Genetic Engineering and Biotechnology

Genetic Engineering and Biotechnology

Editor: Rosanna Mann

www.callistoreference.com

Callisto Reference,
118-35 Queens Blvd., Suite 400,
Forest Hills, NY 11375, USA

Visit us on the World Wide Web at:
www.callistoreference.com

ISBN: 978-1-63239-920-5 (Hardback)

Cataloging-in-Publication Data

Genetic engineering and biotechnology / edited by Rosanna Mann.
     p. cm.
Includes bibliographical references and index.
ISBN 978-1-63239-920-5
1. Genetic engineering. 2. Biotechnology. 3. Genetics. I. Mann, Rosanna.
QH442 .G46 2018
660.65--dc23

Table of Contents

Preface

This book aims to highlight the current researches and provides a platform to further the scope of innovations in this area. This book is a product of the combined efforts of many researchers and scientists, after going through thorough studies and analysis from different parts of the world. The objective of this book is to provide the readers with the latest information of the field.

Genetic engineering has been studied for a number of years for understanding the formation of cells and cell structures as well as the processes involved in evolution. The scientific advancements in the field of genetic engineering and biotechnology have resulted in the manipulation of genes of organisms as well as plants to enhance their traits for commercial purposes. Protein expression and DNA sequencing are key topics of research in this field. This book on genetic engineering and biotechnology discusses the theories and practices related to genes and genetic modification. While understanding the long-term perspectives of the topics, the book makes an effort in highlighting their impact as a modern tool for the growth of the discipline. This book is an essential guide for both academicians and those who wish to pursue this discipline further.

I would like to express my sincere thanks to the authors for their dedicated efforts in the completion of this book. I acknowledge the efforts of the publisher for providing constant support. Lastly, I would like to thank my family for their support in all academic endeavors.

Editor

Microsatellite Tandem Repeats Are Abundant in Human Promoters and Are Associated with Regulatory Elements

Sterling Sawaya[1]*, Andrew Bagshaw[2], Emmanuel Buschiazzo[3], Pankaj Kumar[4], Shantanu Chowdhury[4,5], Michael A. Black[6], Neil Gemmell[1]

1 Centre for Reproduction and Genomics, Department of Anatomy, and Allan Wilson Centre for Molecular Ecology and Evolution, University of Otago, Dunedin, New Zealand, 2 Department of Pathology, University of Otago, Christchurch, New Zealand, 3 School of Natural Sciences, University of California Merced, Merced, California, United States of America, 4 G. N. Ramachandran Knowledge Centre for Genome Informatics, Delhi, India, 5 Proteomics and Structural Biology Unit, Institute of Genomics and Integrative Biology, Council of Scientific and Industrial Research, Delhi, India, 6 Department of Biochemistry, University of Otago, Dunedin, New Zealand

Abstract

Tandem repeats are genomic elements that are prone to changes in repeat number and are thus often polymorphic. These sequences are found at a high density at the start of human genes, in the gene's promoter. Increasing empirical evidence suggests that length variation in these tandem repeats can affect gene regulation. One class of tandem repeats, known as microsatellites, rapidly alter in repeat number. Some of the genetic variation induced by microsatellites is known to result in phenotypic variation. Recently, our group developed a novel method for measuring the evolutionary conservation of microsatellites, and with it we discovered that human microsatellites near transcription start sites are often highly conserved. In this study, we examined the properties of microsatellites found in promoters. We found a high density of microsatellites at the start of genes. We showed that microsatellites are statistically associated with promoters using a wavelet analysis, which allowed us to test for associations on multiple scales and to control for other promoter related elements. Because promoter microsatellites tend to be G/C rich, we hypothesized that G/C rich regulatory elements may drive the association between microsatellites and promoters. Our results indicate that CpG islands, G-quadruplexes (G4) and untranslated regulatory regions have highly significant associations with microsatellites, but controlling for these elements in the analysis does not remove the association between microsatellites and promoters. Due to their intrinsic lability and their overlap with predicted functional elements, these results suggest that many promoter microsatellites have the potential to affect human phenotypes by generating mutations in regulatory elements, which may ultimately result in disease. We discuss the potential functions of human promoter microsatellites in this context.

Editor: Robert W. Sobol, University of Pittsburgh, United States of America

Funding: SS and NG were partially funded by Royal Society of New Zealand Marsden Grant (UOO 0721) and Minority Health International Research Training grant (National Institutes of Health). AB was funded by a Royal Society of New Zealand Marsden Grant (UOO 085). PK was funded by a Sr. Research Fellowship from the Indian Council of Medical Research. The funders had no role in study design, data collection and analysis, decision to publish, or preparation of the manuscript.

Competing Interests: The authors have declared that no competing interests exist.

* E-mail: sterlingsawaya@gmail.com

Introduction

Approximately 3% of the human genome is composed of microsatellites [1], tandem repeats composed of subunits between one and six nucleotides in length. During DNA replication, these sequences change in length at a rate that is many orders of magnitude higher than the average rate of point mutations [2–4]. Because microsatellites are often polymorphic, they have historically been used as markers for parentage and forensic analyses [5,6]. Traditionally, microsatellites and other tandem repeats have been considered to be non-functional, neutral markers. However, there is increasing evidence that this is not always the case [7,8]. For example, in the yeast genome, tandem repeats are frequently found in promoters and are directly responsible for divergence in transcription rates [9]. When tandem repeats within yeast promoters change in length, promoter structure and transcription factor binding can be altered [9,10]. A similar process may occur in the human genome, where tandem repeats can also be found at a high density within promoters [9], defined here as 5 kilobases (kb) upstream and downstream of the transcription start site (TSS).

Recently, we identified human microsatellites that are conserved across vertebrate genomes [11], and later developed a phylogenetic method to measure this conservation [12]. We discovered that highly conserved mammalian microsatellites are over-represented in the promoter regions of various human genes, many of which regulate growth and development [12,13]. Changes in the lengths of microsatellites within promoters can sometimes drastically alter phenotypes [7,13]. For example, expansion of microsatellites in protein coding or 5′ untranslated regions (UTR) is well known to cause disease, including Huntington's disease and fragile-X syndrome [7].

Microsatellites can also affect phenotypes when they are not transcribed [7,13,14]. By altering levels of gene expression, untranslated microsatellites proximal to a TSS can have significant effects on phenotypes. For example, a large body of work has linked variation in human phenotypes with regulatory microsatellites composed of the motif AC/GT [15–34]. Intriguingly, many of these studies focus on genes expressed in neuronal cells [15–21], such as PAX6 expression during eye development [20,21] or NOS1 expression in the brain [15–17]. The promoters of neural development genes such as these contain a striking number of conserved microsatellites [12,35].

Promoter microsatellites have the potential to form various DNA secondary structures, some of which are known to be involved in the regulation of gene expression [13,36]. For example, microsatellites with the motif AC/GT can form Z-DNA, a left-handed spin double helix [37], and microsatellites composed of the motif AG/CT can form H-DNA, a DNA triplex [38–41]. Another DNA secondary structure of interest here is the G-quadruplex (G4, reviewed in [42]). G4 is predicted to form in sequences with the pattern $(G_{3+}N_{1-7})_{3+}(G_{3+})$ which due to its repetitive nature can be composed of microsatellites [43], such as $(TGGG)_{4+}$ [44]. Formation of G4 induces single-strandedness in the complement C-rich strand, which can sometimes form an i-motif [42]. Predicted G4 sequences show a strong preference for promoter regions [45–48]. These structures can regulate transcription by modulating polymerase activity [49,50] or by affecting RNA folding when present in 5′ UTR [51,52].

To better understand how microsatellites are related to promoters and their various regulatory elements we used a wavelet analysis, adapted from ref. [53]. A wavelet decomposition transforms a signal into two components: detail coefficients and smooth coefficients. These coefficients have values at different scales, and these scales increase by a factor of two. The wavelet coefficients can be used to reconstruct the original data. The smoothed coefficients can be seen as similar to a weighted average of the signal, taken at multiple scales. If two signals are compared using smooth coefficients, the result is similar to that which would be found if their average densities were compared. If instead the details coefficients were compared, the result would be similar to comparing covariance between signals, because the detail coefficients measure the change in a signal [53]. Importantly, the wavelet coefficients at any single scale are independent (orthogonal) measures from the coefficients at the other scales [53]. This conveniently allows us to measure correlations between signals at multiple scales [53–55].

Our wavelet analysis included 32 non-continuous regions in the human genome, each 2^{15} kb in length, for a total of 2^{20} kb of DNA (approximately one billion bases). Wavelets are able to easily handle discontinuities in the data, such as those that are present between each of the 32 regions examined here [56]. We measured the densities of various elements across these regions, including those of canonical importance to promoters: GC content, protein coding regions and 5′ UTR. In addition, we examined two other factors known to be associated with promoters: predicted G4 regions [45–48] and CpG islands (CpG dinucleotide rich regions [57]). We focused on G/C rich promoter elements because promoter microsatellites tend to be G/C rich [58]. We examined the pair-wise relationship between all of these variables, and then using a linear model of wavelet coefficients, we examined how these different factors may interact to affect the association between microsatellites and promoters. The intention of the linear model of the wavelet coefficients was to determine if the significant association between microsatellites and promoters was caused by these other elements.

This is the first study to statistically test for an association between microsatellites and promoters. We discovered a highly significant, but complex relationship that depends heavily on microsatellite motif. In addition, we also found associations between microsatellites and the various promoter elements examined in the wavelet analysis. We discuss how microsatellite variation within these promoter elements may modulate gene expression, with a focus on DNA and RNA structure.

Results and Discussion

Microsatellite Motifs in Promoters

The most common microsatellite motifs in the human genome are A/T rich and more than a third of microsatellites in our data set (36.4%) are composed of the motifs A/T or AC/GT (Table 1). These two motifs are also the most common motifs within 5 kb of the TSS (Table 2). The third most common motif within the promoter region is CCG/CGG, but importantly, this motif is very uncommon in the genome, representing less than 1% of the microsatellites in our data set. In fact, of the 3820 CCG/CGG microsatellites we examined, 74% were found within 5 kb of the TSS. A similar motif, CCCG/CGGG, displayed the same preference for promoters, with 62% found within 5 kb of the TSS (Table 2). Intriguingly, microsatellites with the motif CCG/CGG are often very highly conserved in mammals, while the other G/C rich motifs are usually not conserved [12].

Linear Modeling of Distance to TSS

There is a high density of microsatellites around the TSS of human genes (Figure 1). To determine which motifs show the strongest preference for the TSS, we used a linear model. For the response variable in this model we used distance to the nearest TSS, calculated for all microsatellites within 5 kb of the TSS, and we examined this variable in relation to motif for upstream and downstream regions separately. Table 3 displays the motifs with the strongest association to promoters for both upstream and downstream regions. G/C rich motifs have a strong association with promoters. Intriguingly, the most common motifs in the genome, mostly A/T rich, have a strong negative association with promoters. The intent of this model was to uncover the motifs with the strongest positive or negative relationship with distance to the

Table 1. Frequencies of motifs for all simple microsatellites in the human genome.

Motifs	Counts (freqency)
A/T	104,373 (19.4%)
AC/GT	91,786 (17.0%)
AT/TA	37,219 (6.91%)
AAAT/ATTT	30,771 (5.71%)
AAT/ATT	26,782 (4.97%)
AG/CT	23,680 (4.39%)
AAAC/GTTT	21,156 (3.92%)
AAC/GTT	17,974 (3.33%)
AATG/CATT	15,045 (2.79%)
AAAG/CTTT	14,865 (2.75%)
AAAAC/GTTTT	12,610 (2.33%)
AAGG/CCTT	10,681 (1.98%)
AGG/CCT	10,438 (1.93%)
AGGG/CTTT	10,314 (1.91%)
AGC/GCT	6,169 (1.14%)
CCG/CGG	3,820 (0.70%)
CCCG/CGGG	1,098 (0.20%)

The most common motifs in the human genome are shown, along with their counts and frequencies relative to all other microsatellites. A few motifs commonly found in promoters are also shown. The total number of microsatellites examined here is 538,964.

Table 2. Most common motifs found within 5 kb of the TSS and their strand-specific motif results.

Motifs	Counts (on coding strand)	Binom. p-value	KS Test Distance (p-value)
A/T	6559 (2803/3756)	5.2E−32	0.135 (<1E−300)
AC/GT	5072 (2051/3021)	2.1E−42	0.118 (3.1E−15)
CCG/CGG	2833 (1151/1682)	1.7E−23	0.06 (7.2E−3)
AAAT/ATTT	1419 (610/809)	1.4E−7	0.166 (9.1E−9)
AG/CT	1405 (686/719)	0.39	0.07 (0.042)
AGGG/CCCT	1308 (662/646)	0.68	0.07 (0.06)
AAT/ATT	1245 (577/668)	0.011	0.06 (0.15)
AGC/GCT	990 (373/617)	8.36E−15	0.134 (4.7E−4)
AAAC/GTTT	983 (434/549)	2.7E−4	0.188(6.4E−8)
AAC/GTT	952 (460/492)	0.315	0.182 (2.7E−7)
AATG/CATT	876 (452/424)	0.36	0.09 (0.055)
AAAG/CTTT	751 (325/426)	2.6E−4	0.084 (0.146)
AAAAC/GTTTT	651 (304/347)	0.10	0.137 (4.5E−3)
CCCG/CGGG	687 (274/413)	1.28E−7	0.114 (0.027)
AAGG/CCTT	659 (299/350)	0.050	0.092 (0.128)

The most common motifs and their strand-specific counts are displayed. The binomial test (Binom.) p-value is the chance that these strand-specific frequencies deviate from an expected value of 50%. The Kolmogorov-Smirnov (KS) test values provide a measurement of the difference between the distribution of the two different strand-specific motifs, for each motif pair. The p-values shown are not corrected for multiple tests.

TSS. We did not include overlap with functional elements, such as the 5′ UTR, or microsatellite length so that the results could be interpreted simply as the repeated motifs enriched or depleted around the TSS.

Potential Functions of Promoter Microsatellites

As noted in a previous study of a subset of the human genome, there are many G/C rich microsatellites near the TSS of human genes [58]. Here we add that motifs with 100% G/C content are rarely found outside of promoter regions (Table 1) and are usually found very close to the TSS (Table 3). Many of these motifs have the potential to form various secondary structures [43,59]. The G4 secondary structure is of particular interest to this study because there is increasing evidence that G4 elements play an important role in gene regulation [45,46,60]. These structures can be highly conserved in mammals [60], especially in promoter regions [45–47] and have been shown to modulate gene expression levels in microbes [61] and cancer cell lines [62]. Their prevalence in human gene promoters is particularly striking [45,46] and our results support this observation (Figures 2, 3).

Many of the motifs found near the TSS have structural potential (Table 3). For example, the CCG/CGG motif can form secondary structures that are similar but not identical to canonical G4 structures [63], and changes in the length of these microsatellites have the potential to modulate gene expression [64] and cause disease when expanded [65]. A similar motif, CCCG/CGGG, is predicted to form G4 if repeated at least four times, and is similar to the GC-box, a transcription initiation site associated with the transcription factor SP1 [66]. Another motif that is predicted to form G4 DNA is AGGG/CCCT. This motif is common within promoters but is also relatively common elsewhere in the genome. Of the 10,314 AGGG/CCCT microsatellites, 1,308 of them are found within 5 kb of the TSS (Table 2).

G/C rich motifs that contain CpG dinucleotides are potential sites of epigenetic modification. Each of the 100% G/C microsatellites, except for the rare mononucleotide motif C/G, contain CpG dinucleotides [57]. Changes in repeat number for these CpG containing microsatellites would alter the number of potential methylation sites. However, changes in microsatellite length may also affect structural potential, which is important because G4 formation appears to restrict methylation at CpG dinucleotides [67]. So, although longer CpG containing microsatellites may contain more potential methylation sites, this may not directly translate into an increase in methylation because longer microsatellites may also have increased structural potential, and these structures may in turn interfere with methylation [68].

Motifs on the Coding Strand

Transcription is most often uni-directional, with only one strand transcribed into RNA, leading to potential differences in sequence composition between the coding and non-coding strand. Therefore, we wondered if the microsatellite motifs on the coding strand might have different distributions around promoters than their counterparts on the opposite strand. Strand asymmetry exists between all non-palindromic motifs, and these motifs can be broken into pairs of strand-specific motifs. To examine how these strand-specific motifs are related to promoters, we obtained the microsatellite motifs on the coding (non-template) strand for the 37,249 microsatellites found within 5 kb of the TSS (Table 2).

The distributions for the most common strand-specific motif pairs, A/T and AC/GT are shown in Figure 4. These graphs show the smoothed density estimates for both 1 kb and 100 base pair bins. The strand-specific motifs A and AC display a preference for the upstream region and a depletion from the downstream region. Intriguingly, their counterparts, T and GT, display the complete opposite pattern, with their highest densities in the downstream regions. All of these motifs show depletion around the TSS, but this depletion is only clear when fine scale densities (100 base-pair bins) are examined.

Some of these strand-specific motifs have a preference for the coding strand (Table 2). For example, the motifs with 100% G/C (CCG/CGG and CCCG/CGGG) have a preference for the G-

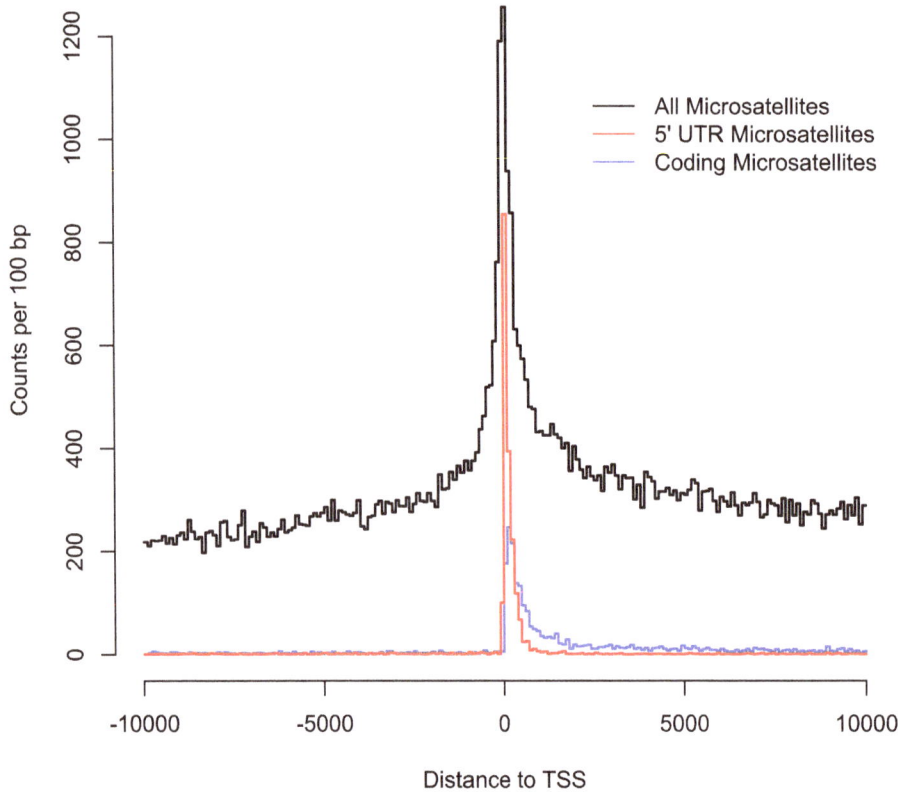

Figure 1. Distribution of microsatellites around promoters. The total number of microsatellites present in each 100 base-pair bin are provided for all microsatellites within 10 kb of the TSS. Also shown are the total number of only coding microsatellites (blue) or only 5′ UTR microsatellites (red).

rich motif to be on the coding strand (59% and 60%, respectively). The binomial test p-values for these observations are $1.7E-23$ and $1.28E-7$, for CCG/CGG and CCCG/CGGG respectively. The other G-rich motifs common in promoters, AGGG/CCCT and AAGG/CCTT, do not show any preference for G-richness on the coding strand.

A strand-specific preference may be due to a selection for G-richness in RNA, and/or G-richness on the coding strand [69]. G-richness on the coding strand is also seen in predicted G4 forming regions around promoters [47]. Therefore, we were surprised that the predicted G4 motif AGGG/TCCC did not show any strong strand preference. The motif AG/CT, which is predicted to form H-DNA [38–41], also displayed no strand preference.

To examine whether the strand-specific distributions are different for each motif pair, we used the Kolmogorov-Smirnov test. The results of this non-parametric test indicate the distributions of many of these motif pairs are dissimilar to each other (Table 2). For example, the strand-specific motifs AC and GT have very different distributions around the TSS (Figure 4), and the Kolmogorov-Smirnov test results indicate this with a large distance value supported by a very low p-value. Notably, some motif pairs do not show any strand differences, such as the poly-

purine/poly-pyrimidine motifs AG/CT, AAAG/CTTT and AAGG/CCTT.

Depletion of the motifs A and AC on the coding strand indicates that they may interfere with transcription (or translation when present in 5′ UTR). Perhaps this is unsurprising for the motif A, which is commonly known as a signal for the end of the transcript in the 3′ UTR, and may be selected against in the 5′ UTR. We are unaware of a similar explanation for the motif AC, which shows particularly strong depletion immediately downstream of the TSS. The Z-DNA structure that can form in AC/GT microsatellites is a left-handed double-helix with no known strand bias [37]. Changes in AC/GT length have been shown to modulate gene expression [70], as seen in the large number of studies associating AC/GT length variation with human phenotypes [15–34]. These strand-specific biases support the hypothesis that microsatellite motif can affect RNA structure [35,71].

Wavelet Analysis: Results on Multiple Scales

To statistically test for an association between microsatellites and promoters, we used a wavelet analysis on approximately one billion bases, a third of the entire genome. G/C rich motifs showed the strongest association with the TSS, so we wondered if the high density of microsatellites at the TSS (Figure 1) was caused by G/C

Table 3. Most significant motifs associated with distance to the TSS from the linear analysis.

Upstream: Motif	Sorted q-values	Reg.coef.
(Intercept): A/T	0.0E+00	−2.2E+03
CCG/CGG	2.7E−195	1.7E+03
CCCCG/CGGGG	2.1E−102	1.9E+03
CCCG/CGGG	1.2E−70	1.7E+03
AGG/CCT	2.7E−26	6.7E+02
CG/CG	5.6E−23	1.8E+03
C/G	3.2E−17	1.0E+03
CCCCCG/CGGGGG	1.3E−12	1.6E+03
AGGG/CCCT	6.7E−12	4.5E+02
CCGCG/CGCGG	7.5E−12	1.9E+03
CCCGG/CCGGG	1.5E−11	1.9E+03
AGCG/CGCT	3.4E−11	1.6E+03
AAAT/ATTT	1.9E−09	−3.7E+02
AT/AT	3.2E−09	−3.7E+02
AAT/ATT	7.9E−08	−3.4E+02

Downstream: Motif	Sorted q-values	Reg.coef.
(Intercept): A/T	0.0E+00	−2.5E+03
CCG/CGG	0.0E+00	2.0E+03
CCCG/CGGG	7.4E−165	1.9E+03
AGC/GCT	1.7E−122	1.3E+03
AGG/CCT	8.8E−71	8.8E+02
CCCCG/CGGGG	3.9E−52	1.8E+03
CCCGG/CCGGG	3.8E−39	2.1E+03
AGCG/CGCT	1.7E−35	2.1E+03
AGGG/CCCT	4.7E−31	6.5E+02
CG/CG	1.0E−21	1.7E+03
CCGG/CCGG	1.2E−21	1.7E+03
CCGCG/CGCGG	7.3E−19	2.0E+03
CCCCGG/CCGGGG	2.5E−17	2.0E+03
AGGGG/CCCCT	5.1E−12	8.8E+02
CCCCCG/CGGGGG	7.4E−12	1.6E+03

The top 10 most significant motifs associated with distance to TSS (in base-pairs), for the upstream and downstream regions, analyzed separately. These factors are sorted by their false discovery rate q-value (Sorted q-values). The size of the regression coefficient (Reg. coef.) indicates the strength of the association, with large positive coefficients belonging to motifs frequently found near the TSS. The full list of significant factors can be found in. Tables S1 and S2.

rich regulatory elements. Therefore, in addition to promoters and microsatellites, we included various factors known to be associated with promoters: 5′ UTR, coding regions, predicted G4 regions [46–48], GC content, and CpG islands [57].

Figure 2 shows the pairwise Kendall rank correlations between each element at each scale for both the smooth and detail coefficients. Red indicates significant positive associations, and blue significant negative associations (p-value < 0.001). The power spectrum is shown on the diagonal, and represents the proportion of total variation explained by variation at each scale. Correlations between the smooth coefficients of these different elements (upper right portion of Figure 2) are functionally

equivalent to correlations between average densities of these elements at various scales. The correlations between detail coefficients (bottom left portion of Figure 2) are more closely related to covariance between the signals [53].

The results of the pairwise comparisons indicate that promoters and microsatellites are significantly associated, but only on fine-scale measurements (Figure 2). At larger scales, microsatellites are negatively associated with promoters. We interpret these results as support for a local association between microsatellites and the TSS, but that microsatellites are, in general, found at higher densities in regions that do not contain promoters. This change in value between fine and coarse scales highlights the importance of examining multiple scales for associations between genomic elements, as processes acting at fine scales can be different from those acting at coarse scales [53]. Intriguingly, microsatellites display the same positive fine-scale and negative large-scale association with every factor examined except G/C content. The negative correlation between microsatellites and GC content highlights the fact that most microsatellites in the human genome are AT rich (Table 1).

Because G/C rich motifs are strongly associated with promoters and because many of these motifs have the potential to act as sites of DNA methylation or structural formation, we hypothesized that CpG islands or G4 forming regions could influence the apparent association between microsatellites and promoters seen in Figure 1. To investigate this we used linear modeling of the wavelet coefficients, again following methods of ref. [53] (Figure 3). This approach used the microsatellite wavelet coefficients as the response variable, and the wavelet coefficients for the other factors as covariates. The $-log_{10}$ p-values are shown for each factor, at each scale. Again, significant positive associations are red, and negative associations are blue.

After controlling for these other factors, the relationship between promoters and microsatellites remained significant, but was again only positive at fine scales. Because fewer of the fine scales showed a significant positive association, the association between microsatellites and promoters at these scales can be partially attributed to the other factors examined. Intriguingly, the positive fine-scale associations between coding regions and microsatellites is absent when these other factors are considered.

The small r^2 values here indicate that the total variance explained by this model is minimal. Therefore, there is a large amount of variation in microsatellite density that is not explained by these factors. Nevertheless, results of this linear model are highly informative and we stress that the intention of the model was not to determine which factors predict microsatellite density. Microsatellites are found throughout the genome, and hypothetically can arise and degrade by entirely neutral mutational processes [5], so we did not expect promoters and promoter-related factors to explain a large amount of variation in the microsatellite signal. We used this model to determine if the association between microsatellites and promoters was the result of a high density of GC rich elements around the TSS. Because the significant positive association between promoters and microsatellites remains when these other factors are included in the model, we can conclude that they are not entirely responsible for the high density of microsatellites found at the TSS (Figure 1).

Relationship between Microsatellites and G4 Elements

The highly significant association between microsatellites and G4 supports the hypothesis that microsatellites sometimes play a role as structural elements [43]. In the pairwise comparison between G4 and microsatellite wavelet coefficients there is a highly

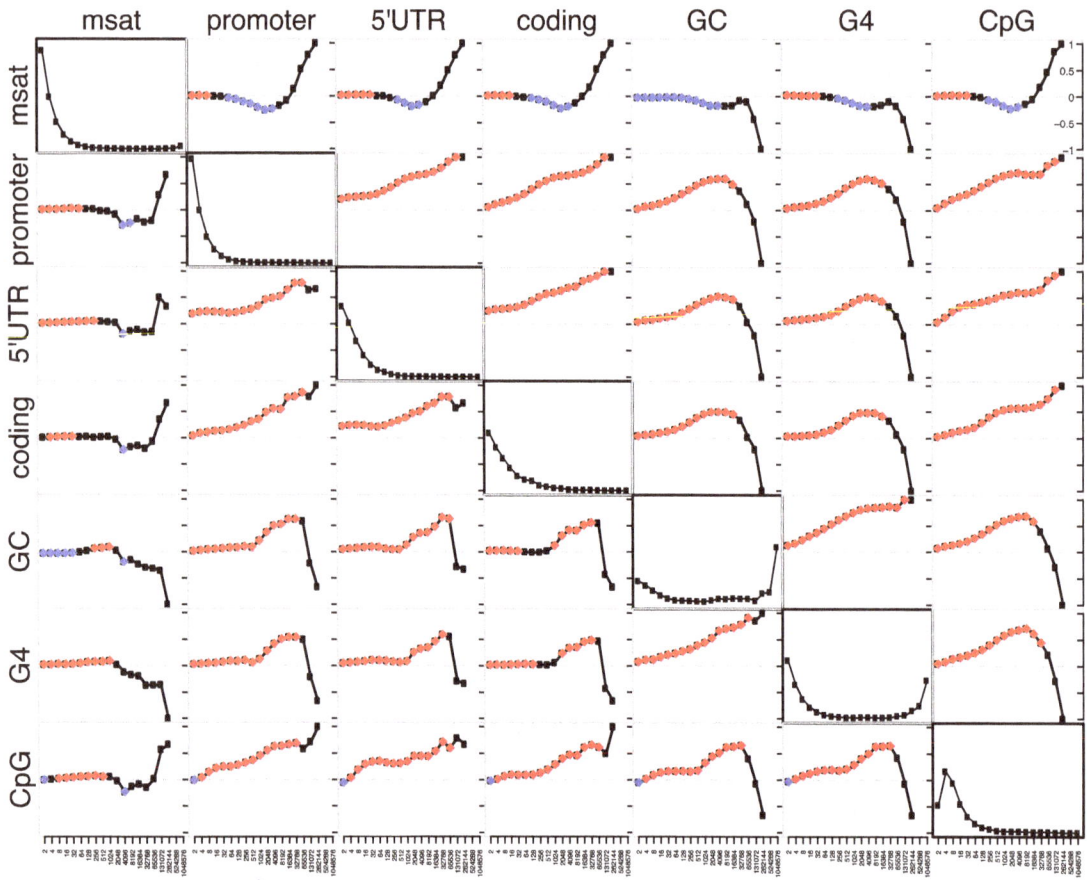

Figure 2. Kendall rank correlations between wavelet coefficients. The pairwise correlations between smooth coefficients are in the top right, and detail coefficients are the bottom left. The diagonal displays the normalized power spectrum for the wavelet coefficients, which can be interpreted as a measure of the variation of each signal at each scale. Note that the majority of factors examined here have most of their variation at the finest scales, while GC content and G4 elements contain a large amount of variation at the largest scales. Abbreviations for each element are "msat" for microsatellite, "G4" for predicted G4 regions, "CpG" for CpG islands, and "GC" for G/C content. Associations with a p-value above 0.001 are shown in red if positive, blue if negative. The smallest scale examined was 1 kb in size, and each successive scale increases by a factor of two.

significant association at fine scales (Figure 2), and this association increases when other factors are considered (Figure 3).

The motifs for microsatellites that overlap with G4 elements are shown in Table 4. Most of these motifs are similar to the canonical G4 definition but not all microsatellites with these G4-like motifs are considered G4 for two reasons. Some of these G4-like microsatellites are too short to have G4 potential (e.g., $(AGGG)_3$). For longer microsatellites, we allow a few point mutations to disrupt the repeating pattern (i.e., they are imperfect repeats). If a point mutation disrupted the runs of adjacent guanines it would disrupt the G4 forming potential. Importantly, expansion of these G4-like microsatellites could result in novel G4 elements that would not present in the reference genome. For example, the G4-like microsatellite $AGTG(AGGG)_3$ contains a point mutation that disrupts the perfect repeat and prevents G4 forming potential. This microsatellites could expand to form $AGTG(AGGG)_4$, a microsatellite with G4 potential.

As discussed above, some motifs have higher rates of expansion and contraction than others [72,73], and therefore, some G4 and G4-like microsatellites will be more polymorphic than others. One motif in particular has a relatively high rate of expansion and contraction, the mononucleotide motif C/G [72]. Intriguingly, there are 1,402 C/G microsatellites in our data set and 961

(68.5%) overlap with a G4 element. G4 elements that overlap with these rare C/G microsatellites are expected to be highly variable.

Less variable G4 microsatellites may also be important because even small changes in repeat number for larger, G-rich motifs have the potential to alter secondary structure. Variation within G/C rich tandem repeats has been shown to affect gene expression and/or be associated with phenotypic differences in humans [64,74–80]. For example, a CGGGGG/CCCCCG microsatellite in the ALOX5 gene has been repeatedly associated with cardiovascular disease [75–77]. Unfortunately, there is limited information about microsatellite variation available [81], even from the 1000 Genomes Project [82], so we are unsure exactly which G4 microsatellites contain variation that might affect structural potential. We expect recent advances in sequencing technology to help resolve this uncertainty [83].

To determine which pathways contain G4 elements that overlap with microsatellites, we used the Genomic Regions Enrichment of Annotations Tool (GREAT, [84]). This tool examines which genes contain a set of elements defined by the user (here G4 that overlap with microsatellites). To control for the fact that a limited sub-set of genes contain G4 elements within their promoters, we used the entire G4 set as a control group. Some of the results can be found in Table 5, and the rest are found in Table S4. Intriguingly, many

Smoothed model: Microsatellites ~ CpG + GC + G4 + Coding + 5' UTR + Promoter -1

	2 kb	4 kb	8 kb	16 kb	32 kb	64 kb	128 kb	256 kb	512 kb	1024 kb	2048 kb	4096 kb
Promoter	11.08	1.12	0.76	5.13	17.46	27.39	30.76	30.9	19.77	13.65	7.41	3.78
5' UTR	62.36	33.9	15.53	4.76	3.67	3.75	3.84	3.19	3.24	2.9	1.66	1.62
Coding	0.8	1.33	4	5.93	5.56	3.79	1.12	0.09	0.27	0.35	0.27	0.22
G4	300	300	151.77	85.19	55.97	35.11	14.91	6.97	4.59	1.53	0.58	7.66
GC	300	211.43	120.19	62.98	33.73	18.2	6.18	3.07	2.45	2.14	1.19	0.27
CpG	49.94	63.45	64.03	44.5	25.03	12.69	4.58	2.25	0.45	0.87	3.21	9.68
Adj r^2	0.01	0.01	0.01	0.01	0.02	0.02	0.03	0.05	0.07	0.09	0.12	0.28

Wavelet model: Microsatellites ~ CpG + GC + G4 + Coding + 5' UTR + Promoter -1

	2 kb	4 kb	8 kb	16 kb	32 kb	64 kb	128 kb	256 kb	512 kb	1024 kb	2048 kb	4096 kb
Promoter	5.02	11.94	3.83	2.23	2.94	0.12	0.83	1.33	8.8	3.05	4.24	3.59
5' UTR	23.52	24.24	20.58	16.08	2.48	0.98	0.32	0.93	0.52	0.93	1.48	0.59
Coding	0.64	0.04	1.87	0.76	0.12	0.19	1.48	1.19	0.22	0.23	0.05	0.24
G4	300	259.35	151.7	63.57	28.21	17.6	21.51	6.36	0.61	0.9	1.25	0.7
GC	300	204.41	108.77	75.65	36.72	8.48	3.47	2.22	3.25	2.78	0.61	1.07
CpG	9.42	0.27	9.89	25.5	25.73	16.99	14.4	2.91	4.58	0.05	0.09	0.4
Adj r^2	0.01	0.01	0.01	0.01	0.01	0.01	0.03	0.03	0.04	0.04	0.05	0.14

Figure 3. Linear model of wavelet results, displaying $-\log_{10}$ **p-values.** The top figure shows the results of the smooth coefficients, the bottom shows the results of the detail coefficients. Positive relationships are shown in red, negative in blue. The r^2 value is shown at the bottom of the figure. The largest scales were not included in this figure for simplicity.

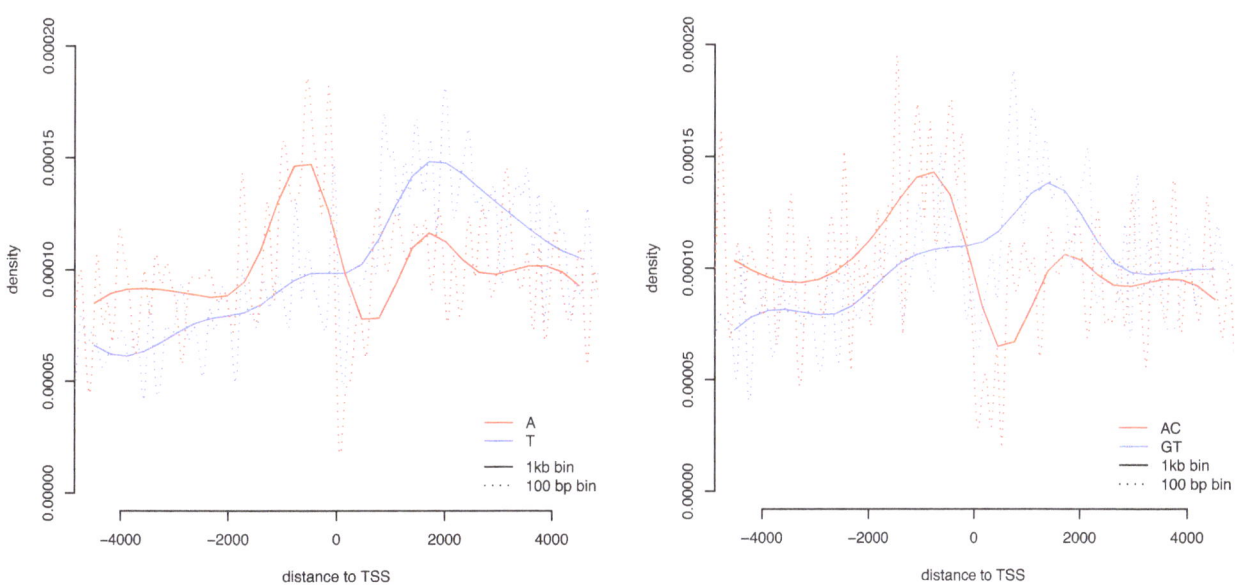

Figure 4. Strand-specific densities for the motifs A/T and AC/GT around promoters. These figures show the cubic spline of the densities of each strand-specific motif for bins of size 1kb (solid) and 100 base-pair (dashed) for the entire 5 kb promoter region.

Table 4. Motifs of microsatellites that overlap with G4.

Motifs	Count	Avg. overlap (bp)	Avg. Overlap fraction
AGGG/CCCT	4610	16.9	0.85
ACCC/GGGT	1417	14.1	0.88
AGGGG/CCCCT	1114	25.9	0.85
C/G	961	18.0	0.98
ACCCC/GGGGT	585	18.6	0.92
CCCG/CGGG	583	14.0	0.86
CCCCG/CGGGG	485	19.6	0.88
AAGGG/CCCTT	427	27.5	0.79
AAGG/CCTT	352	8.4	0.21
AG/CT	306	9.1	0.23
AGCCC/GGGCT	293	19.7	0.87
AGGGC/GCCCT	264	19.4	0.86
AGG/CCT	236	10.7	0.36
ACCCCC/GGGGGT	234	22.1	0.92
AC/GT	176	4.8	0.17
CCG/CGG	157	7.2	0.31
AGCCCC/GGGGCT	154	21.4	0.78
CCCCCG/CGGGGG	116	21.5	0.88
CCCGG/CCGGG	106	19.5	0.88
AGAGGG/CCCTCT	93	24.0	0.70

Of the 13,838 microsatellites that overlap with a G4 element, the most common motifs are shown. For each microsatellite motif, the average base-pair overlap with G4 is shown (Avg. overlap (bp)). The average fraction of each microsatellite that overlaps with the G4 element is also shown (Avg. Overlap fraction). Note that motifs that are dissimilar to the canonical G4 definition, such as AC, usually share only a portion of the microsatellite in the G4 element.

of the genes that contain G4-microsatellites regulate cell signaling and/or development (Table 5).

The relationship between microsatellites and G4 may have implications for quantitative genetics. Single nucleotide substitutions within predicted G4 regions can influence gene expression [85] and changes in microsatellite length within or around predicted G4 may be of equal or greater importance, as they would result in changes that are physically larger than single base changes. G4 microsatellites are potential sources of human phenotypic variation, and would make interesting candidates for association studies or molecular genetics experiments.

Conclusions

The high density of microsatellites in promoters (Figure 1), together with their potential to function as structural elements [43,59], suggests that some microsatellites can function as regulators of gene expression. Microsatellites are present in promoters more often than expected by chance. Promoter microsatellites are often G/C rich, and many promoter microsatellites are within or near 5′ UTR, CpG islands, and G4 structures. Variation within these promoter microsatellites has the potential to affect promoter function, which can ultimately lead to variation in phenotypes. This variation may be selectively beneficial [86,87], and by targeting promoter microsatellites, especially those that are conserved [12,71], we hope to uncover sources of human phenotypic variation.

Materials and Methods

Data

The microsatellite positions, their motifs, conservation and functional region (coding, 3 and 5-UTR, intronic, and intergenic) were taken directly from our previous work [11,12], and we have previously released our data [12]. Our microsatellite definition is a tandem repeat composed of 1–6 base-pair motifs that is at least 12 nucleotides in length for motifs of length 1–4, and at least three uninterrupted repeats for motifs of length 5 and 6. As before, we only examined simple (non-adjacent) microsatellites on the autosomes that are found outside of transposable elements and duplicated regions. The positions for the CpG islands and the TSS (start of unique transcripts from the KnownCanonical table) were obtained from the UCSC genome browser [88]. To obtain the predicted G4 regions, we used the definition of G4 from ref. [60] and scanned the human genome (build 36/hg18) for unique (non-overlapping) G4 regions using the canonical G4 definition, $(G_{3+}N_{1-7})_{3+}(G_{3+})$ [45]. The positions for the 5′ UTR and coding regions of the human genome were obtained from Ensembl [89,90]. The strand-specific motifs were obtained by taking the microsatellites found within 5 kb of the TSS, and analyzing the sequences on the coding strand. We detected microsatellites using SciRoKo [91], using the same parameters as we used in our previous work [11,12].

Linear Regression Analysis

Linear modeling was performed using the R statistical software package [92]. The response variable was the distance to the TSS, for microsatellites within 5 kb of the promoter, as defined by the start of the transcript in the KnownCanonical table from UCSC [88]. The covariate in this model was microsatellite motif (284 types). We corrected for multiple hypothesis testing by controlling the false discovery rate using the R package "fdrtool" and computed the false discovery rate q-value for each regression coefficient [93].

Strand-specific Comparisons

To compare the distributions and counts of each strand-specific motif pair, we used a two tailed binomial test and a Kolmogorov-Smirnov test. Both of these tests were performed in R using default functions [92]. We did not correct for multiple tests here so that researchers interested in specific motifs can extract results independent of the other tests done.

Wavelet Analysis

The methods and R code used for the wavelet analyses were adapted from ref. [53]. The value for each factor examined in the wavelet analysis was measured in 1 kb windows for each of the 32, 2^{15} kb regions. For promoters, this regional measurement was a count of the number of promoters. For the other factors, this measurement was the total coverage in each of the 1 kb windows, as determined using the Galaxy [94–96] overlap tool. By examining coverage in each region, the length of each element is implicitly included in the model.

The regions we used cover 13 chromosomes, and the positions and brief description of each region can be found in Table S3. These regions were chosen because they are well annotated, and because they were used in a previous wavelet analysis on microsatellites [97]. The wavelet coefficients were generated for the entire set of regions, or 2^{20} kb, and were scaled to preserve variance across scales.

To generate the wavelet coefficients, we used the Daubechies 4-tap wavelet transform, a slight variation from ref. [53], in which

Table 5. GO Results for genes with microsatellites that overlap with G4 elements.

Ontology	Category	Hyper FDR Q value	Hyper fold enrichment	Number of genes found
GO Biological Process	Signal release	3.39684e−7	2.1533	52
	Cartilage development	2.28192e−6	2.0690	41
	Negative regulation of B cell activation	5.84903e−5	4.4914	11
	Multicellular organismal homeostasis	1.20456e−4	2.1504	27
	Regulation of ion transmembrane transporter activity	1.87001e−4	2.1324	32
	Camera-type eye morphogenesis	5.21582e−4	2.0009	30
	Neurotransmitter secretion	5.19828e−4	2.1208	29
	Spinal cord anterior/posterior patterning	5.76268e−4	10.1377	1
	Tissue homeostasis	1.08506e−3	2.1131	21
	Regulation of long-term neuronal synaptic plasticity	1.25384e−3	3.0197	13
	Hormone secretion	1.42278e−3	2.2221	22
	Hormone transport	1.76257e−3	2.1627	23
	Negative regulation of synaptic transmission, glutamatergic	3.77337e−3	6.1327	3
	Elevation of cytosolic calcium ion concentration involved in G-protein signaling coupled to IP3 second messenger	3.92996e−3	5.2566	5
	Peptide hormone secretion	4.22613e−3	2.2528	18
PANTHER Pathway	TGF-beta signaling pathway	4.57321e−4	2.0458	32
	General transcription regulation	6.12838e−3	3.1400	35
	Ras Pathway	6.46098e−3	2.0119	22
	Beta2 adrenergic receptor signaling pathway	2.56959e−2	2.0132	15
	Gamma-aminobutyric acid synthesis	2.90416e−2	4.7309	3
	Transcription regulation by bZIP transcription factor	3.47807e−2	2.0310	14

Gene ontology (GO) results for genes that contain microsatellites that overlap with G4 elements in their promoter. Hyper FDR Q-value is the false discovery rate q-value, Hyper fold enrichment is the enrichment of the test set on the overall (control) set for each category. 2,666 genes contain a G4 that overlaps with a microsatellite. For a control set we used genes that contain G4 elements in their promoters, for a total of 14,977 genes. The promoter region here was again 5 kb upstream and down of the TSS.

the Haar wavelet transform (Daubechies 2-tap) was used. Although we found similar results with other values for the Daubechies wavelet bases (results not shown), we chose the 4-tap basis because the results were more consistent between adjacent scales than the 2-tap bases, and it requires less computational time than the higher valued Daubechies transforms.

Gene Ontology Analysis

GREAT 2.0.2 [84] was used for the gene ontology analysis. This web tool allows the user to input a set of genomic regions of interest (here G4 that overlap with microsatellites), and a control set on which to compare these regions (here all G4 regions). GREAT then tests the gene ontology categories which contain the regions of interest against the background set. It also corrects for false discovery rates. We used 5 kb upstream and downstream of the TSS as our promoter region.

Acknowledgments

We would like to acknowledge the helpful comments and suggestions from two anonymous reviewers.

Author Contributions

Conceived and designed the experiments: SS EB NG. Performed the experiments: SS MB. Analyzed the data: SS MB AB PK SC. Contributed reagents/materials/analysis tools: SS MB AB EB PK SC. Wrote the paper: SS MB AB EB NG.

References

1. Warren WC, Hillier LW, Marshall Graves JA, Birney E, Ponting CP, et al. (2008) Genome analysis of the platypus reveals unique signatures of evolution. Nature 453: 175–183.

2. Ellegren H (2004) Microsatellites: simple sequences with complex evolution. Nature Reviews Genetics 5: 435–445.

3. Sun JX, Helgason A, Masson G, Ebenesersdottir SS, Li H, et al. (2012) A direct characterization of human mutation based on microsatellites. Nat Genet 44: 1161–1165.

4. Payseur BA, Jing P, Haasl RJ (2011) A genomic portrait of human microsatellite variation. Mol Biol Evol 28: 303–312.

5. Buschiazzo E, Gemmell N (2006) The rise, fall and renaissance of microsatellites in eukaryotic genomes. Bioessays 28: 1040–1050.

6. Tracey M (2001) Short tandem repeat-based identification of individuals and parents. Croat Med J 42: 233–238.

7. Gemayel R, Vinces MD, Legendre M, Verstrepen KJ (2010) Variable tandem repeats accelerate evolution of coding and regulatory sequences. Annu Rev Genet 44: 445–477.

8. Hannan A (2010) Tandem repeat polymorphisms: modulators of disease susceptibility and candidates for 'missing heritability'. Trends in Genetics 26: 59–65.

9. Vinces MD, Legendre M, Caldara M, Hagihara M, Verstrepen KJ (2009) Unstable tandem repeats in promoters confer transcriptional evolvability. Science 324: 1213–1216.

10. Lee TH, Maheshri N (2012) A regulatory role for repeated decoy transcription factor binding sites in target gene expression. Mol Syst Biol 8: 576.

11. Buschiazzo E, Gemmell NJ (2010) Conservation of human microsatellites across 450 million years of evolution. Genome Biol Evol 2: 153–165.

12. Sawaya SM, Lennon D, Buschiazzo E, Gemmell N, Minin VN (2012) Measuring microsatellite conservation in mammalian evolution with a phylogenetic birth-death model. Genome Biol Evol 4: 636–647.

13. Sawaya S, Bagshaw A, Buschiazzo E, Gemmell N (2012) Promoter microsatellites as modulators of human gene expression. In: Hannan A, editor, Tandem Repeat Polymorphisms: Genetic Plasticity, Neural Diversity and Disease, Austin, Texas: Landes Biosciences, chapter 4.

14. Rockman M, Wray G, Wray G (2002) Abundant raw material for cis-regulatory evolution in humans. Molecular Biology and Evolution 19: 1991–2004.

15. Rife T, Rasoul B, Pullen N, Mitchell D, Grathwol K, et al. (2009) The effect of a promoter polymorphism on the transcription of nitric oxide synthase 1 and its relevance to Parkinson's disease. J Neurosci Res 87: 2319–2325.

16. Reif A, Jacob CP, Rujescu D, Herterich S, Lang S, et al. (2009) Inuence of functional variant of neuronal nitric oxide synthase on impulsive behaviors in humans. Arch Gen Psychiatry 66: 41–50.

17. Kopf J, Schecklmann M, Hahn T, Dresler T, Dieler AC, et al. (2011) NOS1 ex1f-VNTR polymorphism inuences prefrontal brain oxygenation during a working memory task. Neuroimage 57: 1617–1623.

18. Knafo A, Israel S, Darvasi A, Bachner-Melman R, Uzefovsky F, et al. (2008) Individual differences in allocation of funds in the dictator game associated with length of the arginine vasopressin 1a receptor RS3 promoter region and correlation between RS3 length and hippocampal mRNA. Genes Brain Behav 7: 266–275.

19. Itokawa M, Yamada K, Yoshitsugu K, Toyota T, Suga T, et al. (2003) A microsatellite repeat in the promoter of the N-methyl-D-aspartate receptor 2A subunit (GRIN2A) gene suppresses transcriptional activity and correlates with chronic outcome in schizophrenia. Pharmacogenetics 13: 271–278.

20. Okladnova O, Syagailo YV, Tranitz M, Stober G, Riederer P, et al. (1998) A promoter-associated polymorphic repeat modulates PAX-6 expression in human brain. Biochem Biophys Res Commun 248: 402–405.

21. Ng TK, Lam CY, Lam DS, Chiang SW, Tam PO, et al. (2009) AC and AG dinucleotide repeats in the PAX6 P1 promoter are associated with high myopia. Mol Vis 15: 2239–2248.

22. Chen YH, Lin SJ, Lin MW, Tsai HL, Kuo SS, et al. (2002) Microsatellite polymorphism in promoter of heme oxygenase-1 gene is associated with susceptibility to coronary artery disease in type 2 diabetic patients. Hum Genet 111: 1–8.

23. Gao PS, Heller NM, Walker W, Chen CH, Moller M, et al. (2004) Variation in dinucleotide (GT) repeat sequence in the first exon of the STAT6 gene is associated with atopic asthma and differentially regulates the promoter activity in vitro. J Med Genet 41: 535–539.

24. Agarwal AK, Giacchetti G, Lavery G, Nikkila H, Palermo M, et al. (2000) CA-Repeat polymorphism in intron 1 of HSD11B2 : effects on gene expression and salt sensitivity. Hypertension 36: 187–194.

25. Akai J, Kimura A, Hata RI (1999) Transcriptional regulation of the human type I collagen alpha2 (COL1A2) gene by the combination of two dinucleotide repeats. Gene 239: 65–73.

26. Searle S, Blackwell JM (1999) Evidence for a functional repeat polymorphism in the promoter of the human NRAMP1 gene that correlates with autoimmune versus infectious disease susceptibility. J Med Genet 36: 295–299.

27. Yamada N, Yamaya M, Okinaga S, Nakayama K, Sekizawa K, et al. (2000) Microsatellite polymorphism in the heme oxygenase-1 gene promoter is associated with susceptibility to emphysema. Am J Hum Genet 66: 187–195.

28. Shimajiri S, Arima N, Tanimoto A, Murata Y, Hamada T, et al. (1999) Shortened microsatellite d(CA)21 sequence down-regulates promoter activity of matrix metalloproteinase 9 gene. FEBS Lett 455: 70–74.

29. Hough C, Cameron CL, Notley CR, Brown C, O'Brien L, et al. (2008) Inuence of a GT repeat element on shear stress responsiveness of the VWF gene promoter. J Thromb Haemost 6: 1183–1190.

30. Wang B, Ren J, Ooi LL, Chong SS, Lee CG (2005) Dinucleotide repeats negatively modulate the promoter activity of Cyr61 and is unstable in hepatocellular carcinoma patients. Oncogene 24: 3999–4008.

31. Valverde P, Koren G (1999) Purification and preliminary characterization of a cardiac Kv1.5 repressor element binding factor. Circ Res 84: 937–944.

32. Gebhardt F, Zanker KS, Brandt B (1999) Modulation of epidermal growth factor receptor gene transcription by a polymorphic dinucleotide repeat in intron 1. J Biol Chem 274: 13176–13180.

33. Funke-Kaiser H, Thomas A, Bremer J, Kovacevic SD, Scheuch K, et al. (2003) Regulation of the major isoform of human endothelin-converting enzyme-1 by a strong housekeeping promoter modulated by polymorphic microsatellites. J Hypertens 21: 2111–2124.

34. Domart MC, Benyamina A, Lemoine A, Bourgain C, Blecha L, et al. (2012) Association between a polymorphism in the promoter of a glutamate receptor subunit gene (GRIN2A) and alcoholism. Addict Biol 17: 783–785.

35. Riley D, Krieger J (2009) UTR dinucleotide simple sequence repeat evolution exhibits recurring patterns including regulatory sequence motif replacements. Gene 429: 80–86.

36. Kouzine F, Levens D (2007) Supercoil-driven DNA structures regulate genetic transactions. Front Biosci 12: 4409–4423.

37. Wang G, Vasquez KM (2007) Z-DNA, an active element in the genome. Front Biosci 12: 4424–4438.

38. Beaulieu M, Barbeau B, Rassart E (1997) Triplex-forming oligonucleotides with unexpected affinity for a nontargeted GA repeat sequence. Antisense Nucleic Acid Drug Dev 7: 125–130.

39. Rustighi A, Tessari MA, Vascotto F, Sgarra R, Giancotti V, et al. (2002) A polypyrimidine/polypurine tract within the Hmga2 minimal promoter: a common feature of many growth-related genes. Biochemistry 41: 1229–1240.

40. Han YJ, de Lanerolle P (2008) Naturally extended CT.AG repeats increase H-DNA structures and promoter activity in the smooth muscle myosin light chain kinase gene. Mol Cell Biol 28: 863–872.

41. Xu G, Goodridge AG (1998) A CT repeat in the promoter of the chicken malic enzyme gene is essential for function at an alternative transcription start site. Arch Biochem Biophys 358: 83–91.

42. Qin Y, Hurley LH (2008) Structures, folding patterns, and functions of intramolecular DNA G-quadruplexes found in eukaryotic promoter regions. Biochimie 90: 1149–1171.

43. Bacolla A, Larson JE, Collins JR, Li J, Milosavljevic A, et al. (2008) Abundance and length of simple repeats in vertebrate genomes are determined by their structural properties. Genome Res 18: 1545–1553.

44. Gudin A, Gros J, Alberti P, Mergny JL (2010) How long is too long? effects of loop size on g-quadruplex stability. Nucleic Acids Research 38: 7858–7868.

45. Verma A, Halder K, Halder R, Yadav VK, Rawal P, et al. (2008) Genome-wide computational and expression analyses reveal G-quadruplex DNA motifs as conserved cis-regulatory elements in human and related species. J Med Chem 51: 5641–5649.

46. Du Z, Zhao Y, Li N (2009) Genome-wide colonization of gene regulatory elements by G4 DNA motifs. Nucleic Acids Res 37: 6784–6798.

47. Du Z, Zhao Y, Li N (2008) Genome-wide analysis reveals regulatory role of G4 DNA in gene transcription. Genome Res 18: 233–241.

48. Huppert JL, Balasubramanian S (2007) G-quadruplexes in promoters throughout the human genome. Nucleic Acids Res 35: 406–413.

49. Eddy J, Maizels N (2008) Conserved elements with potential to form polymorphic G-quadruplex structures in the first intron of human genes. Nucleic Acids Res 36: 1321–1333.

50. Eddy J, Vallur AC, Varma S, Liu H, Reinhold WC, et al. (2011) G4 motifs correlate with promoter-proximal transcriptional pausing in human genes. Nucleic Acids Res 39: 4975–4983.

51. Kumari S, Bugaut A, Huppert JL, Balasubramanian S (2007) An RNA G-quadruplex in the 5′ UTR of the NRAS proto-oncogene modulates translation. Nat Chem Biol 3: 218–221.

52. Wieland M, Hartig JS (2007) RNA quadruplex-based modulation of gene expression. Chem Biol 14: 757–763.

53. Spencer CC, Deloukas P, Hunt S, Mullikin J, Myers S, et al. (2006) The inuence of recombination on human genetic diversity. PLoS Genet 2: e148.

54. Arneodo A, d'Aubenton Carafa Y, Bacry E, Graves P, Muzy J, et al. (1996) Wavelet based fractal analysis of dna sequences. Physica D: Nonlinear Phenomena 96: 291–320.

55. Dodin G, Vandergheynst P, Levoir P, Cordier C, Marcourt L (2000) Fourier and wavelet transform analysis, a tool for visualizing regular patterns in DNA sequences. J Theor Biol 206: 323–326.

56. Nason GP (2008) Wavelet Methods in Statistics with R. New York: Springer. URL http://www.springer.com/978-0-387-75960-9. ISBN 978-0-387-75960-9.

57. Deaton AM, Bird A (2011) CpG islands and the regulation of transcription. Genes Dev 25: 1010–1022.

58. Lawson MJ, Zhang L (2008) Housekeeping and tissue-specific genes differ in simple sequence repeats in the 5'-UTR region. Gene 407: 54–62.

59. Brahmachari SK, Meera G, Sarkar PS, Balagurumoorthy P, Tripathi J, et al. (1995) Simple repetitive sequences in the genome: structure and functional significance. Electrophoresis 16: 1705–1714.

60. Yadav VK, Abraham JK, Mani P, Kulshrestha R, Chowdhury S (2008) QuadBase: genome-wide database of G4 DNA–occurrence and conservation in human, chimpanzee, mouse and rat promoters and 146 microbes. Nucleic Acids Res 36: D381–385.

61. Rawal P, Kummarasetti VB, Ravindran J, Kumar N, Halder K, et al. (2006) Genome-wide predic-tion of G4 DNA as regulatory motifs: role in Escherichia coli global regulation. Genome Res 16: 644–655.

62. Verma A, Yadav VK, Basundra R, Kumar A, Chowdhury S (2009) Evidence of genome-wide G4 DNA-mediated gene expression in human cancer cells. Nucleic Acids Res 37: 4194–4204.

63. Darlow JM, Leach DR (1998) Secondary structures in d(CGG) and d(CCG) repeat tracts. J Mol Biol 275: 3–16.

64. Roberts RL, Gearry RB, Bland MV, Sies CW, George PM, et al. (2008) Trinucleotide repeat variants in the promoter of the thiopurine S-methyltrans-ferase gene of patients exhibiting ultra-high enzyme activity. Pharmacogenet Genomics 18: 434–438.

65. Nithianantharajah J, Hannan AJ (2007) Dynamic mutations as digital genetic modulators of brain development, function and dysfunction. Bioessays 29: 525–535.

66. Todd AK, Neidle S (2008) The relationship of potential G-quadruplex sequences in cis-upstream regions of the human genome to SP1-binding elements. Nucleic Acids Res 36: 2700–2704.

67. Halder R, Halder K, Sharma P, Garg G, Sengupta S, et al. (2010) Guanine quadruplex DNA structure restricts methylation of CpG dinucleotides genome-wide. Mol Biosyst 6: 2439–2447.

68. Bacolla A, Pradhan S, Larson JE, Roberts RJ, Wells RD (2001) Recombinant human DNA (cytosine-5) methyltransferase. III. Allosteric control, reaction order, and inuence of plasmid topology and triplet repeat length on methylation of the fragile X CGG.CCG sequence. J Biol Chem 276: 18605–18613.

69. Eddy J, Maizels N (2009) Selection for the G4 DNA motif at the 5' end of human genes. Mol Carcinog 48: 319–325.

70. Rothenburg S, Koch-Nolte F, Haag F (2001) DNA methylation and Z-DNA formation as mediators of quantitative differences in the expression of alleles. Immunol Rev 184: 286–298.

71. Riley DE, Krieger JN (2009) Embryonic nervous system genes predominate in searches for dinu-cleotide simple sequence repeats anked by conserved sequences. Gene 429: 74–79.

72. Kelkar YD, Tyekucheva S, Chiaromonte F, Makova KD (2008) The genome-wide determinants of human and chimpanzee microsatellite evolution. Genome Res 18: 30–38.

73. Kelkar YD, Eckert KA, Chiaromonte F, Makova KD (2011) A matter of life or death: How mi-crosatellites emerge in and vanish from the human genome. Genome Res.

74. Whetstine JR, Witt TL, Matherly LH (2002) The human reduced folate carrier gene is regulated by the AP2 and sp1 transcription factor families and a functional 61-base pair polymorphism. J Biol Chem 277: 43873–43880.

75. Allayee H, Baylin A, Hartiala J, Wijesuriya H, Mehrabian M, et al. (2008) Nutrigenetic association of the 5-lipoxygenase gene with myocardial infarction. Am J Clin Nutr 88: 934–940.

76. Dwyer JH, Allayee H, Dwyer KM, Fan J, Wu H, et al. (2004) Arachidonate 5-lipoxygenase promoter genotype, dietary arachidonic acid, and atherosclerosis. N Engl J Med 350: 29–37.

77. Todur SP, Ashavaid TF (2012) Association of sp1 tandem repeat polymorphism of alox5 with coronary artery disease in indian subjects. Clinical and Translational Science : no–no.

78. Wang S, Wang M, Yin S, Fu G, Li C, et al. (2008) A novel variable number of tandem repeats (VNTR) polymorphism containing Sp1 binding elements in the promoter of XRCC5 is a risk factor for human bladder cancer. Mutat Res 638: 26–36.

79. Borel C, Migliavacca E, Letourneau A, Gagnebin M, Bena F, et al. (2012) Tandem repeat sequence variation as causative Cis-eQTLs for protein-coding gene expression variation: The case of CSTB. Hum Mutat 33: 1302–1309.

80. Herdewyn S, Zhao H, Moisse M, Race V, Matthijs G, et al. (2012) Whole-genome sequencing reveals a coding non-pathogenic variant tagging a non-coding pathogenic hexanucleotide repeat expansion in c9orf72 as cause of amyotrophic lateral sclerosis. Human Molecular Genetics.

81. Treangen TJ, Salzberg SL (2012) Repetitive DNA and next-generation sequencing: computational challenges and solutions. Nat Rev Genet 13: 36–46.

82. McIver LJ, Fondon JW, Skinner MA, Garner HR (2011) Evaluation of microsatellite variation in the 1000 Genomes Project pilot studies is indicative of the quality and utility of the raw data and alignments. Genomics 97: 193–199.

83. Koren S, Schatz MC, Walenz BP, Martin J, Howard JT, et al. (2012) Hybrid error correction and de novo assembly of single-molecule sequencing reads. Nat Biotechnol.

84. McLean CY, Bristor D, Hiller M, Clarke SL, Schaar BT, et al. (2010) GREAT improves functional interpretation of cis-regulatory regions. Nature Biotechnol-ogy 28: 495–501.

85. Baral A, Kumar P, Halder R, Mani P, Yadav VK, et al. (2012) Quadruplex-single nucleotide polymorphisms (Quad-SNP) inuence gene expression differ-ence among individuals. Nucleic Acids Res 40: 3800–3811.

86. Rando OJ, Verstrepen KJ (2007) Timescales of genetic and epigenetic inheritance. Cell 128: 655–668.

87. King DG, Kashi Y (2007) Indirect selection for mutability. Heredity (Edinb) 99: 123–124.

88. Fujita PA, Rhead B, Zweig AS, Hinrichs AS, Karolchik D, et al. (2010) The ucsc genome browser database: update 2011. Nucleic Acids Research.

89. Guberman JM, Ai J, Arnaiz O, Baran J, Blake A, et al. (2011) BioMart Central Portal: an open database network for the biological community. Database (Oxford) 2011: bar041.

90. Smedley D, Haider S, Ballester B, Holland R, London D, et al. (2009) BioMart–biological queries made easy. BMC genomics 10: 22+.

91. Koer R, Schlotterer C, Lelley T (2007) SciRoKo: a new tool for whole genome microsatellite search and investigation. Bioinformatics 23: 1683–1685.

92. R Development Core Team (2011) R: A Language and Environment for Statistical Computing. R Foundation for Statistical Computing, Vienna, Austria. URL http://www.R-project.org. ISBN 3-900051-07-0.

93. Strimmer K (2008) A unified approach to false discovery rate estimation. BMC Bioinformatics 9: 303.

94. Goecks J, Nekrutenko A, Taylor J, Afgan E, Ananda G, et al. (2010) Galaxy: a comprehensive approach for supporting accessible, reproducible, and transpar-ent computational research in the life sciences. Genome Biol 11: R86.

95. Blankenberg D, Von Kuster G, Coraor N, Ananda G, Lazarus R, et al. (2010) Galaxy: a web-based genome analysis tool for experimentalists. Curr Protoc Mol Biol Chapter 19: 1–21.

96. Giardine B, Riemer C, Hardison RC, Burhans R, Elnitski L, et al. (2005) Galaxy: a platform for interactive large-scale genome analysis. Genome Res 15: 1451–1455.

97. Brandstrom M, Bagshaw AT, Gemmell NJ, Ellegren H (2008) The relationship between microsatellite polymorphism and recombination hot spots in the human genome. Mol Biol Evol 25: 2579–2587.

2

Age-Specific Signatures of Glioblastoma at the Genomic, Genetic, and Epigenetic Levels

Serdar Bozdag[1,2]*, Aiguo Li[1], Gregory Riddick[1], Yuri Kotliarov[1,3], Mehmet Baysan[1], Fabio M. Iwamoto[1,4], Margaret C. Cam[1], Svetlana Kotliarova[1], Howard A. Fine[1,5]

1 Neuro-Oncology Branch, National Cancer Institute, National Institute of Neurological Disorders and Stroke, National Institutes of Health, Bethesda, Maryland, United States of America, **2** Department of Mathematics, Statistics, and Computer Science, Marquette University, Milwaukee, Wisconsin, United States of America, **3** Center for Human Immunology, Autoimmunity and Inflammation, National Heart Lung and Blood Institute, National Institutes of Health, Bethesda, Maryland, United States of America, **4** The Neurological Institute of New York, College of Physicians and Surgeons, Columbia University, New York, New York, United States of America, **5** New York University Cancer Institute, New York University Langone Medical Center, New York, New York, United States of America

Abstract

Age is a powerful predictor of survival in glioblastoma multiforme (GBM) yet the biological basis for the difference in clinical outcome is mostly unknown. Discovering genes and pathways that would explain age-specific survival difference could generate opportunities for novel therapeutics for GBM. Here we have integrated gene expression, exon expression, microRNA expression, copy number alteration, SNP, whole exome sequence, and DNA methylation data sets of a cohort of GBM patients in The Cancer Genome Atlas (TCGA) project to discover age-specific signatures at the transcriptional, genetic, and epigenetic levels and validated our findings on the REMBRANDT data set. We found major age-specific signatures at all levels including age-specific hypermethylation in polycomb group protein target genes and the upregulation of angiogenesis-related genes in older GBMs. These age-specific differences in GBM, which are independent of molecular subtypes, may in part explain the preferential effects of anti-angiogenic agents in older GBM and pave the way to a better understanding of the unique biology and clinical behavior of older versus younger GBMs.

Editor: Waldemar Debinski, Wake Forest University, United States of America

Funding: This work was supported by the Intramural Research Program of the National Institutes of Health (NIH), National Cancer Institute. The funders had no role in study design, data collection and analysis, decision to publish, or preparation of the manuscript.

Competing Interests: The authors have declared that no competing interests exist.

* E-mail: serdar.bozdag@marquette.edu

Introduction

Glioblastoma multiforme (GBM) is the most common malignant primary brain tumor [1]. GBM patients have a median survival of about fourteen months despite aggressive multimodality treatment [2]. Given the pathological and clinical heterogeneous nature of GBMs, there have been a number of recent attempts to better understand and characterize these tumors at the molecular and genetic level [3–16]. Among these studies, The Cancer Genome Atlas Project (TCGA) has generated a vast amount of high-throughput data for about 500 GBM samples [4,15].

Advanced age has been identified as an independent significant prognostic factor for survival in glioblastoma clinical trials since the 1970s (Table S1). An analysis of three randomized phase III trials conducted by the Radiation Therapy Oncology Group (RTOG) showed that median survival of GBM patients aged 60 or older was 7.5 months compared to 16.2 months in patients younger than 40 years old [17]. Older age as negative prognostic factor for GBM survival was confirmed by other National Cancer Institute sponsored cooperative groups [18,19] and a large meta-analysis of 3,004 patients with high-grade gliomas [20]. The study that established the current standard of care for newly diagnosed glioblastoma with radiation and concurrent temozolomide also showed a shorter median survival of patients older than 60 years (11.4 months) compared to those who were 50 years or younger (17.4 months) [21].

The reasons why older age is such a negative prognostic factor remain unclear. Retrospective data and randomized controlled trials do not suggest that older patients receive less than optimal treatment and/or tolerate treatment less well than younger patients thereby suggesting a potential difference in the biology of GBMs in older patients. Thus, it would be valuable to discover age-specific signatures in GBM biology that might explain this survival difference and allow clinicians to develop age-specific therapeutic clinical trials for GBM.

Noushmehr, *et al.* discovered a glioma-CpG island methylator phenotype (G-CIMP) in GBMs [22]. G-CIMP positive patients (about 11% of GBM samples in TCGA) have significantly longer survival than G-CIMP negative patients. G-CIMP positive patients are also significantly younger than G-CIMP negative patients. Nevertheless, age still turns out to be a significant independent prognostics factor for survival despite the G-CIMP status of the tumor [22].

In this study, we computationally analyzed gene expression, exon expression, microRNA expression, DNA methylation, copy number alteration, somatic mutation derived from whole exome sequence, and SNP data sets of the TCGA GBM samples to discover age-specific signatures at the transcriptional, genetic, and epigenetic levels. In order to avoid the confounding variable of the

G-CIMP status of the tumor, we trained a model to predict G-CIMP status of GBM samples based on gene and exon expression profiles in order to exclude G-CIMP positive patients in our analyses.

Materials and Methods

Determining Old and Young GBM Groups

For two-sample tests, we defined a "young" and "old" group. We hypothesized that if there is an old and young biology then samples with "intermediate" ages might represent a mix of these biologies. Thus, we did not include samples with "intermediate" ages in our old and young groups. In order to define the age boundaries for the old and young groups, we examined the histogram of survival for different age groups (Figure S1) and number of samples in each age group (Figure S2). We assigned patients ≤40 years old to the *young* group and patients ≥70 years old to the *old* group. The number of available samples in the young and old groups changes depending on the data set (Table 1), but overall young and old patients constitute 9% and 21% of all samples, respectively. For linear regression tests, we also used samples with *intermediate* ages (i.e., between 40 and 70) to increase the power of the analysis.

Predicting G-CIMP Status of GBM Patients and Removing G-CIMP Positive Patients

We obtained G-CIMP calls of samples that have methylation data from [22] (Figure 1). To predict G-CIMP calls of samples for which no methylation data is available, we used the k-nearest neighbor algorithm to train models from gene and exon expression profiles of samples with a G-CIMP call. The G-CIMP call prediction results from models of gene and exon expression overlapped by more than 95%. We chose consistent prediction calls as final G-CIMP calls. All analyses were conducted on

Partek® Genomics SuiteTM version 6.5.

G-CIMP positive patients are significantly younger than G-CIMP negative patients and there is a significant difference between G-CIMP positive and negative patients at the transcriptional, genetic, and epigenetic levels [22]. If we compared the old and young groups without eliminating G-CIMP positive samples, some of the G-CIMP-specific signatures would be potentially considered as age-specific (i.e., Type I error). In order to eliminate this error, we excluded the G-CIMP positive samples, which constituted about 11% of the database in our analysis.

Computing Age-specific Differentially Expressed/methylated Genes and microRNAs

We used data sets from Affymetrix U133A, Affymetrix Human Exon 1.0, and Agilent 244 K G4502A platforms in TCGA for differential gene expression analysis (Table 1). We used DNA methylation data set from Illimuna Infinium Human DNA Methylation 27 in TCGA for differential methylation analysis (Table 1). The sample IDs for each data set are listed in Table S2. We used data set from Agilent 8×15 K Human miRNA-specific microarray in TCGA for differential microRNA expression analysis (Table 1). Finally, we used 100 microRNA-specific probes in the Illumina Infinium platform for differential microRNA methylation analysis. We applied two-sample t-test to compute age-specific differentially expressed genes (DEGs) between the old and young groups. Considering age as a continuous variable, we applied linear regression to compute age-specific DEGs. We also applied a nonparametric *ranked-based* linear regression to find age-specific differentially expressed microRNAs and differentially methylated genes (DMGs). We used a ranked-based linear regression on microRNA expression and DNA methylation data, since these data were not normally distributed. We used the samr v1.28 package in R [23] for all tests. For multiple test correction, we applied a permutation-based FDR threshold of 0.05.

Table 1. Number of GBM samples used in this study (downloaded from the TCGA repository on June 29, 2011, Sample IDs are in Table S1).

Data Type	Platform	Level[1]	Institute	# Old[2]	# Young[2]	Total
Gene expression	Affymetrix HT Human Genome U133 Array Plate Set	2	Broad Institute of MIT and Harvard	92	37	422
Exon expression	Affymetrix Human Exon 1.0 ST Array	3	Lawrence Berkeley National Laboratory	80	34	382
Gene expression	Agilent 244K Custom Gene Expression G4502A	2	University of North Carolina	92	37	420
miRNA expression	Agilent 8×15K Human miRNA-specific microarray	3	University of North Carolina	80	34	385
Methylation	Illumina Infinium Human DNA Methylation 27	2	Johns Hopkins/University of Southern California	56	22	256
Copy Number	Agilent Human Genome CGH Microarray 244A	3	Memorial Sloan-Kettering Cancer Center	87	36	406
SNP	Affymetrix Genome-Wide Human SNP Array 6.0	3	Broad Institute of MIT and Harvard	88	32	390
SNP	Illumina 550K Infinium HumanHap550 SNP Chip	3	HudsonAlpha Institute for Biotechnology	78	33	376
Whole Exome Sequence	Illumina Genome Analyzer IIx	N/A	Broad Institute of MIT and Harvard	55	12	202

[1]Level 2 refers to probeset-level data and level 3 refers to gene-level data for expression and methylation data sets. Level 3 refers to segmented data for copy number and SNP data sets. There is no level number for whole exome sequence data set as we just used the mutations derived from this data set.
[2]Old and Young refer to samples ≥70 and ≤40 years old, respectively.

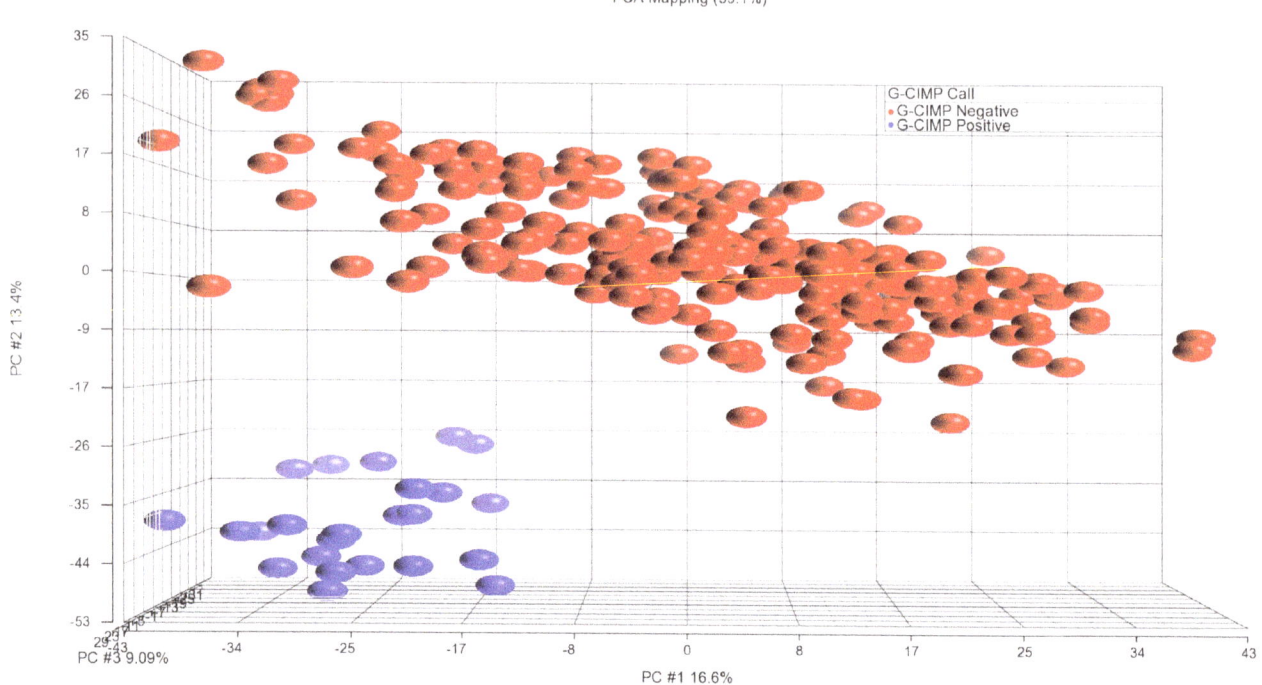

Figure 1. PCA plot of GBM samples with methylation data. Red: G-CIMP negative, Blue: G-CIMP positive. Methylation sites with std. deviation >0.2 are selected to generate this graph.

Computing Age-specific Differentially Altered Genes

We used segmented copy number/SNP data sets in Agilent HG CGH 244 K, Affymetrix Human SNP 6.0, and Illumina 550 K Infinium HumanHap550 platforms in TCGA (Table 1). The samples from Agilent HG CGH 244 K data set cover all samples in Infinium HumanHap550 data set and 95% of the samples in SNP6 data set (Figure S3). We performed the analysis on each data set independently. We generated a project in Nexus v5.1 (BioDiscovery Inc., El Segundo, CA, USA) for each data set and used Nexus' *comparison* function to find differentially altered regions between old and young groups (q-value≤0.05). The *comparison* function compares the frequency of alteration in both groups and finds areas where there is significant difference in frequency [24].

Somatic Mutation Data Analysis

We used somatic mutations derived from whole exome sequences in TCGA (TCGA Analysis Working Group Data Release Package 1, 8/26/2011). We performed Fisher's exact test to find genes that are significantly mutated in old or young GBMs.

Survival Analysis

We applied Cox multivariate analysis on variables namely age, molecular subtypes derived in [15] (i.e., classical, neural, mesenchymal, proneural), gender, and Karnofsky performance score. We used *coxph* function in R [25]. We also generated Kaplan-Meier survival plots in Partek® Genomics Suite™ version 6.6.

Functional Analysis

We applied the gene set enrichment analysis (GSEA) algorithm [26,27] to identify upregulated expression pathways and signatures by comparing old and young groups. GSEA mapped all 3272 gene sets in the functional c2 v3 MsigDB database to ranked genome-wide expression profiles (Affymetrix U133A) of old versus young groups. To compute p-values for enrichment scores, we applied Kolmogorov-Smirnov statistics by constructing a cumulative null distribution with permuting old and young group assignments 1000 times. The significant gene sets were claimed for nominal $p \leq 0.05$. We also used DAVID [28] to create functional annotation charts on age-specific upregulated genes derived from both Affymetrix U133A and Agilent 244 K G4502A data sets via a linear regression method (Fisher's exact test $p \leq 0.05$) and Ingenuity Pathway Analysis (IPA) software (Ingenuity® Systems, www.ingenuity.com) on age-specific angiogenesis related genes to display the interactions among these genes.

Motif Enrichment Analysis

In order to compute enriched motifs in the promoters of the age-specific DEGs, we used PScan [29] to find enriched motifs in the JASPAR database [30]. We checked sequences from 450 bp upstream to 50 bp downstream of the transcription start size for each Refseq transcript of the gene in human genome version hg19. We applied a Benjamini-Hochberg multiple test correction method [31] to correct for multiple testing.

Cross-validation on TCGA Data Set

To validate age-specific DEGs, we applied 10-fold cross-validation on TCGA U133A data set of old and young patients. For each fold, we used the support vector machine (SVM) algorithm to build a model based on training data (i.e., age-specific DEGs obtained from 90% of samples) and used this model to predict the old/young status of the remaining 10% of samples. For comparison, we also used the same algorithm to build a model based on the molecular subtype-specific DEGs of training data. We used Partek® Genomics SuiteTM version 6.5 and tried different parameter choices for the SVM algorithm. To compute DEGs, we used one-way ANOVA.

Validation on External Data Set

To validate the age-specific signatures derived from the TCGA data set, we obtained gene expression profiles of GBM samples from the REMBRANDT database (http://caintegrator-info.nci.nih.gov/rembrandt). We predicted the G-CIMP status of these samples as described above. There were 153 G-CIMP negative samples (27 old, 15 young, and 111 intermediate). Due to the small sample size, we were unable to compute statistically significant age-specific DEGs on this data set. We, therefore, filtered in the age-specific genes at the transcriptional level derived from both TCGA Agilent 244 K G4502A and Affymetrix U133A data sets (hereafter the TCGA age signature) in the REMBRANDT data set to create a *filtered* data set. We clustered the filtered data set via hierarchical clustering to see if old and young samples would be separated by TCGA age signature. We also clustered the unfiltered REMBRANDT data set (i.e., all genes) and compared both results. As a more quantitative approach, we also built an ANOVA model on filtered and unfiltered REMBRANDT data sets to compute how much of the variation could be explained by the age group (i.e., old and young). We built a 3-way ANOVA by using gender, age group, and sample source institute as categorical factors. We used Partek® Genomics Suite^{TM} version 6.6 for clustering and building ANOVA model.

Results

Age is an Independent Significant Prognostic Factor for Survival within G-CIMP Negative GBMs

Age is known to be an independent significant prognostic factor for survival in GBMs [1,22]. Our multivariate Cox regression analysis on G-CIMP negative GBM samples also demonstrated that age is an independent significant factor for survival within G-CIMP negative GBM patients (p-value<5.02e-07, Table S3). The Kaplan-Meier plot also shows that there is significant survival difference between old and young GBM samples (log-rank p≤2.42e-08, Figure 2). Of note, our results show that GBM molecular subtypes are not a significant factor for survival as previously reported [15].

Age-specific Signature at the Transcriptional Level

We computed age-specific DEGs by using both t-test and linear regression. DEGs found by linear regression mostly contain DEGs found by t-test in all three platforms suggesting that the use of all samples gives more power to the analysis (Table 2). Using linear regression, we found 1749, 909, and 91 DEGs in Affymetrix U133A, Agilent G4502A, and Affymetrix Human Exon 1.0 platforms, respectively (FDR<0.05, Table S4). The low number of DEGs in the exon data set is possibly due to the lower sample size compared with the other data sets. There are 334 DEGs found both in Agilent G4502A and Affymetrix U133A platforms. Among these DEGs, seven of them (*SOD2, GTPBP4, TPST1, GNA12, SIM2, ZFP2, and SLC22A5*) have also age-specific differential expression in normal brain tissues based on a data set described in [32]. Two hundred and thirty of these genes (69%) are upregulated in older GBMs. There are fourteen genes that were found by both tests in all three platforms (upregulated in old: PRUNE2, TMEM144, SLC14A1; downregulated in old: H2AFY2, ENOSF1, SFRP1, RANBP17, SVIL, TUSC3, ATF7IP2, FZD6, TSPYL5, DLK1, HIST3H2A). A number of these genes are of apparent interest for GBM biology such as TUSC3, which is a tumor suppressor candidate gene and known to be

hypermethylated in GBMs [33]. Additionally, SFRP1 and FZD6 are in the Wnt signalling pathway [34].

Age-specific microRNA Expression Signature

We applied ranked-based linear regression and found 19 differentially expressed microRNAs (FDR<0.05) (ebv-miR-BART1-5p, hsa-miR-422b, hsa-miR-507, hsa-miR-147, ebv-miR-BHRF1-2, hsa-miR-620, hsa-miR-554, hsa-miR-625, hsa-miR-661, hcmv-miR-UL70-5p, hsa-miR-325, hsa-miR-453, hsa-miR-552, hsa-miR-558, hsa-miR-223, hsa-miR-302c, hsa-miR-142-5p, hsa-miR-649, hsa-miR-142-3p). All these microRNAs are downregulated in older GBMs. We used the mirWalk database [35] to find experimentally validated targets of these microRNAs. We found 172 experimentally validated target genes (Table S5). Two of these target genes are upregulated in older GBMs (LOX, VEGFA). VEGFA is known to be upregulated in older GBMs [16,36]. LOX and HIF-1 act in synergy to help tumor cells adapt to hypoxia [37].

Age-specific Signature at the Epigenetic Level

We found 389 age-specific DMGs by using ranked-based linear regression (Table S6). Ninety-eight percent of these DMGs are hypermethylated in the older GBMs. Seventeen genes that are hypermethylated in the older GBMs are polycomb group protein target genes (PCGTs) (Table S7, Fisher's exact test p-value<1.0e-10). Hypermethylation of PCGTs has been previously shown to be associated with aging [38]. We subtracted out genes that are normally methylated in an age specific manner based on previous data sets [39], and found 184 and four genes that are uniquely hypermethylated in the old and young GBMs, respectively (Table S8). Eighteen of the GBM-specific DMGs exist in the Pubmeth database [40], which stores genes that are known to undergo methylation in cancer (Table S9, Fisher's exact test p-value<1.27e-05). Eleven genes are both differentially expressed (Agilent and Affymetrix U133A platforms) and methylated with respect to age (Table S10). Seven of them are hypermethylated and downregulated in older GBMs (MYO1B, PRKCB1, VRK2, FZD6, DLK1, SLC25A21, MSC). We also found three differentially methylated microRNAs (hsa_miR_196b, miR_34b, and miR_34c), all of which were hypermethylated in the old group.

Age-specific Signature at the Genetic Level

Each copy number/SNP data set was analyzed in Nexus independently. Figure 3a–c shows the whole genome copy number alteration (CNA) profiles of the old and young groups on these data sets. We found 1044 and 455 differentially altered genes (DAGs) in Affymetrix SNP 6 and HG-CGH 244A platforms, respectively (Table S11). The DAG list found in SNP 6 platform covers 88% of DAGs found in HG-CGH 244A. We could not detect any DAGs on HumanHap550 platform possibly due to the low resolution of this data set. We found the largest DAG list on SNP 6 platform possibly because of its high resolution.

Analyzing the SNP 6 data set, we detected differential deletions only on chromosome 10 for 722 genes. We observed that the old group had a higher frequency of deletion than the young group. We found 321 differentially amplified genes on chromosome 7 with a higher frequency in the old group than the young group and one gene on chromosome 1q (CFHR3) with a higher frequency in the young group than the old group. The high frequency of chromosome 10 deletion and chromosome 7 amplification in the old group, and high frequency of chromosome 1q amplification in the young group have also been reported in a study that compared a cohort of pediatric GBMs with adult GBMs [41].

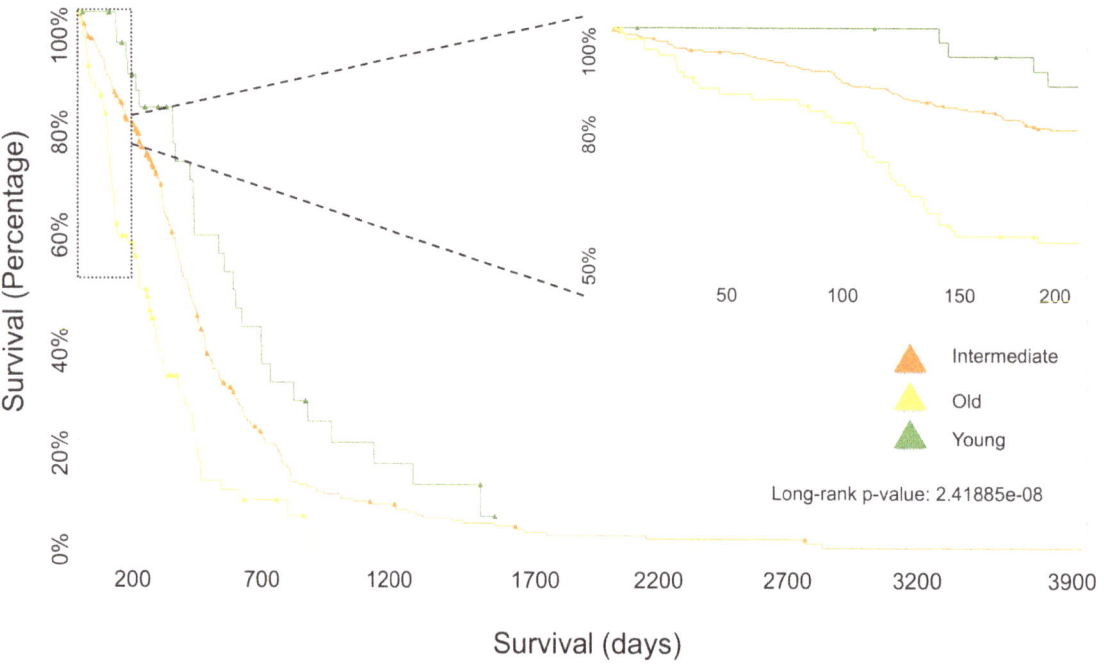

Figure 2. Kaplan-Meier plot between old, young, and middle-aged GBM samples.

We compared the list of DAGs on SNP 6 and the list of DMGs, and found three genes that are both heterozygously deleted and hypermethylated in the old group (HHEX, ITGA8, RASGEF1A). Among these genes, HHEX is downregulated in older GBMs. HHEX is known to downregulate VEGF and VEGF receptors [42]. We also observed a significant negative correlation between HHEX and VEGFA expression levels in the TCGA data set (Table S12).

We also compared the list of DAGs on SNP 6 and the list of DEGs found on both Agilent G4502A and Affymetrix U133A data sets. There are 21 genes that are deleted and downregulated, and 7 genes that are amplified and upregulated in the old group (Table S13).

We also analyzed somatic mutations derived from whole exome sequence data. In general, there are more mutations in the old group (Table 3). There are two genes that stand out: TP53 is mutated in 19 old samples and in only one young sample (Fisher's exact test p-value<0.068). GRM3 is mutated in 3 out of 12 young samples and none of the old samples (Fisher's exact test p-value<0.01645).

Motif Enrichment Analysis

We analyzed the promoter regions of differentially expressed genes appear in Affymetrix U133A and Agilent G4502A data sets. We have found several motifs statistically enriched in the promoter regions of these genes including HIF-1A and MYC (FDR≤0.05 Table S14).

Functional Analysis of Age-specific DEGs

We ran a GSEA on the old and young groups to discover upregulated gene sets for each group (Table S15). GSEA analysis found that the younger GBMs maintain an active regulation of G1 entry checkpoint in cell cycle (p<0.05) and have a quiescent phenotype (p<0.03). Older GBMs uphold a strong oxygen depletion environment (p<0.04) that induces the hypoxia inducible factor signaling as indicated by three up-regulated HIF signatures (p<0.05) (Table S15). Furthermore, carbohydrate metabolism with over-expressed glycolysis and glucagon signaling reactomes are enhanced in older GBMs. Younger GBMs showed enrichment of P38_MAPK signaling (p<0.01) and upregulated targets of MYC (p<0.04), BMYB signature (p<0.02), and enhanced stem cell signatures (p<0.02) (Table S15). Moreover, a premalignant signature driven by hepatic stem cell marker, epithelial cell adhesion molecular (EpCAM) is enhanced in younger GBMs (p<0.04) whereas older GBMs showed more advanced tumor profiles (p<0.01) and more invasive expression signatures regulated by integrin-mediated cell migration (p<0.03). Additionally, glioblastoma tumor in young patients showed an increased TNF signaling (p<0.05) and protein translational

Table 2. Number of differentially expressed genes between Old and Young GBM samples for three transcriptomic platforms.

	T-test[1]	Linear regression[1]	Common
Affymetrix HT HG U133A	630	1749	**595**
Affymetrix Human Exon 1.0 ST	62	91	**40**
Agilent 244K G4502A	348	909	**313**
Common (U133A and G4502A)	130	334	115
Common (all three platforms)[2]	17	40	14

The last row shows the number of differentially expressed genes found in all three platforms.
[1]In each test, FDR≤0.05 threshold is applied.
[2]Shows the number of differentially expressed genes found in all three platforms.

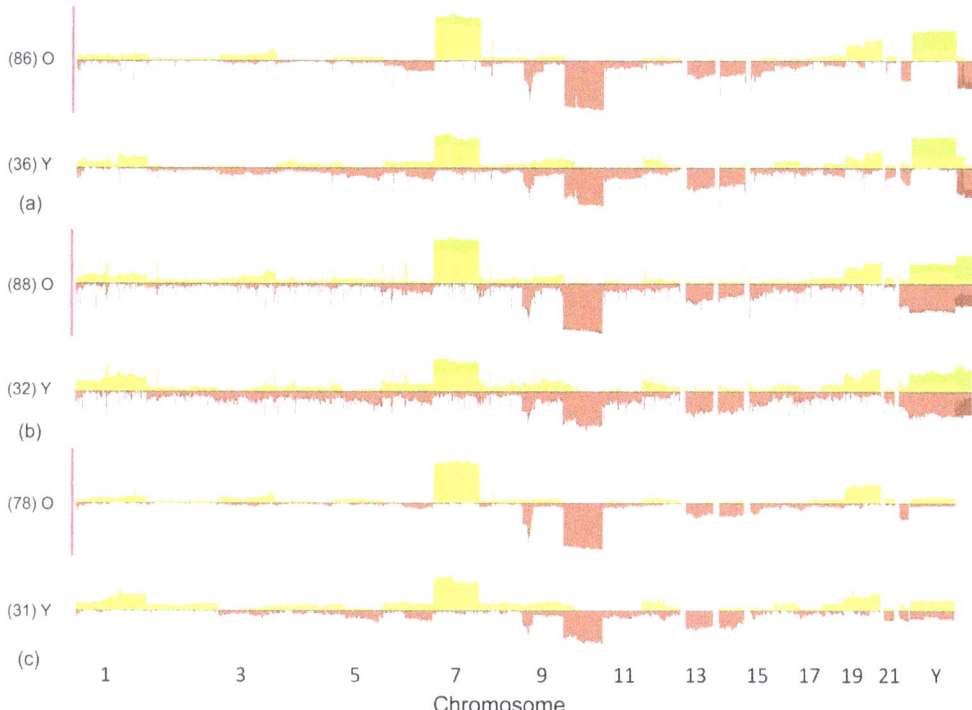

Figure 3. Genome-wide copy number alteration profiles of old and young GBM samples. Data are from (a) Agilent Human Genome CGH Microarray 244A (Memorial Sloan-Kettering Cancer Center), (b) Affymetrix Genome-Wide Human SNP Array 6.0 (Broad Institute of MIT and Harvard), (c) Illumina 550 K Infinium HumanHap550 SNP Chip (HudsonAlpha Institute for Biotechnology) platforms (chr 1–23). Green bars represent amplification and red bars represent deletion. The height of each bar represents the frequency of the alteration in the group. The differentially amplified genes are in chromosome 7 and differentially deleted genes are in chromosome 10.

activities (p<0.03), as indicated by the formation of translation initiation complex involving 43S unit (Table S15).

We also ran DAVID on upregulated genes that appear in the Affymetrix U133A and Agilent G4502A data sets. We found enrichment in several GO terms such as "response to hypoxia" (p-value<0.00123, enriched genes: VEGFA, SOD2, BNIP3, SLC11A2, EGLN3, PLOD2, NOL3, and ALDOC); "vasculogenesis" (p-value<0.088, enriched genes: VEGFA, NTRK2, and QKI) (Table S16).

Cross-validation on TCGA Data Set

We applied 10-fold cross-validation and built a model based on age-specific DEGs on training data and used this model to predict

the old/young status of the remaining test samples. The model achieved over 77% prediction accuracy. For comparison, we also used the same algorithm to build a model based on the molecular subtype-specific DEGs of training data. This model predicted about 64% of prediction accuracy.

Validation on External Data Set

We obtained the gene expression profiles of G-CIMP negative GBM samples from the REMBRANDT database to validate our findings. We created a filtered REMBRANDT data set by filtering in the TCGA age signature (i.e. age-specific differentially expressed genes derived from both TCGA Affymetrix U133A and Agilent 244 K G4502A data sets via linear regression). We

Table 3. Number of mutated genes in old and young GNEG GBMs.

Number of samples[1]	Number of mutated genes in Old	Number of mutated genes in Young
>0	3038	720
>1	561	37
>2	159	3 (GRM3, TTN, PTEN)
>3	49	0
>6	8 (PTEN, EGFR, MUC16, TTN, TP53, RYR2, SLIT3, LRP2)	0
>8	5 (PTEN, EGFR, MUC16, TTN, TP53)	0
>15	2 (PTEN, TP53)	0

[1]Shows number of old of young samples each gene is mutated. For instance, there are 3038 genes that are mutated in at least one old sample (see first row) and PTEN is mutated in more than 15 old samples (see last row).

clustered both filtered (Figure 4-A) and unfiltered (i.e. all genes) REMBRANDT data sets (Figure 4-B). To create a reference point, we also clustered old and young GBMs in TCGA Affymetrix U133A data set based on the TCGA age signature (Figure 4-C). We observed that the separation between old and young groups is more apparent in the cluster on filtered data set than the separation on the unfiltered data set (Figure 4). We also observed that the separation of old and young samples in the clustering of filtered data set are very similar to the separation of old and young samples in TCGA Affymetrix data set (Figure 4). Additionally, we built a 3-way ANOVA model on both filtered and unfiltered REMBRANDT data sets by using the categorical factors of age group (i.e., old and young), gender, sample source institute to compute how much of the variation in gene expression is explained by each factor (Figure 5, 6). The results show that age group explains the majority of the variation in the filtered data set, whereas it could not explain the variation in the unfiltered data set.

We also checked whether TCGA DEGs have the same direction of regulation (up or down) in REMBRANDT data set. We applied age-specific linear regression to compute p-value for TCGA DEGs in REMBRANDT data set and created a gene list for both FDR<0.05 and unadjusted p-value<0.05 thresholds. There were 55 and 148 genes in these gene lists, respectively, which had 100% consistency with respect to the directionality of regulation in TCGA and REMBRANDT data sets.

Discussion

Age has consistently been shown to be one of the most powerful prognostic factors for survival in patients with malignant gliomas

with younger patients generally living much longer than older patients. The negative effects of age seen in a number of systemic cancers have often been ascribed to the physiological stress of metastatic cancer in the setting of concurrent medical problems leading to an increased rate of medical related deaths [43]. Additionally, the poor tolerance of older patients to aggressive toxic systemic chemotherapy often results in either treatment-related complications and/or suboptimal tumor treatment [44]. These clinical variables do not, however, adequately explain the profoundly negative effect of age in patients with GBM since such patients almost never have metastatic disease and do not usually die of concurrent medical problems. Furthermore, the marginal effects of systemic chemotherapy in patients with GBMs means that patients are generally treated less aggressively than patients with systemic cancer and the amount of chemotherapy that GBM patients receive has little impact on overall survival. Furthermore, there are few data to suggest that involved field radiotherapy, the one effective treatment for GBM, is associated with increased mortality in older versus younger patients. The lack of a clinical explanation for the poorer survival of older patients with GBM, together with the growing appreciation of the heterogeneous nature of the disease, leads to the hypothesis that the impact of age on survival may be do to a difference in the biology of GBMs in older patients compared to that in younger patients.

There have been a relatively large number of studies over the last decade demonstrating that GBM is a heterogeneous tumor with the most recent studies suggesting that there are at least four major molecular subtypes of GBM based on gene expression profiling [15]. Those major subtypes, however, do not account for

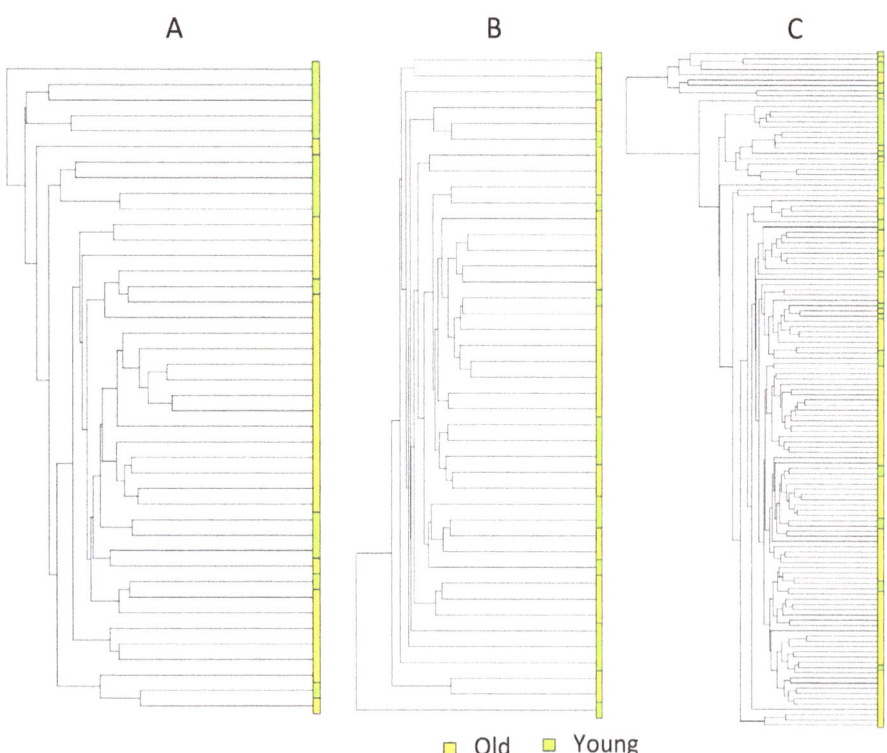

Figure 4. Hierarchical clustering of GBM samples in the REMBRANDT and TCGA data sets. (A) Clustering of old and young REMBRANDT GBM samples based on the expression profiles of age-specific genes derived from both TCGA Affymetrix U133A and Agilent G4502A data sets. (B) Clustering of old and young REMBRANDT GBM samples based on the expression profiles of all genes in the REMBRANDT data set. (C) Clustering of the old and young TCGA GBM samples based on the expression profiles of age-specific genes derived from both TCGA Affymetrix U133A and Agilent G4502A data sets.

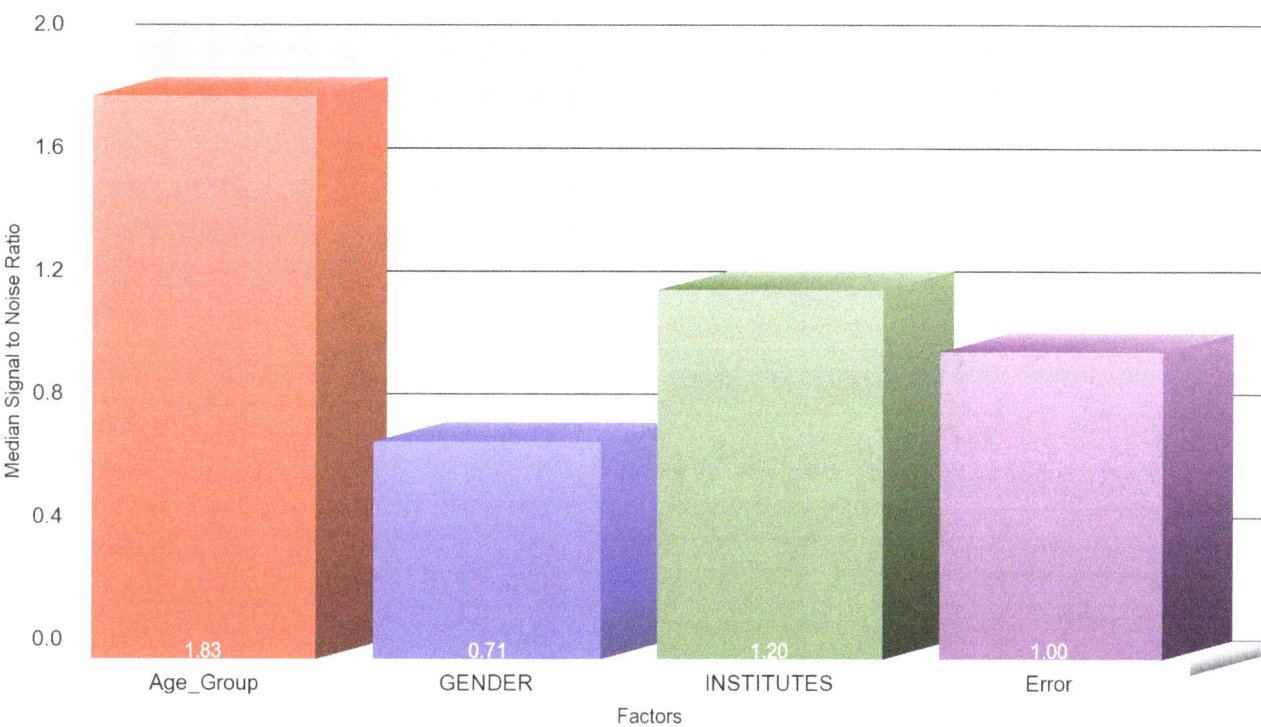

Figure 5. Source of variation of expression profiles of age-specific genes in the REMBRANDT data set. The x-axis shows the components of the 3-way ANOVA model and the y-axis shows the median signal to noise ratio. The ANOVA model is built based on the expression profiles of the TCGA age-specific genes in REMBRANDT data set. The TCGA age-specific genes are the intersection of DEGs computed on TCGA Affymetrix U133A and Agilent G4502A data sets.

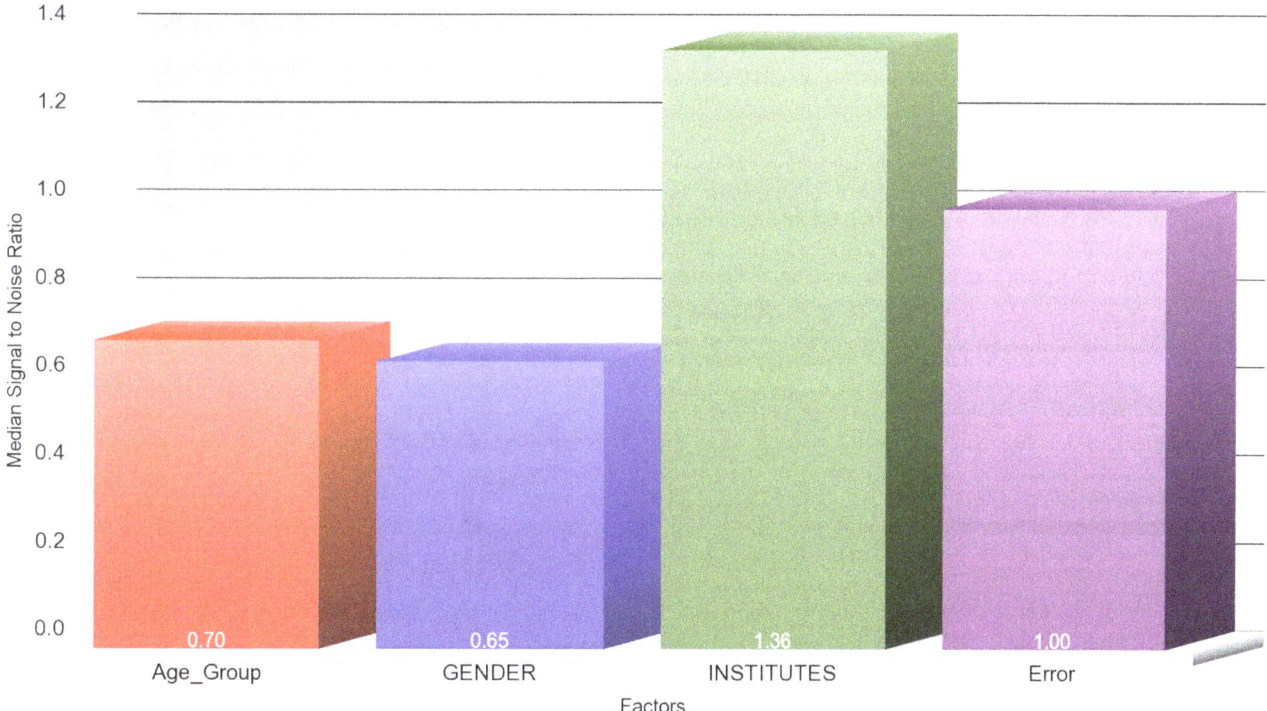

Figure 6. Source of variation of expression profiles of all genes in the REMBRANDT data set. The x-axis shows the components of the 3-way ANOVA model and the y-axis shows the median signal to noise ratio. The ANOVA model is built based on the expression profiles of all genes in REMBRANDT data set.

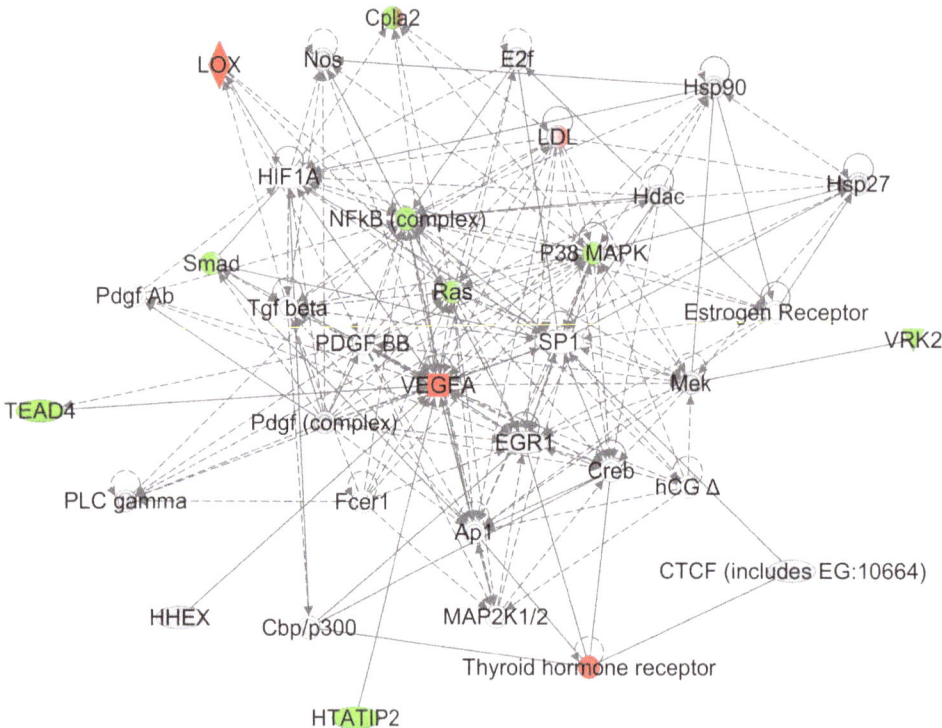

Figure 7. IPA network of angiogenesis-related genes. Node color represents the expression status based on Affymetrix U133A data set (Red: Upregulated in old GBMs, Green: Downregulated in old GBMs, Gray: Baseline, White: Unknown, Mix of green and red: both upregulated and downregulated genes in the complex).

the effect of age on survival. Recently, however, the G-CIMP positive subgroup of the proneural subtype of GBM was described based on a pattern of differential genomic methylation [22]. G-CIMP positive GBM patients tend to be younger and have a significantly longer survival than the G-CIMP negative GBMs thus accounting for some of the age-associated effects on survival [22]. Nevertheless, our data demonstrates that age remains a powerful predictor of survival amongst G-CIMP negative tumors. Thus, we sought to elucidate a biological basis for this age–related effect. To do so, we integrated the high-throughput transcriptomic, genetic, and epigenetic profiles of about 425 GBM samples in the TCGA project to find age-specific signatures at the transcriptional, genetic, and epigenetic levels and found such differences at all levels.

We observed a relative small number of DAGs that consistently differentiated old versus young GBMs. Specifically we found that chromosome 10 deletion and chromosome 7 amplification was found commonly in the old group whereas there was a relatively high frequency of chromosome 1q amplification in the young group, observations that have been previously reported in a study that compared a cohort of pediatric GBMs with adult GBMs [41].

In contrast to the relatively few consistent genomic changes between each group of tumors, we did observed a large age-specific signature at the transcriptional level. We observed a major overlap between the DEGs found in Affymetrix U133A and Agilent G4502A platforms, although not surprisingly, each platform had unique DEGs as described in [45]. We observed fewer DEGs on the Affymetrix Human Exon 1.0 platform possible due to the lower sample size in this platform.

We applied 10-fold cross validation on TCGA U133A data set to validate age-specific DEGs. We applied the SVM algorithm to build a model based on age-specific DEGs of training samples and

applied this model to predict old/young status of the test samples. This model achieved over 77% prediction accuracy whereas a model based on molecular subtype-specific DEGs only achieved 64% accuracy.

We also validated age-specific signatures at the transcriptional level on an external data set. We obtained G-CIMP negative old and young GBM samples from the REMBRANDT database and clustered these samples based on the TCGA age signature genes and all genes. The clustering results showed that the TCGA age signature could separate the old and young REMBRANDT samples as well as it separates the old and young TCGA samples. We also showed that when the TCGA age signature was selected in the REMBRANDT data set, the majority of the variation could be explained by age group (i.e., old and young). The age group, however, could not explain the variation on the entire REMBRANDT data set. We also showed that the upregulated (downregulated) TCGA DEGs are also upregulated (downregulated) in REMBRANDT data set. These findings indicate that the TCGA age-specific signature at the transcriptional level is also age-specific in REMBRANDT data set.

We also observed a large age-specific signature at the epigenetic level. In particular, we found that about 98% of the DMGs between the old and young group are hypermethylated in the old group. There are several studies that show that aging increases methylation of DNA including cancer-related genes [38,46–48]. The hypermethylated genes in the old group enriched for PCGTs that are known to be undergo methylation with aging [38]. It has been also shown that PCGTs in stem cells undergo hypermethylation with aging and this methylation locks the cell in an undifferentiated state. Thus, our results are consistent with the cancer stem cell hypothesis of gliomagenesis being most prevalent in the older GBMs, in part through hypermethylation of PCGTs.

Obviously, such a hypothesis awaits further experimental validation, as do the significance and meaning of our other age-specific epigenetic signatures.

The primary focus of this study was to identify the significant genomic, genetic and epigenetic signatures between young and old GBMs for hypothesis generation and future study and although the annotation and the biological significance of these changes are well beyond the scope of this manuscript, there is one striking observation worth noting. We found a significant number of genes involved in the hypoxic response and in angiogenesis deregulated in older GBMs compared to younger GBMs. In particular, we found a remarkable number of genes involved in the regulation of the proangiogenic protein, VEGF, deregulated in the older GBMs. This is consistent with older pathology-based studies that have demonstrated VEGF to be more highly expressed in older GBMs than in younger GBMs. Examples of the genes we found deregulated at the expression level or through transcriptional factor motif activation in older GBM that contribute to VEGF expression include HIF, HHEX, EGR1, CTCF, HTATIP2, lox, and DLK1.

To better elucidate the potential angiogenesis-related signaling aberrations found in older GBMs, we entered into the IPA network analysis a number of the genes deregulated in older GBMs at either the transcriptional level or at the TF motif enrichment level that have been associated with angiogenesis in the literature. These genes included EGR1 [49], VEGF [50], CTCF [50], Myc [51], Mycn [52], Sp1 [53], MSX1 [54], NDRG2 [55], HTATIP2 [56], VRK2 [57], TEAD4 [58], PKRCB1 [59], HHEX [42], HIF1 [37], DLK1 [60], and Lox [37]. The resulting IPA-generated network (Figure 7) demonstrates a complex network with a number of prototypic deregulated GBM genes (i.e. HIF1A, PDGF, TGF-b, Creb, and HCG) located at key nodes. Most prominently displayed in this network is the central role of VEGF.

Thus, it appears that a key biological difference between older GBMs compared to younger GBMs is the central role of VEGF and angiogenesis signaling. Although we cannot know for certain that the more prominent angiogenic profile of older GBMs is responsible in part for the shorter survival of older patients with these tumors, there is an extensive literature linking greater angiogenic potential with decreased survival in GBMs [61–63]. Additionally, the central role of VEGF in the biology of GBMs, as determined by this computational analysis, may in part explain recently published data showing that older GBM patients benefit more from treatment with the VEGF inhibitor, bevacizumab, than do younger patients [36,64]. These clinical observations have been considered paradoxical because responses to therapy with standard cytotoxic agents were historically always greater in younger GBM patients than in older. Our analysis now gives biological rationale to these previously unexplained clinical results with bevacizumab.

In conclusion, we have demonstrated through computational analyses of high-throughput genomic data from hundreds of tumors that there are substantial and consistent biological differences between GBMs found in older patients compared to those found in younger patients. Although the ultimate biological meaning and clinical significance of many of these findings await experimental validation, it appears clear that the pro-angiogenic phenotype of older GBMs compared to younger GBMs has biological, clinical, and therapeutic significance. This finding demonstrates how computational analysis of high-throughput data of a human tumor can help explain long standing clinical observations and point the way to more rationale therapeutics targeted to a specific biological process in selected patients.

Supporting Information

Figure S1 Histogram of survival days of samples for each age bin.

Figure S2 Histogram of number of samples for each age bin.

Figure S3 The comparison of copy number/SNP samples obtained from different institutes in the TCGA project. MSKCC: Agilent Human Genome CGH Microarray 244A (Memorial Sloan-Kettering Cancer Center), BI: Affymetrix Genome-Wide Human SNP Array 6.0 (Broad Institute of MIT and Harvard), HAIB: Illumina 550 K Infinium HumanHap550 SNP Chip (HudsonAlpha Institute for Biotechnology.

Table S1 Survival analysis of GBM in randomized phase trials.

Table S2 Sample IDs for each data type used in this study.

Table S3 Cox multivariate analysis results for survival. P-values≤0.05 are bold faced.

Table S4 List of differentially expressed genes found by standard linear regression (FDR≤0.05).

Table S5 Experimentally validated targets of the differentially expressed miRs between the old and young GBMs.

Table S6 List of differentially methylated genes found by ranked-based regression (FDR≤0.05).

Table S7 List of differentially methylated genes that are polycomb group protein target genes.

Table S8 List of genes that undergo hypermethylation only in GBMs.

Table S9 List of differentially methylated genes that have been reported to undergo hypermethylation in cancer in Pubmeth database.

Table S10 List of genes that are differentially expressed (Agilent and Affymetrix U133A platforms) and methylated with respect to age.

Table S11 List of differentially altered genes between Old and Young on different data sets.

Table S12 Pearson and Spearman Rank Correlation between HHEX and VEGFA expression on Affymetrix U133A, Agilent G4502A, and Affymetrix Human Exon platforms. For all correlations, p-values are significant (p≤0.05)

Table S13 List of genes that are (a) deleted and downregulated (b) amplified and upregulated in the old group. DAGs are computed from SNP 6 platform. DEGs are the intersection of genes found in Affymetrix U133A and Agilent G4502A platforms.

Table S14 List of enriched motifs in the promoter regions of the differentially expressed genes appear in both Affymetrix U133A and Agilent G4502A platforms. FDR(BH): Benjamini-Hochberg FDR value

Table S15 GSEA results on old vs. young genes. NES: normalized enrichment score. Gene sets with negative (positive) enrichment scores are active in young (old) group.

Table S16 DAVID analysis results on age-specific upregulated genes in old that appear in Affymetrix U133A and Agilent G4502A data sets. Fisher's exact test p-value<0.05

Acknowledgments

The authors would like to thank the TCGA team for providing somatic mutations from whole-exome sequence data.

Author Contributions

Conceived and designed the experiments: SB HAF. Performed the experiments: SB AL GR YK MB FMI MCC SK. Analyzed the data: SB AL GR YK MB FMI MCC SK. Wrote the paper: SB HAF.

References

1. Adamson C, Kanu OO, Mehta AI, Di C, Lin N, et al. (2009) Glioblastoma multiforme: a review of where we have been and where we are going. Expert opinion on investigational drugs 18: 1061–1083.
2. Ohgaki H, Kleihues P (2005) Epidemiology and etiology of gliomas. Acta neuropathologica 109: 93–108.
3. Beroukhim R, Mermel CH, Porter D, Wei G, Raychaudhuri S, et al. (2010) The landscape of somatic copy-number alteration across human cancers. Nature 463: 899–905.
4. Cancer Genome Atlas Research Network (2008) Comprehensive genomic characterization defines human glioblastoma genes and core pathways. Nature 455: 1061–1068.
5. Kotliarov Y, Steed ME, Christopher N, Walling J, Su Q, et al. (2006) High-resolution global genomic survey of 178 gliomas reveals novel regions of copy number alteration and allelic imbalances. Cancer Research 66: 9428–9436.
6. Li A, Walling J, Ahn S, Kotliarov Y, Su Q, et al. (2009) Unsupervised analysis of transcriptomic profiles reveals six glioma subtypes. Cancer Research 69: 2091–2099.
7. Liang Y, Diehn M, Watson N, Bollen AW, Aldape KD, et al. (2005) Gene expression profiling reveals molecularly and clinically distinct subtypes of glioblastoma multiforme. Proceedings of the National Academy of Sciences of the United States of America 102: 5814–5819.
8. Mischel PS, Shai R, Shi T, Horvath S, Lu K, et al. (2003) Identification of molecular subtypes of glioblastoma by gene expression profiling. Oncogene 22: 2361–2373.
9. Murat A, Migliavacca E, Gorlia T, Lambiv WL, Shay T, et al. (2008) Stem cell-related "self-renewal" signature and high epidermal growth factor receptor expression associated with resistance to concomitant chemoradiotherapy in glioblastoma. Journal of clinical oncology: official journal of the American Society of Clinical Oncology 26: 3015–3024.
10. Nutt CL, Mani DR, Betensky RA, Tamayo P, Cairncross JG, et al. (2003) Gene expression-based classification of malignant gliomas correlates better with survival than histological classification. Cancer Research 63: 1602–1607.
11. Phillips HS, Kharbanda S, Chen R, Forrest WF, Soriano RH, et al. (2006) Molecular subclasses of high-grade glioma predict prognosis, delineate a pattern of disease progression, and resemble stages in neurogenesis. Cancer Cell 9: 157–173.
12. Ruano Y, Mollejo M, Ribalta T, Fiaño C, Camacho FI, et al. (2006) Identification of novel candidate target genes in amplicons of Glioblastoma multiforme tumors detected by expression and CGH microarray profiling. Molecular cancer 5: 39.
13. Shai R, Shi T, Kremen TJ, Horvath S, Liau LM, et al. (2003) Gene expression profiling identifies molecular subtypes of gliomas. Oncogene 22: 4918–4923.
14. Tso CL, Freije WA, Day A, Chen Z, Merriman B, et al. (2006) Distinct transcription profiles of primary and secondary glioblastoma subgroups. Cancer Research 66: 159–167.
15. Verhaak RG, Hoadley KA, Purdom E, Wang V, Qi Y, et al. (2010) Integrated genomic analysis identifies clinically relevant subtypes of glioblastoma characterized by abnormalities in PDGFRA, IDH1, EGFR, and NF1. Cancer Cell 17: 98–110.
16. Lee Y, Scheck A, Cloughesy T, Lai A, Dong J, et al. (2008) Gene expression analysis of glioblastomas identifies the major molecular basis for the prognostic benefit of younger age. BMC Medical Genomics 1: 52.
17. Simpson JR, Horton J, Scott C, Curran WJ, Rubin P, et al. (1993) Influence of location and extent of surgical resection on survival of patients with glioblastoma multiforme: results of three consecutive Radiation Therapy Oncology Group (RTOG) clinical trials. International journal of radiation oncology, biology, physics 26: 239–244.
18. Buckner JC, Ballman KV, Michalak JC, Burton GV, Cascino TL, et al. (2006) Phase III trial of carmustine and cisplatin compared with carmustine alone and standard radiation therapy or accelerated radiation therapy in patients with glioblastoma multiforme: North Central Cancer Treatment Group 93-72-52 and Southwest Oncology Group 9503 Trials. Journal of clinical oncology: official journal of the American Society of Clinical Oncology 24: 3871–3879.
19. Halperin EC, Herndon J, Schold SC, Brown M, Vick N, et al. (1996) A phase III randomized prospective trial of external beam radiotherapy, mitomycin C, carmustine, and 6-mercaptopurine for the treatment of adults with anaplastic glioma of the brain. CNS Cancer Consortium. International journal of radiation oncology, biology, physics 34: 793–802.
20. Stewart LA (2002) Chemotherapy in adult high-grade glioma: a systematic review and meta-analysis of individual patient data from 12 randomised trials. Lancet 359: 1011–1018.
21. Gorlia T, van den Bent MJ, Hegi ME, Mirimanoff RO, Weller M, et al. (2008) Nomograms for predicting survival of patients with newly diagnosed glioblastoma: prognostic factor analysis of EORTC and NCIC trial 26981-22981/CE.3. The lancet oncology 9: 29–38.
22. Noushmehr H, Weisenberger D, Diefes K, Phillips HS, Pujara K, et al. (2010) Identification of a CpG island methylator phenotype that defines a distinct subgroup of glioma. Cancer Cell 17: 510–522.
23. R Development Core Team (2008) R: A Language and Environment for Statistical Computing. Vienna Austria R Foundation for Statistical Computing 1: ISBN 3–0.
24. BioDiscovery I (2009) Locating Significantly Different CNV Regions between Two Groups.
25. Andersen P, Gill R (1982) Cox's regression model for counting processes: a large sample study. The annals of statistics: 1100–1120.
26. Mootha VK, Lindgren CM, Eriksson KF, Subramanian A, Sihag S, et al. (2003) PGC-1alpha-responsive genes involved in oxidative phosphorylation are coordinately downregulated in human diabetes. Nature genetics 34: 267–273.
27. Subramanian A, Tamayo P, Mootha VK, Mukherjee S, Ebert BL, et al. (2005) Gene set enrichment analysis: a knowledge-based approach for interpreting genome-wide expression profiles. Proceedings of the National Academy of Sciences of the United States of America 102: 15545–15550.
28. Dennis G, Sherman BT, Hosack DA, Yang J, Gao W, et al. (2003) DAVID: Database for Annotation, Visualization, and Integrated Discovery. Genome Biology 4: P3.
29. Zambelli F, Pesole G, Pavesi G (2009) Pscan: finding over-represented transcription factor binding site motifs in sequences from co-regulated or co-expressed genes. Nucleic acids research 37: W247–252.
30. Portales-Casamar E, Thongjuea S, Kwon AT, Arenillas D, Zhao X, et al. (2010) JASPAR 2010: the greatly expanded open-access database of transcription factor binding profiles. Nucleic acids research 38: D105–110.
31. Benjamini Y, Yekutieli D (1995) Controlling the false discovery rate: a practical and powerful approach to multiple testing. Annals of Statistics 29: 1165–1188.
32. Gibbs JR, van der Brug MP, Hernandez DG, Traynor BJ, Nalls MA, et al. (2010) Abundant Quantitative Trait Loci Exist for DNA Methylation and Gene Expression in Human Brain. PLoS Genet 6: e1000952.
33. Laffaire J, Everhard S, Idbaih A, Crinière E, Marie Y, et al. (2011) Methylation profiling identifies 2 groups of gliomas according to their tumorigenesis. Neuro-Oncology 13: 84–98.
34. Katoh M (2005) WNT/PCP signaling pathway and human cancer (review). Oncology reports 14: 1583–1588.
35. Dweep H, Sticht C, Pandey P, Gretz N (2011) miRWalk - Database: Prediction of possible miRNA binding sites by "walking" the genes of three genomes. Journal of biomedical informatics.
36. Nghiemphu PL, Liu W, Lee Y, Than T, Graham C, et al. (2009) Bevacizumab and chemotherapy for recurrent glioblastoma: a single-institution experience. Neurology 72: 1217–1222.
37. Pez F, Dayan F, Durivault J, Kaniewski B, Aimond G, et al. (2011) The HIF-1-inducible lysyl oxidase activates HIF-1 via the Akt pathway in a positive

regulation loop and synergizes with HIF-1 in promoting tumor cell growth. Cancer Research 71: 1647–1657.

38. Teschendorff AE, Menon U, Gentry-Maharaj A, Ramus SJ, Weisenberger D, et al. (2010) Age-dependent DNA methylation of genes that are suppressed in stem cells is a hallmark of cancer. Genome Research 20: 440–446.

39. Hernandez D, Nalls M, Gibbs J, Arepalli S, Van Der Brug M, et al. (2011) Distinct DNA methylation changes highly correlated with chronological age in the human brain. Human molecular genetics 20: 1164–1172.

40. Ongenaert M, Van Neste L, De Meyer T, Menschaert G, Bekaert S, et al. (2008) PubMeth: a cancer methylation database combining text-mining and expert annotation. Nucleic acids research 36: D842–846.

41. Paugh B, Qu C, Jones C, Liu Z, Adamowicz-Brice M, et al. (2010) Integrated Molecular Genetic Profiling of Pediatric High-Grade Gliomas Reveals Key Differences With the Adult Disease. Journal of Clinical Oncology 28: 3061–3068.

42. Noy P, Williams H, Sawasdichai A, Gaston K, Jayaraman PS (2010) PRH/Hhex controls cell survival through coordinate transcriptional regulation of vascular endothelial growth factor signaling. Molecular and cellular biology 30: 2120–2134.

43. Stein S, Linn MW, Stein EM (1989) Psychological correlates of survival in nursing home cancer patients. The Gerontologist 29: 224–228.

44. Wedding U, Honecker F, Bokemeyer C, Pientka L, Höffken K (2007) Tolerance to chemotherapy in elderly patients with cancer. Cancer control : journal of the Moffitt Cancer Center 14: 44–56.

45. Tan PK, Downey TJ, Spitznagel EL, Xu P, Fu D, et al. (2003) Evaluation of gene expression measurements from commercial microarray platforms. Nucleic acids research 31: 5676–5684.

46. Ahuja N, Issa JP (2000) Aging, methylation and cancer. Histology and histopathology 15: 835–842.

47. Ahuja N, Li Q, Mohan AL, Baylin SB, Issa JP (1998) Aging and DNA methylation in colorectal mucosa and cancer. Cancer Research 58: 5489–5494.

48. Fraga MF, Esteller M (2007) Epigenetics and aging: the targets and the marks. Trends in genetics: TIG 23: 413–418.

49. Kim JH, Choi DS, Lee OH, Oh SH, Lippman SM, et al. (2011) Antiangiogenic antitumor activities of IGFBP-3 are mediated by IGF-independent suppression of Erk1/2 activation and Egr-1-mediated transcriptional events. Blood 118: 2622–2631.

50. Tang M, Chen B, Lin T, Li Z, Pardo C, et al. (2011) Restraint of angiogenesis by zinc finger transcription factor CTCF-dependent chromatin insulation. Proceedings of the National Academy of Sciences of the United States of America 108: 15231–15236.

51. Baudino TA, McKay C, Pendeville-Samain H, Nilsson JA, Maclean KH, et al. (2002) c-Myc is essential for vasculogenesis and angiogenesis during development and tumor progression. Genes & development 16: 2530–2543.

52. Kang J, Rychahou PG, Ishola TA, Mourot JM, Evers BM, et al. (2008) N-myc is a novel regulator of PI3K-mediated VEGF expression in neuroblastoma. Oncogene 27: 3999–4007.

53. Li ZY, Zhu F, Hu JL, Peng G, Chen J, et al. (2011) Sp1 inhibition-mediated upregulation of VEGF 165 b induced by rh-endostatin enhances antiangiogenic and anticancer effect of rh-endostatin in A549. Tumour biology: the journal of the International Society for Oncodevelopmental Biology and Medicine 32: 677–687.

54. Lee H-H, Park K, Choi K (2009) Novel agent that inhibits angiogenesis and metastasis targeting mtor signaling pathway.

55. Liu S, Yang P, Kang H, Lu L, Zhang Y, et al. (2011) NDRG2 induced by oxidized LDL in macrophages antagonizes growth factor productions via selectively inhibiting ERK activation. Biochimica et biophysica acta 1801: 106–113.

56. NicAmhlaoibh R, Shtivelman E (2001) Metastasis suppressor CC3 inhibits angiogenic properties of tumor cells in vitro. Oncogene 20: 270–275.

57. Blanco S, Santos C, Lazo PA (2007) Vaccinia-related kinase 2 modulates the stress response to hypoxia mediated by TAK1. Molecular and cellular biology 27: 7273–7283.

58. Lux A, Salway F, Dressman HK, Kröner-Lux G, Hafner M, et al. (2006) ALK1 signalling analysis identifies angiogenesis related genes and reveals disparity between TGF-beta and constitutively active receptor induced gene expression. BMC cardiovascular disorders 6: 13.

59. He LF, Wang TT, Gao QY, Zhao GF, Huang YH, et al. (2011) Stanniocalcin-1 promotes tumor angiogenesis through up-regulation of VEGF in gastric cancer cells. Journal of biomedical science 18: 39.

60. Rodríguez P, Higueras MA, González-Rajal A, Alfranca A, Fierro-Fernández M, et al. (2011) The non-canonical NOTCH ligand DLK1 exhibits a novel vascular role as a strong inhibitor of angiogenesis. Cardiovascular research 93: 232–241.

61. Chi AS, Sorensen AG, Jain RK, Batchelor TT (2009) Angiogenesis as a therapeutic target in malignant gliomas. The oncologist 14: 621–636.

62. Kaur B, Tan C, Brat DJ, Post DE, Van Meir EG (2004) Genetic and hypoxic regulation of angiogenesis in gliomas. Journal of neuro-oncology 70: 229–243.

63. Kesari S (2011) Understanding glioblastoma tumor biology: the potential to improve current diagnosis and treatments. Seminars in oncology 38 Suppl 4: S2–10.

64. Kreisl TN, Kim L, Moore K, Duic P, Royce C, et al. (2009) Phase II trial of single-agent bevacizumab followed by bevacizumab plus irinotecan at tumor progression in recurrent glioblastoma. Journal of clinical oncology: official journal of the American Society of Clinical Oncology 27: 740–745.

3

High Fidelity Copy Number Analysis of Formalin-Fixed and Paraffin-Embedded Tissues Using Affymetrix Cytoscan HD Chip

Yan P. Yu[1], Amantha Michalopoulos[1], Ying Ding[2], George Tseng[2], Jian-Hua Luo[1]*

1 Department of Pathology, University of Pittsburgh School of Medicine, Pittsburgh, Pennsylvania, United States of America, 2 Department of Statistics, University of Pittsburgh School of Medicine, Pittsburgh, Pennsylvania, United States of America

Abstract

Detection of human genome copy number variation (CNV) is one of the most important analyses in diagnosing human malignancies. Genome CNV detection in formalin-fixed and paraffin-embedded (FFPE) tissues remains challenging due to suboptimal DNA quality and failure to use appropriate baseline controls for such tissues. Here, we report a modified method in analyzing CNV in FFPE tissues using microarray with Affymetrix Cytoscan HD chips. Gel purification was applied to select DNA with good quality and data of fresh frozen and FFPE tissues from healthy individuals were included as baseline controls in our data analysis. Our analysis showed a 91% overlap between CNV detection by microarray with FFPE tissues and chromosomal abnormality detection by karyotyping with fresh tissues on 8 cases of lymphoma samples. The CNV overlap between matched frozen and FFPE tissues reached 93.8%. When the analyses were restricted to regions containing genes, 87.1% concordance between FFPE and fresh frozen tissues was found. The analysis was further validated by Fluorescence In Situ Hybridization on these samples using probes specific for BRAF and CITED2. The results suggested that the modified method using Affymetrix Cytoscan HD chip gave rise to a significant improvement over most of the previous methods in terms of accuracy in detecting CNV in FFPE tissues. This FFPE microarray methodology may hold promise for broad application of CNV analysis on clinical samples.

Editor: Renato Franco, Istituto dei tumori Fondazione Pascale, Italy

Funding: This work was funded by the National Cancer Institute, Amercain Cancer Society and University of Pittsburgh Cancer Institute. The funders had no role in study design, data collection and analysis, decision to publish, or preparation of the manuscript.

Competing Interests: The authors have declared that no competing interests exist.

* E-mail: luoj@msx.upmc.edu

Introduction

Genome abnormalities are the hallmark of human malignancies [1]. These include chromosome deletion, amplification, translocation, inversion and isochromosome formation. Analysis of genome abnormality is critical in making diagnosis of human malignancies, congenital birth defects and a variety of inheritable diseases. Array comparative genome hybridization (aCGH) or Affymetrix SNP array has been frequently applied to clinical samples to examine loss of heterozygosity and to detect amplification or deletion of genome fragments in the chromosomes [2–8]. The current methodologies using aCGH or Affymetrix SNP6.0 require high quality genome DNA from fresh frozen tissues. However, most of the samples for pathological evaluation are formalin-fixed and paraffin-embedded (FFPE) tissue blocks. Suboptimal and high background results are obtained when tissues from FFPE tissue blocks are analyzed due to the fragmenting nature of genome DNA in FFPE tissues. The low quality of genome copy number analysis from FFPE tissues practically precludes the application of whole genome copy number variation (CNV) analyses in clinical setting. Thus, a new method that can reproducibly generate high quality CNV analysis from FFPE tissues is needed to make high throughput genome CNV analysis applicable to clinical setting.

Lymphoma is one of the human malignancies that are frequently associated with large number of structural genome abnormalities. The classification and treatment of lymphomas are based on their genotypes in the tumor cells. Thus, lymphoma is an ideal human malignancy to investigate whether FFPE tissue is suitable for CNV analysis using Affymetrix Cytoscan HD chip. In this report, we describe a method that is adapted to the genomic DNA extracted from FFPE tissues to prepare the DNA cocktail for Affymetrix Cytoscan HD analysis. To validate this method, frozen genome DNA samples from matched lymphoma cases were also analyzed on Affymetrix Cytoscan HD chips. The results showed close overlaps in CNV profiles between FFPE and matched frozen tissues.

Materials and Methods

Tissue samples

Fresh frozen tissues of eight cases of human malignant lymphomas, including 5 diffused large B cell lymphomas, 2 follicular lymphomas and 1 T cell non-Hodgkins lymphoma, were obtained from clinical services. These tissues were dissected to have at least 70% purity of tumor cells. The study was approved by University of Pittsburgh Medical Center Quality Insurance Committee and Institutional Review Board, and exempted from

Table 1. Genotyping concordance between FFPE and Frozen tissues.

Case	Chr	Frozen genotype Locus	Copy	Def	Size	FFPE genotype Locus	Copy	Def	Size	Karyotype
Case 1	2	2p16.1	1	Loss	0.7 MB	2p16.1	1	Loss	0.5 MB	−2
		2q33.2-34	2.3	Gain	7.2 MB	2q33.2-34	2.3	Gain	7.2 MB	
		2q35-37.3	3	Gain	24 MB	2q35	2.4	Gain	24 MB	
	5	5q13.2-14.3	1.3	Loss	18.4 MB	5q13.2-14.3	1.4	Loss	17 MB	−5
		5q21.1-22.1	1.4	Loss	12 MB	5q21.1-22.1	1.5	Loss	10.5 MB	
		5q33.3-35.1	1.3	Loss	12.4 MB	5q33.3-35.1	1.45	Loss	11.8 MB	
	7	7p14.3-14.1	1.3	Loss	6.5 MB	7p14.3-14.1	1.5	Loss	6.7 MB	der(7)t(7;?)(p11.2;?)
		7q11.22-22.1	4	Gain	27.8 MB	7q11.22-22.1	4	Gain	24.4 MB	add(7)(q22) [8]
		7q22.1-36.3	3	Gain	38.3 MB	7q22.1-36.3	3	Gain	33.5 MB	der(7;16)(p10;q10)
	8	8p23.3-11.21	1.3	Loss	42.6 MB	8p23.3-11.22	1.4	Loss	39.6 MB	−8 [17]
		10q23.1-25.1	1.3	Loss	21.7 MB	10q23.1-25.1	1.5	Loss	21.3 MB	
	11	11p15.3-15.1	1.5	Loss	10.1 MB	11p15.4-14.3	1.5	Loss	14.2 MB	
		11p14.2-11.2	2.5	Gain	22 MB	11p14.2-11.12	2.4	Gain	23.6 MB	
	13	13q33.3-34	1.3	Loss	6.6 MB	13q33.3-34	1.3	Loss	6.7 MB	add(13)(q34) [15]
	16	16p11.2-11.1	4	Gain	0.8 MB	16p11.2-11.1	3	Gain	1.2 MB	
		16p11.1-q21	1.3	Loss	30.4 MB	16p11.1-q22.1	1.5	Loss	33 MB	−16
		16q23.1	4	Gain	2.7 MB	16q23.1	4	Gain	2.6 MB	
		16q23.1-23.2	1.2	Loss	3.7 MB	16q23.1-23.2	1.2	Loss	3.7 MB	
		16q23.2-23.3	4	Gain	1.2 MB	16q23.2-23.3	4	Gain	1.2 MB	
		16q23.3-24.1	1.1	Loss	2.5 MB	16q23.3-24.1	1.1	Loss	2.5 MB	
		16q24.1	4	Gain	2.0 MB	16q24.1	4	Gain	2.0 MB	
		16q24.1-24.3	1.2	Loss	2.8 MB	16q24.1-24.3	1.2	Loss	2.8 MB	
	17	17p13.3-13.1	2.5	Gain	8.5 MB	17p13.3-13.1	2.5	Gain	8.7 MB	−17
		17p12-q11.2	3	Gain	14.5 MB	17p12-q11.2	2.7	Gain	16.7 MB	
		17q21.33-25.33	2.7	Gain	31.8 MB	17q21.33-25.32	2.7	Gain	32.2 MB	
	X	Xp22.33-q28	2	NP	154.3 MB	Xp22.33-q28	3	Gain	154.3 MB	+X [6]
Case 2	1	1p36.33-36.21	1.4	Loss	16.7 MB	1p36.33-36.13	1.5	Loss	18.2 MB	
		1q21.2-q44	2.7	Gain	102 MB	1q21.2-q44	2.6	Gain	102 MB	der(1)(qter>1q21::p36.3>qter)
	2	2p25.3-11.2	2.8	Gain	89.1 MB	2p25.3-11.2	2.6	Gain	88.9 MB	
	6	6q21-27	1.2	Loss	58.9 MB	6q21-27	1.4	Loss	58.7 MB	
	17	17p13.3-11.2	1.3	Loss	16.9 MB	17p13.3-11.2	1.5	Loss	19.2 MB	del(17)(p11.2)
		17p11.2-q25.3	2.8	Gain	59.7 MB	17p11.2-q25.3	2.7	Gain	60.3 MB	i(17)(q10)
	18	18p11.32-q21.33	2.9	Gain	60.5 MB	18p11.32-q21.33	2.8	Gain	60.1 MB	+der(18)(t(14;18)(q32;q21) [4]
	22	22q11.22	1	Loss	0.9 MB	22q11.22	1	Loss	1.0 MB	
	X	Xp22.33-q28	3	Gain	156 MB	Xp22.33-q28	3	Gain	156 MB	+X
Case 3	1	1q24.3-42.11	2.9	Gain	55.1 MB	1q24.3-42.11	2.7	Gain	52 MB	+1, der (1)
	5	5p15.33-q35.3	2.6	Gain	180.6 MB	5p15.33-q35.3	2.6	Gain	180.6 MB	5
	8	8p23.3-21.3	1.4	Loss	22.8 MB	8p23.3-21.3	1.4	Loss	22.3 MB	del(8)(p21p23)
	9	9p24.1-23	2.5	Gain	4.7 MB	9p24.2-23	2.4	Gain	7.1 MB	add(9)(p22)
	10	10p15.3	1.4	Loss	0.8 MB	10p15.3	1.5	Loss	0.8 MB	dic(1;10)(10qter>10p15::1p13>1q25::1q21>1q32::1q25>1qter)
		10p15.1-11.21	2.5	Gain	31.6 MB	10p15.1-11.21	2.5	Gain	31 MB	
	14	14q32.33	1	Loss	0.7 MB	q32.33	1	Loss	1.0 MB	add(14)(q32)

Table 1. Cont.

Case	Chr	Frozen genotype Locus	Copy	Def	Size	FFPE genotype Locus	Copy	Def	Size	Karyotype
	18	18p11.32-q11.2	3	Gain	22 MB	18p11.32-q11.2	3	Gain	22 MB	18
		18q11.2-23	4	Gain	55 MB	18q11.2-23	4	Gain	55 MB	
	19	19q13.2-13.43	2.7	Gain	18.1 MB	19q13.2-13.43	2.6	Gain	19.4 MB	
	X	Xp22.33-q28	3	Gain	156 MB	Xp22.33-q28	3	Gain	156 MB	+X
Case 4	1	1p31.1	1.2	Loss	13.7 MB	1p31.1	1.4	Loss	13.2 MB	78–79,inc[cp5]*
		1q23.3-32.1	2.3	Gain	31.4 MB	1q23.3-32.1	2.3	Gain	29.5 MB	
		1q42.2-44	1.2	Loss	17.2 MB	1q42.2-43	1.5	Loss	10 MB	
	2	2p25.3-24.3	2.7	Gain	13.3 MB	2p25.3-24.3	2.3	Gain	14.3 MB	
		2p24.3-16.3	1.5	Loss	34 MB	2p24.3-16.3	1.6	Loss	34.1 MB	
	3	3p26.3-q13.13	1.6	Loss	108.1 MB	3p25.1-12.2	1.7	Loss	66.2 MB	
	4	4p16.3-q24	1.6	Loss	106.6 MB	4p16.3-q24	1.6	Loss	104.6 MB	
		4q25-34.1	2.7	Gain	64.7 MB	4q25-34.1	2.5	Gain	65.5 MB	
		4q34.3-35.2	2.6	Gain	7.4 MB	4q34.3-35.2	2.5	Gain	9.3 MB	
	5	5p15.33-15.31	2.6	Gain	8.7 MB	5p15.33	3	Gain	0.7 MB	
		5q15-22	1.5	Loss	54.6 MB	5q23.3-33.1	1.6	Loss	22.5 MB	
	7	7p22.3-21.1	1.1	Loss	17.8 MB	7p22.3-21.1	1.4	Loss	17.3 MB	
		7q11.22-36.3	3	Gain	93.5 MB	7q11.22-36.3	2.8	Gain	93.6 MB	
	8	8p23.3-12	1.6	Loss	32.9 MB	8p23.3-12	1.6	Loss	30.9 MB	
	9	9p23.1-22.3	0.8	Loss	8.5 MB	9p23.1-22.3	1.2	Loss	8.1 MB	
		9p21.3-21.1	0.7	Loss	7 MB	9p21.3-21.1	1	Loss	6.9 MB	
		9p21.1	0.8	Loss	1 MB	9p21.1	1.3	Loss	1 MB	
		9p13.3	1	Loss	1.3 MB	9p13.3	1	Loss	1.3 MB	
		9p13.3-q21.33	0.9	Loss	54.9 MB	9p13.3-q21.33	1.3	Loss	53.2 MB	
		9q22.2-22.31	0.7	Loss	2.5 MB	9q22.2-22.31	1	Loss	2.4 MB	
		9q22.31-22.33	1.2	Loss	4.4 MB	9q22.31-22.33	ND	NA	NA	
		9q31.1-31.2	0.8	Loss	6.3 MB	9q31.1-31.2	1.1	Loss	5.6 MB	
		9q32-34.11	1	Loss	15.1 MB	9q32-34.11	1.3	Loss	17 MB	
	10	10p15.3-12.1	3	Gain	28.1 MB	10p15.3-12.1	2.8	Gain	28.4 MB	
		10p11.22	0.3	Loss	2.6 MB	10p11.22	0.7	Loss	3.6 MB	
	11	11q13.1-13,4	1.6	Loss	8.7 MB	11q13.1-13,4	1.7	Loss	7 MB	
	13	13q12.11-34	1.2	Loss	96.7 MB	13q12.11-34	1.3	Loss	96.7 MB	
	14	14p11.2	0.3	Loss	0.5 MB	14p11.2	0.7	Loss	0.6 MB	
		14q32.32	0.7	Loss	0.6 MB	14q32.32	1.2	Loss	0.6 MB	
	17	17p13.3-13.2	1.2	Loss	5.9 MB	17p13.3-13.2	1.5	Loss	5.8 MB	
		17p13.2-13.1	3	Gain	1.7 MB	17p13.2-13.1	3	Gain	0.9 MB	
		17p13.1-11.2	1.2	Loss	14.4 MB	17p13.1-11.2	1.3	Loss	13.4 MB	
		17q11.1-25.3	3.2	Gain	55.7 MB	17q11.1-25.3	2.9	Gain	54.2 MB	
	20	20q11.22-13.33	3.2	Gain	30.4 MB	20q11.22-13.33	2.7	Gain	31.1 MB	
	21	21q11.2-22.3	1.5	Loss	33.1 MB	21q21.2-22.3	1.7	Loss	23 MB	
Case 5	10	10q22.3-24.32	1.4	Loss	21.7 MB	10q22.3-24.32	1.5	Loss	22.3 MB	del(10)(q24q26)
	13	13q14.13-22.1	1.5	Loss	28.4 MB	13q14.13-22.1	1.5	Loss	27.4 MB	de(13)(q14q22)
Case 6	1	1p11.2-q21.2	2.6	Gain	27.4 MB	1p11.2-q21.2	2.5	Gain	33.3 MB	

Table 1. Cont.

Case	Frozen genotype				FFPE genotype				Karyotype
	Chr	Locus	Copy	Def	Size	Locus	Copy	Def	Size	
	4	4p16.3-q36.2	1.4	Loss	190.9 MB	4p16.3-q36.2	1.6	Loss	188.2 MB	−4
	5	5p15.33-q11.2	3.2	Gain	58.8 MB	5p15.33-q11.2	2.5	Gain	57.5 MB	+5, i(5)(p10)
	6	6p25.3-21.1	3.1	Gain	45 MB	6p25.3-21.1	2.5	Gain	45.3 MB	
		6p12.3-q27	1.4	Loss	121 MB	6p12.3-q27	1.6	Loss	97 MB	−6
	7	7p22.3-14.1	2.7	Gain	41.9 MB	7p22.3-14.1	2.5	Gain	45.2 MB	
	13	13q31.1-34	3	Gain	35.8 MB	13q31.1-34	2.5	Gain	35.8 MB	add(13)(q34)
	18	18q21.11-21.33	3.2	Gain	28.4 MB	18q21.11-21.33	2.7	Gain	28.2 MB	add18(q21)
		18q21.33-23	1	Loss	17.5 MB	18q21.33-23	1.3	Loss	17.6 MB	
	20	20p13-q11.23	2.6	Gain	38.1 MB	20p13-q11.23	2.5	Gain	38.1 MB	
Case 7	1	1p12-q44	2.4	Gain	128.8 MB	1p12-q44	2.4	Gain	128.2 MB	+1,add(1)(p32)
	2	2p25.3-16.1	1.6	Loss	58.6 MB	2p25.3-16.1	1.6	Loss	58.1 MB	−2
		2p16.1-14	3.5	Gain	8.7 MB	2p16.1-14	3	Gain	8.7 MB	
	3	3p14.1-13	3.9	Gain	7.6 MB	3p14.1-13	3.5	Gain	7.6 MB	3
		3q25.2-26.1	1.7	Loss	8.3 MB	3q25.2-26.1	1.7	Loss	6.0 MB	
	4	4p16.3-q35.2	1.6	Loss	190.9 MB	4p16.3-q35.2	1.6	Loss	190.9 MB	−4
	6	6p26.3-11.2	2.4	Gain	58.3 MB	6p26.3-11.2	2.5	Gain	57.3 MB	+6,i(6)(p10)
	7	7p22.3-36.3	2.4	Gain	159 MB	7p22.3-36.3	2.4	Gain	159 MB	7
	10	10p16.3-15.2	1.6	Loss	3.6 MB	10p16.3	1.7	Loss	1.8 MB	dic(1;10)(q10;p13)
	17	17p13.3-q11.1	1.6	Loss	25.6 MB	17p13.3-q11.1	1.6	Loss	25.9 MB i(17)(q10)	i(17)(q10)
	20	20p12.3-12.1	1.5	Loss	5.9 MB	20p12.3-12.1	1.6	Loss	5.9 MB	−20
		20q11.1-13.33	1.6	Loss	33.3 MB	20q11.1-13.33	1.6	Loss	33.1 MB	
Case 8	1	1p36.33-12	1.6	Loss	96.7 MB	1p36.33-12	1.7	Loss	96.7 MB	i(1)(q10)
		1q12-44	2.5	Gain	102.3 MB	1q12-44	2.4	Gain	102.1 MB	
	2	2p25.3-25.2	1.7	Loss	6.3 MB	2p25.3-25.2	1.7	Loss	5.0 MB	
		2p21-13.3	2.4	Gain	22.1 MB	2p16.3-13.3	2.4	Gain	17.2 MB	add(2)(p11.2)
		2q23.1-37.3	2.4	Gain	44.3 MB	2q23.1-37.3	2.3	Gain	37.1 MB	
	3	3p26.3-12.1	1.7	Loss	86.1 MB	3p26.3-12.1	1.7	Loss	84.1 MB	del(3)(p13p25)
	4	4p16.3-q35.2	1.6	Loss	188.5 MB	4p16.3-q35.2	1.7	Loss	188.5 MB	−4
	6	6q25.2-27	2.8	Gain	17.3 MB	6q25.2-27	2.5	Gain	18.5 MB	
	7	7q21.11-36.3	2.5	Gain	78.3 MB	7q21.11-36.3	2.4	Gain	79.2 MB	add(7)(q36)
	8	8p23.3-q23.1	1.6	Loss	108.8 MB	8p23.3-q21.11	1.7	Loss	78 MB	del(8)(q13q24)
	9	9p24.3-q34.3	2.5	Gain	140.8 MB	9p24.3-q34.3	2.4	Gain	140.8 MB	9
	10	10p15.3-q26.3	1.7	Loss	136.3 MB	10p15.3-q26.3	1.7	Loss	136.3 MB	−10
	11	11q12.1-13.5	1.7	Loss	19.2 MB	11q12.1-13.5	ND	NA	NA	−11
	15	15q11.2-26.3	1.7	Loss	79.6 MB	15q11.2-26.3	1.7	Loss	74 MB	−15
	18	18p11.32-q23	1.6	Loss	77.8 MB	18p11.31-q23	1.7	Loss	72.3 MB	−18
	X	Xp22.33-21.2	1.5	Loss	30.1 MB	Xp22.33-21.2	ND	NA	NA	XXX
		Xp21.2-q21.33	2.5	Gain	63 MB	Xp21.2-q28	2.5	Gain	133 MB	

NP-normal ploidy; ND-Not detected; NA-Not applicable; Def-copy number variation definition;
*-No analysis was reported due to low number of cells survived.

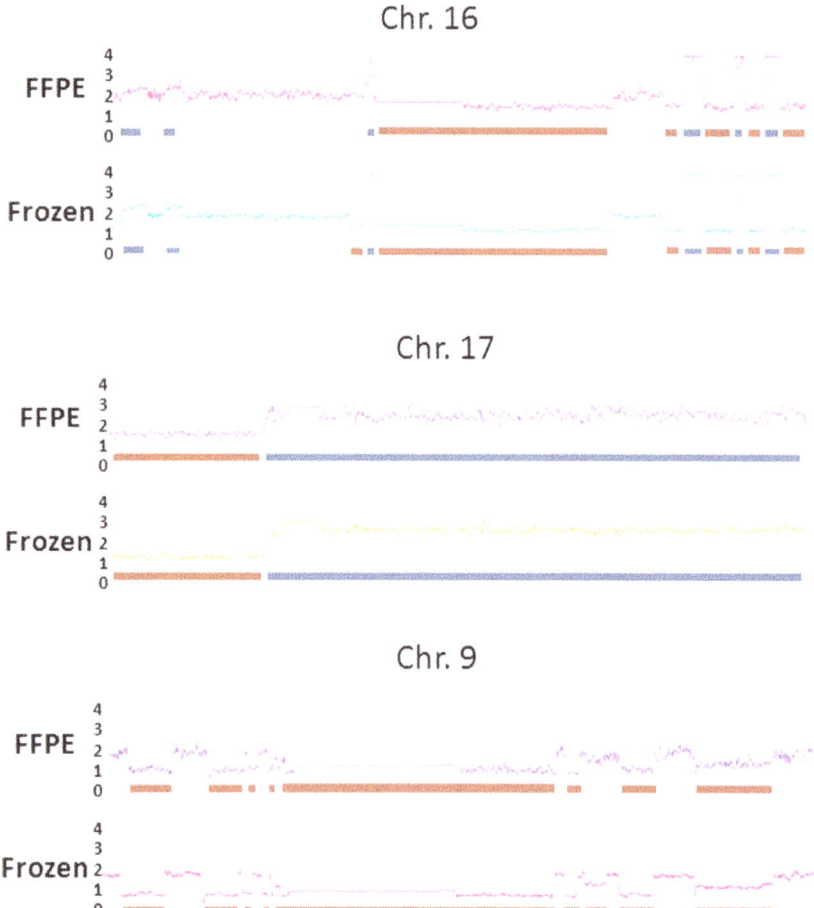

Figure 1. Histograms of matched FFPE and Fresh frozen samples on selected chromosomes. Top panel: FFPE and frozen histograms of chromosome 16 of Case 1; Mid panel: FFPE and frozen histograms of chromosome 17 of Case 2; Lower panel: FFPE and frozen histograms of chromosome 9 of Case 4. Red bar denotes deletion. Blue bar denotes amplification.

informed-consent. The matched FFPE tissues from the same patients were also frozen sectioned onto slides, fixed and dehydrated with 100% ethanol and similarly micro-dissected to obtain tumor cells. Karyotyping analyses were performed on all these cases to detect chromosome abnormalities.

Affymetrix CytoScan HD chip analysis of copy number variation of tumor cells

For macrodissected frozen tissues, DNA was extracted using QIAamp blood and tissue kit (Qiagen, Valencia, CA). Five hundred nanograms of genome DNA were digested with Nsp1 for 2 hours at 37°C. The digested DNA was purified and ligated with primer/adaptors at 16°C for 12–16 hours. Amplicons were generated by performing PCR using primers provided by the manufacturer (Affymetrix, CA) on the ligation products using the following program: 94°C for 3 min, then 35 cycles of 94°C 30 second, 60°C for 45 sec and 65°C for 1 minute. This was followed by extension at 68°C for 7 min. The PCR products were then purified and digested with DNAseI for 35 min at 37°C to fragment the amplified DNA. The fragmented DNA was then labeled with biotinylated nucleotides through terminal deoxynucleotide transferase for 4 hours at 37°C. Two hundred fifty micrograms of fragmented DNA were hybridized with a pre-equilibrated Affymetrix chip Cytoscan HD chip at 50°C for 18 hours. The procedures of washing and scanning of Cytos-

canHD chips followed the manuals provided by Affymetrix, Inc. Cel files were generated from AGCC software from Affymetrix, Inc. (Santa Clara, CA). For FFPE tissues, micro-dissected tumor cells were treated with xylene for 12 hours. DNA was then extracted using QiaAmp FFPE DNA extraction kit. Gel purification of DNA sizes ranging from 200 to 1000 bp was performed. Two hundred fifty nanograms of purified DNA was then digested with NSP1, and similarly processed as frozen tissues.

Statistical analysis

Sixteen cel files were analyzed with Genotyping console for quality control analysis. Samples with QC call above 80% were admitted into the analysis. To analyze CNV, cel files were imported into Chromosome Analysis Suite 1.2 (Affymetrix, Inc) to generate copy number from raw intensity. For frozen tissues, Cytoscan HD files from fresh frozen tissues of 380 healthy individuals provided by Affymetrix were used as a baseline. For FFPE tissues, Cytoscan HD files from FFPE tissue of 100 healthy individuals from Affymetrix were used. Deletions or amplifications of genomes were analyzed by first limiting to the regions with p-value less than 0.05/total number of regions detected, i.e. family-wise error rate (FWER) is controlled using Bonferroni's correction [9]. The selected regions were filtered by limiting to the regions with at least 25 markers and 500 kb. For genome fragment gain determination, a mean of >2.3 for autosomal chromosomes or

Table 2. CNV Overlaps between Frozen tissues and matched formalin-fixed paraffin-embedded tissues.

Case	Loss by Frozen	loss by FFPE	Overlap	Gain by Frozen	Gain by FFPE	Overlap
Case 1	170.4 MB	170.3 MB	91.70%	180.8 MB	177.3 MB*	92.70%
Case 2	93.4 MB	97.1 MB	95.60%	311.3 MB	311.3 MB	99.60%
Case 3	24.3 MB	24.1 MB	98.80%	523.1 MB	523.1 MB	98.60%
Case 4	614.9 MB	545 MB	82.10%	334.9 MB	327.5 MB	94.90%
Case 5	50.1 MB	49.7 MB	96.80%	N/D	N/D	N/A
Case 6	329.4 MB	302.8 MB	91.90%	275.4 MB	283.4 MB	95.90%
Case 7	326.2 MB	321.7 MB	98.40%	362.7 MB	360.8 MB	99.60%
Case 8	829.4 MB	734.9 MB	88.70%	468.1 MB	527.9 MB	82%

N/D-Not detected; N/A-Not applicable.
*Excluding X chromosome.

>1.5 for sex chromosomes of male was required, while for genome fragment loss determination, a mean of <1.7 for autosomal chromosomes or <0.5 of sex chromosomes of male was required. Loss of heterozygosity was not analyzed due to lack of matched normal tissues.

Fluorescence In-situ Hybridization (FISH)

Tissue slides (5 microns) were placed in 2×SSC at 37°C for 30 min. Slides were then removed and dehydrated in 70% and 85% ethanol for 2 min each at room temperature, and air dried. The probes for CITED2 and BRAF FISH analysis were obtained from BACPAC Resource Center, Oakland, CA. The DNA from the selected clone was extracted using Nucleobond Ax kit (Macherey-Nagel, Easton, PA). The probe was prepared by combining 7 μl of biotin-labeled genomic sequence containing CITED2 or BRAF (150 Kb)/50% formamide with 1 μl of direct-labeled CEP7 spectrum green for BRAF or CEP6 for CITED2 (Vysis, Downers Grove, IL). The probe was denatured for 5 min at 75°C. Sections of formalin-fixed tissues were denatured in 70% formamide for 3 min, and dehydrated in 70%, 85%, and 100% ethanol for 2 min each at room temperature. The denatured probe was placed on the slide, cover-slipped, sealed with rubber cement, placed in a humidified chamber and hybridized overnight at 37°C. Coverslips were removed and the slides were washed in 2×SSC/0.3% NP-40 at 72°C for 2 min. Slides were then held in phosphate buffered saline (PBS) at room temperature in the dark for 2 min. The biotin label was visualized by conjugation with Avidin-spectrum orange (Zymed, San Francisco, CA), cover-slipping and incubating in a moist chamber in the dark at 37°C for 20 min. Slides were washed 3 times for 2 min each in fresh PBS. Slides were then air-dried in the dark and counterstained with DAPI. Analysis was performed using a Nikon Optiphot-2 and Quips Genetic Workstation equipped with Chroma Technology 83000 filter set with single band exitors for Texas Red/ Rhodamine, FITC and DAPI (uv 360 nm). Only individual and well delineated cells with two hybridization signals were scored. Overlapping cells were excluded from the analysis. Fifty to 100 cells per sample were scored to obtain an average of signals.

Karyotyping analysis

Cytogenetic analysis was performed on cell cultures of lymphoma samples. The cell cultures were stimulated with phytohemagglutinin (PHA) for 72 hours, and harvested and treated briefly with a hypotonic solution. The cells were then fixed with Carnoy fixative. GTG-banding was carried out using standard protocols [10].

Results

Affymetrix Cytoscan HD contains 2.6 million markers covering all RefSeq genes and 750K SNPs of human genome, and has >99% genotype accuracy. It has been widely applied to ascertain genome abnormalities of a variety of human diseases. However, it is rarely applied to FFPE tissues. To evaluate the usage of Cytoscan HD on clinical FFPE samples, eight cases of lymphoma FFPE tissues were micro-dissected and analyzed through Affymetrix Cytoscan HD chips. The frozen counterparts of these cases were similarly analyzed. Karyotype analyses were performed on all these cases for validation purpose. As shown in Table 1, karyotype abnormalities in 24 of 51 loci had complete matches with frozen tissue copy number analyses. Twenty-two loci from Karyotype analyses had significant overlapped regions detected by frozen tissue CNV analysis. Overall, this represents 90.2% (46/51) concordance between Karyotype and Cytoscan HD analyses. Five loci of Karyotype analyses, however, did not concord with CNV analysis from both frozen and FFPE tissues analyses. These differences may result from heterogeneity of the tumor samples that some genome abnormalities appear in only a fraction of tumor cells.

Most of the clinical samples are in the form of formalin-fixed and paraffin-embedded tissue blocks. The dependence of array analysis on fresh frozen tissues significantly limits its application in clinical setting. To investigate whether FFPE tissues are suitable for cytoscan HD analysis, DNA from the matched FFPE tissue samples was extracted. Similar Affymetrix Cytoscan HD analyses were performed. Cel files of FFPE tissues from 100 normal individuals were used as baseline to calculate the copy number of genome fragments of these FFPE samples. CNV was determined by p<0.05 with at least 25 markers and a minimal length of 500 Kb. As shown in Tables 1–2 and figure 1, the concordant

Table 3. Correlation of CNV callings between FFPE and matched fresh frozen tissues by segment number.

CNV	fresh gain	fresh loss
FFPE loss	68	18
FFPE gain	13	352

$P < 2.2 \times 10^{-16}$; Pearson correlation coefficient = 0.72.

A

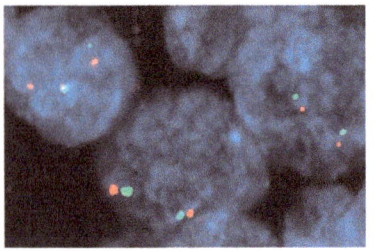

Normal controls   Lymphoma cells

B

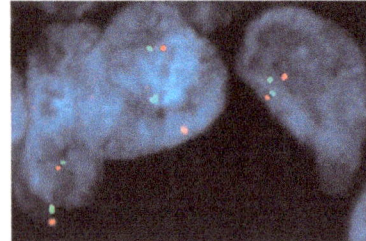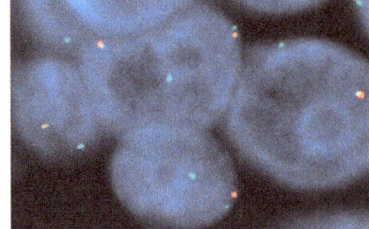

Normal controls   Lymphoma cells

Figure 2. FISH validation of copy number variation detected by Cytoscan HD analysis. (A) Representative images of FISH analysis using probes specific for BRAF (7q34, red) and centromere of chromosome 7 (green). Left, normal diploid control, right-case 4 lymphoma (BRAF amplified). (B) Representative images of FISH analysis using probes specific for CITED2 (6q23.3, red) and centromere of chromosome 6 (green). Left, normal diploid control, right-case 2 lymphoma (CITED2 deleted).

Table 4. Correlation of CNV callings and FISH on BRAF and CITED2.

Genes	Cases	CN by frozen tissue	CN by FFPE	CN by FISH
BRAF	1	3	3	2.8*
	2	Unchanged	Unchanged	2.1
	3	Unchanged	Unchanged	1.9
	4	3	3	2.9*
	5	Unchanged	Unchanged	2.2
	6	Unchanged	Unchanged	1.9
	7	3	2.5	2.7*
	8	3	3	3.1*
CITED2	1	Unchanged	Unchanged	2.1
	2	1	1	1.3*
	3	Unchanged	Unchanged	1.9
	4	Unchanged	Unchanged	2
	5	Unchanged	Unchanged	2.2
	6	1.4	Unchanged	1.6*
	7	Unchanged	Unchanged	1.8
	8	Unchanged	Unchanged	2

CN-copy number;
*-p<0.01.

rates based on CNV length between arrays from the matched FFPE and frozen tissues ranged from 82% to over 99%. Seven of these FFPE blocks are over 1 year old and one is 3 years. The average concordant rate for 1 year old samples is 94.1%, while the rate for the 3 year old sample is 92%. This suggests that the quality of the assays is stable, and is probably not adversely affected by the age of the fixed tissues for at least 1–3 years. When CNV calling was limited to regions that cover at least one gene, we found that 482 genome segments were either deleted or amplified in at least one of the 16 samples. Among these segments, 68 of 86 segments that were determined as deletion in FFPE samples matched the same callings from fresh frozen tissue counterparts, while 352 of 365 segments that were determined as amplification matched those from the fresh tissues (Table 3). The results confirmed a strong correlation between FFPE and fresh frozen tissues (87.1%, Pearson correlation coefficient = 0.72, $p < 2.2 \times 10^{-16}$). Interestingly, when CNV results of FFPE were matched with Karyotype studies, the results were extremely similar to those between frozen CNV and Karyotype. There is no statistically significant difference in terms of accuracy. In fact, a slight improvement (25/51 complete match versus 24/51 frozen) was seen due to detection of X chromosome gain in a case that was missed in frozen tissue.

To validate the CNV analysis at individual gene level, FISH assays were performed on all 8 cases of lymphoma using probes specific for BRAF and CITED2. Four samples of lymphoma were found to have gain of BRAF gene by CNV analyses in both frozen and FFPE tissues. Each of these 4 cases was found to have similar amplification of BRAF in the FISH assays, while the other 4 CNV neutral samples were found to have copy number near diploid condition in FISH. The concordant rate between the CNV calls from FFPE or frozen samples and the FISH results from the

Table 5. Comparison of methods analyzing CNV using FFPE tissue.

	Thompson et al	Little et al	Tuefferd et al	Lips et al	Oosting et al	Soroush et al	Yu et al
Platform	Affymetrix	GenomePlex	Affymetrix	Illumina	Illumina/Affymetrix	Agilent/Affymetrix	Affymetrix
Array Type	10K	aCGH	SNP 6.0	BeadArray	BeadArray/10K	aCGH/SNP6.0	CytoscanHD
Number of Loci	10000	5623	1.85 million	5861	5861/10000	40161/1.85 million	2.6 million
CNV FFPE vs. Frozen	NE	NE	53%	NA	NE	NA	93.80%
CNV FFPE vs. karyotype	NE	NE	NE	NA	NE	NA	91%
FISH validation rate	NE	NE	NE	NA	NE	NA	93.80%

NE-Not examined; NA-Not applicable.

matched samples reaches 100% (8/8, Table 4 and figure 2). CITED2, a transcription modulator essential for glycolytic metabolism for adult hematopoietic stem cells and potential tumor suppressor [11,12], was found deleted in 2 cases of lymphoma by CNV analysis in frozen tissues. On the other hand, CNV analysis in the matched FFPE tissues suggests loss of CITED2 in only one sample. FISH assays using a probe specific for CITED2 found loss of one copy of CITED2 gene in both cases where deletions were also indicated by Cytoscan HD chip analyses of frozen tissues (Table 4 and figure 2). However, 55% (117/212) of the counted cells in one of the cases (case 6) in the FISH assay do not contain deletion of CITED2. The large dilution by the diploid cells in this sample may result in a negative result in FFPE CNV analysis.

Despite the high level statistical CNV concordance between FFPE and frozen tissues, a moderately higher level of fluctuation was readily detected in FFPE array analysis (figure 1). The average of copy number for genome deletion for FFPE tissues is 1.44. This magnitude of deletion is significantly less than that from frozen tissues (1.29, p = 6.7×10^{-11}). As a result, CNV analysis from FFPE tissues could be less sensitive. Despite this mild drawback, excellent CNV concordance between FFPE and frozen samples was evident in most of the CNV regions, particularly those CNV loci with large DNA fragment abnormalities (figure 1).

Discussion

The current methodologies to detect chromosome abnormalities include Karyotyping using Giemsa staining of chromosomes, high throughput array analysis of single nucleotide polymorphism and DNA copy number, and whole genome sequencing. The first two are the most commonly used methods to determine the copy number of genome fragments, while the last one might be highly precise but expensive. Array genome copy number analysis offers a high resolution alternative to Karyotyping assay. However, its clinical application is limited due to the requirement of high quality DNA. Most clinical specimens, however, are stored in the form of formalin-fixed and paraffin-embedded tissue blocks. The DNA from FFPE is highly fragmented and cross-linked. This produces a significant challenge in using FFPE tissues for high resolution genotyping analysis. Our method using FFPE tissues to analyze chromosomes shows an average of concordance of 93.8% between FFPE and fresh frozen tissues. It suggests that FFPE Cytoscan HD analysis is readily applicable to clinical setting.

Studies using FFPE tissues to analyze copy number variation had been peviously attempted on Affymetrix 10K and SNP6.0 chips, GenomePlex aCGH, Illumina beadArray and Agilent 244K chips (Table 5) [13–19]. However, many of these studies lack direct validation using other methodologies. In one study, high noise level on CNV analysis was found in FFPE samples when using Affymetrix 6.0 chip. Only 53% concordant rate was found between the matched frozen and FFPE samples [15]. In another study, only selected concordance analyses were performed on selected regions of FISH and Agilent 244K chips using FFPE tissues [14]. The study concluded complementary roles of between aCGH and FISH analyses. Both analyses were performed on FFPE tissues. Recently, a FFPE OncoScan service was developed by Affymetrix Inc to use Molecular Inversion Probe technology to detect CNV of oncogenic hot spots [20,21]. The signal-to-noise ratios were reduced even using FFPE tissues more than 5 years old. However, these studies lacked direct validation comparison between FFPE and matched frozen tissues. Thus, the fidelity of CNV callings was not determined. Another study using optimization of universal linkage system labeling to analyze 3 cases FFPE and matched frozen tissues. They found a good correlation in 2 cases (Pearson correlations 0.54–0.58), but found poor correlation in the other case [22]. To our knowledge, this is the first report that shows high concordance in CNV analysis between FFPE and frozen tissues using Affymetrix Cytoscan HD chip. The CNV results from FFPE not only matched well with those from frozen tissues, but were also largely validated by cytogenetic karyotyping and FISH analyses. The Cytoscan HD FFPE analysis holds promise for being widely used in solid tumor and hematological diseases diagnosis. It may be also useful in differential diagnoses of hereditary diseases.

Acknowledgments

We thank Song-Yang Zheng and Kathleen Cieply for technical support in the study. Sarah Gibson provided pathology support in identifying lymphoma regions. We also thank Steven Swedlow and George Michalopoulos for frank and constructive comments on the manuscript.

Author Contributions

Conceived and designed the experiments: JHL. Performed the experiments: YPY AM YD GT. Analyzed the data: YPY YD GT. Contributed reagents/materials/analysis tools: YD GT. Wrote the paper: JHL.

References

1. Barigozzi C, Cusmano L (1947) Chromosome numbers in cancer cells. Nature 159: 505.
2. Ren B, Yu G, Tseng GC, Cieply K, Gavel T, et al. (2006) MCM7 amplification and overexpression are associated with prostate cancer progression. Oncogene 25: 1090–1098.
3. Yu G, Tseng GC, Yu YP, Gavel T, Nelson J, et al. (2006) CSR1 suppresses tumor growth and metastasis of prostate cancer. American Journal of Pathology 168: 597–607.
4. Luo JH, Ren B, Keryanov S, Tseng GC, Rao UN, et al. (2006) Transcriptomic and genomic analysis of human hepatocellular carcinomas and hepatoblastomas. Hepatology (Baltimore, Md 44: 1012–1024.
5. Yu YP, Luo JH (2007) Pathological factors evaluating prostate cancer. Histology and histopathology 22: 1291–1300.
6. Liu W, Laitinen S, Khan S, Vihinen M, Kowalski J, et al. (2009) Copy number analysis indicates monoclonal origin of lethal metastatic prostate cancer. Nature medicine 15: 559–565.
7. Yu YP, Song C, Tseng G, Ren BG, Laframboise W, et al. (2012) Genome abnormalities precede prostate cancer and predict clinical relapse. The American journal of pathology 180: 2240–2248.
8. Nalesnik MA, Tseng G, Ding Y, Xiang GS, Zheng ZL, et al. (2012) Gene deletions and amplifications in human hepatocellular carcinomas: correlation with hepatocyte growth regulation. The American journal of pathology 180: 1495–1508.
9. Strassburger K, Bretz F (2008) Compatible simultaneous lower confidence bounds for the Holm procedure and other Bonferroni-based closed tests. Statistics in medicine 27: 4914–4927.
10. Hu J, Surti U (1991) Subgroups of uterine leiomyomas based on cytogenetic analysis. Human pathology 22: 1009–1016.
11. Du J, Li Q, Tang F, Puchowitz MA, Fujioka H, et al. Cited2 Is Required for the Maintenance of Glycolytic Metabolism in Adult Hematopoietic Stem Cells. Stem cells and development.
12. Cheung KF, Zhao J, Hao Y, Li X, Lowe AW, et al. (2013) CITED2 is a novel direct effector of peroxisome proliferator-activated receptor gamma in suppressing hepatocellular carcinoma cell growth. Cancer 119: 1217–1226.
13. Thompson ER, Herbert SC, Forrest SM, Campbell IG (2005) Whole genome SNP arrays using DNA derived from formalin-fixed, paraffin-embedded ovarian tumor tissue. Human mutation 26: 384–389.
14. Wang L, Rao M, Fang Y, Hameed M, Viale A, et al. (2013) A Genome-Wide High-Resolution Array-CGH Analysis of Cutaneous Melanoma and Comparison of aCGH to FISH in Diagnostic Evaluation. J Mol Diagn.
15. Tuefferd M, De Bondt A, Van Den Wyngaert I, Talloen W, Verbeke T, et al. (2008) Genome-wide copy number alterations detection in fresh frozen and matched FFPE samples using SNP 6.0 arrays. Genes, chromosomes & cancer 47: 957–964.
16. Little SE, Vuononvirta R, Reis-Filho JS, Natrajan R, Iravani M, et al. (2006) Array CGH using whole genome amplification of fresh-frozen and formalin-fixed, paraffin-embedded tumor DNA. Genomics 87: 298–306.
17. Nasri S, Anjomshoaa A, Song S, Guilford P, McNoe L, et al. (2010) Oligonucleotide array outperforms SNP array on formalin-fixed paraffin-embedded clinical samples. Cancer genetics and cytogenetics 198: 1–6.
18. Oosting J, Lips EH, van Eijk R, Eilers PH, Szuhai K, et al. (2007) High-resolution copy number analysis of paraffin-embedded archival tissue using SNP BeadArrays. Genome research 17: 368–376.
19. Lips EH, Dierssen JW, van Eijk R, Oosting J, Eilers PH, et al. (2005) Reliable high-throughput genotyping and loss-of-heterozygosity detection in formalin-fixed, paraffin-embedded tumors using single nucleotide polymorphism arrays. Cancer research 65: 10188–10191.
20. Krijgsman O, Israeli D, Haan JC, van Essen HF, Smeets SJ, et al. (2012) CGH arrays compared for DNA isolated from formalin-fixed, paraffin-embedded material. Genes, chromosomes & cancer 51: 344–352.
21. Wang Y, Cottman M, Schiffman JD (2012) Molecular inversion probes: a novel microarray technology and its application in cancer research. Cancer genetics 205: 341–355.
22. Salawu A, Ul-Hassan A, Hammond D, Fernando M, Reed M, et al. (2012) High quality genomic copy number data from archival formalin-fixed paraffin-embedded leiomyosarcoma: optimisation of universal linkage system labelling. PloS one 7: e50415.

4

On the Power and the Systematic Biases of the Detection of Chromosomal Inversions by Paired-End Genome Sequencing

José Ignacio Lucas Lledó[1], Mario Cáceres[1,2]*

1 Institut de Biotecnologia i de Biomedicina, Universitat Autònoma de Barcelona, Bellaterra, Barcelona, Spain, **2** Institució Catalana de Recerca i Estudis Avançats, Barcelona, Spain

Abstract

One of the most used techniques to study structural variation at a genome level is paired-end mapping (PEM). PEM has the advantage of being able to detect balanced events, such as inversions and translocations. However, inversions are still quite difficult to predict reliably, especially from high-throughput sequencing data. We simulated realistic PEM experiments with different combinations of read and library fragment lengths, including sequencing errors and meaningful base-qualities, to quantify and track down the origin of false positives and negatives along sequencing, mapping, and downstream analysis. We show that PEM is very appropriate to detect a wide range of inversions, even with low coverage data. However, $\geq 80\%$ of inversions located between segmental duplications are expected to go undetected by the most common sequencing strategies. In general, longer DNA libraries improve the detectability of inversions far better than increments of the coverage depth or the read length. Finally, we review the performance of three algorithms to detect inversions —SVDetect, GRIAL, and VariationHunter—, identify common pitfalls, and reveal important differences in their breakpoint precisions. These results stress the importance of the sequencing strategy for the detection of structural variants, especially inversions, and offer guidelines for the design of future genome sequencing projects.

Editor: Zhanjiang Liu, Auburn University, United States of America

Funding: This work was supported by the European Research Council Starting Grant 243212 (INVFEST) under the European Union Seventh Research Framework Programme (FP7) to MC and a Beatriu de Pinós Postdoctoral fellowship from the Generalitat de Catalunya to JILL. The funders had no role in study design, data collection and analysis, decision to publish, or preparation of the manuscript.

Competing Interests: The authors have declared that no competing interests exist.

* E-mail: mcaceres@icrea.cat

Introduction

In the last several years, genomic techniques have discovered an unprecedented degree of structural variation (SV) in multiple species, including humans [1–3]. This advance has spurred a renovated interest in the study of all kinds of SV, both in normal situations and in disease. Currently, one of the most used techniques for SV detection is paired-end mapping (PEM), which has been associated to high-throughput DNA sequencing methods [4–7]. Millions of pairs of short reads of DNA are sequenced from a DNA library of a target genome with a known length distribution. The two reads in a pair are the sequenced ends of a template molecule from the DNA library. When a pair of reads are mapped to a reference genome, they are expected to lay at a certain distance, and in a specific relative orientation. Deviations from these expectations are then interpreted as structural variations between the target and the reference genomes.

Several algorithms have been developed to translate the PEM data into a list of structural variants (reviewed in [8,9]), and some studies have successfully applied them to whole human genomes [4,10–14]. However, the proportions of false positives produced by PEM-based methods are high [4,15], if not unknown [13,16]. False negatives are also suspected to be many, especially in repetitive regions of the genome analysed [17]. The most common source of errors in PEM-based SV detection is probably mismapping, that is, the spurious alignment of reads to non-orthologous positions of the reference genome.

In principle, it is possible to analytically derive the probability of detecting a structural variant as a function of the sequencing strategy (e.g., template length, and sequencing effort; [18]), which could be used to estimate the likelihood of a candidate structural variant. However, these theoretical expectations are overly optimistic because they do not take into account the repetitive structure of the sequenced genome, nor the ambiguously mapped reads. A more realistic alternative is to use genome-specific simulations and empirical models of the SV-detection process. Recently, genome-specific simulations are being used to evaluate the performance of SV-detection software and to estimate rates of false positives and false negatives [19–21]. Most of such simulations lack a realistic distribution of sequencing errors, which is essential when researching mapping-related issues. Furthermore, simulated structural variation is either distributed randomly or copied from known variants. Neither strategy represents the real, unknown distribution of SV in the human genome. In particular, the tendency of SV to happen in repetitive regions is elusive for the most common detection methods, largely ignored by simulation studies, and overlooked in databases.

One particular type of SV that is especially problematic is chromosomal inversions, which simply change the orientation of a fragment of DNA. Polymorphic inversions have been studied in the species of *Drosophila* for decades [22], and they are also known to exist in human populations [23]. Interestingly, inversions could have important consequences on the genome both through the effect on nearby genes or the inhibition of recombination within the inverted region in heterozygotes. As such, they have been shown to be involved in phenotypic characteristics [24], susceptibility to genetic disorders [23], and evolution [25]. Traditionally inversions have been very difficult to detect and validate across a whole genome in a high-throughput manner, and are one of the less well characterized type of SV. PEM has the advantage of being able to discover balanced or dosage-invariant rearrangements, such as inversions and translocations, and has been used to predict several hundreds of inversions in different human individuals [4,10–14]. Nevertheless, as mentioned above, very little is known about the proportion of inversions that are missed or incorrectly predicted by different mapping and sequencing methods, and current PEM predictions could be giving us an inaccurate and incomplete view of the inversions in the human genome.

Inversions are expected to produce a very specific and distinct pattern of discordantly mapped reads, consisting on one of the ends being mapped in the unexpected orientation. Because this signature is known with absolute precision, in contrast with the expected distance between two mapped reads across an insertion or a deletion, inversions should be easier to detect by PEM methods. However, inversions are frequently located where it is most difficult to map reads uniquely. Inversions have been proposed to originate by two main types of mechanisms: non-homologous end-joining of random breaks in more or less simple sequences, or non-allelic homologous recombination between inverted repeats. Although each mechanism relative contribution to inversion generation is discussed and varies depending on the detection method [4,10,26], a big fraction of polymorphic inversions in humans, especially the largest ones (>100 kb), are flanked by highly identical segmental duplications [23]. Therefore, many reads sequenced across inversion breakpoints are mapped concordantly (Figure 1), and the power of PEM methods to detect them is significantly reduced. As an example, the pilot study of the 1000 genomes project described several new big insertions and deletions in human populations, but neglected inversions because 'methods capable of discovering inversions […] in low coverage data […] remain to be developed' [27]. In a similar study, 80 individuals from natural populations of *Arabidopsis thaliana* were sequenced to a depth of 10–20 each with paired-end reads, but inversions were not reported yet [28].

In this study, we use computer simulations to estimate the sensitivity and the specificity of different sequencing strategies in detecting chromosomal inversions of different sizes and in different sequence contexts. We use human chromosome 1 as a model of the human genome, and we simulate realistic paired-end sequencing experiments with meaningful base qualities, sequencing errors, and sequence divergence between the target and the reference genomes. Simulated inversions are located either randomly or between inverted repeats, in order to represent two types of mechanisms of origin, either mediated by homology or not. We explain why the discovery of inversions from PEM experiments have had limited success, and make recommendations for future experiments. We also predict the expected levels of false positives and false negatives for each kind of inversion, under different strategies, and we compare the performance of three different SV-detecting algorithms: SVDetect [29], VariationHunter [16], and GRIAL (Martinez-Fundichely, S. Casillas, and M. Cáceres, unpublished data).

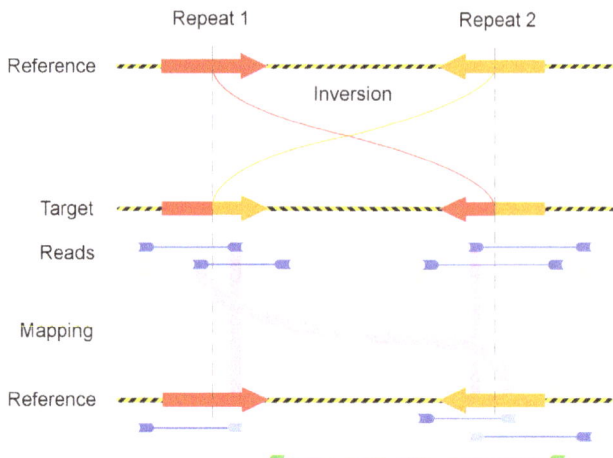

Figure 1. Inversion between the reference and the target genomes. The breakpoints (dashed lines) are located inside inverted repeats (red or orange arrows). Four pairs of reads that span the breakpoints are depicted in blue, with their sequenced ends in opposite orientations. Yellow bands indicate the correct mappings in the reference genome of ends located in unique sequences. The reads sequenced from a repeat are erroneously mapped to the alternative copy (pink bands), because concordant alignments are favored by the aligner. The mapped reads at the bottom are displayed in dark blue if correctly mapped or in light blue otherwise. The only discordant pair of reads that report the inversion is shown in green.

Methods

Simulation of inversions

From the reference sequence of human chromosome 1 (hg19), we simulated two target chromosomes: the *inversionless* chromosome, colinear to the reference, and the *inversionful* chromosome, including 948 inversions. Gaps in the reference genome (9.6% of its length) were substituted by random sequence of equivalent length and composition in both simulated chromosomes. We then introduced two main kinds of inversions: 424 randomly located inversions, and 524 inversions located between inverted repeats (Table 1). To generate the latter, we first identified all pairs of inverted repeats not more than 200 kb apart present in chromosome 1. We used three databases of such repeats: segmental duplications, self-alignments, and repeat-masked sequences, all downloaded from the UCSC Genome Browser ftp site. Inversions were distributed as evenly as possible between the three kinds of inverted repeats, making sure that they did not overlap and that they were separated by at least 50 kb. In all, 29 inversions were located between segmental duplications, 97 between other alignable regions, and 398 between repeat-masked fragments of the same type (Table 1). The distribution of inversion lengths is roughly linearly decreasing between 200 bp and 200 kb.

The breakpoints of inversions between segmental duplications were located in the middle point of the longest tract of perfect identity between them, which is a good approximation of real inversions produced by non-allelic homologous recombination (M. Cáceres, unpublished data). To determine what was the longest tract of perfect identity between two copies, we performed either a global exhaustive alignment, if possible, or a local heuristic alignment with the program Exonerate and parsed the output. The identity between the two copies was recorded as the number of identical residues divided by the average length of the two copies. Similarly, the breakpoints of inversions situated between other alignable regions were chosen in the middle of the longest

Table 1. Characteristics of the inversions simulated in four sequence contexts.

Sequence context	Number of inversions	Inversion size (bp)		Repeat length (bp)		Repeat identity (%)	
Random	424	61,756	(46,449)	NA		NA	
Repeat-masked	399	75,544	(57,549)	347	(383)	81.6	(7.3)
Alignable	98	16,716	(39,069)	201	(131)	94.2	(4.4)
Segmental dup.	29	102,469	(65,558)	19,962	(29,027)	89.2	(15.0)

Number and average size of inversions simulated in each type of sequence context, and average length and average percentage of identity between the two inverted copies flanking the breakpoints (repeat-masked and segmental duplications) or within their largest alignment blocks (alignable). The standard deviations are shown in parentheses.

block of ungapped alignment between the two copies. The length of the chosen block and its percentage of identity were recorded. For inversions between masked repeats, breakpoints were located in the middle points of the copies. Their average length and the percentage of identity between them were also recorded.

Sequencing

Because divergence between the sequenced and the reference genomes affects the ability to map the reads, we also introduced a non-trivial proportion of 0.005 mutations in the simulated copies of human chromosome 1 (hg19), including point mutations (80%) and indels (20%) of 1–4 bp. Several paired-end sequencing experiments of the two target genomes were simulated with the wgsim utility distributed with SAMtools [30]. To correct the homogeneity of sequencing errors along the reads produced by wgsim, we originally simulated the reads without any error, and then used a custom perl script to assign stochastic base qualities and add sequencing errors with a probability corresponding to the assigned quality. Base qualities were distributed along each read independently, according to a generalization of the empirical error models available in the package MetaSim [31].

We simulated three different read lengths that are representative of the available data in current and past paired-end genome sequences generated by the most popular sequencing technologies: 36, 75, and 150 bp. These read lengths were combined with five commonly used library template lengths from 250 bp to 40 kb (except when the template was shorter than twice the read length), generating 14 realistic sequencing experiments in each simulated chromosome (Table 2). Standard deviations were proportional to the template lengths, according to a linear regression estimated from empirical data from different types of real DNA libraries [4,10,27].

When sequencing the inversionless chromosome, the number of simulated reads in each experiment was determined to generate an expected sequencing depth of 20. When sequencing the inversionful chromosome, though, we aimed at a sequencing depth of at least 20 and to a physical coverage of at least 50 (Table 2). We call 'physical coverage' what others have called 'clonal coverage' [18] or 'span coverage' [19], namely the number of times that a site lays between the two sequenced ends of a pair. We define the expected physical coverage as $n \cdot (t - 2r)/g$, where n is the number of templates sequenced, t is the average template length, r is the read length, and g is the genome size. This assumes that the length needed to map the pair is not available to detect a breakpoint, which lets us focus on the problem of detecting inversions by different PEM strategies, and set aside the complementary approach of detecting them by the use of split reads (but see Discussion).

Mapping

We used Novoalign (http://www.novocraft.com) to map the sequenced reads to the reference genome. We allowed for a score difference of 5 (default) between alternative alignments to consider the read ambiguously mapped, and we kept up to 100 alignments for each ambiguously mapped read. The alignments were done in paired-end mode, using the information of the expected distance between the two ends of a pair to find the most likely mapping and to determine if a pair is concordant or discordant. To favor concordant mappings over discordant ones, we set an SV penalty, which represents how much more likely a discordant mapping must be, relative to its concordant alternative, for it to be preferred (it is equivalent to the phred-scaled a priori probability of a breakpoint being covered by a read). Higher values of SV-penalty are expected to increase the specificity of SV detection and to reduce the sensitivity. We tested SV-penalty values between 0 and 70.

Table 2. Sequencing strategies and sequencing efforts of the inversionful chromosome.

Template (bp)	Read (bp)	Num. reads	Seq. depth	Phys. cov.
250 (15)	36	70,014,219	20.22	50.00
250 (15)	75	124,625,311	75.00	50.00
450 (27)	36	69,236,284	20.00	105.00
450 (27)	75	41,541,770	25.00	50.00
450 (27)	150	83,083,540	100.00	50.00
2,500 (150)	36	69,236,284	20.00	674.44
2,500 (150)	75	33,233,416	20.00	313.33
2,500 (150)	150	16,616,708	20.00	146.67
10,000 (599)	36	69,236,284	20.00	2,757.78
10,000 (599)	75	33,233,416	20.00	1,313.33
10,000 (599)	150	16,616,708	20.00	646.67
40,000 (2394)	36	69,236,284	20.00	11,091.11
40,000 (2394)	75	33,233,416	20.00	5,313.33
40,000 (2394)	150	16,616,708	20.00	2,646.67

Sequencing strategies tested, defined by the template and read lengths. Standard deviations of template lengths are shown in brackets. The number of reads simulated from the inversionful chromosome and their corresponding expected sequencing depth and physical coverage are shown.

SV-detection algorithms

Three SV-detection algorithms were used to identify common difficulties in the post-mapping stage of PEM data analysis: SVDetect [29], VariationHunter [16], and GRIAL (A. Martnez-Fundichely, S. Casillas, and M. Cáceres, unpublished data), which is available in http://grupsderecerca.uab.cat/cacereslab/grial. Care was taken to offer the same paired ends mapped in discordant orientation to all programs, while respecting their specific requirements. Because there is a trade off between template length and throughput of current sequencing technologies, we considered more realistic to downsample the reads from experiments with an expected physical coverage larger than 50 (see Table 2). Thus, we evened up the physical coverage, rather than the sequencing depth, across experiments before using the SV-detection algorithms.

SVDetect uses a sliding-window approach to first identify pairs of windows (links) connected by one or more discordant read. Redundant links are purged and reads within them are filtered. Finally, the program defines clusters of reads and identifies their corresponding structural variation. We set the minimum number of reads required to call a cluster to be 3, and followed the author's suggestion to set the lengths of both the window and its sliding step in order to be able to detect large SVs [29]. A mapping quality threshold of 20 was applied to the input reads, which proved to reduce the number of false positives significantly. GRIAL only predicts inversions. It relies on the average template length and on its standard deviation to apply some geometric rules and define a minimum range where the breakpoints must be (A. Martnez-Fundichely, S. Casillas, and M. Cáceres, unpublished data). As before, the minimum number of reads required to call a cluster was also 3, and a mapping quality threshold of 20 was used. Both GRIAL and SVDetect are hard-clustering algorithms, meaning that they assume a unique mapping for each read. In contrast, VariationHunter takes as input all possible mappings of each read, being aware of their mapping qualities. It applies a sophisticated algorithm to find the minimal set of compatible structural variants collectively supported by all discordant reads, so that each read only gives support to one variant [32]. Although VariationHunter usually work with all the potential read mappings provided by its companion aligner, MrFast [33], we instead parsed the SAM files produced by Novoalign into VariationHunter's native format, including only up to 100 alternative mappings for each read.

All three programs produce a set of chromosomal intervals where the breakpoints of the inversions are predicted to be. Predictions other than inversions (or 'inverted segments' and the like) were discarded. The length of the interval is the precision of each breakpoint prediction. If any program produced overlapping predictions, we merged them in one larger interval that included all of them. This was necessary for 90% of predictions across experiments by SVDetect, but infrequent for GRIAL (0.04%) or VariationHunter (0.2%). Then, the predictions were compared with the true locations of the breakpoints, and the numbers of true positives, false positives and false negatives were recorded for each program and sequencing strategy. The breakpoints predicted were also compared among programs.

In order to determine if false breakpoints were predicted on inverted repeats more often than expected, we counted the overlaps between false breakpoints and all the inverted repeats present in chromosome 1 (segmental duplications, repeat-masked regions, and other alignable segments). Then, we counted all the positions in the genome where a breakpoint prediction of certain length would have overlapped with one, two… or any number of repeats of each kind. From them, we determined the total expected number of overlaps that false breakpoint predictions by

each program could have produced with each kind of repeat if they were randomly located. Finally, we used this number as the λ parameter of the Poisson distribution to test if the number of overlaps observed was higher or lower than the random expectation.

Results

Sequencing and mapping

Two simulated target genomes were generated: the inversionless and the inversionful (948 inversions of different types, see Table 1), derived from human chromosome 1 in the hg19 assembly. Each of them was paired-end sequenced 14 times, with different combinations of template and read lengths (Table 2). After sequencing, we mapped the reads to the reference genome using Novoalign. In all the experiments ~10% of the reads were not mappable, due to the presence of gaps in the reference sequence.

Because the original positions of the reads from the inversionless chromosome were known, we were able to measure their distances to the mapped positions and evaluate the true quality of the alignments. Table 3 shows some statistics of the performance of the aligner in the different experiments. We counted as correct all mappings within a distance to their expected position not larger than the length of the sequenced end, in order to account for potential deviations due to either small indels or alignment clipping. Between 1 and 3% of all mapped reads had at least one alternative mapping. In the majority of ambiguous mappings, the primary alignment is incorrect (Table 3), and the true alignment is to be found, if at all, among the secondary mappings.

Both the template length and the length of the reads have positive effects on the mapping quality, with some nuances. It is remarkable that when the length of the read is shorter than 150 bp, templates of 40 kb produce more mapping errors than templates of 10 kb. This increase in the number of erroneously mapped reads is paralleled by a similar decrease in the number of unmapped reads. We interpret this as a result of the over-zealous alignment of unmappable reads, the presence of gaps in the reference genome, and the proportionality between the average template length and its standard deviation (Table 2). Short reads proceeding from regions not represented in the reference genome are more likely to have spurious concordant alignments when the length of the template is known with less precision.

The application of a mapping quality threshold of 20 (mapping error probability <0.01), reduces the mapping error rate by about 2 orders of magnitude. Such an improvement in average mapping quality comes at the cost of removing more well mapped reads than erroneously mapped ones. Remarkably, all ambiguously mapped reads, that is, all reads with at least two possible mappings within 5 score points from each other, are removed by this filter.

Mapping specificity in inversion detection

Because we simulated Illumina reads, expected to map in forward-reverse orientation, only reads with forward-forward or reverse-reverse orientations are informative of the presence of inversions. These discordant orientations may also arise from mismapping. We used the inversionless chromosome to determine the probability of finding spurious inversion-like orientations in paired-end mappings to human chromosome 1, using different combinations of template length and structural variation (SV) penalty (see Methods).

If ends are mapped independently of each other (that is, with a null SV penalty), about 2% of all pairs with 36 bp reads are mapped in discordant orientations, suggesting the spurious presence of inversions (data not shown). A positive SV penalty

Table 3. Summary statistics of the mapping of reads from the inversionless chromosome.

| | | | Uniquely mapped (%) | | | | Ambiguously mapped (%) | | |
| | | | MAPQ ≥ 20 | | MAPQ <20 | | (all MAPQ <20) | | |
Read (bp)	Template (bp)	Total simulated	correct	wrong	correct	wrong	correct	wrong	unmapped (%)
36	250	138,472,568	85.41	0.005	1.62	0.14	1.16	2.05	9.61
36	450	138,472,568	85.70	0.005	1.50	0.12	1.15	1.92	9.61
36	2,500	138,472,568	86.24	0.003	1.37	0.06	1.15	1.58	9.60
36	10,000	138,472,568	86.40	0.004	1.56	0.05	1.10	1.36	9.52
36	40,000	138,472,568	85.96	0.009	2.01	0.13	1.12	1.55	9.23
75	250	66,466,832	87.64	0.005	0.73	0.04	0.79	1.18	9.62
75	450	66,466,832	87.71	0.006	0.71	0.03	0.79	1.14	9.61
75	2,500	66,466,832	87.97	0.007	0.62	0.03	0.77	1.02	9.58
75	10,000	66,466,832	88.24	0.012	0.55	0.03	0.70	0.97	9.50
75	40,000	66,466,832	88.26	0.041	0.58	0.09	0.66	1.23	9.14
150	450	33,233,416	88.77	0.002	0.32	0.01	0.56	0.71	9.63
150	2,500	33,233,416	88.87	0.002	0.30	0.01	0.54	0.65	9.63
150	10,000	33,233,416	89.01	0.002	0.30	0.01	0.50	0.56	9.63
150	40,000	33,233,416	89.03	0.002	0.33	0.01	0.51	0.50	9.63

For each experiment, defined by the length of the reads and the average length of the templates, we show the total number of reads simulated from the inversionless chromosome and the percentages thereof that have been: mapped uniquely or ambiguously, with a mapping quality (MAPQ) of at least 20 or lower, correctly mapped or not, or unmapped. An ambiguous mapping is considered correct if the primary alignment is correct.

rapidly decreases this proportion, which asymptotically approaches 1.0×10^{-5} by SV-penalty 70. Different template lengths do not significantly change this figure. Longer reads were mapped only with an SV-penalty of 70.

Most of the discordant paired ends from the inversionless chromosome are assigned a low mapping quality. If reads are 36 bp long, and only paired ends having both a mapping quality of at least 20 are considered, around 95% of the orientation-discordant mappings are removed, while only ∼5% of concordant reads are affected by the filter. The effect of the mapping quality threshold is equivalent for all values of SV penalty and template length tested (data not shown). Overall, with 36 bp reads, the combination of a mapping quality ≥ 20 and an SV penalty of 70 reduces the frequency of false orientation discordant paired-ends about 5 orders of magnitude, to between 5.6×10^{-7} and 9.8×10^{-7}. Longer reads from the inversionless chromosome, with a mapping quality of at least 20 (and mapped with an SV-penalty of 70), include proportions of orientation-discordant pairs always lower than 2.0×10^{-7}. If reads from the inversionful chromosome have similar rates of mismapping, from a physical coverage of 50 we expect between less than 1 (reads longer than 36 bp and templates longer than 250 bp) and 39 (36 bp reads, 250 bp templates) spurious orientation-discordant pairs with a mapping quality of at least 20, that would suggest the presence of false inversions.

Mapping sensitivity in inversion detection

To determine the ability of PEM experiments to detect inversions, we computationally sequenced the inversionful chromosome using different combinations of read and template lengths (Table 2). Before the application of any SV-detection software, we determined the performance of the alignment software at providing evidence of the breakpoints. We applied an SV penalty

of 70, necessary to remove most false positives (see above). A pair of ends sequenced from alternative sides of a breakpoint is potentially informative of the existence of the breakpoint. For every breakpoint, we counted the potentially informative pairs obtained with each sequencing strategy, and how many of them were mapped correctly, erroneously mapped and unmapped.

The informative physical coverage of a breakpoint depends on the sequencing strategy and on the length of the inversion. The expected informative physical coverage can be expressed as the product of the total number of templates sequenced and the probability that a template encompasses a single breakpoint between its two sequenced ends. Assuming that, as it is the case in our experiments, average template lengths are larger than twice the read length, and inversions are larger than the reads, then:

$$E(c) = \frac{\min(t-2r, i-r)}{g} \cdot n \qquad (1)$$

where c is the physical coverage, r is the length of the reads, t is the average length of the templates, i is the length of the inversion, n is the sequencing effort in number of templates sequenced, and g is the length of the genome.

Equation 1 describes well the number of reads actually sequenced across breakpoints. However, a variable portion of those reads are either unmapped or, more often, erroneously mapped. Figure 2 shows the average proportion of pairs of reads sequenced across a breakpoint that are correctly mapped in each experiment for inversions located in 4 different contexts: 1) randomly, 2) between inverted repeat-masked sequences, 3) between other inverted alignable regions, and 4) between inverted segmental duplications. The rest are erroneously mapped elsewhere, many as concordant (a small, and rather constant fraction of unmapped reads are not counted there).

As shown in Figure 2, reads of 36 bp (green points) with templates ≤ 450 bp perform quite badly for all inversion types and have a probability of being well mapped around inverted repeats well below 0.6. It can also be seen that long templates are instrumental to correctly map reads across repeats. While short repeats, such as those identified by RepeatMasker and other alignable regions, are effectively bypassed with 2500 bp-long templates, most segmental duplications are challenging even for 40 kb-long templates.

The values shown in Figure 2 are averages across inversions. For each inversion, we estimated the probability of mapping a pair of ends correctly across one of its breakpoints, if at least 50 pairs had been simulated covering its breakpoints. These probabilities where then used to calculate the expected number of breakpoints detected by at least 2 paired ends with a given physical coverage and a given sequencing strategy (Figure 3). Our results show that low physical coverages can detect very efficiently randomly generated inversions. However, for inversions located between inverted repeats, maximal inversion detection requires quite different amounts of physical coverage depending on the PEM conditions, and suboptimal sequencing strategies are predicted to fail to detect a substantial amount of inversions located between segmental duplications, irrespectively of the sequencing effort. In particular, a physical coverage of ~ 50 can achieve sensitivities higher than 90% in all sequencing contexts, with average template length of 40 kb and reads of 150 bp. Notice that such an experiment would produce a sequence coverage (i.e. sequencing depth) of only 0.4. In addition, easier to obtain libraries of 2.5 kb perform also very well for most types of inversions, except those mediated by segmental duplications.

The probability of correctly mapping a pair of reads across a breakpoint is expected to depend on the characteristics of the inverted repeats present around the breakpoints, if any. At least, the length of the repeats and the similarity between them must affect directly the fraction of templates spanning a breakpoint that are mapped either discordantly or concordantly across the breakpoint. The expected value of that proportion (discordant over the sum of concordant and discordant) in a candidate breakpoint would be useful to determine the likelihood of that candidate. Thus, we attempted to fit a generalized linear model of the proportion of templates sequenced across a breakpoint that are mapped across that breakpoint either concordantly or discordantly, using characteristics of the inversion and of the sequencing strategy as predictors. We failed to correct the overdispersion present in all the models that we tested. We suspect that the specific distribution of mismatches along the alignment of the two inverted copies, and the amount and distribution of gaps thereof, which were not characterized, significantly affect the chances of a read being mapped to the correct copy. In any case, we captured part of the pattern of variation with two compound variables (interactions) using the data from the 14 experiments (Figure 4). First, the interaction between the length of the repeat and the length of the template is apparent in Figure 1: only templates longer than the repeats may have ends with unique sequences, that can be correctly mapped. And second, the interaction between inverted copies identity and the read length represents that longer reads are more likely to contain a difference between repeat copies than shorter reads from the same repeat. The logarithmic transformation of all lengths and the squaring of the identity improved the quality of the relationship.

Sensitivity, specificity and precision of SV-detecting algorithms

The post-mapping analysis of PEM data to discover inversions may introduce its own biases. We used three different algorithms designed to detect inversions and other SV from PEM data to identify common sources of false positives and false negatives: SVDetect [29], VariationHunter [16], and GRIAL (A. Martnez-Fundichely, S. Casillas, and M. Cáceres, unpublished data). Figure 5 represents the percentage of breakpoints of each type of inversion detected by each program with different sequencing strategies. In this comparison, the physical coverage was kept at 50 across experiments, that is, longer templates entail fewer paired ends sequenced (see Methods).

The sensitivity (proportion of true breakpoints correctly predicted) of the three algorithms (Figure 5) is in general close to, but sometimes lower than, what expected from the mappability of the reads around breakpoints (Figure 3, for a physical coverage of 50). As predicted by Figure 4, sensitivity of shorter template libraries decreases with highly identical inverted repeats, especially segmental duplications. In addition, templates of 40 kb recall fewer breakpoints than 10 kb templates in most sequence contexts. This is due to a higher proportion of inversions being shorter than the template, and therefore receiving lower useful physical coverage (see Equation 1), than if sequenced with shorter templates. This effect is very pronounced in the case of inversions located between alignable regions, because they are on average the shortest (Table 1). If the number of paired ends sequenced or the sequencing depth, instead of the physical coverage, was kept constant across experiments, longer templates would always outperform shorter ones, as suggested by Figure 3 (data not shown).

Inversion detection for random and repeat-masked inversions with VariationHunter and GRIAL is almost 100%. In contrast, SVDetect does not reach the same sensitivity with short templates (250 and 450 bp). A careful inspection of these cases showed that SVDetect is calling inversion breakpoints some base pairs off the true breakpoints, thus producing false positives (see below). On the other hand, using 150 bp reads and 40 kb templates, GRIAL detected 51/58 breakpoints within segmental duplications, while SVDetect and VariationHunter detected 47/58 and 44/58, respectively. According to Figure 3, 53.5 breakpoints were expected to be detectable, on the bases of the mappability of the reads. Across experiments, GRIAL was about 6% more sensitive than SVDetect, and 0.4% more sensitive than VariationHunter.

We also measured the average precision attained by each program in their breakpoint predictions. In all cases, the ranges of positions where breakpoints are predicted to lay are larger than the theoretical expectation derived by Bashir et al. [18] for large inversions, assumed to be randomly located (Figure 6). As expected, the breakpoints of inversions smaller than the template length cannot be detected with a precision better than the difference between the length of the template and the length of the inversion. This reflects the fact that for a pair of ends to be informative, the read sequenced from outside of a small inversion must be at a certain distance, such that its partner is sequenced from inside the inversion. For both size classes, but in particular for big inversions, GRIAL achieves finer precision than the other two programs, and VariationHunter offers the coarsest precision. Note that both axes in Figure 6) are in logarithmic scale to appreciate how substantial the differences are.

In addition to the true positives shown, all three programs produce a number of false positives (Table 4), which in general are higher for short templates. We compared the positions spanned by the false predictions among the three programs (Figure 7). The

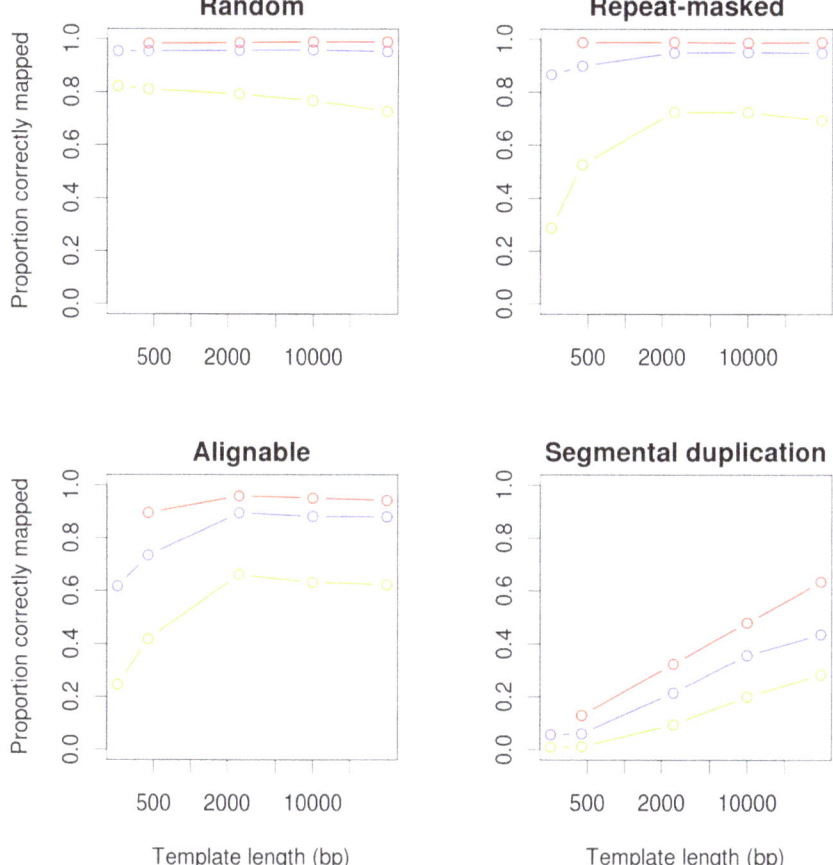

Figure 2. Average portion of potentially informative reads that are correctly mapped across a breakpoint. Informative reads are represented as a function of the template length for inversions located in four different sequence contexts. Colors represent the three read lengths: green, 36 bp; blue, 75 bp; and red, 150 bp.

false positives predicted by VariationHunter tend to be different from those predicted by either GRIAL or SVDetect. We attribute these differences to the fact that VariationHunter uses ambiguously mapped reads with low mapping quality, that neither GRIAL nor SVDetect use.

We carefully looked at the origin of false positives and distinguished three different types. First, a small number of reads originally colinear with the reference (no more than 50 per experiment) were erroneously mapped in discordant orientation and gave support to false breakpoints, at least in SVDetect and in GRIAL (we could not keep track of what reads supported each prediction from VariationHunter). This is in agreement with the small number of false positives expected from erroneous mappings (see section on the Mapping specificity in inversion detection). We do not observe this kind of false positives when the template lengths are at least 2500 bp long or if the reads are at least 150 bp long. In principle, an SV-penalty higher than 70 during mapping could also reduce the number of this kind of false positives (not tested).

Second, truly discordant reads, originated across true breakpoints and mapped in the correct (discordant) orientation, but to an erroneous location, gave rise to false predictions. In the experiment with reads of 75 bp and templates of 250 bp, GRIAL predicted 10 false inversions (involving 17 false breakpoints, and 3 true breakpoints assigned to wrong inversions) and 7 of them are also predicted by SVDetect. These common false inversions are

due to mismapped reads, and they are larger than 20 Mb. In contrast, VariationHunter filters out inversion predictions larger than 1 Mb, although it does predict individual, unpaired breakpoints in other locations.

And third, there are correctly mapped reads that are not well interpreted by the SV-detection algorithm, and give rise to 'false' breakpoints that do not overlap true breakpoints, but lay close. Across experiments, 88% of SVDetect's false breakpoints (see Table 4 and Figure 7) lay within 50 bp of a true breakpoint, and they predict inversions that do overlap with real inversions. Also the two false positives predicted by GRIAL with reads of 150 bp and template lengths of 10 kb and 40 kb are very close (at 23 and 62 bp, respectively) to real breakpoints. These two false positives may be due to random departures from the expected template lengths, upon which GRIAL predictions heavily depend (A. Martnez-Fundichely, S. Casillas, and M. Cáceres, unpublished data).

Just as inverted repeats are hotspots of false negatives (see section on Mapping sensitivity in inversion detection), they can also generate false positives, due to the possibility of mapping reads to either copy. To understand better the origin of false positives, we compared their positions with those of all the segmental duplications, RepeatMasker-filtered regions and other alignable regions present in chromosome 1. In all the experiments with average template lengths of 250 or 450, the false breakpoints predicted by GRIAL or by SVDetect overlap with either repeat-

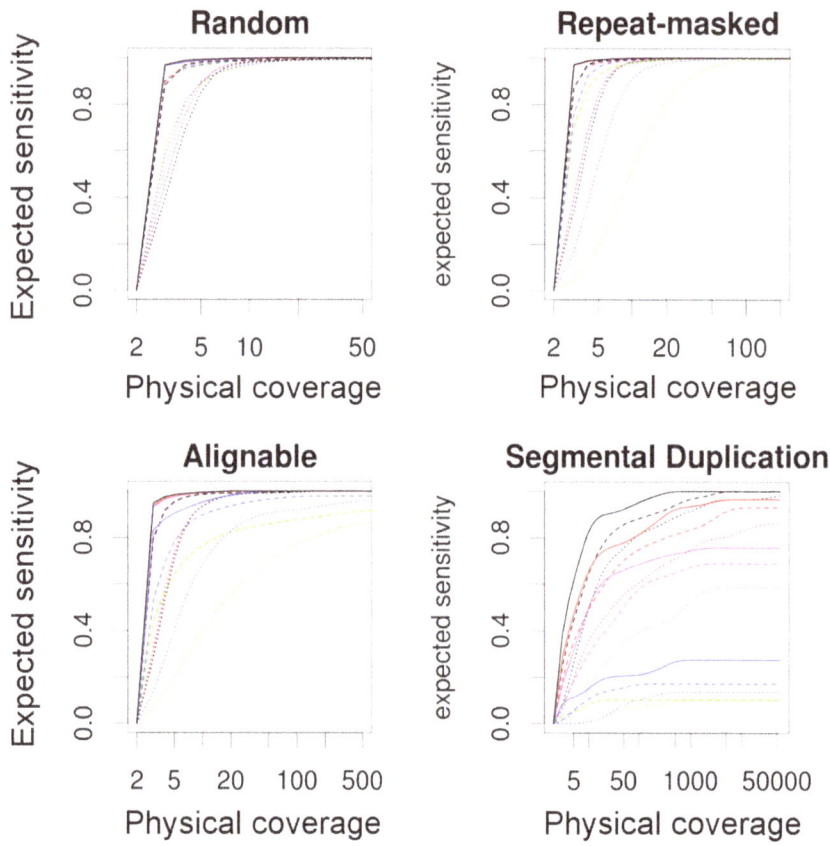

Figure 3. Relationship between physical coverage and the expected sensitivity of different sequencing strategies to detect inversions. The expected sensitivity is based on the probability of correctly mapping paired ends across inversion breakpoints in four different sequence contexts. Inversions are assumed to be longer than the templates. The sequencing strategy is defined by the read length: dotted lines, 36 bp; dashed lines, 75 bp; solid lines, 150 bp; and by the template length: green, 250 bp; blue, 450 bp; purple, 2.5 kb; red, 10 kb; and black, 40 kb. Notice the different ranges of physical coverage among plots.

masked regions or with other alignable regions more often than expected by chance; but they did not overlap segmental duplications more frequently than expected (data not shown). In contrast, the false positives predicted by VariationHunter in 4 experiments did overlap segmental duplications more often than expected by chance (at 0.01 significance level, Poisson test), and they also overlapped other alignable regions (not repeat-masked segments) more often than expected by chance in some experiments.

Discussion

Currently there is a great interest in the complete characterization of SV at a genome level, with multiple projects to sequence whole genomes using different PEM strategies. However, it is not clear to what extent these projects are giving us an adequate picture of the SV present in the human genome. Therefore, it is important to have quantitative estimates as realistic as possible of the amount of variants that we may be missing or describing incorrectly.

The mapping stage of PEM data analysis caps the sensitivity of any SV-detection program. For well understood reasons, inversions between segmental duplications may be undetectable, under some experimental designs. Unfortunately, most PEM experiments performed to date were done with very small templates (e.g., [11,12,27]; but see [10]) that are not suited to detect inversions

between inverted repeats (Figure 5). Around 90% of the paired-end sequencing experiments (~80% of the reads) generated by the 1000 genomes project have template lengths below 500 bp (according to the sequence indexes downloaded from their ftp site on October 9th, 2012). These template lengths, combined with modest coverages, are expected to miss more than 80% of the inversions between segmental duplications and around 5–50% of the inversions between repeat-masked or other alignable regions. Neither an increase in coverage, nor an improvement in SV-detection algorithms can prevent false negatives completely. It is also important to note that our sensitivity estimates may be overly optimistic, since we have simulated inversions between inverted repeats with identities as low as 60%, whereas real inversions are probably enriched in highly identical repeats. Thus, PEM studies have been systematically missing most of the inversions present between inverted repeats, and a similar problem may affect other types of structural variants. The actual relative abundance of inversions between inverted repeats is impossible to evaluate with current data from massively parallel paired-end sequencing studies, precisely due to the ascertainment bias against them.

It is known that longer templates improve the assembly in *de novo* sequencing projects [34], and extend the range of insertions that can be discovered by PEM [8]. However, little emphasis has been put on the importance of template length for inversion discovery. When detecting inversions, longer templates always improve sensitivity (Figure 2) and specificity (Table 4). If longer templates

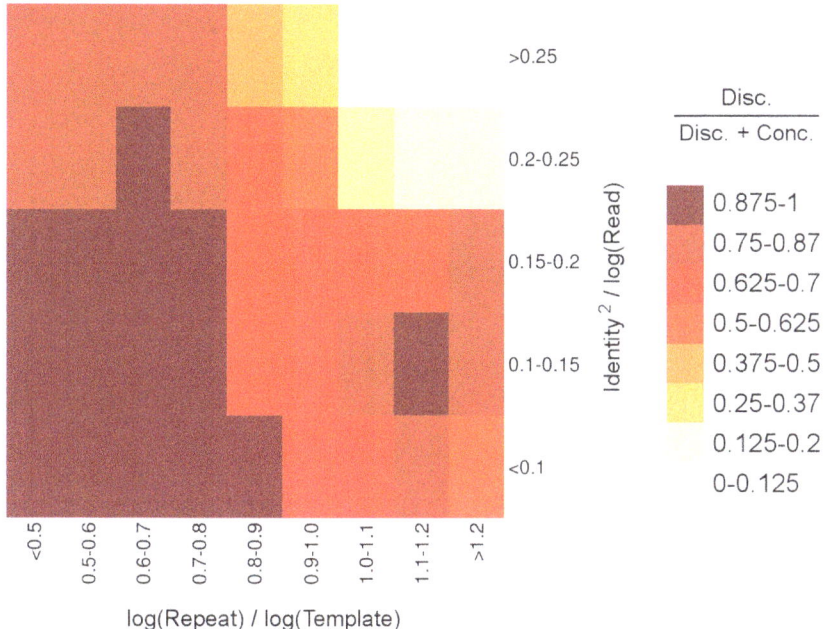

Figure 4. Average proportion of pairs of reads mapped across a breakpoint that are correctly mapped as discordant. Correct discordant read pairs are expressed relative to all reads mapped across the same breakpoint, as a function of the length of the repeat (relative to the length of the template) and the identity between the copies (relative to the read length). Data from all 14 simulated paired-end sequencing experiments are used. Cells may have different standard errors, due to differences in the total number of reads used to calculate the proportion of discordant pairs in each situation.

were used, the bias against inversions between inverted repeats could be traded for a bias against short inversions, but only as long as current technologies impose a trade off between the template length and the throughput. When designing an experiment, one could give priority to inversions between inverted repeats, and choose first the longest average template length available and then, the affordable sequencing effort. For example, Kidd et al. [10] used average template lengths of ∼40 kb (fosmid genomic libraries), and sequenced about 400 bp of each end. They reached a sequencing depth of about 0.3 per individual, which, according to Equation 1, implies that the breakpoints of inversions shorter than 3 kb were expected to be physically covered less than once. In addition, Bashir et al. [18] reported a trade-off between detectability (template length) and breakpoint precision for large inversions in random locations, and they recommended a mixture of long and short template lengths to optimize both. Although this trade-off progressively vanishes with increasing physical coverage, it has an important corollary: the longer the templates, the higher the proportion of inversions that are shorter than the templates. In Figure 6, it can be seen that for a physical coverage of 50 the loss of precision due to longer templates in large inversions is small compared to that in inversions shorter than the templates. Thus, a coarse precision may be the price to pay for the detection of inversions between inverted repeats.

From our results (figures 2, 4, and 5), it is apparent that both longer reads and longer templates improve inversion detectability. In most of the genome, sequenced ends of 150 bp perform almost as well as possible. The constant development of sequencing technologies offers ever longer reads, going up to several kilobases in the case of Pacific Biosciences or Illumina's Moleculo technology. Eventually, long enough reads with high enough quality could override the need for paired-ends, and inversions would be detected by direct sequencing. However, increasingly

longer reads will not help much for the detection of inversions located between large segmental duplications, but longer DNA libraries would. The breakpoints of an inversion located between inverted repeats are virtually invisible at the sequence level within the repeat, what renders split reads useless in this context. Longer sequences could be useful to map reads more accurately in the two inverted repeats, although in highly identical regions the mapping would rely at most in a few base differences between copies. These differences are known to vary between individual genomes and make the mappings that are not based in unique or quite divergent sequences unreliable.

In terms of sensitivity, the three programs tested perform similarly (Figure 5), stressing the importance of the sequencing design and the mapping stage. However, the programs differ significantly in terms of precision, GRIAL being the program with the most accurate breakpoints (Figure 6). In terms of false positives, the apparently high false discovery rate by SVDetect (Table 4, and Figure 7) is mostly due to the predictions missing the actual breakpoints by a few base pairs. VariationHunter is the only tested algorithm producing an excess of false positives in segmental duplications, that we attribute to its usage of low quality, secondary mappings. To avoid those false positives, either the discordantly mapped reads must be further filtered, or their mapping qualities should be more accurate.

Even low rates of false positives can produce high posterior error probabilities if inversions are rare [35]. On the other hand, if inversions (and maybe other kinds of SV) are frequent between the target and the reference genomes, as in the case of cancer genomes, the rate of false positives could be even higher. High levels of SV could produce large numbers of false positives because pairs of reads that span a breakpoint have a higher chance of being mapped to a wrong location *and* in a discordant orientation than those colinear to the reference. Thus, part of the false positives

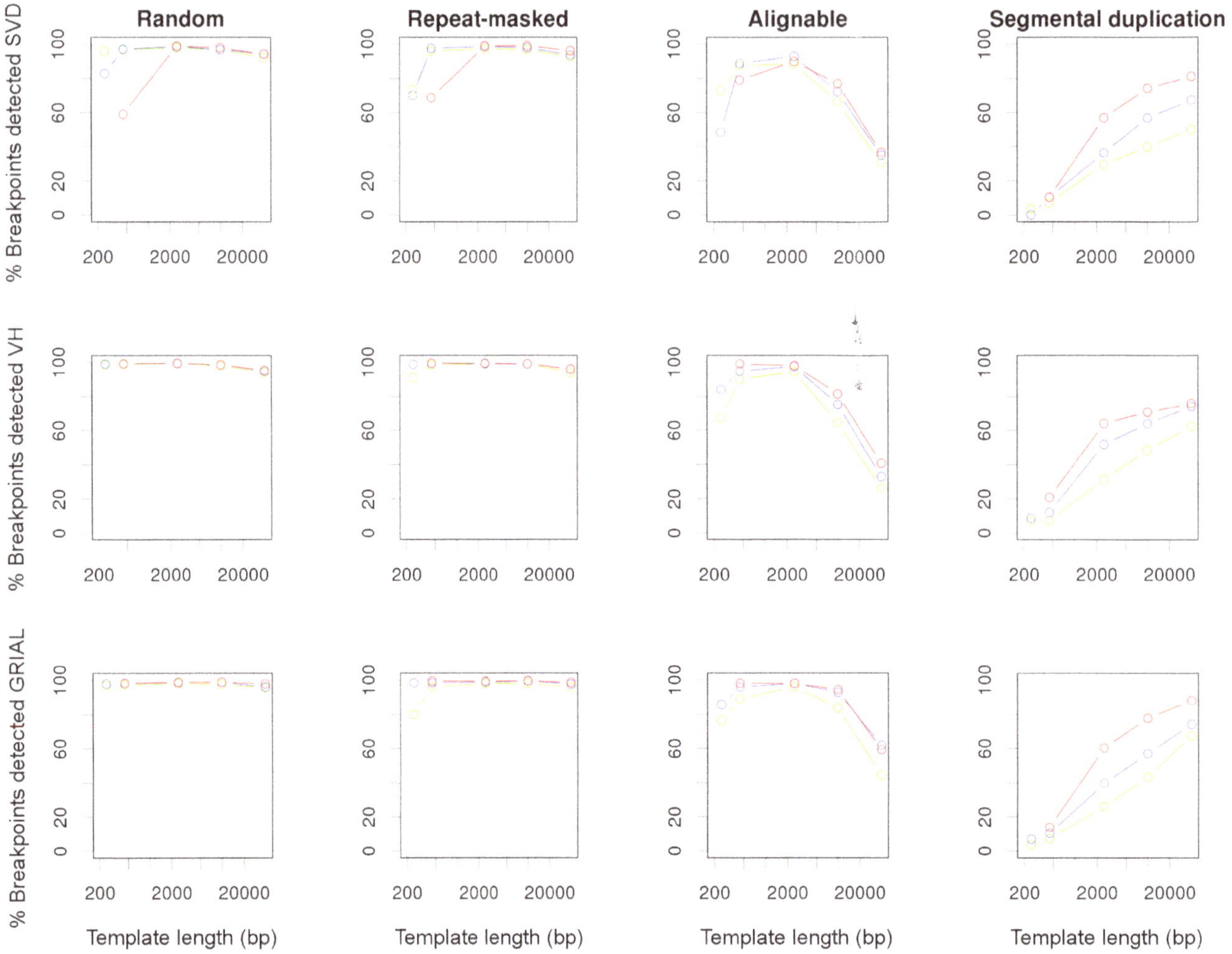

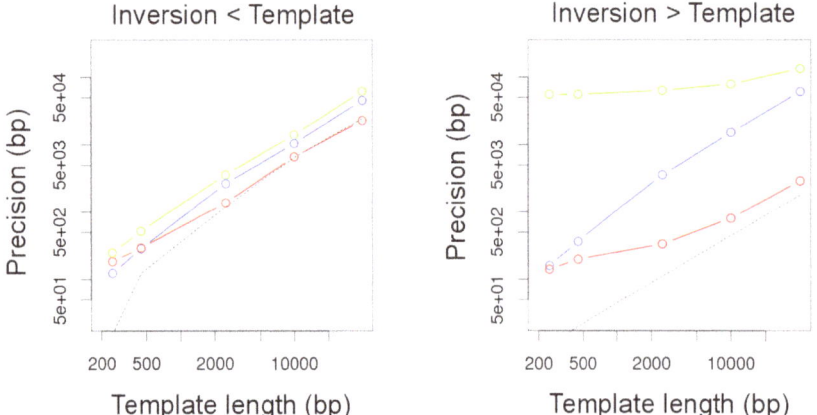

Figure 5. Percentage of inversion breakpoints from each sequence context that are successfully detected by different programs. Results from SVDetect (SVD, upper row), VariationHunter (VH, middle row), or GRIAL (bottom row) are plotted against the template length used. Colors correspond to the length of the reads: green, 36 bp; blue, 75 bp; and red, 150 bp.

Figure 6. Precision of breakpoint prediction plotted against the length of the template. The average size of the predicted range of a breakpoint is represented separately for inversions smaller (left) or larger (right) than the template. Colors correspond to the programs used to predict the breakpoints: green, VariationHunter; blue, SVDetect; and red, GRIAL. The dashed lines correspond to the theoretical expected precisions, obtained either from equation 3 in reference [18] for large inversions, or from the average difference between the inversion size and the template length for small inversions.

SVDetect   GRIAL

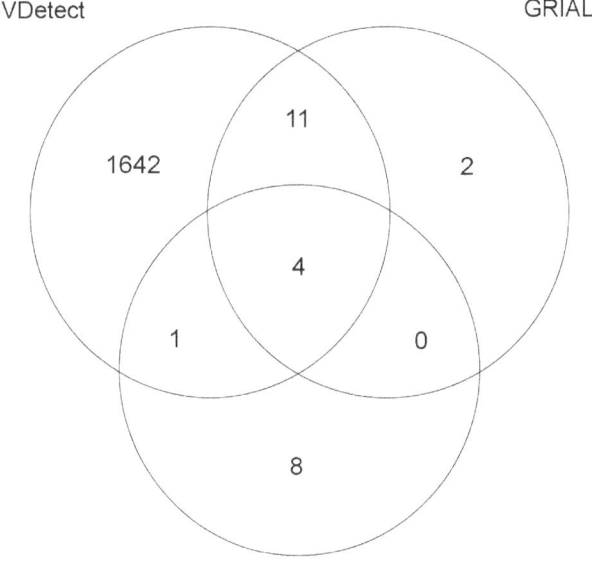

VariationHunter

Figure 7. Comparison of the false breakpoints predicted by the three programs. Templates of 250 bp and 75 bp reads were used. The sharing of a breakpoint between two programs imply that their predictions overlap in at least one base and are of the same kind, namely, either the first or the second breakpoint of an inversion.

Table 4. False inversion breakpoints called by three SV-detection algorithms.

Read (bp)	Template(bp)	SVD	VH	GRIAL
36	250	67 (0.0418)	11 (0.0065)	10 (0.0062)
36	450	6 (0.0034)	10 (0.0055)	4 (0.0023)
36	2,500	1 (0.0006)	5 (0.0027)	0 (0.0000)
36	10,000	0 (0.0000)	3 (0.0017)	0 (0.0000)
36	40,000	1 (0.0006)	2 (0.0012)	0 (0.0000)
75	250	1,658 (0.5508)	13 (0.0072)	17 (0.0095)
75	450	215 (0.1080)	7 (0.0038)	4 (0.0022)
75	2,500	6 (0.0033)	3 (0.0016)	0 (0.0000)
75	10,000	5 (0.0028)	4 (0.0022)	0 (0.0000)
75	40,000	0 (0.0000)	6 (0.0036)	0 (0.0000)
150	450	1,527 (0.5591)	10 (0.0054)	13 (0.0071)
150	2,500	15 (0.0081)	12 (0.0064)	0 (0.0000)
150	10,000	17 (0.0093)	8 (0.0044)	1 (0.0005)
150	40,000	4 (0.0024)	8 (0.0047)	1 (0.0006)

Number of false inversion breakpoints predicted by SVDetect (SVD), VariationHunter (VH) or GRIAL under each sequencing strategy, defined by the template and the read lengths. In parentheses, the proportion that these false positives represent among all the predictions.

predicted here are due to pairs of reads sequenced across true breakpoints and mismapped, and they could be considered an artifact of the high density of the simulated inversions. Our results suggest that false positives can be kept low by using a stringent SV-penalty during mapping, filtering out low quality reads, choosing an appropriate algorithm, and using templates of at least 2.5 kb (Table 4). However, our simulations represent a best-case scenario, any departure from which will make it more difficult to detect the true inversions and to avoid the false ones. For example, the presence of other types of SV, and especially the presence of complex rearrangements, are expected to increase the rate of false positives, as mentioned earlier.

Polymorphic inversions are more likely to be detected where they are less likely to happen, namely in non-repetitive sequences; and difficult to detect where they are more likely to be, that is, between inverted repeats (Figure 5). As a result, the frequency of polymorphic inversions in the human genome could be underestimated in one hand, and overestimated due to false positives in the other. Supposedly simple tasks such as comparing the frequency of inversions among chromosomes, or estimating the total number of inversions in one genome, are not supported by any SV-detection algorithm to date, because the unknown numbers of false positives and false negatives would bias the results. Yet, with the information contained in PEM data, it should be easier to estimate the total number of inversions, than to enumerate all of them. Thus, we think that SV-detection algorithms will keep evolving to implement sound statistical models with estimates of both false positives and false negatives.

One step in this direction is the recent appearance of GASVPro, an SV-detection algorithm that implements a probabilistic model to determine the most likely set of structural variants supported by PEM data from one individual [36]. GASVPro uses multiple possible alignments of discordant reads, and approximates the posterior probabilities of the mappings. Although GASVPro does

not explicitly estimate the total number of SVs, nor it reports the probability that a prediction is false, its probabilistic formulation would allow such extensions. Instead, GASVPro follows the trend of reporting a list of variants, biased as it may be. Therefore, it is not surprising that even GASVPro has a very low rate of recall of known inversions from two sequenced individuals, and apparently high rates of false positives, just as all other programs tested by the authors (Tables 2 and 3 in [36]). Other recent developments in SV-detection algorithms tend to use evidence from both paired-ends and split reads to improve the definition of breakpoints [37,38]. These methods take the most of the data at hand and improve the sensitivity and the specificity in some circumstances. However, they fail to address the main concern raised by our results, namely the overlooking of inversions between inverted repeats, where split reads do not add any information.

In summary, current SV-detection algorithms fail to account for the heterogeneous distribution of SV, and in particular of inversions, along the genome; and they fail to account for the also heterogeneous probability of false positives. In order to study their mechanisms of origin and to perform population genetic analyses of inversions, we need to estimate parameters of an explicit model of SV distribution, rather than an incomplete and biased list of differences between two genomes, and they will have to pay attention to the genome-specific repetitive structure. Future improvements of both algorithms and sequencing strategies are expected to give us a better idea of the genomic landscape of SVs in general, and inversions in particular.

Acknowledgments

We thank Alexander Martnez-Fundichely and Sònia Casillas for help with the implementation of the GRIAL algorithm, and the standard deviation estimates of libraries of different template lengths; and Meritxell Oliva for useful advice and comments.

Author Contributions

Conceived and designed the experiments: MC JILL. Performed the experiments: JILL. Analyzed the data: JILL. Contributed reagents/materials/analysis tools: MC. Wrote the paper: MC JILL.

References

1. Iafrate AJ, Feuk L, Rivera MN, Listewnik ML, Donahoe PK, et al. (2004) Detection of large-scale variation in the human genome. Nat Genet 36: 949–951.

2. Sebat J, Lakshmi B, Troge J, Alexander J, Young J, et al. (2004) Large-scale copy number polymorphism in the human genome. Science 305: 525–528.

3. Tuzun E, Sharp AJ, Bailey JA, Kaul R, Morrison VA, et al. (2005) Fine-scale structural variation of the human genome. Nat Genet 37: 727–732.

4. Korbel JO, Urban AE, Affourtit JP, Godwin B, Grubert F, et al. (2007) Paired-end mapping reveals extensive structural variation in the human genome. Science 318: 420–426.

5. Campbell PJ, Pleasance ED, Stephens PJ, Dicks E, Rance R, et al. (2008) Subclonal phylogenetic structures in cancer revealed by ultra-deep sequencing. Proc Natl Acad Sci U S A 105: 13081–13086.

6. Stephens PJ, McBride DJ, Lin ML, Varela I, Pleasance ED, et al. (2009) Complex landscapes of somatic rearrangements in human breast cancer genomes. Nature 462: 1005–1010.

7. Hillmer AM, Yao F, Inaki K, Lee WH, Ariyaratne PN, et al. (2011) Comprehensive long-span paired-end-tag mapping reveals characteristic patterns of structural variations in epithelial cancer genomes. Genome Res 21: 665–675.

8. Medvedev P, Stanciu M, Brudno M (2009) Computational methods for discovering structural variation with next-generation sequencing. Nat Methods 6: S13–S20.

9. Xi R, Kim TM, Park PJ (2011) Detecting structural variations in the human genome using next generation sequencing. Brief Funct Genomics 9: 405–415.

10. Kidd JM, Cooper GM, Donahue WF, Hayden HS, Sampas N, et al. (2008) Mapping and sequencing of structural variation from eight human genomes. Nature 453: 56–64.

11. Wang J, Wang W, Li R, Li Y, Tian G, et al. (2008) The diploid genome sequence of an Asian individual. Nature 456: 60–65.

12. Ahn SM, Kim TH, Lee S, Kim D, Ghang H, et al. (2009) The first Korean genome sequence and analysis: full genome sequencing for a socio-ethnic group. Genome Res 19: 1622–1629.

13. McKernan KJ, Peckham HE, Costa GL, McLaughlin SF, Fu Y, et al. (2009) Sequence and structural variation in a human genome uncovered by short-read, massively parallel ligation sequencing using two-base encoding. Genome Res 19: 1527–1541.

14. Pang AW, MacDonald JR, Pinto D, Wei J, Rafiq MA, et al. (2010) Towards a comprehensive structural variation map of an individual human genome. Genome Biol 11: R52.

15. Chen K, Wallis JW, McLellan MD, Larson DE, Kalicki JM, et al. (2009) BreakDancer: an algorithm for high-resolution mapping of genomic structural variation. Nat Methods 6: 677–681.

16. Hormozdiari F, Alkan C, Eichler EE, Sahinalp SC (2009) Combinatorial algorithms for structural variation detection in high-throughput sequenced genomes. Genome Res 19: 1270–1278.

17. Onishi-Seebacher M, Korbel JO (2011) Challenges in studying genomic structural variant formation mechanisms: the short-read dilemma and beyond. Bioessays 33: 840–850.

18. Bashir A, Bansal V, Bafna V (2010) Designing deep sequencing experiments: detecting structural variation and estimating transcript abundance. BMC Genomics 11: 385.

19. Korbel JO, Abyzov A, Mu XJ, Carriero N, Cayting P, et al. (2009) PEMer: a computational framework with simulation-based error models for inferring genomic structural variants from massive paired-end sequencing data. Genome Biol 10: R23.

20. Suzuki S, Yasuda T, Shiraishi Y, Miyano S, Nagasaki M (2011) ClipCrop: a tool for detecting structural variations with single-base resolution using soft-clipping information. BMC Bioinformatics 12: S7.

21. Zhang ZD, Du J, Lam H, Abyzov A, Urban AE, et al. (2011) Identification of genomic indels and structural variations using split reads. BMC Genomics 12: 375.

22. Krimbas CB, Powell JR, editors (1992) Drosophila inversion polymorphism. CRC Press, 1–560 pp.

23. Feuk L (2010) Inversion variants in the human genome: role in disease and genome architecture. Genome Med 2: 11.

24. Stefansson H, Helgason A, Thorleifsson G, Steinthorsdottir V, Masson G, et al. (2005) A common inversion under selection in Europeans. Nat Genet 37: 129–137.

25. Hoffmann AA, Rieseberg LH (2008) Revisiting the impact of inversions in evolution: from population genetic markers to drivers of adaptive shifts and speciation? Annu Rev Ecol Evol Syst 39: 21–42.

26. Kidd JM, Graves T, Newman TL, Fulton R, Hayden HS, et al. (2010) A human genome structural variation sequencing resource reveals insights into mutational mechanisms. Cell 143: 837–847.

27. The 1000 Genomes Project Consortium (2010) A map of human genome variation from population scale sequencing. Nature 467: 1061–1073.

28. Cao J, Schneeberger K, Ossowski S, Günther T, Bender S, et al. (2011) Whole-genome sequencing of multiple Arabidopsis thaliana populations. Nat Genet 43: 956–963.

29. Zeitouni B, Boeva V, Janoueix-Lerosey I, Loeillet S, Legoix-né P, et al. (2010) SVDetect: a tool to identify genomic structural variations from paired-end and mate-pair sequencing data. Bioinformatics 26: 1895–1896.

30. Li H, Handsaker B, Wysoker A, Fennell T, Ruan J, et al. (2009) The Sequence Alignment/Map format and SAMtools. Bioinformatics 25: 2078–2079.

31. Richter DC, Ott F, Auch AF, Schmid R, Huson DH (2008) MetaSim—A sequencing simulator for genomics and metagenomics. PLoS One 3: e3373.

32. Hormozdiari F, Hajirasouliha I, Dao P, Hach F, Yorukoglu D, et al. (2010) Next-generation VariationHunter: combinatorial algorithms for transposon insertion discovery. Bioinformatics 26: i350–i357.

33. Alkan C, Kidd JM, Marques-Bonet T, Aksay G, Antonacci F, et al. (2009) Personalized copy number and segmental duplication maps using next-generation sequencing. Nat Genet 41: 1061–1067.

34. Roach JC, Boysen C, Wang K, Hood L (1995) Pairwise end sequencing: a unified approach to genomic mapping and sequencing. Genomics 26: 345–353.

35. Manly KF, Nettleton D, Hwang JTG (2004) Genomics, prior probability, and statistical tests of multiple hypotheses. Genome Res 14: 997–1001.

36. Sindi SS, Onal S, Peng LC, Wu HT, Raphael BJ (2012) An integrative probabilistic model for identification of structural variation in sequencing data. Genome Biol 13: R22.

37. Jiang Y, Wang Y, Brudno M (2012) PRISM: paired read informed split read mapping for base-pair level detection of insertion, deletion and structural variants. Bioinformatics 28: 2576–2583.

38. Rausch T, Zichner T, Schlattl A, Stütz AM, Benes V, et al. (2012) DELLY: structural variant discovery by integrated paired-end and split-read analysis. Bioinformatics 28: i333–i339.

5

Profiling of Olfactory Receptor Gene Expression in Whole Human Olfactory Mucosa

Christophe Verbeurgt[2]'ʼ, Françoise Wilkin[1]*ʼ, Maxime Tarabichi[3], Françoise Gregoire[4], Jacques E. Dumont[3], Pierre Chatelain[1]

1 ChemCom S.A., Brussels, Belgium, **2** Department of Otorhinolaryngology, Erasme University Hospital, Brussels, Belgium, **3** Institute of Interdisciplinary Research in human and molecular Biology, Free University of Brussels, Brussels, Belgium, **4** Laboratory of Pathophysiological and Nutritional Biochemistry, Department of Biochemistry, Free University of Brussels, Brussels, Belgium

Abstract

Olfactory perception is mediated by a large array of olfactory receptor genes. The human genome contains 851 olfactory receptor gene loci. More than 50% of the loci are annotated as nonfunctional due to frame-disrupting mutations. Furthermore haplotypic missense alleles can be nonfunctional resulting from substitution of key amino acids governing protein folding or interactions with signal transduction components. Beyond their role in odor recognition, functional olfactory receptors are also required for a proper targeting of olfactory neuron axons to their corresponding glomeruli in the olfactory bulb. Therefore, we anticipate that profiling of olfactory receptor gene expression in whole human olfactory mucosa and analysis in the human population of their expression should provide an opportunity to select the frequently expressed and potentially functional olfactory receptors in view of a systematic deorphanization. To address this issue, we designed a TaqMan Low Density Array (Applied Biosystems), containing probes for 356 predicted human olfactory receptor loci to investigate their expression in whole human olfactory mucosa tissues from 26 individuals (13 women, 13 men; aged from 39 to 81 years, with an average of 67 ± 11 years for women and 63 ± 12 years for men). Total RNA isolation, DNase treatment, RNA integrity evaluation and reverse transcription were performed for these 26 samples. Then 384 targeted genes (including endogenous control genes and reference genes specifically expressed in olfactory epithelium for normalization purpose) were analyzed using the same real-time reverse transcription PCR platform. On average, the expression of 273 human olfactory receptor genes was observed in the 26 selected whole human olfactory mucosa analyzed, of which 90 were expressed in all 26 individuals. Most of the olfactory receptors deorphanized to date on the basis of sensitivity to known odorant molecules, which are described in the literature, were found in the expressed olfactory receptors gene set.

Editor: Richard David Newcomb, Plant and Food Research, New Zealand

Funding: M. Tarabichi is supported by FRIA/FNRS grant. URL: http://www.fnrs.be/. C. Verbeurgt is supported by Fonds Erasme, Université Libre de Bruxelles, Brussels, Belgium. URL: http://www.fondserasme.org/fonds-erasme-pour-la-recherche-medicale. ChemCom S.A. is supported by Innoviris (Brussels Institute for Research and Innovation) URL: http://www.innoviris.be/site/. The funders had no role in study design, data collection and analysis, decision to publish, or preparation of the manuscript.

Competing Interests: Two authors are employed by a commercial company (ChemCom S.A.): P. Chatelain and F. Wilkin. The authors have declared that no competing interests exist.

* E-mail: fwi@chemcom.be

ʼ These authors contributed equally to this work.

Introduction

Analysis of published mammalian genomes indicates that olfactory receptor (OR) genes constitute by far the largest gene family. Initially, Buck and Axel identified this extremely large multigene family based on the observation that OR genes were expressed in olfactory epithelium of rat [1]. Later, other members of this family were identified by sequence homology with the first set of OR genes [2–4]. Currently it is accepted that the human genome contains 851 OR loci. More than 50% of the loci are annotated as nonfunctional due to frame-disrupting mutations, leaving approximately 400 potentially functional OR genes.

In spite of this rather accurate genomic characterization, very little is known of the functional and integrative mechanisms of human olfactory receptor in odorant perception. To date, the responses of only 48 human ORs with one or more odorant

molecules have been reported [5–20] and less than ten of these receptors have been reliably associated with olfactory perception of an odorant stimuli [8–10,13,14].

In the quest to develop industrial applications based on the use of human odorant receptors, ChemCom is committed to the systematic identification of ligands for these chemoreceptors. To fulfill this ambitious deorphanization project, considering the huge number of anticipated functional OR genes, it is mandatory to obtain clues about the involvement of the targeted ORs in the olfactory perception. Moreover, the expression of several predicted OR genes has been detected in non-olfactory tissues, suggesting that a subset of predicted OR genes could have functions unrelated to olfaction. Indeed, expression of OR transcripts has been described in various tissues, including testis and spermatozoa [19,21–26], prostate [27–30], enterochromaffin cells [6], pulmonary neuroendocrine cells [31], brain [32–35], tongue [36–38],

Table 1. Patient's data.

Patient	Age (years)	Sex	RIN*	Smoking**	Cause of death	Delay***	Origin
1	46	F	7.7	–	Digestive hemorrhage	25	Africa
2	50	F	6.8	Φ	Cerebral hemorrhage	49	European
3	59	F	7.8	+	Septic shock	27	European
4	61	F	8.1	+	Cerebral hemorrhage	16	European
5	61	F	7.4	Φ	Cerebral hemorrhage	51	European
6	68	F	5.9	–	Septic shock	26	European
7	70	F	7.4	Φ	Cerebral hemorrhage	48	European
8	72	F	8.5	–	Aortic dissection	9	European
9	75	F	6.8	Δ	Cardiogenic shock	11	European
10	75	F	7.5	–	Digestive hemorrhage	23	European
11	79	F	6.3	–	Septic shock	21	European
12	79	F	7.5	–	Cardiac infarction	24	European
13	81	F	6.7	–	Cardiac infarction	23	European
14	39	M	7.2	–	Cardiopulmonary stop on hypoxemia	17	European
15	50	M	6.5	–	Cerebral hemorrhage	24	European
16	52	M	8.4	+	Septic shock	27	European
17	53	M	7.9	+	Cerebral hemorrhage	16	European
18	54	M	6.6	Δ	Cryptococcal meningitis	22	European
19	65	M	7.7	–	Respiratory failure	43	European
20	67	M	6.5	+	Septic shock	24	European
21	69	M	7	+	Respiratory failure	48	European
22	70	M	7.3	–	Digestive hemorrhage	25	European
23	74	M	6.8	Φ	Aortic dissection	46	European
24	74	M	9	+	Septic shock lung	14	European
25	75	M	6.4	–	Septic Shock	19	European
26	79	M	6.8	+	Cardiac infarction	30	European

*RNA integrity number.
**Smoking: –, never; Δ, past; +, current; Φ, unknow.
***Delay in hours between death and sample collection.

erythroid cells [39], placenta [40], breast [41] and kidney [42]. In addition, systematic expression profiling of ORs in non-olfactory tissues using EST data, microarray or deep sequencing analysis [43–45] have shown that a large number of putative human OR genes are expressed in these tissues. The analysis of the entire olfactory subtranscriptome in a variety of different human tissues provides a list of several OR genes that are highly expressed in non-olfactory tissues [44]. At least some of these ORs could play a role in spermatozoa chemotactism [19], in muscle regeneration [46] or in blood pressure regulation [47]. Although, it cannot be excluded that OR may present double olfactory and non-olfactory functions; it remains possible that some members of the reported odorant receptors family could be solely non-olfactory G protein-coupled receptors.

Another issue pertaining to ORs deorphanization results from the significant allelic variation observed for human ORs. A recent data mining of the sequence repository of the 1000 Genomes Project, has estimated that the number of variants per OR locus is on average about ten. However, some variants may be nonfunctional missense haplotypic alleles [48]. Furthermore, as it has been demonstrated in mice that functional olfactory receptors are required for proper targeting of olfactory neuron axons to their corresponding glomeruli in the olfactory bulb [49], one may suppose that alleles of OR genes predominantly expressed in the olfactory epithelium correspond to functional haplotypes.

Taken together, a study of OR gene expression in the whole human olfactory mucosa (WHOM) provides an opportunity to define ORs specifically involved in olfaction, allowing choosing frequently expressed and potentially functional ORs for deorphanization campaigns. OR gene expression in WHOM has been seldomly studied, probably due to the difficult access to human material. Two publications have reported the characterization of the expression of the human OR gene family in 3 individuals only using DNA microarray and only in one individual using deep sequencing [45,50]. Therefore, we designed an innovative approach based on a TaqMan Low Density Array (TLDA) containing probes for 356 predicted OR loci to investigate more thoroughly the OR gene expression profile in human olfactory mucosa. Real-time reverse transcription PCR (qRT-PCR) is frequently used for gene expression quantification, at the transcriptional level, due to its reproducibility and sensitivity. The method has also become the preferred method for validating results obtained by other techniques, such as microarrays or deep sequencing.

Herein we present our data obtained using an innovative high throughput transcriptome profiling approach of human OR genes, in WHOM of much larger set of 26 individuals.

Materials and Methods

Ethics statement

This project was approved by the Erasme Hospital ethics committee (ULB, Brussels, Belgium: P2011/135 and A2013/050).

Patients and tissues specimens

WHOM were collected 27 ± 12 hours post-mortem from 26 individuals (13 women and 13 men; aged from 39 to 81 years, with an average of 67 ± 11 years for women and 63 ± 12 years for men). Most individuals were of European origin. For each patient, the clinical information is summarized in Table 1. Patients with a history of dysosmia or rhinologic diseases, including allergic rhinitis and chronic sinusitis, were excluded. We also investigated the history of smoking, associated with smell's disorder probably related to alterations of the olfactory mucosa [51,52]. Amongst the

26 subjects, 8 were smokers and the smoking status was unknown for 4 of them.

The WHOM was accurately dissected from the septum, the cribriform plate, the middle and the superior turbinates. As the boundaries between the olfactory and the respiratory epithelium are not clearly defined in humans [53], the septal mucosa was dissected up to the lower limit of the middle turbinate. A control tissue was taken from the mucosa of the inferior turbinate. The dissected tissue samples (about 3.5×5 cm from each side of the olfactory cleft mucosa) were collected, frozen in liquid nitrogen and stored immediately at $-80°C$.

Total RNA isolation

Frozen WHOM was crushed in liquid nitrogen. Total RNA was purified and treated with DNase using the RNeasy kit (Qiagen) according to manufacturer instructions. DNase treatment is mandatory as intron spanning primers is not possible due to lack of introns in the OR genes. Quantitative and qualitative assessment of RNA samples (pooled from each side) was performed by NanoDrop spectophotometry (Thermo Scientific) and by microfluidic analysis using a 2100 Bioanalyser (Agilent Technologies). This latter technique produces an electropherogram allowing the evaluation of the integrity of the 18S and 28S ribosomal RNAs (Figure S1A and S1B) and the algorithm assigns a RNA integrity number (RIN) ranging from 1 to 10, where 10 corresponds to ideally intact RNA and 1 to highly degraded RNA (Table 1).

cDNA synthesis

Total RNA (1 µg per sample-loading port of the 48 PCR reaction channels) was used in the reverse transcription (RT) Quantiscript reaction (Qiagen), performed with a combination of oligo-dT primer and random hexamers following the manufacturer's protocol. Each RNA sample was additionally run on one port (feeding 48 PCR assays) of the TaqMan Low Density Array (TLDA) in the absence of reverse transcriptase (RT-) to assess its potential contamination by genomic DNA. The latter, resulted for all samples, in a borderline amplification for a small subset of the large panel of intronless genes tested. On average, 90% of the PCR yielded a quantification cycle (C_q) value labeled as undetermined or above 35 cycles. The remaining 10% gave an average C_q of 34.1 ± 1.1 indicating a potential low residual genomic DNA contamination.

TLDA design and preparation

A customized TLDA was designed in collaboration with Applied Biosystems. The design process for the assays is described in the White Paper TaqMan Gene Expression Assays from Applied Biosystems. The software TaqExpress was used for the design. Whenever possible, the assays were designed to amplify part of the gene coding sequence. The context sequence determines approximate assay position and the assay IDs allows retrieving the details from Applied Biosystems website (Table S1).

The 384 wells of the TLDA contain FAM dye-labeled NFQ probes and primers for an internal control (GAPDH, 4 wells), 10 endogenous control genes exhibiting low differential expression across tissues (MRPL19, CASC3, POLR2A, CDKN1B, TBP, RPL30, PSMC4, YWHAZ, UBC, PPIA), 356 human OR genes, and 6 reference genes specifically expressed in olfactory epithelium (CNGA2, GNAL, ADCY3, RIC8B, RTP1, OBP2A&2B) [50,54]. cDNA (pre-mixed with TaqMan Universal PCR Master Mix) was loaded onto the TLDA and PCR amplifications were performed in a 7900HT Thermocycler (Applied Biosystems). Thermal cycling conditions used were: 2 min at 50°C, 10 min at 94.5°C, followed by 40 cycles at 97°C for 30 sec, and 59.7°C for 1 min.

Real-time PCR with genomic DNA

One TLDA card was run with 150 ng (per port) of a pooled human genomic DNA from Clontech to evaluate the efficacy of the assays.

Real-time PCR with plasmid DNA

One TLDA card was run with 30 pg (per port) of a pool of 30 OR coding plasmids cloned by ChemCom to evaluate the specificity of the assays. The receptors chosen to perform this experiment were spread throughout the different families of OR genes represented by an unrooted tree based on similarity of amino acid properties. One pg of each plasmid represents about 3000 molecules of specific plasmid per PCR.

TLDA analysis and Statistical analysis

The real-time PCR focuses on the exponential phase, where amplification doubles target templates, following the exponential amplification (2^n where n is the number of cycles). The real-time PCR instrument calculates a C_q value representing the PCR cycle at which the reaction reaches a fluorescent intensity threshold above background. For C_q calculation, the threshold was manually set at $\Delta Rn = 0.1$ for all samples and all targets (threshold set within the 2^n exponential amplification phase). The results were analyzed using the Sequence Detection Systems (SDS) version 2.4 and Qbase$^+$ software packages [55]. Determination of the optimal number of reference genes for the normalization of qPCR data was performed using the geNorm algorithm [56].

Association analyses were performed with R 2.14.1 [R Development Core Team (2008). R: A language and environment for statistical computing. R Foundation for Statistical Computing, Vienna, Austria. ISBN 3-900051-07-0, URL http://www.R-project.org/.]. Significance Analysis of Microarray (SAM) [http://www.pnas.org.gate1.inist.fr/content/98/9/5116.full] was performed using the samr package v2.0 Genes with q-value below 0.05 were considered significant. For each variable (i.e. age, sex and smoking status) SAM was performed to find the receptors presenting expressions individually associated with each variable. To assess whether the expressions of all receptors were globally associated with one of these variables, the sum of the square of scores of association (Pearson correlation coefficients for age, t-scores for the two other variables) of all the receptors expressions was compared to a null-distribution of the sum of the scores of associations obtained after 10.000 permutations of the patient labels. Heatmap visualization was obtained with the heatmap.2 function within the gplots v2.10.1 package [gplots: Various R programming tools for plotting data (2011), Gregory R. Warnes, URL http://CRAN.R-project.org/package = gplots].

Results

A 384-customized TLDA was designed to investigate the gene expression of a large array of OR genes from 26 WHOM samples. All experiments were performed according to the MIQE (minimum information for publication of quantitative real-time PCR experiments) guidelines [57].

Analysis of RNA purity and integrity

The 260/280 and 260/230 OD ratios were measured for all RNA samples to assess respectively the purity of RNA with respect to protein contamination and residual organic solvent. All samples used showed a 260/280 and 260/230 OD ratios between 1.8 and

2.0, indicative of good quality RNA with minimal contaminations. RNA integrity was also assessed, and samples characterized by RIN (RNA integrity number) ranging from 5.9 to 9.0 were used, the average $\pm SD$ being 7.3 ± 0.8 (Figure S1A, S1B, Table 1). These RIN values are usually considered acceptable for qRT-PCR experiments [58]. No correlation between the RIN and the delay between death and sample collection was observed (Statistical analysis reveal a p value = 0.36 for a Pearson's correlation with a $R^2 = 0.035$).

Selection of candidate reference genes

Classical endogenous control genes, exhibiting minimal differential expression across different tissues (MRPL19, CASC3, POLR2A, CDKN1B, TBP, RPL30, PSMC4, YWHAZ, UBC, PPIA) were added to the TLDA, in order to perform a first technical normalization and to compare expression of genes from different tissues as for example WHOM and inferior turbinate. The structure of WHOM is often patchy and contains a significant proportion of respiratory epithelium [54,59]. Therefore, to compare expression of genes from different WHOM, assays for tissue specific olfactory epithelium reference genes were also added to the TLDA for a second biological normalization purpose. The six selected genes were CNGA2, GNAL, ADCY3, RIC8B, RTP1 and OBP2A&2B.

Expression profiling and stability analysis of candidate reference genes

Average C_q values of classical endogenous reference genes was 17.1 ± 0.7 for UBC (mean $\pm$ SD), 17.6 ± 0.5 for GAPDH, 18.0 ± 0.6 for PPIA, 20.0 ± 0.6 for CDKN1B, 20.4 ± 0.6 for CASC3, 21.5 ± 0.7 for PSMC4, 21.7 ± 0.6 for POLR2A, 22.1 ± 0.8 for YWHAZ, 22.6 ± 0.9 for RPL30, 22.8 ± 3.6 for MRPL19 and 24 ± 0.6 for TBP (Table S1). Their stabilities were evaluated by the geNorm algorithm [56] and the geometric mean of 3 stably expressed classical endogenous reference genes (CASC3, PSMC4, CDKN1B) were selected to technically normalize the results. C_q of the olfactory epithelium-specific reference genes (Ric8B, GNAL, RTP1, CNGA2, ADCY3, OBP) are shown in the box plots of Figure 1. The distribution of the olfactory epithelium-specific reference genes C_q provides a global representation of the variation of reference gene expression as well as information on their relative abundance. More highly expressed genes are associated with lower C_q. Average C_q values ranged from 21.3 ± 0.8 (ADCY3) to 30.6 ± 1.1 (OBP, means $\pm SD$, n = 26) (Table S1). As suggested in Khan et al. [54], in view of obtaining a biologically relevant normalization, the geometric mean of the six specific reference genes provided a normalizing factor, rather than a factor from a single reference gene.

Real-time PCR with genomic DNA

One TLDA card was run with a pool of human genomic DNA to evaluate the assays (individual gene efficiency amplification). The C_q average value of 351 detected OR genes is 24.5 ± 0.8 (Table S2) suggesting similar amplification rates for all the genes, as expected as the number of targets is identical for all genes in a human genome. Furthermore 21 assays gave an expected undetermined C_q values because they correspond either to reference genes or to 4 OR genes for which the assays are designed with intron spanning primers. These assays are identified by a suffix '_m' in the assayID. One out of 4 GAPDH assays gave a non-expected value of 36 and PPIA gave a non-expected value of 27 whereas these assays are designed with intron spanning primers. This reflects a slight non-significant amplification

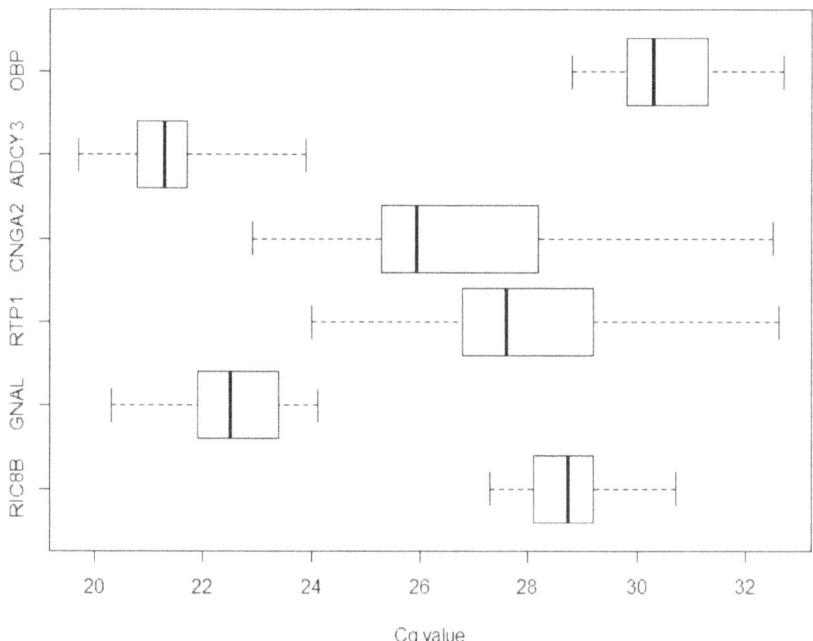

Figure 1. Expression profiling of candidate reference genes in whole human olfactory mucosa. Box plot graph on C_q obtained for the reference genes specific for the olfactory epithelium across the 26 individuals samples. Left and right box limits are first and third quartiles. The inner line conventionally marks the median. Whiskers show the extreme of the series.

compared with results obtained on RNA (delta-C_q of 18 for GAPDH and 9 for PPIA). Finally, one target (OR2A14) showed an abnormal C_q value of 12.4, therefore this assays has not been taken into account for the qRT-PCR analysis.

Real-time PCR with plasmid DNA

One TLDA card was run with a pool of 30 OR coding plasmids to evaluate the specificity of the assays. The expected specific PCR amplification of the 30 targets gives an average C_q value of 25.4 ± 1.1, (mean $\pm$SD, n = 30). However, we observe a non-specific amplification for 15 additional receptors. For 11 of them, the average C_q value is 32.5 ± 1.8, (mean $\pm$SD, n = 11) which reflects a delta-C_q value of 7 as compared to specific amplification. In this case, the non-specific amplification is then negligible. For 4 of them, the average C_q value is 25.0 ± 0.8, (mean $\pm$SD, n = 4) which reflects no difference as compared to specific amplification. These 4 couples of genes OR2L3 and OR2L8 (97% identity on the entire DNA sequence), OR52E6 and OR52E8 (90% identity), OR52I1 and OR52I2 (97% identity) and OR5D16 and OR5D18 (83% identity) cannot be discriminated by this TLDA card. Then for the 30 OR coding plasmids, 88% of the PCR amplifications are specific for the target.

Inter-run calibration

Three different experiments were conducted to run the 26 samples, to correct for possible run-to-run variation whenever all samples are not analyzed in the same run, identical sample have been tested in all runs. Figure 2 shows the correlation between C_q values for all detected targets (ie. C_q average <35) from the same ARN sample in two different runs. C_q values above 35 are not reliable because duplicates are not reproducible. The correlation coefficient reaches 0.83 and the intercept is 0.91 for the detected OR genes. The correlation coefficient reaches 0.99 for the

olfactory epithelium-specific reference genes and for classical endogenous control genes.

Expression profiling of olfactory receptors genes in WHOM

On average, for the 26 samples, C_q values computed from amplification plots of 355 OR genes range between 25.8 and 39.8 (Table S1). These results reflect a low expression of the OR genes compared to other genes involved in the olfactory cascade. One target (OR2A14) shows an abnormal amplification plot with a C_q value of 16.6 ± 7.3 (mean $\pm$SD); this assays will not be taken into account for the analysis. On average, 62 ± 29 (mean $\pm$SD) OR genes per sample gave an undetermined C_q value which was arbitrarily assigned to 40 cycles to allow the calculation of an average C_q values. 74 ± 34 (mean $\pm$SD) OR genes per sample gave a C_q value above 35 were considered as expressed at very low level or not expressed at all.

To make a more quantitative analysis, the C_q values of each OR were converted into normalized relative quantities (NRQ) following the method previously described [55]. Briefly, we apply the delta-C_q quantification model using the average C_q obtained for all ORs in the 26 individuals as calibrator (here 32.7) which is transformed into relative quantities using the exponential function, so results are fully equivalent and thus only rescaled. Then the normalization of relative quantities was performed with the geometric mean of the multiple stably expressed classical endogenous reference genes (CASC3, PSMC4, CDKN1B) defined by the geNorm algorithm [56] and followed by the normalization with the geometric mean of the six reference genes specific for olfactory epithelium. Results obtained on genomic DNA and on specific OR coding plasmids allowed to calculate the approximate number of copies of target for 20 ng RNA engaged in the RT-PCR reaction (Table S3).

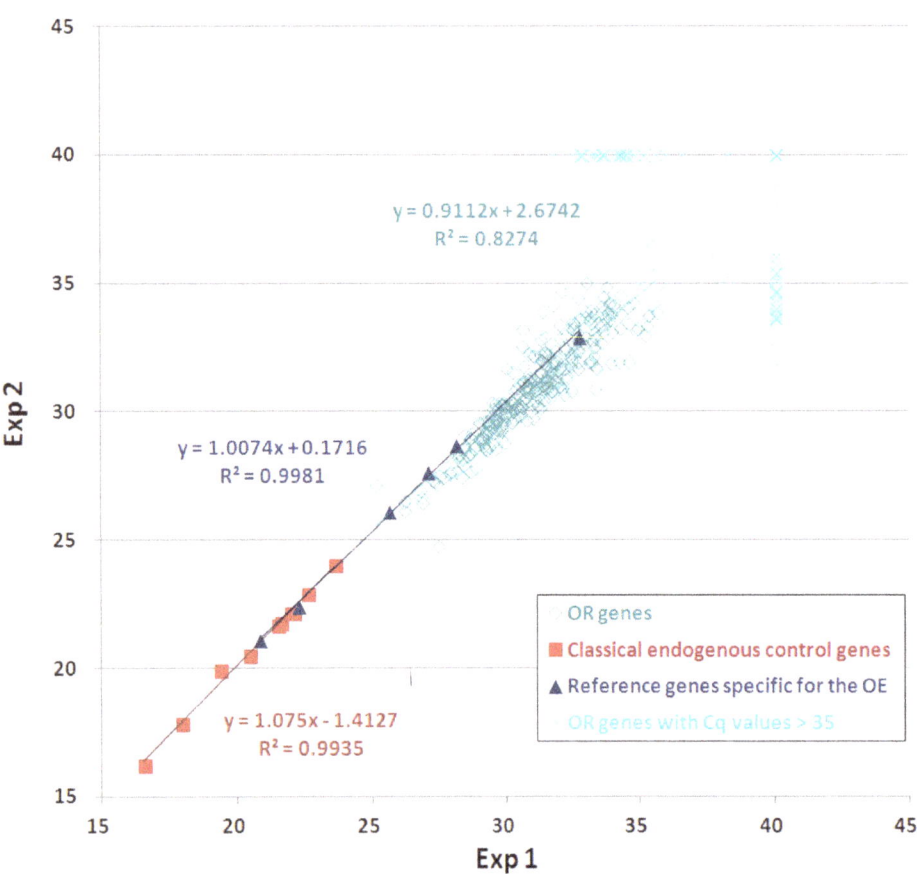

Figure 2. Inter-run calibration analysis. Plotted expression pattern correlation for all detected targets (C_q average below 35) from the same RNA WHOM sample in two different runs. OR genes (green ◇), reference genes specific for the olfactory epithelium (blue Δ) and classical endogenous control genes (red ■). C_q values above 35 are not reproducible (turquoise x). R^2 is the coefficient of correlation.

For each individual, the detected OR gene number (>5 copies/20 ng RNA) was counted (Figure 3). This cut-off value corresponds to a C_q value ≥35.3. On average, on the 355 OR gene targets, the detected OR gene number is 232 ± 28 for women and 238 ± 28 for men.

There is a substantial difference in the expressed OR gene repertoire of each of the samples. A set of 90 human OR genes were detected (>5 copies/20 ng RNA) in all tested individuals. Another set of 140 human OR genes were detected not in all tested individuals but in more than half of the population (in 13 individuals and more on 26) and a third set composed of 125 human OR genes were more rarely detected (in less than 13 individuals on 26) (Figure 4).

Globally, the OR gene expression was not associated with age (p value = 0.19), sex (p value = 0.23) or smoking (p value = 0.66, Pearson's correlation). Individually, 22 OR genes showed a decreased profile and 7 OR genes showed an increased profile related with age (Figure 5). There is no significant association between individual OR gene expression and sex or smoking.

Figure 6 shows OR genes ranked in function of their expression level, from the highest to lowest. It shows 273 (77%) human OR genes above the cut-off value of 5 copies/20 ng RNA. No significant enrichment in class I or class II ORs is observed in the expressed set. Indeed, 17.6% of OR genes expressed belong to class I while 15.2% of the OR genes tested belong to this class.

Interestingly, most of the published deorphanized olfactory receptors [5–11,13–20] are found into the set of expressed OR genes (Figure 6). Indeed, we count 43 expressed OR genes among the 47 deorphanized receptors described in the literature which are tested in this study. In other words, 16% of expressed OR genes are deorphanized while this percentage drops down to 4.8% for non-expressed OR genes (p value = 0.009, Fisher's exact test).

An inverse distribution is observed with potentially non-functional OR genes. Expression levels of 52 OR genes with mutations affecting positions in the consensus amino acid motifs specific for OR genes [60] were analyzed. These receptors, although regarded as intact OR genes, harbor a mutation affecting P or Y in the LHT**PMY** motif, or affecting M, R or the second A in the **MAYDRYVA**IC motif, or Y in the S**Y** motif or finally, on H in the FSTCSS**H** motif. All known variants of these OR genes, correspond to a mutated haplotype of these highly conserved positions [48]. Presumably, these receptors are no longer functional (highlighted in blue in the Figure 6). We observed that 25% of non-expressed OR genes are potentially non-functional whereas 11.3% from the expressed set are potentially non-functional (p value = 0.002, Fisher's exact test). In addition, the average RNA copy numbers of the 47 deorphanized receptors (174 ± 247) and of the 52 potentially non-functional ORs (48 ± 115) are significantly different (p value = 0.002, Student's t-Test, two-tailed distribution, two-samples with unequal variance).

As shown in Table 2, the average RNA copy number varies drastically among the 47 reported deorphanized receptors. The most expressed OR gene corresponds to OR7C1 with an estimate average copy number of about 1108/20 ng RNA. A huge

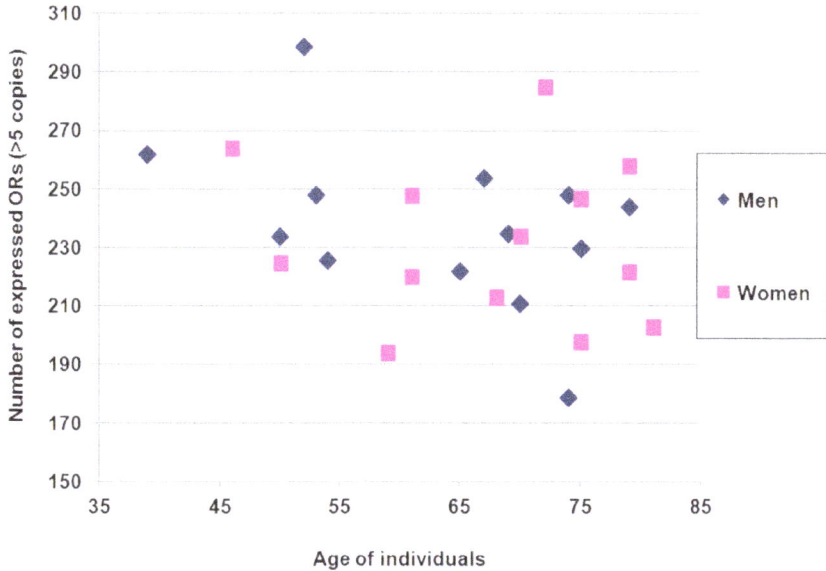

Figure 3. Number of OR genes expressed for each individual in function of age. Scatterplot of the number of OR genes expressed at a level above 5 copies/20 ng RNA for each individuals. Women are colored in pink and men in blue.

difference has also been noted in the expression of the four closely related paralogs, OR10G3, OR10G4, OR10G7 and OR10G9 that respond to ethyl vanillin and eugenol [5]. Indeed OR10G3 is well expressed in 25/26 WHOM samples with an average of 487 copies/20 ng RNA. OR10G4 and OR10G7 are moderately expressed (with an average of 29 and 13 copies/20 ng RNA respectively) and OR10G9 is detected only in 3 WHOM samples above the cut-off of 5 copies (with an average of 2 copies/20 ng RNA). We can count 19 ORs expressed by the 26 individuals among the 47 genes. In others words, 21% of the group of 90

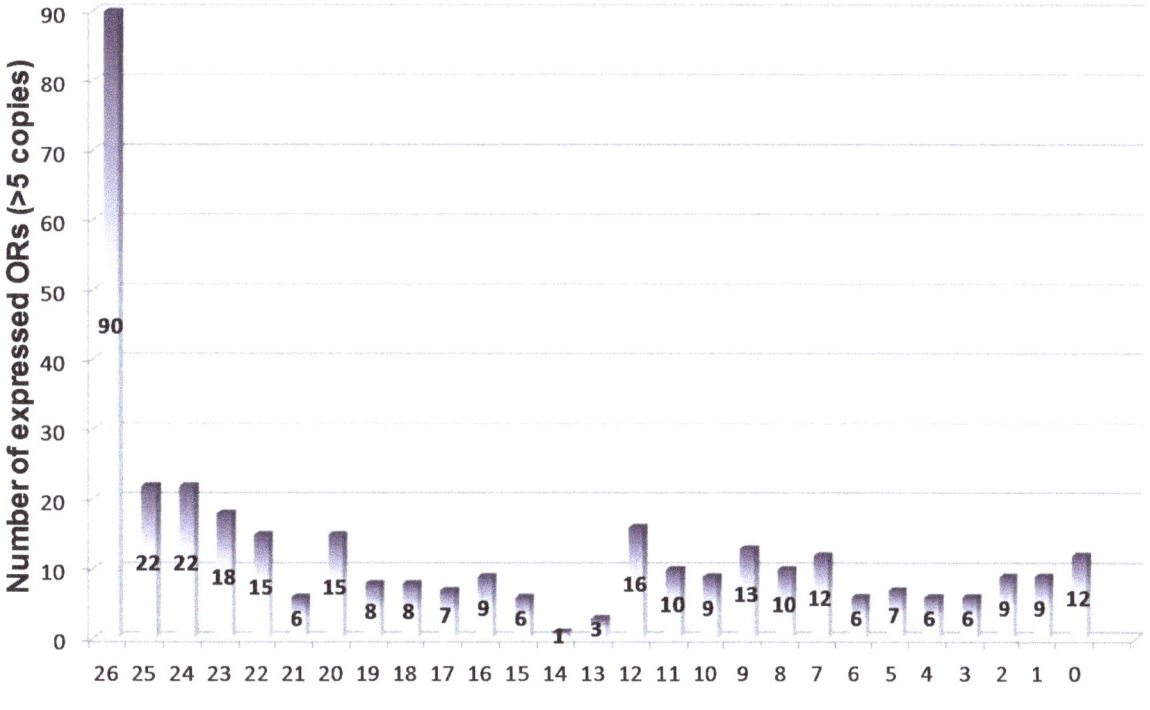

Figure 4. Expression frequency of OR genes in the population of 26 individuals. The bar chart represents the number of expressed OR genes (>5 copies/20 ng RNA) as a function of the number of expressing individuals, e.g. the number of expressed OR genes in all tested individuals (26) corresponds to 90.

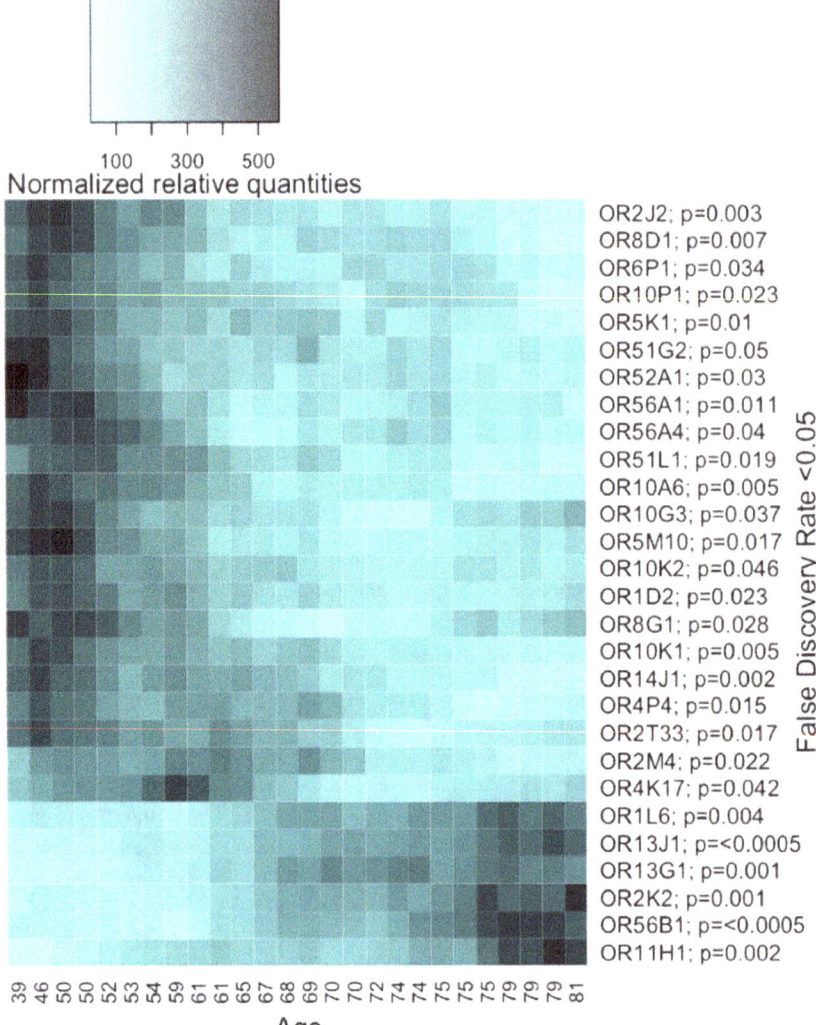

Normalized relative quantities

Figure 5. The expression of several OR genes is statistically related with age. 22 OR genes showed a decreased expression profile and 7 OR genes showed an increased expression profile. The False discovery rate (FDR) calculated by SAM is <0.05 for all represented genes and the p value of the Spearman's correlation is <0.05 are indicated next to the name of the ORs in the heatmap.

receptors expressed in every individual were deorphanized. The 4 non-expressed ORs (OR1A2, OR2M7, OR5D18, OR10G9) are expressed only in 1, 4, 0 and 3 individuals respectively.

Expression profiling of olfactory receptor genes in inferior turbinate sample

One sample of inferior turbinate (IT) has been tested in a TLDA and gives 210 undetermined C_q values compared to 62 on average in WHOM tissues. This observation reflects the non-detection of OR gene expression, expected for a non-olfactory tissue. To make a more quantitative analysis, the C_q values were converted into normalized relative quantities with classical endogenous stably expressed reference genes (CASC3, PSMC4, CDKN1B) defined by the geNorm algorithm. Figure 7 shows results obtained for OR gene expression ratio between WHOM and inferior turbinate. We observe 250 OR genes (70%) more expressed in WHOM than in inferior turbinate (ratio WHOM/IT≥2) (Table S4). Some reference genes specific for olfactory sensory neurons (CNGA2 and Ric8B) are significantly expressed more in WHOM than in

IT; RTP1 and ADCY3 are expressed a little more in WHOM than in IT while OBP and GNAL are detected in equivalent amount in both tissues.

Discussion

Although the OR gene family was discovered over 20 years ago by Buck and Axel, few data are available on their expression in human olfactory mucosa, contrasting with the recent significant increase of results on the genetic polymorphism of OR genes [48]. This probably reflects the difficulty to acquire human tissue and obtain good quality RNA from WHOM.

We report here the first extensive high throughput transcriptome profiling of OR gene expression directly by real-time reverse transcription PCR performed furthermore on WHOM from a relatively large population of 26 patients. Indeed, our study was focused on the expression of 356 predicted functional OR genes among the 851 OR loci scattered throughout the human genome.

Only a small percentage of the olfactory mucosa consists of olfactory sensory neurons. Moreover, the boundaries of the

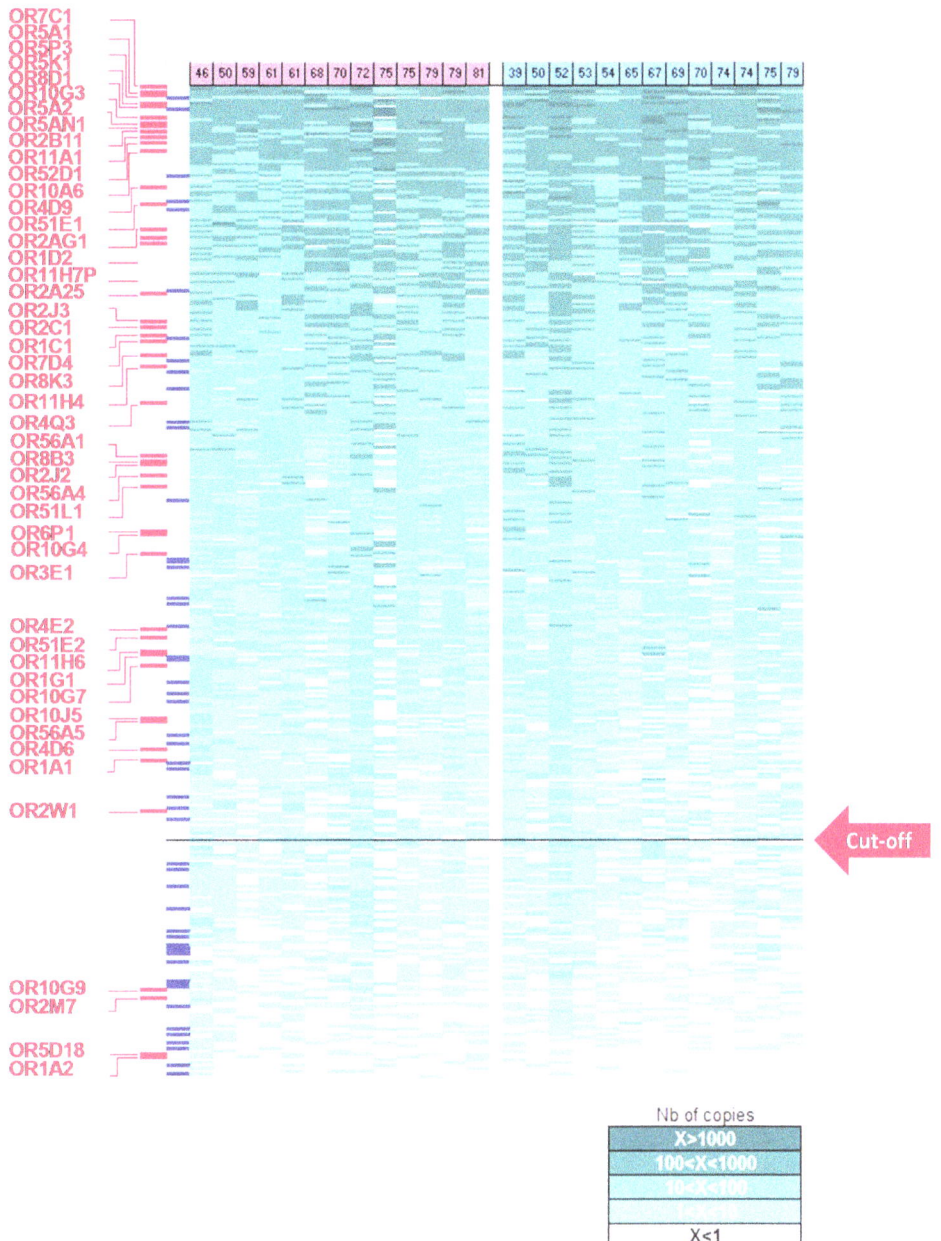

Figure 6. Expression profiling of OR genes in whole human olfactory mucosa. For each of the 355 OR genes (rows), the RNA copies number were estimated from normalized relative quantities obtained for each of the 26 individuals (columns). OR genes have been ranked according to their expression, from higher to lower. RNA copies number obtained for each individual are also indicated according to the green color code to show the good consistency of the inter-individual expression. OR genes with an average copies number below a cut-off of 5 copies/20 ng RNA (red arrow; right) are considered as to low or non-expressed. Age of the individuals are shown above the figure, women are colored in pink and men in blue. Published deorphanized receptors are highlighted in red on the left. Potentially non-functional OR genes are highlighted in blue on the left as well.

WHOM are unclear and the tissue is sometimes replaced by respiratory epithelium. Furthermore, the fraction of olfactory sensory neurons can vary significantly from one sample to another. Therefore, to allow OR gene expression comparisons between individual WHOM samples, normalization is mandatory. Consequently, we therefore normalized our expression data from the individual WHOM, with 6 so called tissue specific reference genes, that are expressed specifically by olfactory sensory neurons as previously described by Khan et al. [54].

Our results show that 77% of human intact OR gene repertoire are expressed with an average level above 5 copies/20 ng RNA. A set of 90 human OR messengers were detected in all tested individuals. In addition, 70% of the human OR gene repertoire were found more expressed in WHOM than in the inferior turbinate (ratio WHOM/IT ≥2). Along with the widespread genetic variation reported for human OR protein coding regions, that correlates to individual differences in odorous perception [8,9,13,14] and with genetic variations in auxiliary olfactory genes [50], this differential expression of ORs in the WHOM could

Table 2. RNA copies number average and number of individuals that express reported deorphanized olfactory receptors.

OR name	Agonist	Reference	Copies number average (per 20 ng RNA)	Number of individuals that express
OR1A1	Dihydrojasmone	[15]	7±8	12
OR1A2	Citronellal	[17]	0.6±1	1
OR1C1	Linalool	[5]	77±64	26
OR1D2	Bourgeonal	[19]	146±117	26
OR1G1	1-nonanol	[16]	14±48	11
OR2A25	Geranyl acetate	[5]	103±95	26
OR2AG1	Amyl butyrate	[11]	157±94	26
OR2B11	Coumarin	[5]	365±615	26
OR2C1	Octanethiol	[15]	78±71	26
OR2J2	1-octanol	[15]	37±46	23
OR2J3	Cis-3-hexen-1-ol	[13]	82±118	25
OR2M7	Citronellol	[15]	2±4	4
OR2W1	1-octanol	[15]	6±13	7
OR3A1	Helional	[20]	26±33	23
OR4D6	β-ionone	[8]	8±13	11
OR4D9	β-ionone	[8]	190±216	26
OR4E2	Amyl acetate	[10]	17±15	20
OR4Q3	Eugenol	[10]	50±50	22
OR5A1	β-ionone	[8]	737±1018	25
OR5A2	β-ionone	[8]	435±386	26
OR5AN1	Muscone	[18]	377±290	26
OR5D18	Eugenol	[6]	0.7±1	0
OR5K1	Eugenol methyl ether	[5]	598±412	26
OR5P3	(+)-carvone	[15]	752±642	26
OR6P1	Anisaldehyde	[10]	29±59	16
OR7C1	Androstadienone	[10]	1108±849	26
OR7D4	Androstenone	[9]	74±98	24
OR8B3	(+)-carvone	[10]	37±50	22
OR8D1	4,5-dimethyl-3-hydroxy-2,5-dihydrofuran-2-one	[5]	558±572	25
OR8K3	(+)-menthol	[5]	66±80	21
OR10A6	3-phenyl propyl propionate	[10]	286±273	26
OR10G3	Ethyl vanillin	[5]	487±443	25
OR10G4	Ethyl vanillin	[5]	29±35	22
OR10G7	Eugenol	[5]	13±12	19
OR10G9	Ethyl vanillin	[5]	2±3	3
OR10J5	Lyral	[15]	10±9	19
OR11A1	2-ethyl fenchol	[5]	324±266	26
OR11H4	Isovaleric acid	[14]	64±51	26
OR11H6	Isovaleric acid	[14]	15±12	22
OR11H7P	Isovaleric acid	[14]	142±100	26
OR51E1	Nonanoic acid	[7]	170±266	26
OR51E2	Propionic acid	[15]	16±20	20
OR51L1	4-allylphenylacetate	[15]	35±33	22
OR52D1	Ethyl isobutyrate	[16]	317±272	26
OR56A1	Decyl aldehyde	[5]	39±41	26
OR56A4	Decyl aldehyde	[5]	37±39	25
OR56A5	Decyl aldehyde	[5]	9±11	15

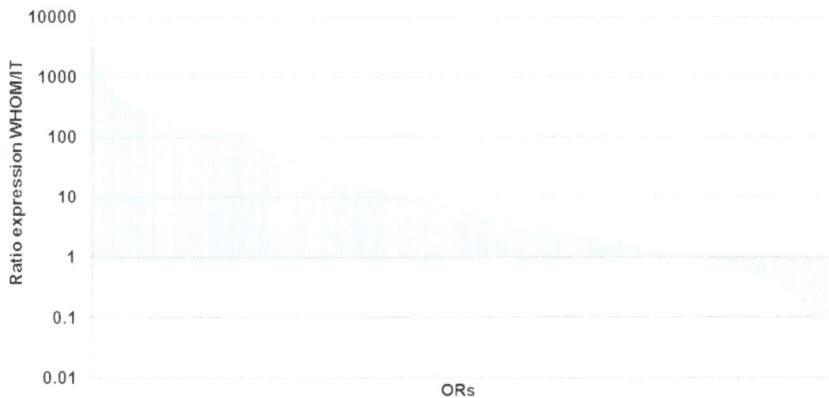

Figure 7. Ratio between normalized relative quantities of RNA obtained for olfactory mucosa and for inferior turbinate. For each of the 355 OR genes, the ratio is calculated from the mean of normalized relative quantities obtained for the 26 individuals for whole human olfactory mucosa (WHOM) and from the normalized relative quantities obtained for inferior turbinate (IT; n = 1).

establish the basis to a well-documented inter-individual variation in olfactory sensitivity.

Statistical analysis was performed for each clinical variable (i.e. age, sex and smoking status) to assess if receptor expressions were globally or individually associated. Our results indicate that OR gene expression is globally not associated with age, sex or smoking status. However, we are not able to detect if these clinical factors may reduce the absolute amount of olfactory sensory neurons. As we have normalized our data with specific references genes of olfactory sensory neurons, we have not taken into account the absolute amount of olfactory tissue. Therefore, the OR gene expression is described relatively to olfactory sensory neurons. Furthermore, at this point, our results cannot explain the decrease of olfactory performance related to age or smoking [51,52,61].

Nevertheless, our results show that individually, the expression of 22 OR genes seems to decrease significantly with age, and the expression of 7 OR genes seems to increase significantly. These results can be compared to those of Khan et al. [54] where the majority of OR gene expression (58.4%) in mice remained stable during aging, while 32.8% presented downward profiles, 7.2% upward profiles and 1.7% of convex or concave profiles. We found no correlation between individual OR gene expression and sex or smoking, although some clinical observations show that these two conditions may influence smell abilities. These differences in smell abilities may occur at another level of the olfactory system than the OR gene expression.

Prior to the present work, only two studies had focused on the expression of the human OR gene family. In these studies, DNA microarray [45] or deep sequencing [50] were used as experimental approaches. Latter report focused on accessory proteins and presented results in the supplementary data only for one human olfactory epithelium biopsy. Over the 261 intact ORs overlapping with our study, 174 OR genes were found to be expressed in this olfactory epithelium biopsy (threshold set at a FPKM≥0.1) whereas we found 202 OR genes expressed in the WHOM (threshold set at a number of copies ≥5). 145 genes (i.e.72% of our expressed OR gene set) turned out to be common to both studies and 30 ORs are not expressed in WHOM according to both studies (Table S5). Therefore, our approach is in agreement with the previous study by Keydar et al. concerning expressed OR genes (p value = 0.0012, hypergeometric test) and for the expression levels of each OR genes (p value<0.001, Spearman's correlation) [50]. With respect to the Zhang et al.

study [45], there are 319 intact ORs overlapping with our study. 202 OR genes were found to be more expressed in human olfactory epithelium than in other tissues by Zhang et al. (threshold set at a p value<0.01) whereas we found 225 OR genes preferentially expressed in WHOM (threshold set at a ratio WHOM/IT≥2). Furthermore, 142 genes (i.e. 63% of our expressed OR gene set) turned out to be common to both studies and 34 ORs are not expressed in WHOM according to both studies (Table S6). The overlap of 63% is exactly what would be expected if the two datasets are completely random with respect to each other (p value = 0.3, hypergeometric test). Moreover, no correlation between the expression levels of each OR genes could be observed between the two studies (p value = 0.96, Spearman's correlation). It is noteworthy that the non-olfactory tissues used to estimate a difference in gene expression are not the same. We compared the expression in the WHOM to the one of inferior turbinate, while human liver, lung, kidney, heart and testis were used in the study by Zhang et al. The latter therefore must be taken with caution, as non-olfactory expression of ORs have been reported for these different tissues, and could therefore bias the comparative results they obtained. We tried to exclude the set of non-olfactory expressed ORs from our comparison. Even excluding these receptors, we could not reveal a correlation between the current data and the Zhang et al. data. Another source of discrepancy between both studies relies on the proportion of the olfactory epithelium used to determine OR expression. In the publication by Zhang et al., it is not clear whether the analyzed tissues cover the entire olfactory mucosa or only a determined anatomical section. This difference might be highly significant as the distribution of olfactory receptors is not homogeneous in the olfactory epithelium of rodents [62,63]. Importantly, though no data currently exist in humans. Another difficulty relies on the fact that the boundaries between the olfactory and the respiratory epithelium are not clearly defined in humans. Different publications report that the human olfactory epithelium is located on the nasal septum, the cribriform plate, the superior and the middle turbinate [53,64,65]. Consequently, this motivated us collecting all these anatomical regions. The mucosa of the nasal septum was dissected along the projection of the middle turbinate. Thus, our samples represent practically the totality of the olfactory mucosa. Finally, our results are in good agreement with those of Keydar et al., but not with Zhang et al. Moreover, we found no correlation between Keydar et al. and Zhang et al. (p value = 0.341, hypergeometric test and p value = 0.13, Spearman's correlation).

Interestingly, we observe an enrichment of functional deorphanized receptors in the set of expressed OR genes and an enrichment of potentially non-functional receptors into the set of non-expressed OR genes. This corroborates the observations of Zhang et al. The latter reports that 80% of intact OR genes and 67% of OR pseudogenes were found to be expressed in WHOM and moreover intact ORs appear to be expressed at a higher level on average than OR pseudogenes.

Taken together, these observations support the hypothesis predicting that if a gene is expressed, it is more likely to be functional. Indeed, a non-functional OR can lead to a defective targeting of olfactory sensory neurons in the olfactory bulb and therefore reduces the survival of these neurons [66]. More precisely, the proper targeting seems more related to OR-derived cAMP signals rather than the OR ability to bind an odorant [67]. The relation between OR genes functionality and expression could be further explored by studying the variants of expressed OR genes. Indeed, for known ORs already deorphanized, both functional and non-functional haplotypes have been described; consequently, it would be worth determining whether expressed allelic variants correspond preferentially to functional haplotypes.

A systematic study of OR gene expression profiles, expressed in non-olfactory tissues using deep sequencing analysis has been recently reported and provides a list of highly expressed OR genes [44]. From this list, 32 intact OR genes are common with our study. We confirmed that the majority of the non-olfactory tissues expressed OR genes (28/32; 87%) are also expressed in WHOM (more than 5 copies/20 ng RNA). 24 out of 32 (75%) non-olfactory tissues expressed OR genes are expressed more in WHOM than in inferior turbinate (ratio WHOM/IT ≥2). The remaining 8 OR genes are detected similarly in both tissues (0.96< ratio WHOM/IT1 <2). From this comparison, it does not seem that non-olfactory tissues expressed OR genes make up a separate group, with a putative non-olfactory function, that would clearly segregate from ORs expressed in the WHOM. However, for one particular receptor, OR2W1, we observed a very low expression level in the WHOM whereas it is well detected in pulmonary neuroendocrine cells [31]. Upon its deorphanization, this receptor was found have a broad spectrum of stimuli [15]. This OR is activated by more than 200 molecules and interacts with a large variety of chemical structures eliciting very different odors [Veithen et al., unpublished data]. Together, our results and those of Gu et al. [31] suggest a role for OR2W1 in the detection of volatile irritants in the human airways. Therefore, this receptor may offer an example of an OR family member that would actually not be only an olfactory mucosa odorant receptor.

Although our study represents the most extensive analysis of human OR expression in the olfactory mucosa, it does present some limitations. The considered population is relatively old. Indeed, because of the difficulty to obtain human material, samples from patients presenting characteristics that may affect the olfactory mucosa such as age and smoking, have not been discarded. However, these conditions do not appear to change the OR genes expression. On another hand, our study includes almost exclusively subjects of European origin and therefore does not explore the possible ethnic related variations of OR gene expression. Logically, therefore, we acknowledge that it would be worth extending the analysis to samples from other origins, if available.

Notwithstanding these points and since the majority of human olfactory receptors are not deorphanized, the information on the expression of OR genes in WHOM collected in this study offers an essential preliminary and lacking understanding that will allow focusing future research on frequently expressed and potentially functional olfactory receptors identified.

Supporting Information

Figure S1   Analysis of RNA integrity. A. Electropherogram showing the integrity of 9 human olfactory epithelium RNA samples (lanes 1 to 9). **B.** Example of profile showing a RIN of 8.5 (sample 8, RNA from a woman of 72 years old) and the integrity of the 18S and 28S ribosomal RNA.

Table S1 C_q values obtained for all OR genes and reference genes in all WHOM samples including assays ID from Applied Biosystems.

Table S2 C_q values obtained for all OR genes and reference genes on genomic DNA and average C_q values obtained on WHOM RNA.

Table S3   Copies number estimated for all OR genes and reference genes in all WHOM samples.

Table S4   Ratio between normalized relative quantities obtained for WHOM and for inferior turbinate and the average copies number for all OR genes and reference genes.

Table S5   Comparison between the 2 studies by Verbeurgt et al. and by Keydar et al.: listing of 145 ORs that are expressed in the WHOM according to both studies and listing of 30 ORs that are not expressed in the WHOM according to both studies.

Table S6   Comparison between the 2 studies by Verbeurgt et al. and by Zhang et al.: listing of 142 ORs that are expressed in the WHOM according to both studies and listing of 34 ORs that are not expressed in the WHOM according to both studies.

Acknowledgments

The authors thank J. Perret from the Laboratory of Pathophysiological and Nutritional Biochemistry, Department of Biochemistry, Free University of Brussels, Brussels, Belgium, for discussions on q-RTPCR and experimental design, use of the Qbase+ Software, help in analyzing data and manuscript critical reviewing as well as for spelling and grammar. We thank C. Degraef and F. Libert from the Institute of Interdisciplinary Research in human and molecular Biology, Free University of Brussels, Brussels, Belgium, for technical assistance and fruitful discussions respectively. We also thank A. Veithen and S. Patiny from ChemCom S.A., Brussels, Belgium for valuable comments on the manuscript.

Author Contributions

Conceived and designed the experiments: CV FW FG JED PC. Performed the experiments: CV FW. Analyzed the data: CV FW MT FG. Contributed reagents/materials/analysis tools: CV FW MT FG. Wrote the paper: CV FW MT FG JED PC.

References

1. Buck L, Axel R (1991) A novel multigene family may encode odorant receptors: a molecular basis for odor recognition. Cell 65: 175–187.
2. Ben-Arie N, Lancet D, Taylor C, Khen M, Walker N, et al. (1994) Olfactory receptor gene cluster on human chromosome 17: possible duplication of an ancestral receptor repertoire. Hum Mol Genet 3: 229–235.
3. Glusman G, Bahar A, Sharon D, Pilpel Y, White J, et al. (2000) The olfactory receptor gene superfamily: data mining, classification, and nomenclature. Mamm Genome 11: 1016–1023.
4. Rouquier S, Taviaux S, Trask BJ, Brand-Arpon V, van den Engh G, et al. (1998) Distribution of olfactory receptor genes in the human genome. Nat Genet 18: 243–250.
5. Adipietro KA, Mainland JD, Matsunami H (2012) Functional evolution of Mammalian odorant receptors. PLoS Genet 8: e1002821.
6. Braun T, Voland P, Kunz L, Prinz C, Gratzl M (2007) Enterochromaffin cells of the human gut: sensors for spices and odorants. Gastroenterology 132: 1890–1901.
7. Fujita Y, Takahashi T, Suzuki A, Kawashima K, Nara F, et al. (2007) Deorphanization of Dresden G protein-coupled receptor for an odorant receptor. J Recept Signal Transduct Res 27: 323–334.
8. Jaeger SR, McRae JF, Bava CM, Beresford MK, Hunter D, et al. (2013) A mendelian trait for olfactory sensitivity affects odor experience and food selection. Curr Biol 23: 1601–1605.
9. Keller A, Zhuang H, Chi Q, Vosshall LB, Matsunami H (2007) Genetic variation in a human odorant receptor alters odour perception. Nature 449: 468–472.
10. Mainland JD, Keller A, Li YR, Zhou T, Trimmer C, et al. (2014) The missense of smell: functional variability in the human odorant receptor repertoire. Nat Neurosci 17: 114–120.
11. Mashukova A, Spehr M, Hatt H, Neuhaus EM (2006) Beta-arrestin2-mediated internalization of mammalian odorant receptors. J Neurosci 26: 9902–9912.
12. Matarazzo V, Clot-Faybesse O, Marcet B, Guiraudie-Capraz G, Atanasova B, et al. (2005) Functional characterization of two human olfactory receptors expressed in the baculovirus Sf9 insect cell system. Chem Senses 30: 195–207.
13. McRae JF, Mainland JD, Jaeger SR, Adipietro KA, Matsunami H, et al. (2012) Genetic Variation in the Odorant Receptor OR2J3 Is Associated with the Ability to Detect the "Grassy" Smelling Odor, cis-3-hexen-1-ol. Chem Senses 37: 585–593.
14. Menashe I, Abaffy T, Hasin Y, Goshen S, Yahalom V, et al. (2007) Genetic elucidation of human hyperosmia to isovaleric Acid. PLoS Biol 5: 2462–2468.
15. Saito H, Chi Q, Zhuang H, Matsunami H, Mainland JD (2009) Odor coding by a Mammalian receptor repertoire. Sci Signal 2: ra9.
16. Sanz G, Schlegel C, Pernollet JC, Briand L (2005) Comparison of odorant specificity of two human olfactory receptors from different phylogenetic classes and evidence for antagonism. Chem Senses 30: 69–80.
17. Schmiedeberg K, Shirokova E, Weber HP, Schilling B, Meyerhof W, et al. (2007) Structural determinants of odorant recognition by the human olfactory receptors OR1A1 and OR1A2. J Struct Biol 159: 400–412.
18. Shirasu M, Yoshikawa K, Takai Y, Nakashima A, Takeuchi H, et al. (2013) Olfactory Receptor and Neural Pathway Responsible for Highly Selective Sensing of Musk Odors. Neuron 81: 165–178.
19. Spehr M, Gisselmann G, Poplawski A, Riffell JA, Wetzel CH, et al. (2003) Identification of a testicular odorant receptor mediating human sperm chemotaxis. Science 299: 2054–2058.
20. Wetzel CH, Oles M, Wellerdieck C, Kuczkowiak M, Gisselmann G, et al. (1999) Specificity and sensitivity of a human olfactory receptor functionally expressed in human embryonic kidney 293 cells and Xenopus Laevis oocytes. J Neurosci 19: 7426–7433.
21. Goto T, Salpekar A, Monk M (2001) Expression of a testis-specific member of the olfactory receptor gene family in human primordial germ cells. Mol Hum Reprod 7: 553–558.
22. Parmentier M, Libert F, Schurmans S, Schiffmann S, Lefort A, et al. (1992) Expression of members of the putative olfactory receptor gene family in mammalian germ cells. Nature 355: 453–455.
23. Vanderhaeghen P, Schurmans S, Vassart G, Parmentier M (1997) Molecular cloning and chromosomal mapping of olfactory receptor genes expressed in the male germ line: evidence for their wide distribution in the human genome. Biochem Biophys Res Commun 237: 283–287.
24. Vanderhaeghen P, Schurmans S, Vassart G, Parmentier M (1993) Olfactory receptors are displayed on dog mature sperm cells. J Cell Biol 123: 1441–1452.
25. Vanderhaeghen P, Schurmans S, Vassart G, Parmentier M (1997) Specific repertoire of olfactory receptor genes in the male germ cells of several mammalian species. Genomics 39: 239–246.
26. Volz A, Ehlers A, Younger R, Forbes S, Trowsdale J, et al. (2003) Complex transcription and splicing of odorant receptor genes. J Biol Chem 278: 19691–19701.
27. Fuessel S, Weigle B, Schmidt U, Baretton G, Koch R, et al. (2006) Transcript quantification of Dresden G protein-coupled receptor (D-GPCR) in primary prostate cancer tissue pairs. Cancer Lett 236: 95–104.
28. Neuhaus EM, Zhang W, Gelis L, Deng Y, Noldus J, et al. (2009) Activation of an olfactory receptor inhibits proliferation of prostate cancer cells. J Biol Chem 284: 16218–16225.
29. Weng J, Wang J, Hu X, Wang F, Ittmann M, et al. (2006) PSGR2, a novel G-protein coupled receptor, is overexpressed in human prostate cancer. Int J Cancer 118: 1471–1480.
30. Xu LL, Stackhouse BG, Florence K, Zhang W, Shanmugam N, et al. (2000) PSGR, a novel prostate-specific gene with homology to a G protein-coupled receptor, is overexpressed in prostate cancer. Cancer Res 60: 6568–6572.
31. Gu X, Karp PH, Brody SL, Pierce RA, Welsh MJ, et al. (2014) Chemosensory functions for pulmonary neuroendocrine cells. Am J Respir Cell Mol Biol 50: 637–646.
32. Otaki JM, Yamamoto H, Firestein S (2004) Odorant receptor expression in the mouse cerebral cortex. J Neurobiol 58: 315–327.
33. Raming K, Konzelmann S, Breer H (1998) Identification of a novel G-protein coupled receptor expressed in distinct brain regions and a defined olfactory zone. Receptors Channels 6: 141–151.
34. Vanti WB, Nguyen T, Cheng R, Lynch KR, George SR, et al. (2003) Novel human G-protein-coupled receptors. Biochem Biophys Res Commun 305: 67–71.
35. Yuan TT, Toy P, McClary JA, Lin RJ, Miyamoto NG, et al. (2001) Cloning and genetic characterization of an evolutionarily conserved human olfactory receptor that is differentially expressed across species. Gene 278: 41–51.
36. Durzynski L, Gaudin JC, Myga M, Szydlowski J, Gozdzicka-Jozefiak A, et al. (2005) Olfactory-like receptor cDNAs are present in human lingual cDNA libraries. Biochem Biophys Res Commun 333: 264–272.
37. Gaudin JC, Breuils L, Haertle T (2001) New GPCRs from a human lingual cDNA library. Chem Senses 26: 1157–1166.
38. Gaudin JC, Breuils L, Haertle T (2006) Mouse orthologs of human olfactory-like receptors expressed in the tongue. Gene 381: 42–48.
39. Feingold EA, Penny LA, Nienhuis AW, Forget BG (1999) An olfactory receptor gene is located in the extended human beta-globin gene cluster and is expressed in erythroid cells. Genomics 61: 15–23.
40. Itakura S, Ohno K, Ueki T, Sato K, Kanayama N (2006) Expression of Golf in the rat placenta: Possible implication in olfactory receptor transduction. Placenta 27: 103–108.
41. Huang E, Cheng SH, Dressman H, Pittman J, Tsou MH, et al. (2003) Gene expression predictors of breast cancer outcomes. Lancet 361: 1590–1596.
42. Pluznick JL, Zou DJ, Zhang X, Yan Q, Rodriguez-Gil DJ, et al. (2009) Functional expression of the olfactory signaling system in the kidney. Proc Natl Acad Sci U S A 106: 2059–2064.
43. Feldmesser E, Olender T, Khen M, Yanai I, Ophir R, et al. (2006) Widespread ectopic expression of olfactory receptor genes. BMC Genomics 7: 121.
44. Flegel C, Manteniotis S, Osthold S, Hatt H, Gisselmann G (2013) Expression profile of ectopic olfactory receptors determined by deep sequencing. PLoS One 8: e55368.
45. Zhang X, De la Cruz O, Pinto JM, Nicolae D, Firestein S, et al. (2007) Characterizing the expression of the human olfactory receptor gene family using a novel DNA microarray. Genome Biol 8: R86.
46. Pavlath GK (2010) A new function for odorant receptors: MOR23 is necessary for normal tissue repair in skeletal muscle. Cell Adh Migr 4: 502–506.
47. Pluznick JL, Protzko RJ, Gevorgyan H, Peterlin Z, Sipos A, et al. (2013) Olfactory receptor responding to gut microbiota-derived signals plays a role in renin secretion and blood pressure regulation. Proc Natl Acad Sci U S A 110: 4410–4415.
48. Olender T, Waszak SM, Viavant M, Khen M, Ben-Asher E, et al. (2012) Personal receptor repertoires: olfaction as a model. BMC Genomics 13: 414.
49. Feinstein P, Bozza T, Rodriguez I, Vassalli A, Mombaerts P (2004) Axon guidance of mouse olfactory sensory neurons by odorant receptors and the beta2 adrenergic receptor. Cell 117: 833–846.
50. Keydar I, Ben-Asher E, Feldmesser E, Nativ N, Oshimoto A, et al. (2013) General olfactory sensitivity database (GOSdb): candidate genes and their genomic variations. Hum Mutat 34: 32–41.
51. Katotomichelakis M, Balatsouras D, Tripsianis G, Davris S, Maroudias N, et al. (2007) The effect of smoking on the olfactory function. Rhinology 45: 273–280.
52. Vennemann MM, Hummel T, Berger K (2008) The association between smoking and smell and taste impairment in the general population. J Neurol 255: 1121–1126.
53. Escada PA, Lima C, da Silva JM (2009) The human olfactory mucosa. Eur Arch Otorhinolaryngol 266: 1675–1680.
54. Khan M, Vaes E, Mombaerts P (2013) Temporal patterns of odorant receptor gene expression in adult and aged mice. Mol Cell Neurosci 57: 120–129.
55. Hellemans J, Mortier G, De PA, Speleman F, Vandesompele J (2007) qBase relative quantification framework and software for management and automated analysis of real-time quantitative PCR data. Genome Biol 8: R19.
56. Vandesompele J, De Preter K, Pattyn F, Poppe B, Van Roy N, et al. (2002) Accurate normalization of real-time quantitative RT-PCR data by geometric averaging of multiple internal control genes. Genome Biol 3: RESEARCH0034.
57. Bustin SA, Benes V, Garson JA, Hellemans J, Huggett J, et al. (2009) The MIQE guidelines: minimum information for publication of quantitative real-time PCR experiments. Clin Chem 55: 611–622.
58. Schroeder A, Mueller O, Stocker S, Salowsky R, Leiber M, et al. (2006) The RIN: an RNA integrity number for assigning integrity values to RNA measurements. BMC Mol Biol 7: 3.

59. Witt M, Bormann K, Gudziol V, Pehlke K, Barth K, et al. (2009) Biopsies of olfactory epithelium in patients with Parkinson's disease. Mov Disord 24: 906–914.

60. Mombaerts P (1999) Molecular biology of odorant receptors in vertebrates. Annu Rev Neurosci 22: 487–509.

61. Kobal G, Klimek L, Wolfensberger M, Gudziol H, Temmel A, et al. (2000) Multicenter investigation of 1,036 subjects using a standardized method for the assessment of olfactory function combining tests of odor identification, odor discrimination, and olfactory thresholds. Eur Arch Otorhinolaryngol 257: 205–211.

62. Ressler KJ, Sullivan SL, Buck LB (1993) A zonal organization of odorant receptor gene expression in the olfactory epithelium. Cell 73: 597–609.

63. Vassar R, Ngai J, Axel R (1993) Spatial segregation of odorant receptor expression in the mammalian olfactory epithelium. Cell 74: 309–318.

64. Leopold DA, Hummel T, Schwob JE, Hong SC, Knecht M, et al. (2000) Anterior distribution of human olfactory epithelium. Laryngoscope 110: 417–421.

65. Nibu K, Li G, Zhang X, Rawson NE, Restrepo D, et al. (1999) Olfactory neuron-specific expression of NeuroD in mouse and human nasal mucosa. Cell Tissue Res 298: 405–414.

66. Fuss SH, Zhu Y, Mombaerts P (2013) Odorant receptor gene choice and axonal wiring in mice with deletion mutations in the odorant receptor gene SR1. Mol Cell Neurosci 56: 212–224.

67. Imai T, Suzuki M, Sakano H (2006) Odorant Receptor-Derived cAMP Signals Direct Axonal Targeting. Science 314: 657–661.

6

Genome-Wide Analysis of Copy Number Variation Identifies Candidate Gene Loci Associated with the Progression of Non-Alcoholic Fatty Liver Disease

Shamsul Mohd Zain[1,2]*, Rosmawati Mohamed[3]*, David N. Cooper[4], Rozaimi Razali[5], Sanjay Rampal[6], Sanjiv Mahadeva[3], Wah-Kheong Chan[3], Arif Anwar[5], Nurul Shielawati Mohamed Rosli[5], Anis Shafina Mahfudz[7], Phaik-Leng Cheah[8], Roma Choudhury Basu[9], Zahurin Mohamed[1,2]

1 The Pharmacogenomics Laboratory, Faculty of Medicine, University of Malaya, Kuala Lumpur, Malaysia, 2 Department of Pharmacology, Faculty of Medicine, University of Malaya, Kuala Lumpur, Malaysia, 3 Department of Medicine, Faculty of Medicine, University of Malaya, Kuala Lumpur, Malaysia, 4 Institute of Medical Genetics, School of Medicine, Cardiff University, Cardiff, United Kingdom, 5 Sengenics Sdn Bhd, High Impact Reseach Building, University of Malaya, Kuala Lumpur, Malaysia, 6 Julius Centre University of Malaya, Department of Social and Preventive Medicine, Faculty of Medicine, University of Malaya, Kuala Lumpur, Malaysia, 7 Medical Imaging Unit, Faculty of Medicine, University of Technology MARA, Sungai Buloh Campus, Selangor, Malaysia, 8 Department of Pathology, Faculty of Medicine, University of Malaya, Kuala Lumpur, Malaysia, 9 Clinical Investigation Centre, University Malaya Medical Centre, Kuala Lumpur, Malaysia

Abstract

Between 10 and 25% of individuals with non-alcoholic fatty liver disease (NAFLD) develop hepatic fibrosis leading to cirrhosis and hepatocellular carcinoma (HCC). To investigate the molecular basis of disease progression, we performed a genome-wide analysis of copy number variation (CNV) in a total of 49 patients with NAFLD [10 simple steatosis and 39 non-alcoholic steatohepatitis (NASH)] and 49 matched controls using high-density comparative genomic hybridization (CGH) microarrays. A total of 11 CNVs were found to be unique to individuals with simple steatosis, whilst 22 were common between simple steatosis and NASH, and 224 were unique to NASH. We postulated that these CNVs could be involved in the pathogenesis of NAFLD progression. After stringent filtering, we identified four rare and/or novel CNVs that may influence the pathogenesis of NASH. Two of these CNVs, located at 13q12.11 and 12q13.2 respectively, harbour the exportin 4 (XPO4) and phosphodiesterase 1B (PDE1B) genes which are already known to be involved in the etiology of liver cirrhosis and HCC. Cross-comparison of the genes located at these four CNV loci with genes already known to be associated with NAFLD yielded a set of genes associated with shared biological processes including cell death, the key process involved in 'second hit' hepatic injury. To our knowledge, this pilot study is the first to provide CNV information of potential relevance to the NAFLD spectrum. These data could prove invaluable in predicting patients at risk of developing NAFLD and more importantly, those who will subsequently progress to NASH.

Editor: Yanqiao Zhang, Northeast Ohio Medical University, United States of America

Funding: This work was supported by High Impact Research Ministry of Higher Education (HIRMOHE) of Malaysia Grant E000049-20001. The funders had no role in study design, data collection and analysis, decision to publish, or preparation of the manuscript.

Competing Interests: Sengenics Sdn Bhd is not a funder and has no competing interest and financial disclosure along with any other relevant declarations relating to employment, consultancy, patents, products in development or marketed products. The authors certify that the submission is an original work and is not under review at any other journals.

* E-mail: soulz712@gmail.com (SMZ); ros@ummc.edu.my (RM)

Introduction

Non-alcoholic fatty liver disease (NAFLD) has emerged as a silent epidemic, with its worldwide prevalence continuing to increase with the growing incidence of obesity [1]. NAFLD comprises a spectrum of diseases ranging from simple steatosis, which is essentially benign fatty infiltration of the liver, to its inflammatory counterpart non-alcoholic steatohepatitis (NASH) [2]. The pathogenesis of NAFLD is based on the "two hit hypothesis" [3]. The "first hit" is the development of steatosis and involves the accumulation of triglycerides in the liver due to insulin resistance. Insulin resistance prepares the hepatocytes for the second insult. The "second hit" is often due to adipocytokines and oxidative stress, which further damage the liver thereby promoting progression to steatohepatitis and fibrosis. A significant proportion of individuals with NAFLD develop hepatic fibrosis, a key feature of the condition which is associated with progression of the disease to cirrhosis and its related complications, including hepatic failure and hepatocellular carcinoma [4]. The fibrotic progression of NAFLD is identified histologically by the presence of NASH. A high prevalence of NASH is found among those with insulin resistance-related comorbidities such as obesity and type 2 diabetes [5]. The mortality rate among NASH patients has been found to be much higher than for patients with simple fatty liver (simple steatosis) [6].

In addition to environmental factors such as high calorific food intake and a sedentary lifestyle, there is mounting evidence of a genetic component to the complex etiology of NAFLD [7]. This is reflected by marked differences in the prevalence of NAFLD across diverse populations [8–9]. The high heritability of NAFLD was evident in a familial aggregation study, with estimates of 59%

Table 1. Histopathological data in patients with NAFLD.

	NASH (n = 39)	Simple steatosis (n = 10)
Steatosis grade		
1	11	9
2	20	1
3	8	0
Inflammatory activity		
0	1	3
1	19	7
2	19	0
3	0	0
Ballooning		
0	0	10
1	21	0
2	18	0
Fibrosis stage		
0	0	10
1	12	0
2	19	0
3	6	0
4	2	0

NASH, non-alcoholic steatohepatitis.

in siblings and 78% in parents with NAFLD [10]. Until recently, genome-wide association studies (GWAS) and the candidate gene approach have both utilised single nucleotide polymorphisms (SNPs) to explain the genetic component of NAFLD [7,11–12].

The wide distribution of copy number variants (CNVs) in the human genome has underscored the importance of CNVs in relation to genetic diversity, phenotypic variability and disease susceptibility [13–14]. It has been estimated that approximately 12% of the human genome is copy number variable [15] with over 1000 genes having been mapped within or close to regions that are affected by structural variation [16]. A global increase in CNV burden has also been observed in polygenic traits such as schizophrenia [17], autism [18] and attention deficit hyperactivity disorder [19]. Given these findings, the sheer scale of CNVs means that they are likely to make a significant contribution to the 'missing heritability' of some of these conditions [20]. However, despite some success in identifying CNVs responsible for metabolic phenotypes including obesity and diabetes mellitus [21–22], there are as yet no data available to suggest whether or not CNVs might be involved in the etiology of the NAFLD spectrum.

Here, we describe a pilot study designed to detect rare or novel CNVs associated with NAFLD and/or NASH. Predicting NASH non-invasively is very important since this condition is potentially progressive and liver biopsy is currently the gold standard for the diagnosis of NASH. We interrogated the CNVs associated with NASH and ascertained the biological processes associated with those genes covered by the CNVs in order to assess their possible role in the progression of the disease. To this end, we used a high-resolution Agilent aCGH platform to perform genome-wide copy number analysis in patients with both simple steatosis and NASH, which are representative of the clinical spectrum of NAFLD.

Materials and Methods

Ethics Statement

The study protocol was approved by the Medical Ethics Committee of UMMC and all subjects provided their written informed consent to participate.

Subjects

Genome-wide copy number profiling was performed using array comparative genomic hybridization (aCGH) on a total of 49 NAFLD patients (39 with NASH and 10 with simple steatosis) and 49 fatty liver-free controls that were matched both for age and gender. All subjects were, as far as could ascertain, genetically unrelated to each other. All NAFLD patients were consecutively recruited from the University of Malaya Medical Centre (UMMC). NAFLD was confirmed through liver histology and evaluated according to the NASH Clinical Research Network criteria [23–24]. All liver biopsy specimens were on average 1.5 cm long and contained at least six portal tracts. Subjects were excluded if they met any of the following criteria: (i) alcohol consumption >10g/day [25]; (ii) hepatitis B or C infection; (iii) autoimmune hepatitis; (iv) exposure to drugs known to cause steatosis or (v) Wilson's disease. The controls were genetically unrelated healthy subjects with a body mass index (BMI) <25 kg/m^2, a fasting plasma glucose of <110 mg/dL, a normal lipid profile and normal liver enzymes. NAFLD was actively excluded in the controls by ultrasonography according to the absence of the following criteria: (i) slight diffuse increase in bright homogeneous echoes in the liver parenchyma with normal visualization of the diaphragm and portal and hepatic vein borders, and normal hepatorenal echogenicity contrast; (ii) diffuse increase in bright echoes in the liver parenchyma with slightly impaired visualization of the

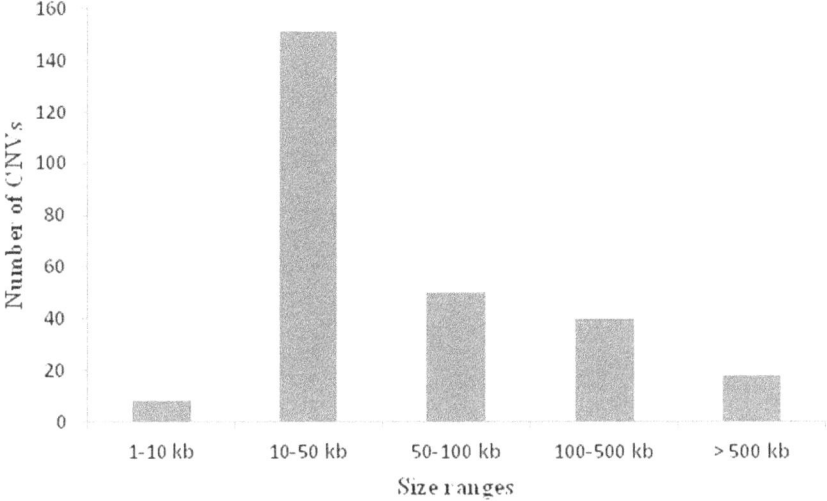

Figure 1. Size range distribution of the CNVs. The CNVs ranged in size from 5.77 kb to 8.15 Mb with a mean size of 194.94 kb and a median size of 38.33 kb.

peripheral portal and hepatic vein borders; (iii) marked increase in bright echoes at a shallow depth with deep attenuation, impaired visualization of the diaphragm and marked vascular blurring [26]. Subsequent magnetic resonance imaging (MRI) to further confirm the fatty liver free status was performed.

Array CGH

Array-CGH was performed according to the protocol established by the manufacturer (Oxford Gene Technology, Begbroke, UK). It was carried out using the SurePrint G3 Human CGH 2×400 K array (Agilent Technologies, Santa Clara, CA, USA) for genome-wide identification of putative disease-associated CNVs. Each oligonucleotide-based microarray slide contained 410,739 probes that enabled the profiling of molecular genomic imbalances with a mean resolution of 5.3 kb. Probes on the array were 60-mers and covered both coding and non-coding regions of the human genome. A total of 1.0 μg genomic DNA from patients and controls was labeled with Cy3 and Cy5 dyes respectively using the CytoSure Genomic DNA labeling kit (Oxford Gene Technology). Probes were then purified using Microcon Centrifugation Filters, Ultracel YM-30 (Millipore, Billerica, MA, USA) and mixed thoroughly. This was followed by denaturation and pre-annealing with 50 μg human Cot-1 DNA (Invitrogen, California). Hybridization of the mixture to the array slide was executed at a constant rotation at 65°C for 40 hours. The slide was then washed with Agilent wash buffers 1 and 2, and scanned immediately using an Agilent Microarray scanner (Agilent Technologies, Santa Clara, CA, USA). Data were extracted from scanned images using Feature Extraction Software, version 10.7.3.1 (Agilent Technologies, USA). The raw data obtained thereafter were uploaded into the CytoSure Interpret software version 4.2.5 (Oxford Gene Technology), normalized and converted into.cgh files. Data normalization software was used to improve inconsistencies in dye incorporation. The data were segmented using a modified Circular Binary Segmentation (CBS) algorithm [27]. Genomic aberrations were identified by applying log2 intensity ratios of sample to reference (Cy3/Cy5: log2-ratios above 0.3 for duplications and below −0.6 for deletions). Chromosomal aberrations were

reported in accordance with the human genome sequence assembly Build 37, hg 19 (http://www.ncbi.nlm.nih.gov). The microarray data have been deposited in the Gene Expression Omnibus (GEO) database (accession number 55645).

CNV Calling and Functional Enrichment Analysis

CNVs were called for the segments with at least 5 consecutive probes. Rare CNVs were defined as those which overlapped by <50% with reported CNVs from the Database of Genomic Variants (DGV; http://dgv.tcag.ca/dgv/app/home). CNVs were deemed to be novel if they did not appear in the DGV database. Gene content within the identified CNVs was retrieved from the *Homo sapiens* (GRCh27) assembly using the Biomart-Ensembl (http://www.ensembl.org). By default, the lists contained both gene and non-gene entities; the latter were removed through a process of cross-checking and verification of gene symbols using the HUGO Gene Nomenclature Committee (HGNC) database (http://www.genenames.org/). To investigate the functional impact of rare and/or novel CNVs, the Database for Annotation, Visualization and Integrated Discovery (DAVID; http://david.abcc.ncifcrf.gov/) was utilised to assess the Gene Ontology (GO; http://www.geneontology.org/) and Kyoto Encyclopedia of Genes and Genomes (KEGG) pathway (http://www.genome.jp/kegg/) annotations between the genes encompassing the rare and/or novel CNVs and the genes associated with NAFLD. The list of genes associated with NAFLD was identified using the MalaCards database (http://www.malacards.org/) – an integrated searchable database of human disease states and their annotations, in association with the GeneCards relational database. Initially, a total of 200 genes associated with NAFLD were identified. Since the gene-disease association was based on a text mining algorithm, a manual verification of the biological processes associated with each of the 200 genes was performed. Only genes that had previously been described as being associated with NAFLD by either expression studies, genotyping or protein array work were selected, thereby lowering the number of genes implicated in NAFLD to 70 (see Table S1 for the complete list of genes).

Table 2. Top regions of copy number gains and losses in NASH.

Cytoband	Sample frequency (%)	DGV coverage (%)	Start	End	Size (kb)	Number of genes within demarcated region	Candidate gene (s)
					Amplification		
14q11.2	53.8	100	19,728,641	20,420,849	692.21	15	*OR*
5p15.33	41	100	723,194	820,424	97.23	3	–
11:p15.4	35.9	100	5,893,184	5,935,144	41.96	3	*OR*
12:p13.31	33.3	100	9,637,323	9,718,846	81.52	0	–
					Deletion		
11:q11	35.9	100	55,368,154	55,450,788	82.63	6	*OR*
14:q24.3	33.3	100	74,001,651	74,022,324	20.67	3	*ACOT1*
16q12.2	33.3	100	55,832,511	55,853,358	20.85	1	*CES1*
4:q13.2	33.3	100	69,392,545	69,483,277	90.73	1	–
					Amplification & Deletion		
12:p13.2	38.5	100	11,219,788	11,249,210	29.42	4	*TASR*

Start = first base-pair location in the copy number region, End = last base-pair location in the copy number region.

Table 3. Rare and novel CNVs in NASH.

Cytoband	Sample frequency (%)	DGV coverage (%)	Start	End	Size (kb)	Number of genes within demarcated region	Candidate Gene (s)
					Amplification		
Rare							
12q24.33	12.8	<50	131,432,076	131,460,728	28.65	1	–
13q12.11	10.3	<50	21,475,933	21,494,311	18.38	1	*XPO4*
Novel							
21p11.1–11.2	10.3	0	10,347,806	10,944,060	596.25	5	–
12q13.2	10.3	0	54,962,801	54,983,141	20.34	2	*PDE1B*

Start = first base-pair location in the copy number region, End = last base-pair location in the copy number region.

Quantitative PCR Validation of CNV Calls

A duplex TaqMan real-time quantitative polymerase chain reaction (qPCR) was performed to validate the CNV regions using a Step One Plus (Applied Biosystems) on three of the samples from two selected regions (11q11: Assay Hs02799097_cn, and 13q12.11: Assay Hs03857719_cn). Each reaction (20 μL) contained 10 μL master mix, 1 μL TaqMan Copy Number Assay, 1 μL TaqMan Copy Number Reference Assay, 4 μL nuclease free water, and 4 μL 5 ng/μL genomic DNA, and was run in quadruplicate. The PCR cycling conditions consisted of 1 PCR cycle at 95°C for 10 min, followed by 40 cycles at 95°C for 15 sec and 60°C for 1 min.

Results

In the aCGH method adopted, DNA samples pooled from multiple subjects (patients and controls) were cross-compared so as to remove/normalise any common copy number changes in the normal control sample. Since the principle of aCGH is to compare the DNA copy number from patient samples against those of normal controls, CNV calls were designed to be patient-specific.

Subjects and Identification of CNVs in the NASH Genome

All DNA samples passed quality control (QC) after a rigorous sample preparation process and a QC check during sample processing. Sets of 39 NASH samples and 39 matched controls were run in parallel on an array CGH platform that allowed the ratio of DNA copy number between a test (patient) and a reference (control) to be simultaneously assessed. From a total of 39 samples, 51.3% (n = 20) were females, 48.7% (n = 19) were males; the mean age of the 39 subjects was 50.4 years. The histopathological data are presented in Table 1. Seven percent of CNV calls were attributable to the sex chromosomes (with a frequency of at least 10%), but we opted to exclude these chromosomes from further analysis owing to the evolutionary biases due to small imbalances of the sex chromosomes [28]. Analysis of copy number variants, on the basis of log ratio and probe incidence filtering, yielded a total of 267 autosomal CNVs (the ratio of the fluorescence intensities between the patients and controls is a measure of the relative DNA copy number), amounting to an average of 6.84 autosomal CNVs per individual. The 267 CNVs detected spanned between 5.77 kb and 8.15 Mb in size, with a mean size of 194.94 kb and a median size of 38.33 kb, covering a total of 52.05 Mb or 1.63% of the genome (Fig. 1). Most chromosomal arms harboured both copy number gains and losses, but copy number gains were more commonly observed than losses (estimated ratio of 1.7:1). However, only 55 CNVs (20.6%) out of the 267 CNVs detected had a frequency of >10%.

Molecular genomic profiling identified 14q11.2 as the most frequently amplified region, which occurred in 53.8% of the NASH samples and contained a clutch of olfactory receptor (*OR*) family genes (Table 2; see Table S2 for the full list of *OR* genes). The most frequently deleted genomic region in the NASH samples, 12p13.2, is enriched in taste receptor (*TASR*) family genes (see Table S2 for the full list of *TASR* genes), and exhibited similar frequencies of losses and gains (38.5%) suggesting a generally unstable region. Several other frequently deleted regions were also observed including one at 16q12.2 harbouring the carboxylesterase 1 (*CES1*) gene and one at 14q24.3 spanning the acyl-CoA thioesterase 1 (*ACOT1*) gene; importantly, both genes are known to promote hepatic steatosis via the action of regulation of hepatic lipid metabolism [29–30]. There were nine CNVs present in at least 33% of the samples whilst only one was present

Table 4. Enriched GO terms associated with NASH.

Category	Term	Count	Involved genes/ total genes (%)	P-Value*
GO_CC	GO:0005576~extracellular region	34	44.16	1.2E-10
GO_BP	GO:0006006~glucose metabolic process	6	7.79	2.0E-04
GO_BP	GO:0019318~hexose metabolic process	6	7.79	6.8E-04
GO_BP	GO:0005996~monosaccharide metabolic process	6	7.79	1.4E-03
GO_BP	GO:0015980~energy derivation by oxidation of organic compounds	4	5.19	8.3E-03
GO_BP	GO:0007186~G-protein coupled receptor protein signaling pathway	12	15.58	2.0E-02
GO_BP	GO:0007166~cell surface receptor linked signal transduction	16	20.78	4.6E-02
GO_BP	GO:0006091~generation of precursor metabolites and energy	5	6.49	3.0E-02
GO_BP	GO:0008219~cell death	8	10.39	4.5E-02

GO_BP, Gene Ontology biological process; GO_CC, Gene Ontology cellular component.
*Modified Fisher's Exact test, P-Value ≤ 0.05.

in at least half of the samples. Six CNVs occurred with a frequency of at least 10% in samples which contained copy number duplications in the chromosomal regions 16p12.2 (27.5%), 12q24.33 (12.8%), 22q13.2 (12.8%), 12q13.2 (10.3%), 2q37.1 (10.3%) and 21p11.2-p11.1 (10.3%), implying that these regions could be involved in the development of NASH. Overall, nearly 50% of genomic regions were reported only as copy number gains; however, only 6 of these regions were present at a frequency of more than 10%. By contrast, about 18% of the genomic regions presented only as losses; however, none had a frequency greater than 10%.

Integrative Analysis of CNVs and Functional Enrichment to Identify Candidate Genes for Involvement in NASH

To identify unique CNVs in NASH patients that could be involved in the pathogenesis of this condition, we performed a cross-comparison with known CNVs from the DGV database. Conservative assessment of the overlap between reported CNVs from the DGV database with the CNVs identified in this study revealed four rare and/or novel CNVs (DGV coverage <50%) that were present in at least 10% of the NASH samples (Table 3). Two of these CNVs were classed as rare (DGV coverage <50%: 12q24.33 and 13q12.11), whereas the other two were novel (DGV coverage 0%: 21p11.1–11.2 and 12q13.2). A Chi-square test confirmed the significance of the association of these CNVs with NASH ($P<0.05$) as compared to simple steatosis. To further assess the likelihood of the involvement of these CNVs in NASH, the genes located within these regions were identified and their involvement in those biological processes shared with known NAFLD genes assessed. First, we profiled the genes within the chromosomal regions that are bounded by the four rare and/or novel CNVs, where genes such as exportin 4 (*XPO4*) and phosphodiesterase 1B (*PDE1B*) are located. A list of genes known to be associated with NAFLD was then obtained (see Table S1). Subsequently, we performed GO enrichment and KEGG pathway analysis using the DAVID gene annotation tool for the two sets of genes (genes within the four unique regions and known genes associated with NAFLD). We observed a number of shared biological processes (Table 4)

between the two sets of genes including those that could be linked to NAFLD progression such as glucose metabolism, cell surface receptor-linked signal transduction and cell death [3].

Identification of CNVs in the Simple Steatosis Genome

Given the greater number of NASH samples (~80%) and the progressive nature of NASH (about one third of NASH patients tend to develop cirrhosis over a 5–10 year period; by contrast, simple steatosis patients tend to be clinically stable over time) [31] in the disease spectrum, the main focus of this study was placed on NASH. However, we were also interested in understanding the progression of simple steatosis to NASH. Unfortunately, we were only able to obtain DNA samples from 10 simple steatosis patients and 10 fatty-liver free controls. Seven of the samples were male and the mean age (all samples) was 47.9. The histopathological data are shown in Table 1. A total of 56 CNVs (simple steatosis patient-specific) were identified, including three (5.4%) which were located on one of the sex chromosomes. All CNVs were present with a frequency of at least 10%. Fifty-three autosomal CNVs were selected for further analysis. Of these, 11 were unique to simple steatosis whereas 42 were found to be shared with NASH. The former 11 CNVs could conceivably play a role in the development of hepatic steatosis, whereas the latter 42 CNVs could be involved in progression to steatohepatitis. Intriguingly, the four rare and/or novel CNVs identified earlier in NASH patients were not found in simple steatosis patients, and remain unique to NASH.

The top scoring regions in terms of copy number gains and losses in simple steatosis are listed in Table 5. The most commonly amplified region, 12p13.31 (50%), was also among the most highly amplified regions observed in NASH patients. A CNV at the 10q11.22 locus that occurred in 40% of the simple steatosis samples contains the neuropeptide Y receptor 4 (*NPYR4*) gene, which is known to be important in obesity through the regulation of appetite and energy metabolism [32]. Three CNVs (located at 4q13.2, 15q11.2 and 11q11) shared the most deleted region at a frequency of 40%, in which two of the CNVs (4q13.2 and 11q11) were also among the most highly deleted regions observed in NASH. These CNVs were enriched

Table 5. Top regions of copy number gains and losses in simple steatosis.

Cytoband	Sample frequency (%)	DGV coverage (%)	Start	End	Size (kb)	Number of genes within demarcated region	Candidate gene (s)
					Amplification		
12p13.31	50	100	9,637,323	9,718,846	81.52	0	–
10q11.22	40	100	46,971,647	47,394,442	422.80	14	NPY4R
16p13.11	40	100	15,048,676	15,120,666	71.99	2	–
8p11.23	40	100	39,234,992	39,386,158	151.17	0	–
					Deletion		
11q11	35.9	100	55,368,154	55,450,788	82.63	6	OR
15q11.2	33.3	100	20,481,702	22,578,630	2096.93	32	OR, IGH
					Amplification & Deletion		
4q13.2	40	100	69,392,545	69,483,277	90.73	1	–

Start = first base-pair location in the copy number region, End = last base-pair location in the copy number region.

for *OR* genes (11q11) and immunoglobulin heavy chain (*IGH*) (15q11.2) family genes (Table 4; see Table S2 for the full list of *OR* and *IGH* genes). However, all CNVs identified in the simple steatosis patients were common (DGV coverage 100%).

qPCR Validation

We validated three samples (each is patient-control matched pair) for each CNV region identified. We selected two CNV regions that represented different statuses of copy number change (the CNV at 13q12.11 was a copy number gain and was rare, 11q11 was a copy number loss). All CNVs were confirmed through qPCR validation. Amplifications and deletions of the genomic regions were defined on the basis of differences between patient's copy number and the wild-type copy number (i.e. a copy number around 2). Fig. 2 illustrates the qPCR results of the validated CNVs.

Discussion

Studies on CNVs are becoming increasingly important in studies of inherited disease, with growing evidence attesting to the substantial impact that they can have on human phenotypic variability and genetic susceptibility. Here, we present a pilot analysis of CNVs in a series of NAFLD patients. We identified four CNVs that are either rare or novel to NASH patients in our study that could potentially contribute to clinical outcome.

In patients with NASH, the most frequently amplified region was 14q11.2, which is enriched in *OR* family genes, while an abundance of *TASR* family genes were found at 12p13.2, the most frequently deleted region. Although the *OR* and *TASR* families play roles in the olfactory and gustatory systems respectively, a search of the database of Expressed Sequence Tags, NCBI dbEST, revealed *OR* and *TASR* gene expression in many tissues and organs, including the liver. Impairment of olfactory and gustatory function has been reported in chronic liver disease including cirrhosis; chemosensory function however improved after liver transplantation [33]. In the early 2000s, a comprehensive database of the human olfactory subgenome was completed using a highly automated data mining system [34]. Glusman et al. (2001) reported the presence of 906 potential coding regions for *OR* genes that cover almost all human chromosomes with the exception of chromosomes 20 and Y, in which 2/3 of the regions have not been reported. Subsequently, new databases termed respectively the Olfactory Receptor Microarray Database (ORMD) which includes microarray gene expression data from the *ORs* [35], and the Database of Chemosensory Receptor Gene Families (CRDB), were developed [36]. The size of these databases highlights the importance of *OR* and *TASR* gene families not only in the olfactory and gustatory systems, but also in tissues and organs throughout the body.

A deletion CNV was noted at the 16q12.2 locus; it includes the *CES1* gene, which is primarily important in the metabolism of fatty acids and cholesterol [37]. Expression of *CES1* has been found to be higher in human NAFLD hepatic tissue as compared to non-NAFLD [38]. A role for *CES1* in lipolysis was evidenced by a positive correlation between *CES1* expression and triglyceride lipase activity as well as with adiposity [30]. On the other hand, *CES1* knockout mice are characterized by a gain in weight, hepatic steatosis and hyperinsulinemia, thereby supporting a role for *CES1* in the regulation of fatty acids [37]. Interestingly, the 16q12.2 locus is known to harbour genetic variants (SNPs) associated with BMI [39]. Although *CES1* has been implicated in hepatic steatosis

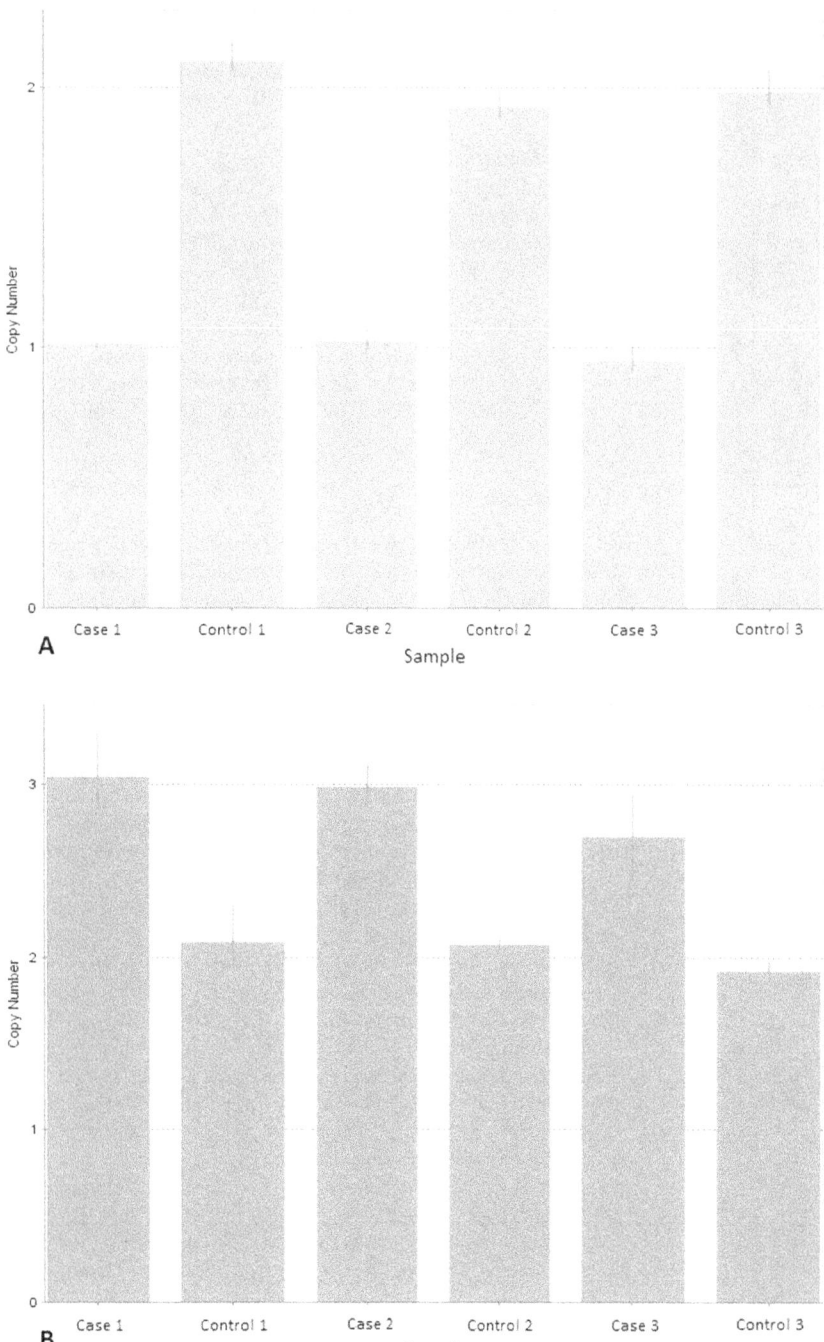

Figure 2. qPCR validation performed on selected genomic regions. (**A**) Results for the region 11q11. (**B**) Results for the region 13q12.11. A copy number of around 2 was deemed to be indicative of wild-type status (i.e. no CNV), a copy number of 1 was indicative of one copy lost, whereas a copy number of 3 or above was held to indicate copy number gain(s). The error bars represent the standard error among four replicates.

[37], a recent study has shown that CES1 may have potential as a biomarker to distinguish hepatocellular carcinoma (HCC) from cirrhosis [40]. Also notable among the highly deleted regions in NASH patients was a copy number loss at the 14q24.3 locus, where the acyl-CoA thioesterase 1 (*ACOT1*) gene resides. Acyl-CoA thioesterase 1 promotes the cellular balance between free fatty acids and acyl-CoAs to maintain cellular processes including lipid metabolism [29]. Compared to other *ACOT* subfamily genes, *ACOT1* is unique in that it is highly

expressed only in association with a high fat diet but not in association with a normal diet [41].

Although determining the CNV frequencies and their gene content are important, most of the CNVs detected here are considered to be common (DGV coverage 100%) and hence may have little or no impact on the pathogenesis of NASH. The definition of 'common' here is however debateable given that reported CNVs from the DGV are (i) from non-NAFLD studies and (ii) unlikely to be from the Malay population.

Caution should therefore be exercised when offering functional interpretation of these CNVs until more comprehensive studies on larger numbers of patients are conducted. To achieve our main goal in this pilot study, which was to identify candidate CNV loci that could play a role in the etiology of NASH, we filtered out common CNVs and identified four CNVs (DGV coverage <50%) that have the potential to be involved in the pathogenesis of NASH; two of these are rare (12q24.33 and 13q12.11) whilst two are novel (12q13.2 and 21p11.1–11.2). We were able to establish the potential significance of these loci by performing a Chi-square test against other loci and validating the findings by qPCR; in this way, we were able to confirm that, despite the relatively small sample size, our analysis has the potential to yield biologically meaningful and reproducible results. We postulate that these CNVs could provide new insights into the biology of NASH. Of particular note was an aberration at the 13q12.11 locus that could serve as a potential copy number biomarker for NASH. This region contains the tumor suppressor gene exportin 4 (*XPO4*), the inactivation of which promotes HCC in mice [42]. On the other hand, increased expression of *XPO4* in human HCC is associated with better prognosis and a better survival rate [43–44]. The phosphodiesterase 1B, calmodulin-dependent (*PDE1B*) gene spanning the 12q13.2 region is important in many signal transduction pathways, and has been found to be downregulated in cirrhotic liver [45]. The 12q13.2 locus was identified as a clear-cut amplification (no deletion event), thereby supporting its candidacy as a potential risk marker CNV associated with the disease. However, there are a limited number of published reports on *XPO4* and *PDE1B* and their putative role in liver disease. Thus, additional comprehensive studies focussed on these two genes will be necessary to confirm or refute this finding.

To assess the plausibility of our results, it was important to verify the functional role of these CNVs (rare/novel) and their potential impact on NASH. In order to explore the possible association between these CNVs and NASH, we extended our analysis to GO functional enrichment and KEGG pathway analysis for genes residing at these CNV loci and known NAFLD genes. The results yielded several shared biological processes between the two sets of genes. Of primary importance are glucose metabolism and cell surface receptor-linked signal transduction and cell death, all of which have been shown to be important in the pathogenesis of NASH [4]. However, no related KEGG pathway was observed.

As for simple steatosis, the most frequently amplified region (12p13.31) also happened to be among the most highly amplified regions in NASH. The 10q11.22 region, which harbours a CNV that occurred in 40% of the simple steatosis samples, contains the *NPY4R* gene. This gene is involved in the regulation of appetite and energy metabolism [32]. The pancreatic peptide, a high affinity ligand for the neuropeptide Y receptor 4 (Y4), has been suggested to have anti-obesity potential [46–47]. Long term antagonism of Y4 causes significant reduction in body weight and adiposity via effects on metabolic rate and energy distribution [48].

We readily acknowledge the small number of simple steatosis samples in the present study. This limitation was due to the lack of availability of simple steatosis patients from our previous study that comprised three major ethnic groups [49–50]. These patients were recruited from the UMMC, a tertiary referral center, which could explain the greater number of NASH patients as compared to those with simple steatosis. In order to minimise ethnicity as a potential confounding factor, we selected samples taken from only one specific ethnic group, namely the Malays, for both the NASH and the simple steatosis group. Under these conditions, the number of simple steatosis patients that we were able to obtain was only 10. Despite the limited numbers of patients available, several of our findings were statistically significant. Importantly, chromosome 11q11 which was one of the most frequently deleted CNVs in our study, was also frequently deleted from 10 hepatic steatosis patients from the study by Royo et al. [51]. It should be noted that the Royo et al. study did not include any NASH patients. This notwithstanding, our pilot study was designed to provide an initial screen of the structural genomic aberrations present in NAFLD samples. Simple steatosis patients mostly presented with either a copy number gain or a loss event at one locus, unlike the NASH group which tended to exhibit both events. In addition, a greater number of CNVs were identified in the NASH group as compared to the simple steatosis group. This could be explained by the complex pathogenesis of NAFLD especially at the NASH stage, involving not only the 'first hit' mechanism but also the 'second hit' [4]. In this study, we were mainly concerned with identifying CNVs that were common to both simple steatosis and NASH, particularly when the CNV frequency was higher in NASH than in simple steatosis ($n = 2$), as they could indicate involvement in the progression of the disease. Surprisingly, histological data from the samples harbouring these CNVs (12p13.2 and 11p15.4) showed a higher frequency (53.3% and 71.4% respectively) of fibrosis score ≥ 2, thereby supporting the disease progression model.

The ethical issue that precluded the use of liver biopsy for the classification of controls (non-NAFLD) required us to adopt a stringent definiton of controls in order to rule out fatty liver in the control subjects; biochemical tests, ultrasonography and MRI evaluations were therefore used to minimise misclassification of our controls. To the best of our knowledge, this is the first study to investigate a genome-wide profile of copy number variation in the NAFLD spectrum; hence, determination of the CNV total number, frequency, genomic location and gene content, is challenging. The use of aCGH technology allows CNV discovery at high resolution and hence allows confidence in CNV detection. The use of 60mer probes provides high sensitivity and specificity to accurately detect both known and *de novo* CNVs as compared to shorter oligonucleotide probes [52]. The source of genes known to be related to NAFLD was Malacard, which is known to use a text-mining approach [53]. Hence, a manual verification of the gene functions was performed that included only genes that have been shown to be associated in either expression studies, genotyping or protein array work. However, we cannot rule out the possibility that other genes could be of importance in NAFLD, as more comprehensive studies are still ongoing. Indeed, it was also difficult for us to assess the significance of such CNVs given that multiple genes often reside within the CNV intervals. We attempted to overcome this limitation by performing a functional enrichment analysis that covered all the genes residing within the CNV regions.

Taken together, the results of our whole genome copy number analysis have documented four rare and/or novel CNV loci that are unique to NASH, and to the best of our knowledge, have not previously been reported. This study nevertheless falls into the hypothesis generating category rather than the hypothesis testing category; hence, our results remain to be substantiated by additional studies on larger patient groups. Moreover, additional functional studies on the genes residing within these loci will be needed to fully characterize the function of the genes and their relationship, if any, to NASH.

Acknowledgments

The authors thank all the patients and volunteers for their assistance in performing the above-described studies.

References

1. Fabbrini E, Sullivan S, Klein S (2010) Obesity and nonalcoholic fatty liver disease: biochemical, metabolic, and clinical implications. Hepatology 51: 679–689.
2. Malaguarnera M, Di Rosa M, Nicoletti F, Malaguarnera L (2009) Molecular mechanisms involved in NAFLD progression. J Mol Med (Berl) 87: 679–695.
3. Jou J, Choi SS, Diehl AM (2008) Mechanisms of disease progression in nonalcoholic fatty liver disease. Semin Liver Dis 28: 370–379.
4. Farrell GC, van Rooyen D, Gan L, Chitturi S (2012) NASH is an inflammatory disorder: pathogenic, prognostic and therapeutic implications. Gut Liver 6: 149–171.
5. Chalasani N, Younossi Z, Lavine JE, Diehl AM, Brunt EM, et al. (2012) The diagnosis and management of non-alcoholic fatty liver disease: practice guideline by the American Gastroenterological Association, American Association for the Study of Liver Diseases, and American College of Gastroenterology. Gastroenterology 142: 1592–1609.
6. Ekstedt M, Franzen LE, Mathiesen UL, Thorelius L, Holmqvist M, et al. (2006) Long-term follow-up of patients with NAFLD and elevated liver enzymes. Hepatology 44: 865–873.
7. Hernaez R (2012) Genetic factors associated with the presence and progression of nonalcoholic fatty liver disease: a narrative review. Gastroenterol Hepatol 35: 32–41.
8. Farrell GC, Wong VW, Chitturi S (2013) NAFLD in Asia–as common and important as in the West. Nat Rev Gastroenterol Hepatol 10: 307–318.
9. Vernon G, Baranova A, Younossi ZM (2011) Systematic review: the epidemiology and natural history of non-alcoholic fatty liver disease and non-alcoholic steatohepatitis in adults. Aliment Pharmacol Ther 34: 274–285.
10. Schwimmer JB, Celedon MA, Lavine JE, Salem R, Campbell N, et al. (2009) Heritability of nonalcoholic fatty liver disease. Gastroenterology 136: 1585–1592.
11. Romeo S, Kozlitina J, Xing C, Pertsemlidis A, Cox D, et al. (2008) Genetic variation in PNPLA3 confers susceptibility to nonalcoholic fatty liver disease. Nat Genet 40: 1461–1465.
12. Speliotes EK, Yerges-Armstrong LM, Wu J, Hernaez R, Kim LJ, et al. (2011) Genome-wide association analysis identifies variants associated with nonalcoholic fatty liver disease that have distinct effects on metabolic traits. PLoS Genet 7: e1001324.
13. Iafrate AJ, Feuk L, Rivera MN, Listewnik ML, Donahoe PK, et al. (2004) Detection of large-scale variation in the human genome. Nat Genet 36: 949–951.
14. Sebat J, Lakshmi B, Troge J, Alexander J, Young J, et al. (2004) Large-scale copy number polymorphism in the human genome. Science 305: 525–528.
15. Redon R, Ishikawa S, Fitch KR, Feuk L, Perry GH, et al. (2006) Global variation in copy number in the human genome. Nature 444: 444–454.
16. Pinto D, Marshall C, Feuk L, Scherer SW (2007) Copy-number variation in control population cohorts. Hum Mol Genet 16 Spec No.2: R168–173.
17. Stone JL, O'Donovan MC, Gurling H, Kirov GK, Blackwood DH, et al. (2008) Rare chromosomal deletions and duplications increase risk of schizophrenia. Nature 455: 237–241.
18. Pinto D, Pagnamenta AT, Klei L, Anney R, Merico D, et al. (2010) Functional impact of global rare copy number variation in autism spectrum disorders. Nature 466: 368–372.
19. Lionel AC, Crosbie J, Barbosa N, Goodale T, Thiruvahindrapuram B, et al. (2011) Rare copy number variation discovery and cross-disorder comparisons identify risk genes for ADHD. Sci Transl Med 3: 95ra75.
20. Clarke AJ, Cooper DN (2010) GWAS: heritability missing in action? Eur J Hum Genet 18: 859–861.
21. Bae JS, Cheong HS, Kim JH, Park BL, Park TJ, et al. (2011) The genetic effect of copy number variations on the risk of type 2 diabetes in a Korean population. PLoS One 6: e19091.
22. Wheeler E, Huang N, Bochukova EG, Keogh JM, Lindsay S, et al. (2013) Genome-wide SNP and CNV analysis identifies common and low-frequency variants associated with severe early-onset obesity. Nat Genet 45: 513–517.
23. Brunt EM, Kleiner DE, Wilson LA, Belt P, Neuschwander-Tetri BA (2011) Nonalcoholic fatty liver disease (NAFLD) activity score and the histopathologic

Author Contributions

Conceived and designed the experiments: SMZ RM ZM. Performed the experiments: SMZ NSMR. Analyzed the data: SMZ RR SR. Contributed reagents/materials/analysis tools: SM WKC RCB ASM PLC. Wrote the paper: SMZ. Interpretation of data: SMZ RR AA DNC. Drafting the article or revising it critically for important intellectual content: SMZ DNC RM ZM.

diagnosis in NAFLD: distinct clinicopathologic meanings. Hepatology 53: 810–820.
24. Kleiner DE, Brunt EM, Van Natta M, Behling C, Contos MJ, et al. (2005) Design and validation of a histological scoring system for nonalcoholic fatty liver disease. Hepatology 41: 1313–1321.
25. Ruhl CE, Everhart JE (2005) Joint effects of body weight and alcohol on elevated serum alanine aminotransferase in the United States population. Clin Gastroenterol Hepatol 3: 1260–1268.
26. Sanyal AJ (2002) AGA technical review on nonalcoholic fatty liver disease. Gastroenterology 123: 1705–1725.
27. Venkatraman ES, Olshen AB (2007) A faster circular binary segmentation algorithm for the analysis of array CGH data. Bioinformatics 23: 657–663.
28. Nguyen DQ, Webber C, Hehir-Kwa J, Pfundt R, Veltman J, et al. (2008) Reduced purifying selection prevails over positive selection in human copy number variant evolution. Genome Res 18: 1711–1723.
29. Chang ML, Yeh CT, Chen JC, Huang CC, Lin SM, et al. (2008) Altered expression patterns of lipid metabolism genes in an animal model of HCV core-related, nonobese, modest hepatic steatosis. BMC Genomics 9: 109.
30. Nagashima S, Yagyu H, Takahashi N, Kurashina T, Takahashi M, et al. (2011) Depot-specific expression of lipolytic genes in human adipose tissues - association among CES1 expression, triglyceride lipase activity and adiposity. J Atheroscler Thromb 18: 190–199.
31. Caldwell S, Argo C (2010) The natural history of non-alcoholic fatty liver disease. Dig Dis 28: 162–168.
32. Herzog H (2003) Neuropeptide Y and energy homeostasis: insights from Y receptor knockout models. Eur J Pharmacol 480: 21–29.
33. Bloomfeld RS, Graham BG, Schiffman SS, Killenberg PG (1999) Alterations of chemosensory function in end-stage liver disease. Physiol Behav 66: 203–207.
34. Glusman G, Yanai I, Rubin I, Lancet D (2001) The complete human olfactory subgenome. Genome Res 11: 685–702.
35. Liu N, Crasto CJ, Ma M (2007) Integrated olfactory receptor and microarray gene expression databases. BMC Bioinformatics 8: 231.
36. Dong D, Jin K, Wu X, Zhong Y (2012) CRDB: database of chemosensory receptor gene families in vertebrate. PLoS One 7: e31540.
37. Quiroga AD, Li L, Trotzmuller M, Nelson R, Proctor SD, et al. (2012) Deficiency of carboxylesterase 1/esterase-x results in obesity, hepatic steatosis, and hyperlipidemia. Hepatology 56: 2188–2198.
38. Ashla AA, Hoshikawa Y, Tsuchiya H, Hashiguchi K, Enjoji M, et al. (2010) Genetic analysis of expression profile involved in retinoid metabolism in non-alcoholic fatty liver disease. Hepatol Res 40: 594–604.
39. Peters U, North KE, Sethupathy P, Buyske S, Haessler J, et al. (2013) A systematic mapping approach of 16q12.2/FTO and BMI in more than 20,000 African Americans narrows in on the underlying functional variation: results from the Population Architecture using Genomics and Epidemiology (PAGE) Study. PLoS Genet 9: e1003171.
40. Na K, Jeong SK, Lee MJ, Cho SY, Kim SA, et al. (2013) Human liver carboxylesterase 1 outperforms alpha-fetoprotein as biomarker to discriminate hepatocellular carcinoma from other liver diseases in Korean patients. Int J Cancer 133: 408–415.
41. Almon RR, Dubois DC, Sukumaran S, Wang X, Xue B, et al. (2012) Effects of high fat feeding on liver gene expression in diabetic goto-kakizaki rats. Gene Regul Syst Bio 6: 151–168.
42. Zender L, Xue W, Zuber J, Semighini CP, Krasnitz A, et al. (2008) An oncogenomics-based in vivo RNAi screen identifies tumor suppressors in liver cancer. Cell 135: 852–864.
43. Liang XT, Pan K, Chen MS, Li JJ, Wang H, et al. (2011) Decreased expression of XPO4 is associated with poor prognosis in hepatocellular carcinoma. J Gastroenterol Hepatol 26: 544–549.
44. Zhang H, Wei S, Ning S, Jie Y, Ru Y, et al. (2013) Evaluation of TGFbeta, XPO4, eIF5A2 and ANGPTL4 as biomarkers in HCC. Exp Ther Med 5: 119–127.
45. Lee S, Kim S (2007) Gene regulations in HBV-related liver cirrhosis closely correlate with disease severity. J Biochem Mol Biol 40: 814–824.

46. Lin S, Shi YC, Yulyaningsih E, Aljanova A, Zhang L, et al. (2009) Critical role of arcuate Y4 receptors and the melanocortin system in pancreatic polypeptide-induced reduction in food intake in mice. PLoS One 4: e8488.

47. Liu YL, Semjonous NM, Murphy KG, Ghatei MA, Bloom SR (2008) The effects of pancreatic polypeptide on locomotor activity and food intake in mice. Int J Obes (Lond) 32: 1712–1715.

48. Zhang L, Bijker MS, Herzog H (2011) The neuropeptide Y system: pathophysiological and therapeutic implications in obesity and cancer. Pharmacol Ther 131: 91–113.

49. Zain SM, Mohamed R, Mahadeva S, Cheah PL, Rampal S, et al. (2012) A multi-ethnic study of a PNPLA3 gene variant and its association with disease severity in non-alcoholic fatty liver disease. Hum Genet 131: 1145–1152.

50. Zain SM, Mohamed Z, Mahadeva S, Rampal S, Basu RC, et al. (2013) Susceptibility and gene interaction study of the angiotensin II type 1 receptor (AGTR1) gene polymorphisms with non-alcoholic fatty liver disease in a multi-ethnic population. PLoS One 8: e58538.

51. Royo F, Zabala A, Paz N, Acquadro F, Echevarria JJ, et al. (2013) Genome-wide analysis of DNA copy number changes in liver steatosis. Br J Med Med Res 3: 1773–1785.

52. Curtis C, Lynch AG, Dunning MJ, Spiteri I, Marioni JC, et al. (2009) The pitfalls of platform comparison: DNA copy number array technologies assessed. BMC Genomics 10: 588.

53. Rappaport N, Nativ N, Stelzer G, Twik M, Guan-Golan Y, et al. (2013) MalaCards: an integrated compendium for diseases and their annotation. Database (Oxford) 2013: bat018.

7

Conserved Noncoding Elements Follow Power-Law-Like Distributions in Several Genomes as a Result of Genome Dynamics

Dimitris Polychronopoulos[1,2], Diamantis Sellis[3], Yannis Almirantis[1]*

1 Institute of Biosciences and Applications, National Center for Scientific Research "Demokritos", Athens, Greece, **2** Department of Biochemistry and Molecular Biology, Faculty of Biology, National and Kapodistrian University of Athens, Athens, Greece, **3** Department of Biology, Stanford University, Stanford, California, United States of America

Abstract

Conserved, ultraconserved and other classes of constrained elements (collectively referred as CNEs here), identified by comparative genomics in a wide variety of genomes, are non-randomly distributed across chromosomes. These elements are defined using various degrees of conservation between organisms and several thresholds of minimal length. We here investigate the chromosomal distribution of CNEs by studying the statistical properties of distances between consecutive CNEs. We find widespread power-law-like distributions, i.e. linearity in double logarithmic scale, in the inter-CNE distances, a feature which is connected with fractality and self-similarity. Given that CNEs are often found to be spatially associated with genes, especially with those that regulate developmental processes, we verify by appropriate gene masking that a power-law-like pattern emerges irrespectively of whether elements found close or inside genes are excluded or not. An evolutionary model is put forward for the understanding of these findings that includes *segmental or whole genome duplication* events and *eliminations (loss)* of most of the duplicated CNEs. Simulations reproduce the main features of the observed size distributions. Power-law-like patterns in the genomic distributions of CNEs are in accordance with current knowledge about their evolutionary history in several genomes.

Editor: Christos A. Ouzounis, The Centre for Research and Technology, Hellas, Greece

Funding: Publication fees for this work were covered in the framework of "Target Identification for Disease Diagnosis and Treatment (DIAS)" project within General Secretariat for Research and Technology (GSRT) Development Proposals of Research Organisations-KRIPIS action, funded by Greece and the European Regional Development Fund of the European Union under the O.P. Competitiveness and Entrepreneurship, National Strategic Reference Framework (NSRF) 2007-2013. The funders had no role in study design, data collection and analysis, decision to publish, or preparation of the manuscript.

Competing Interests: The authors have declared that no competing interests exist.

* E-mail: yalmir@bio.demokritos.gr

Introduction

The sequencing and comparative analysis of many mammalian genomes has indicated that at least 5.5% of the human genome is under selective constraint; of that, 1.5% is estimated to code for proteins, 3.5% displays known regulatory functions, while for the function of the rest there is little or no information available [1]. One of the most interesting findings that have arisen from comparative analysis among mammalian genomes is the discovery of hundreds of ultraconserved elements (UCEs) of more than 200 bp in length that show absolute conservation among human, mouse and rat genomes [2]. One out of four of UCEs overlaps known protein-coding genes. However, such a high degree of conservation (100%) is not expected even in exons, due to the degeneration of the genetic code. Since the discovery of UCEs, there have been efforts to identify conserved elements based on lower thresholds of sequence similarity over whole genome alignments of two or more species. Several thresholds of minimal length of conserved sequence have been used as well as the exclusion of elements inside protein-coding genes [3,4]. Through-out this article, we use the term CNE(s) for Conserved Noncoding Elements to describe all such elements despite their specific characterization as UCEs, UCNEs, HCNEs, CNGs, CNEs etc in the related literature. We here use the specific name only when we refer to the corresponding class of elements.

CNEs are not a vertebrate innovation but are also found in invertebrate and plant genomes [5–7]. The vertebrate, insect and nematode CNEs are not related to each other at the sequence level [6,8,9]. However, a recent study has identified two elements conserved between vertebrates and invertebrates [10] and it is possible that more will be identified in the near future with the advent on new sequencing methodologies and the increasing availability of sequenced genomes. In the relatively recent evolution of vertebrates, the mean length and conservation of CNEs found therein are the highest observed [11] regarding all taxonomic groups, while the conjectured roles they have acquired are particularly important [12].

CNEs are often clustered in the vicinity of genes involved in transcriptional regulation and/or development [13–15]. Using microarray analysis it was reported that a large fraction of noncoding UCEs have tissue-specific expression levels and are deregulated in human cancer [16,17]. When such elements are located in the vicinity of genes, these genes are invariably found in conserved synteny in all vertebrates, possibly due to the fact that the surrounding genomic environment of a regulation-dependent gene has to be maintained intact [18]. Gene deserts are usually enriched in CNEs [19,20] while, in mammalian genomes, the vast majority of those elements are found at long distances from the closest genes, exceeding in some cases 2 Mb, which is the limit for any known cis regulatory element [18,21,22]. Little is known or could be speculated about what those distant CNEs actually do. Published studies tend to support the idea that they might be an essential part of Gene Regulatory Blocks (GRBs) and that they could function in a cooperative way alongside with their target genes [4,23–25].

There is a corpus of literature suggesting that CNEs are selectively constrained and not mutational cold spots [26,27]. Studies showing that CNEs might act as transcriptional regulators, e.g. enhancers or insulators, have been published [28,29], although *in vivo* experiments of elimination of some of these elements yield viable mice [30]. A CNE from one species may drive expression in another species as shown by transgenics experiments [31,32], although this is not a demonstration of whether a particular CNE drives conserved expression. Experiments that have been performed in order to test the same CNE in multiple species or the same CNE from multiple species in one species, have shown that although CNEs can be identified using sequence conservation criteria, the expression patterns they drive across species may show little conservation [33–36]. Another aspect not directly addressed herein is the existence of paralogous CNEs in vertebrate genomes. These are believed to often remain conserved having the possibility of controlling overlapping expression patterns of their adjacent paralogous protein-coding genes [37]. Paralogous CNEs are involved in the gene expression pattern of the vertebrate brain [38].

The alternative hypothesis that CNEs are horizontally transferred between lineages and accumulate during the course of long-term evolution has also been expressed [39]. Furthermore, a study has suggested that CNEs might act as Matrix-Attachment Regions (MARs) by serving as sequences that regulate the architecture of chromatin through specific binding of particular proteins [40]. An association between CNEs and phenotypic variation and disease has also been reported [41–43].

Long-range correlations were reported in the nucleotide sequence of the non-protein-coding part of eukaryotic genome soon after such large sequences became available [44–46]. In previous works, we explored the large-scale features of several classes of genomic elements, such as protein coding segments [47,48] and transposable elements [49,50], by studying the size distribution of inter-exon and inter-repeat distances. In most cases we found power-law-like size distributions, fractality and self-similarity, often spanning several orders of magnitude. We here apply the same methodology for the analysis of inter-CNE distances. We use published datasets, which are characterized by different degrees of evolutionary conservation, identified in a wide variety of organisms spanning vertebrates and invertebrates. We detect power-law-like inter-CNE size distributions in most cases studied. A previous study from Salerno *et al.* [51] reported the existence of a power-law distribution in the length of "perfectly conserved" sequence from mouse/human whole-genome intersection and alignment. The work we present here focuses on the *distances* of consecutive CNEs (inter-CNE spacers) for which we also propose an explanatory model. The model that we propose cannot apply to the length distribution of CNEs themselves, thus the finding of Salerno *et al.* appears to be the expression of an independent phenomenon.

Given the aforementioned detection of long-range correlations in the nucleotide juxtaposition in non-constrained sequences of the eukaryotic genome, simple molecular dynamics have been used in attempts to explain this emergent pattern. A simple expansion -

modification system is shown to generate long-range correlations through the interplay of symbol duplication and symbol elimination events [52]. More recent findings on strand slippage during replication combined with point mutations shed light into homonucleotide tracts and microsatellites' evolution and may be a realistic implementation of the above model to genome dynamics (see Athanasopoulou *et al.* [48], where a short discussion about evolutionary scenarios generating long-range correlations is included). Here we implement a model (initially proposed in [47]) for the generation of the observed power-law-like distribution pattern of distances between evolutionary constrained genomic elements in general (protein-coding segments and CNEs). This evolutionary scenario is based on an earlier model accounting for the explanation of power-law size distributions appearing in aggregative growth of particles in physicochemical systems [53]. This mechanism, as applied in genome evolution herein, mainly involves segmental duplication (including whole genome duplication events) and loss of most of the duplicated CNEs, alongside a moderated loss of non-duplicated CNEs in some cases.

Methods and Materials

Datasets

We systematically investigate the chromosomal distribution of various CNEs. We include in our analysis a phylogenetically wide collection of datasets, ranging from human to elephant shark and from vertebrates to invertebrates:

(i) 13,736 CNEs mapped on the human genome (hg18), of various lengths, that are identical over at least 100 bp in at least 3 of 5 placental mammals (human, mouse, rat, dog and cow) [20]. The whole set is named EU100+. Specific subsets are also considered for our purposes as follows (data kindly provided by J.S. Mattick, see also Table 1): (**ia**) 8,332 EU100+ elements that are not present in fish (Fugu). These appeared during tetrapod evolution (present in frog, chicken and/or mammals) and are named EU-FR. (**ib**) 5,404 elements from EU100+ set with orthologs in fish (ancient). These are named FR. (**ic**) 1,665 elements that are present in frog but not in fish (tetrapod speciation). These are named XT-FR. (**id**) 980 elements that are present in chicken but not in frog or fugu (amniote speciation). These are named GG-XT-FR. (**ie**) 600 elements that are not present in chicken, or frog, or fugu (mammalian speciation). These are named EU-GG-XT-FR.

(ii) 82,335 Mammalian CNEs (conserved within mammals but not found in chicken or fish) and 16,575 Amniotic CNEs (conserved in mammals and chicken but not found in fish) respectively, mapped on the human genome (hg17) [19].

(iii) 4,386 UCNEs (Ultraconserved Noncoding Elements, longer than 200bp) mapped on the human genome (hg19) that display sequence identity which is consistently greater or equal to 95% between human and chicken whole genome alignments [24].

(iv) 3,124 Human – Fugu conserved noncoding elements mapped on the human genome (hg17) with 70% identity and a score of match-mismatch up to 60 [54].

(v) 2,833 Human – Zebrafish CNEs mapped on the human genome (hg17) that display identity greater than 70% over at least 80 bp [32].

(vi) 4,782 Human – Elephant Shark CNEs mapped on the human genome (hg17), with identity ranging from 71% to 98% [55].

(vii) 4,519 PCNEs (Phylogenetically CNEs) mapped on the zebrafish genome (genome-build Ensembl 42) that are conserved across amphioxus, zebrafish, mouse and fugu [56]. These elements are unique due to the way of their identification, which is not biased by rearrangement and duplication. In addition to that, local similarity searches (versus whole genome alignments) in the genomic regions surrounding phylogenetically defined gene families have been employed in order to detect them.

(viii) 23,651 *D. melanogaster* – *D. pseudoobscura* (insect) CNEs of 50 bp or more that are 100% conserved between these two species, mapped on the *D. melanogaster* genome (dm1) [6].

(ix) 2,082 Nematode (worm) CNEs with mean identity of 96% between *C. elegans* and *C. briggsae* mapped on genome WS140 [5].

(x) 2,614 Noncoding elements marked by extreme human-mouse-rat constraint (mapped on hg17), a subset of which act as developmental enhancers [57].

For specific details about the used data sets and the subsequent treatment see File S1. In most cases, the suite of utilities BEDTools has been used for the computational analysis [58]

Gene and CDS masking

We proceed to a complete masking of the regions characterized as genic in the human genome (hg17 and hg18). In addition, we mask flanks surrounding every gene: 5 kb at the 5' end and 2 kb at the 3' end, in order to exclude cis-regulatory elements the localization of which may be principally determined by the positioning of the regulated gene. The region located upstream of transcription start sites is usually particularly enriched in such regulatory sequences. Extended flanks of 10 kb and 100 kb have also been masked in a similar manner (see Results section). We use custom scripts and BEDTools in order to perform the masking. In the case of *D. melanogaster*, when we refer to masked CNEs of insect origin, we refer to elements that do not overlap exonic sequences and splice sites, as adopted from the supplementary material of Glazov *et al.* [6]. For masked genes' genomic coordinate data (file format and availability) see in the File S1.

We do not proceed to the masking of other genomic components, such as transposable elements (TEs), for which there are indications that they do follow power-law-like distributions, because there is no evidence about CNE – TE functional interaction or systematic co-localization. Only a tiny proportion of TEs is reported to have been exapted to the role of a CNE, but they are too few to influence and reshape the whole CNE distribution [59].

Size distributions

Suppose there is a large collection of n objects (in our case spacers between CNEs), each characterized by its length S. In typically random such collections (like runs of heads in a coin tossing experiment) we can approximate the distribution of sizes with an exponential distribution. Let p(S) the probability of a spacer having length between S-s/2 and S+s/2 (where s is the size of the bin width) and N*(S) the number of spacers:

Table 1. Summary characterization of genomes for several datasets: Power-law-like distributions of inter-CNE distances at chromosomal scale.

Dataset	Class of CNEs	Unmasked		Masked		Reference genome
		Average Extent (avg E)	Average Extent of five 'best' chr. (avg E-5)	Average Extent (avg E)	Average Extent of five 'best' chr. (avg E-5)	
i	EU100+	2	2.38	2.48	2.98	hg18
ia	EU-FR	1.97	2.46			hg18
ib	FR	2.2	2.72			hg18
ic	XT-FR	2.32	2.32			hg18
id	GG-XT-FR	2.14	2.14			hg18
ie	EU-GG-XT-FR	1.96	1.96			hg18
iia	Mammalian	1.49	1.9	1.59	2.04	hg17
iib	Amniotic	2.2	2.86	2.25	2.91	hg17
iii	Human/Chicken	2.35	2.63			hg19
iv	Human/Fugu	2.78	3.26			hg17
v	Human/Zebrafish	2.46	2.98	2.31	2.31	hg17
vi	Human/El. shark	2.36	2.69	2.42	2.68	hg17
vii	*D. rerio* PCNEs	2.43	3.01			#
viii	Insect CNEs	1.23	1.23	1.42	1.42	dm1
ix	Worm CNEs	1.7	1.7			WS140
x	Human/Rodents	2.15	2.43			hg17

Propensity for the formation of power-law-like size distributions of the inter-CNE distances as quantified by the extent (E) of linearity in log-log scale. Average values of E for all chromosomes (avg E) and average values of E for 5 chromosomes with the largest E (avg E-5) in each genome are presented. Gene-masked genomes are also included when available (for details see in the text).
#: genome-build Ensembl 42 (zebrafish).

$$\mathbf{N^*(S) = np(S) \propto e^{-\alpha S}} \quad \alpha > 0 \tag{1}$$

When scale-free clustering appears, long-range correlations extend to several length scales (ideally, in our case for the whole examined genomic length) and the spacers' size distributions follow a so-called power-law, which corresponds to a linear graph in a double logarithmic scale:

$$\mathbf{N^*(S) = np(S) \propto S^{-\zeta} = S^{-1-\mu}} \quad \mu > 0 \tag{2}$$

In this article we use the "cumulative size distribution", more precisely: the complementary cumulative distribution function [60], defined as follows:

$$\mathbf{P(S) = \int_S^\infty p(r)dr} \tag{3}$$

where p(r) is the original spacers' size distribution. The cumulative distribution has in general better statistical properties, as it forms smoother "tails", less affected by statistical fluctuations. Also, by definition it is independent of any binning choice: in a cumulative curve the value of P(S) for length S is not associated with the subset of spacers whose length falls in the same bin, as in the original distribution, but it corresponds to the number of all spacers longer than S. For reviews on power-law size distributions, their properties and alternative forms see e.g. [60–64].

The cumulative form of a power-law size distribution is again a power-law characterized by an exponent (slope) equal to that of the original distribution minus 1: if $\mathbf{p(r) \propto r^{-1-\mu}}$, then

$$\mathbf{N(S) = nP(S) \propto \int_S^\infty (r^{-1-\mu})dr \propto S^{-\mu}} \tag{4}$$

where N(S) is the number of spacers longer or equal to S. All the distribution plots presented in this article depict complementary cumulative size distributions of distances (spacers) between consecutive CNEs. The logarithms of these spacers' length (S) are shown in the horizontal axis and the logarithms of the number N(S) of all the spacers longer or equal to S are shown on the vertical axis.

The slope for a typical power-law does not exceed the value of $\mu = 2$, as $\mu < 2$ is a condition leading to a non-convergent standard deviation. In the power-law-like linearities reported in what follows the value of μ is always below 2. Power-law-like distributions in nature always have an upper and a lower cutoff, which determine the linear region in log-log scale, where self-similarity and fractality is observed. The extent of the linear region (E) measures the orders of magnitudes that the fractal geometry spans. Linearity has been determined by linear regression and the associated value of r^2 is in all cases higher than 0.97 and in more than 90% of the cases higher than 0.98.

Additionally to genomic spacers' size distributions, all figures also include a bundle of ten surrogate simulated size distributions (continuous lines) where markers representing CNEs are randomly positioned in a sequence. The number of the randomly positioned markers and the length of the simulated sequence are equal to the number of CNEs and the size of the considered chromosome respectively. The inclusion of these random (surrogate) data sets in the figures visualizes the difference between observed distribution patterns and the ones expected on the grounds of pure randomness. Note that, whenever gene-masking methodology is applied, corresponding surrogates are being made that exclude the masked space from the random positioning of markers.

Simulations using the genomic duplications – CNE loss model

Simulations using an ample choice of parameter values reproduce the observed genomic distributions. In the last figure, we show characteristic cases, while in the appendix of Plot S1, some more examples are also included. Initially, 1000 markers (representing CNEs) are randomly inserted in a sequence 2 Mbp long. Part (a) of the last figure shows snapshots of the emerging power-law-like pattern as it develops through time. Complementary cumulative size distributions of distances (spacers) between consecutive CNEs are computed every 50 segmental duplication events. Each segmental duplication (SD) event involves a region with length sampled from a uniform distribution with maximum the 5% of the actual length of the simulated sequence. In all these simulations, after each SD event, a number of CNEs equal to 90% of the number of the duplicated CNEs are eliminated (denoted as: fr = 0.9). In the part (b) of last figure, three distribution curves are presented produced after numerical simulations where the fraction fr takes the values 0.8, 0.9 and 1. In part (c) of the same figure, three distribution curves are presented again. In these simulations the fraction fr remains constant and equal to 0.9, while, in two of them, additional eliminations of CNEs are allowed, one and two after each event of segmental duplication respectively.

Results

Occurrence of power-law-like size distribution between inter-CNEs' distances

The main finding of this study is the widespread occurrence of power-law-like size distribution of the distances between consecutive CNEs. In our analysis we include CNE datasets from various taxonomic groups and also compare CNE populations exapted at different evolutionary stages. The studied CNEs are mapped on different genomes (human, *D. melanogaster*, *C. elegans*, *D. rerio*).

In Figure 1 we present the size distributions of distances between consecutive CNEs in a double logarithmic plot in some typical cases. We also report the linear region E of the distribution, and the slope μ. The full set of plots is presented in Plot S1, while a complete quantitative description of our results is given in Table S1. In Table 1 we summarize the results per organism and report the average value of the linear extent E in log-log scale for all chromosomes (avg E) and for the five chromosomes with the largest E (avg E-5) (including only linear regressions with $r^2 > 0.97$, see in the Methods). The extent E captures the orders of magnitude that the power-law-like distribution spans. Throughout this work we use the quantity E for measuring the existence of a self-similar chromosomal geometry and for assessing the accordance of the observed genomic distributions with the evolutionary model we propose (see Discussion).

Power-law-like patterns are not only found in alignments of closely related genomes but are widespread. Elements identified from mammalian and amniotic whole genome alignments [19] were among the first non-coding constrained elements found in quantities allowing statistical analysis of their chromosomal distribution. The complete set includes 16,575 Amniotic and 82,335 Mammalian CNEs (see Datasets iia,b, File S1 & Table S1). We observe power-law-like patterns also in collections of CNEs

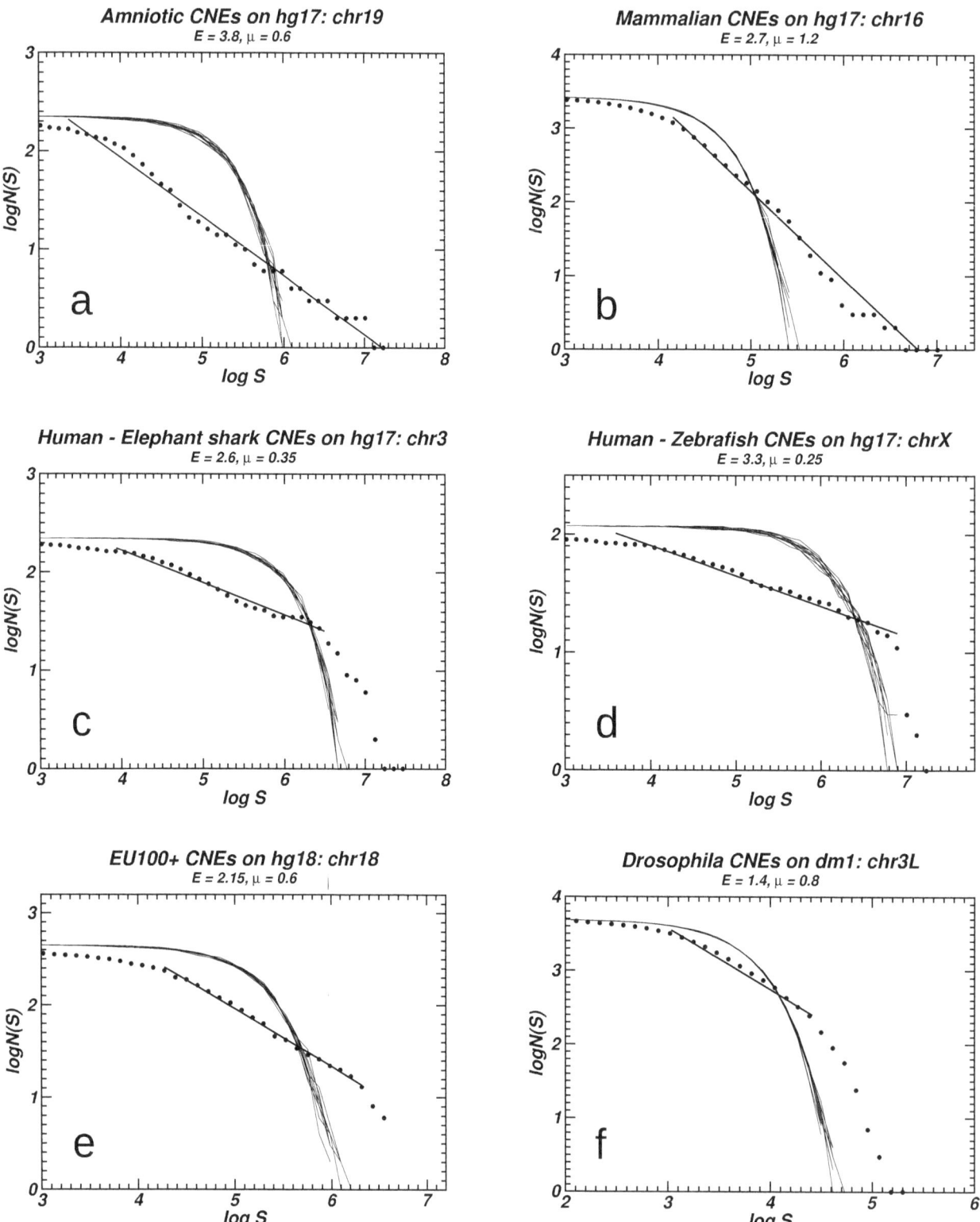

Figure 1. Examples of power-law-like size distributions. Twelve plots of inter-CNE spacers' cumulative size distributions in whole chromosomes. Genomic curves are accompanied in each plot by 10 curves of surrogate data (continuous lines), corresponding to randomly distributed markers. The linear segments are inferred by linear regression. Whenever we mention CNEs in the plots, we refer to the distances between consecutive CNEs.

derived from alignments including mammals along with teleosts, their last common ancestor dated ~450 MYA and cartilaginous fish (elephant shark) that diverged ~530 MYA [65]. The size distribution of inter-CNE spacers in invertebrate genomes, such as *D. melanogaster* and *C. elegans* also follow a similar pattern. It is known that vertebrate and invertebrate CNEs share similar sequence characteristics but are not identical [5], hence indicating in combination with our results that their distributions are shaped by common mechanisms.

Power-law-like distributions are typically characterized by overrepresentation of large spacers. Thus, one could expect that extended power-law-like linearity would be favored in scarce data sets. However, this is not the case. Based on our data, we deduce that the power-law-like size distribution of inter-CNEs's spacers is inherent to the studied system and is not dependent on the population sizes (instances of CNEs). This is evidenced by the fact that when we reduce the numbers of mammalian CNEs (~80,000), by random downsampling to similar numbers as the amniotic ones (~16,000), which are characterized by more extended power-law-like linearity, the extent of linearity is not increased. Instead, linearity in double-log scale disappears as a consequence of the alteration of the studied genomic landscape. Similarly, linearity disappears when we study the merged populations of amniotic and mammalian CNEs. A description of this methodology and the related plots are included in the last section of Plot S1. Thus, we argue that our results are characteristic of each CNE class studied and are not dependent on CNE population sizes provided that the existing populations of constrained elements are sufficient for statistical analysis.

The observed distribution of CNEs is not a mere consequence of the localization of genes in the same chromosome

Power-law-like distributions are found in the chromosomal distribution of protein-coding segments [47]. As it is known from the literature, CNEs are somehow spatially associated with genes coding for transcription factors and developmental regulators (also known as trans-dev genes) [13,15,18]. To rule out the possibility that the observed CNE distributions are a consequence of power-law-like patterns followed by inter-genic distance distributions, we mask all protein coding genes and extended flanking regions, where usually most of the known regulatory elements are located. By masking, we mean excluding all the elements that fall within genes and flanking regions and not removing the genes themselves, as the latter would alter the inter-CNE distance size distribution. Linearity in log-log plots is not only preserved but in most cases improved, as shown by the increase of the linear region extent. This shows that even if we exclude from our study the CNEs that might be bound to be close to genes (thus following their distribution), the remaining CNEs still follow a power-law-like chromosomal distribution. Our principal aim here is to show that the dynamics creating the power-law-like pattern is not a mere consequence of the genic distribution, although the two distributions are expected to influence one another, a fact which is not taken into account in our simple model. Examples of such plots are given in Figure 2. The full set of these plots and the related quantitative description are also included in Plot S1, Plot S2 and Table S1. In Table 1 the results concerning "gene-masked" chromosomes per organism are also given for a direct comparison.

We choose to perform the masking methodology in the human genome for the most abundant sets of CNEs [19,20] as well as for the most ancient elements conserved between human and zebrafish [32] or elephant shark [55]; datasets (**iia,b**), (**i**), (**v**) and (**vi**) respectively. In all cases studied, we observe power-law-

like size distributions of inter-CNEs' spacers that are extended over several orders of magnitude (see Table S1 & Table 1). A similar methodology is applied to the genome of *D. melanogaster*; dataset (**viii**). The possible functions of many CNEs (individually or in blocks) through their interactions with specific genes within the nucleus, by means of chromatin looping and other conformations, has recently received a direct experimental verification through the work of Viturawong *et al.* [66]. These authors have demonstrated, in a collection of 193 UltraConserved Elements, the frequent cis action of (distant to a gene) CNEs through chromatin looping. The scope of our gene masking applied herein is not to exclude CNEs acting as distant regulatory elements through such a mechanism. Thus, we have chosen to present a moderated (5 kb upstream of the 5′ end and 2 kb downstream of the 3′ end) gene-flank masking, in order to only exclude elements, which are probably limited to act as close (e.g. promoter-like) regulators and consequently may be spatially linked to nearby genes. To further validate our claim we also performed gene-masking with extensive flanks (10 kb and 100 kb) in the EU dataset for six chromosomes (the five largest ones and chromosome 10, which is particularly abundant in CNEs). Linearity in log-log scale in the distributions of inter-CNE distances is still evident and extends at several length scales (see various statistics and plots in Table S1/sheets EU100+ _masked10/100 kb and Plot S2 correspondingly). Even in the case of 100 kb flanks, such linearities are preserved, despite the few CNEs left after masking at such a large scale.

Discussion

An evolutionary model reproducing the observed power-law-like distributions based on genomic (segmental or whole-genome) duplications and CNE loss

Segmental duplication events occurred continuously in the evolutionary past of virtually all eukaryotes [67–70]. At least 10% of the non-repetitive human genome consists of identifiable (i.e. relatively recent) segmental duplication events [71]. It is estimated that 50% of all genes in a genome are expected to duplicate, giving an "offspring" at least once on time scales of 35 to 350 million years [72]. Additionally, most extant taxa have experienced paleopolyploidy during their evolution (i.e. duplication of the whole genome and subsequent reduction to diploidy), see e.g. [73,74] and references therein. Segmental duplication and polyploidization generate copies of some or all the genes of an organism, but also of other functional genomic elements, such as CNEs. As all authors agree, see e.g. [72,74,75], the fate of most duplicated genes is that one copy is silenced, losing the ability to be transcribed, and then disintegrates progressively by random mutations, while it is also exposed to the possibility of excision due to recombination driven eliminations. The fate of duplicated CNEs is expected to be similar, therefore a duplicated CNE often can become superfluous and stop to be under purifying selection, being thus gradually decomposed and lost. The existence of another source of CNE loss can be supported by current findings of comparative genomics, as many of the CNEs found to be conserved between elephant shark and human are not recognized in the fugu genome (see next section for further discussion). In that case, not only duplicates of CNEs are lost but also the population of CNEs present in an ancestral organism is considerably reduced.

Occasional CNE loss of function and subsequent degradation, complete genome duplications and repeated segmental duplications alongside with other forms of genomic dynamics (e.g. insertions of transposable elements and of other parasite sequences) can be combined in an evolutionary model the propensity of which to generate power-law-like chromosomal distributions is

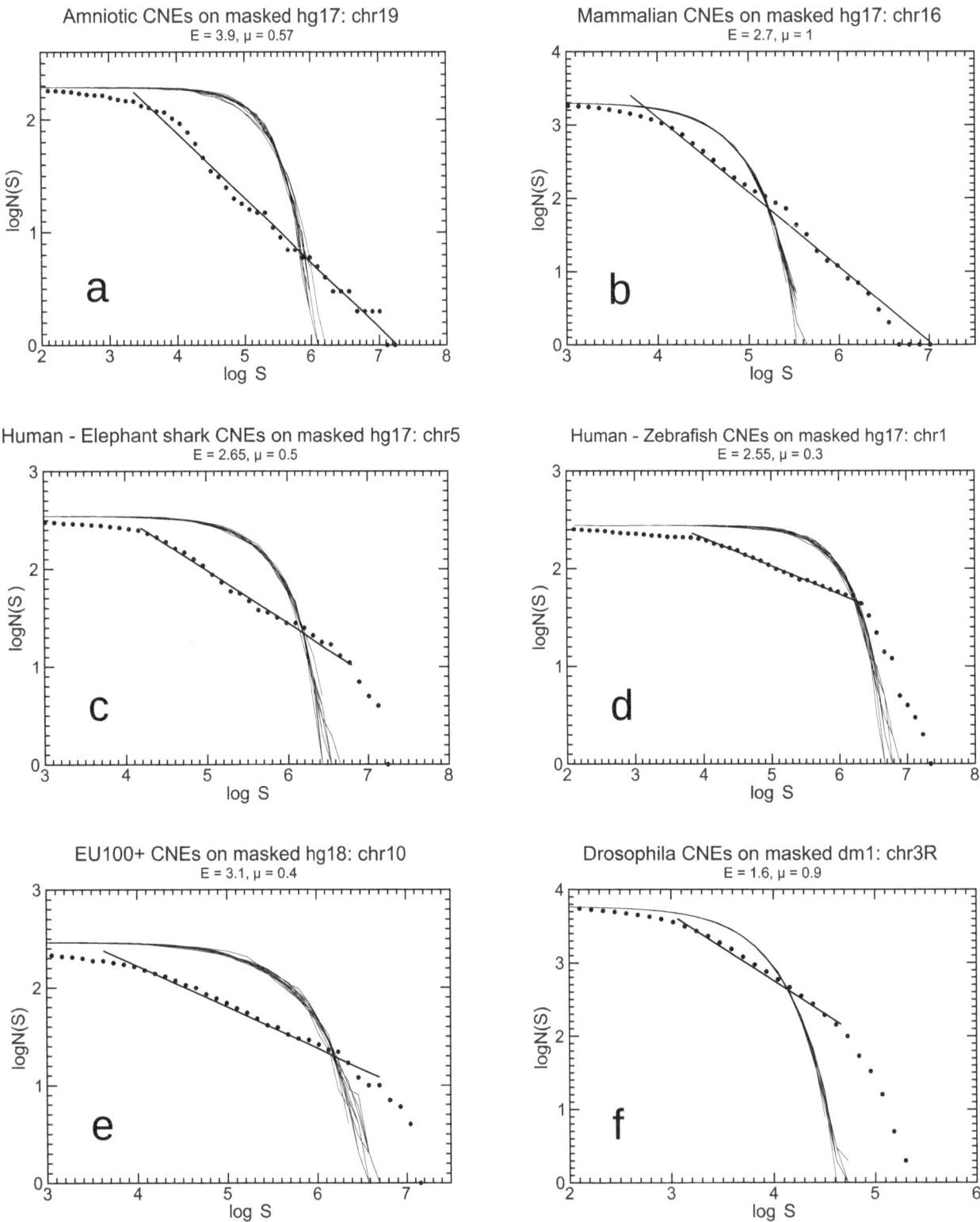

Figure 2. Examples of power-law-like size distributions after gene-masking. Six plots of inter-CNE spacers' cumulative size distributions in whole chromosomes after masking genes and flanks (for further details see in the text). Surrogate curves and regression as in figure 1. Whenever we mention CNEs in the plots, we refer to the distances between consecutive CNEs.

testable through computer simulations. We implement such a model and name it "genomic duplications – CNE loss model" (see link at the bottom of File S1 where we provide the code in Fortran and a detailed description of the model). The genomic events included are:

i   Segmental duplications of extended regions of chromosomes. This step may include as limiting case whole genome duplications, although not considered in the examples shown.

ii   Random eliminations of a number of CNEs which is lower or equal to the number of the duplicated ones.

iii   Occasionally, additional eliminations of non-duplicated CNEs.

iv   Insertions of sequences increasing the total chromosomal length (these could be transposable elements, retroviruses, microsatellite expansions etc).

v   Deletions of sequence stretches (which usually are under weak or no purifying selection).

The proposed evolutionary model reproduces power-law-like distributions of the sizes of inter-CNE distances, see Figure 3 and additional examples of simulations in the appendix of Plot S1. This property is proven numerically to be robust to quantitative modifications of all the involved types of molecular events. Only events *i* and *ii* are indispensable for the appearance of the power-law-like pattern. This dynamics has close parallels with the one described earlier for the explanation of an analogous distributional pattern followed by protein-coding genes [47]. In a completely different genomic framework (i.e. when non-conserved elements, e.g. interspersed repeats or microsatellites are being studied), event types *iii* and *iv* (i.e. insertions of TE families more recent than the studied one) are required instead of *i* and *ii* [49,50]. Events *iv* and *v* are numerically shown not to be required for the emergence of power-law patterns in computer simulations described herein. Inclusion of events of the type *iv* tests the robustness of the model, as for many organisms important parts of the genome represent repeat proliferation. Events of type *v* represent either deletion of sequence regions, usually due to unequal recombination or gradual shrinkage by a balance of indel events, favoring decrease of the sequence length. These types of events are of importance in genomes getting more compact (evolution occurred e.g. in the recent past of *Drosophila melanogaster* or in the case of *Takifugu rubripes*). Examples of simulations including all these types of genomic dynamics can be found in [47].

These evolutionary scenarios are based on an analytically solvable model introduced by Takayasu *et al.* [53] for the appearance of power-law size distributions in aggregative growth of particles in physicochemical systems. Notice that the model presented herein, conceived to describe the genomic dynamics of CNEs, is not analytically solvable and thus no universal exponents (slopes for the linear segment in log-log scale) may be reached. This is verified by all our computer simulations and is in accordance with the variety of slope values met in the study of genomic CNE distributions. Thus, our data deviate from the typical power-law not only because they always have the linearity in log-log scale truncated at a lower and an upper cut-off (in fact, this is a feature common to all cases of naturally occurring "power-laws") but mainly because they lack any universal exponent (slope). This is the principal reason why we call the pattern we have found "power-law-like" throughout this article. For a recent in depth view of the requirements for having a power-law, see Stumpf and Porter [63]. These authors state that these requirements include a statistically sound power-law (extended linearity in log-log scale

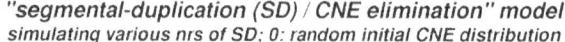

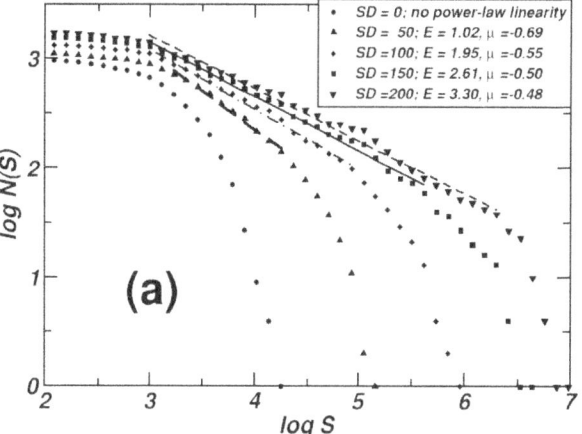

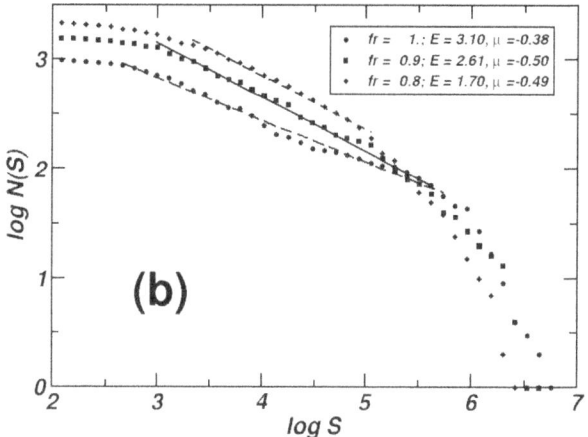

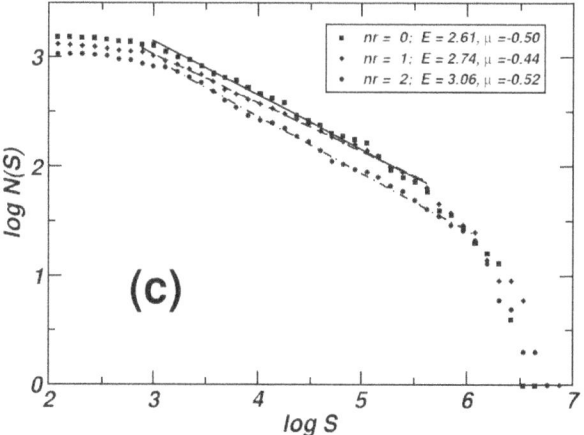

Figure 3. Simulations using the "genomic duplications – CNE loss model". The dependence of the extent of the linearity in log-log scale for the distances between consecutive simulated CNEs on several parameters is shown: **(a)** The number of Segmental Duplications (SD). **(b)** The fraction of the duplicated CNEs eliminated after each SD (fr). **(c)** The number of additional, non-duplicated, CNE eliminations. In **(a)** we

are able to follow the evolution of the emerging power-law-like pattern, as the four curves correspond to consecutive snapshots taken from the same numerical experiment. The curve depicted by squares (■) is common in all three plots, representing a simulation including 150 segmental duplications, where 90% (fr = 0.9) of the number of duplicated CNEs are lost. No additional eliminations are supposed here. Linear segments are computed by linear regression and in all cases $r^2 > 0.98$.

with indications of convergence to a universal exponent) and a concrete underlying theory to support it. In our case we clearly show that the log-log linearities we observe in our genomic data (often being quite extended) lack a universal exponent (slope). On the other hand, this deviation from universality is a characteristic feature shared between the genomic inter-CNE distributional patterns we describe herein and the simulations of the proposed evolutionary model. This feature, along with the common dependence on the evolutionary parameters shared between model and genomic distributions, corroborates the hypothesis that the evolutionary dynamics described by this model is at the origin of the observed genomic patterns.

Under a wide range of parameter combinations, our model reproduces the transient power-law-like distributions we observe in genomic data. In the simulation of Figure 3a & b, events of the types *i* and *ii* are only included: i.e. segmental duplications (SD) followed by CNE losses. In Figure 3a, a simulated chromosome is monitored using consecutive snapshots taken every 50 SD, starting from an initial (at "time zero") random distribution of markers representing genomic CNEs. We see that, gradually, a power-law-like linear region in log-log scale appears in the cumulative distributions of inter-CNE distances, as the ones observed in real chromosomes. This plot also shows the positive dependence of the observed extent of the linearity on the number of the occurred SD. In Figure 3b, the positive dependence of the extent of linearity on the number of the CNEs eliminated (lost after each segmental duplication) is shown. Here, the number of the eliminated CNEs is expressed as a fraction (fr) of the duplicated ones, because of the segmental duplication events. Finally, in Figure 3c additional eliminations of not duplicated CNEs are simulated (events of type *iii*), and the positive dependence of the extent of linearity on the number of non-duplicated lost CNEs is also shown. As we discuss in more detail later, this finding is compatible with the extended linearities found in the distributions of teleosts' CNEs, where important losses of ancestral CNEs are reported. Such ancient CNEs are absent in the teleost genome while retained in the tetrapod lineage.

Evolutionary origin and implications of CNE chromosomal distributions

In the Results section we have seen that the power-law-like distributional pattern reported in the present work is proper to CNEs, UCEs and other highly conserved elements, not being a mere consequence of genic spatial distributions (see Figure 2 and Table 1). The complete study of gene-masked chromosomes has been conducted in the human genome for the most abundant sets of CNEs, as well as for the most ancient elements. In five out of six examined cases (see Table 1, datasets. **i**, **iia**, **iib**, **vi**, **viii**), inspection of both average extent of linearity for all chromosomes (avg E) and for the five chromosomes with the more extended linearity (avg E-5) reveals that the extent of the linear region in log-log scale of the original distribution follows an increasing trend (or is at least preserved) after masking of the CNEs positioned next to, or inside genes (for details see "Methods"). In the remaining one

case (dataset **v**, Human/Zebrafish) well-shaped power-law-like distributions are still present after masking, with reduction of the length of the linearity. In Table S1 more information on the related statistics is given. The persistence of the linearity in log-log plots after gene-masking shows the independence of the two patterns. The fact that, in most cases, the average extent is not only preserved, but increased, further strengthens this conclusion. The frequent improvement of the power-law features when gene-masking applies, might indicate that, when the whole CNE chromosomal population is studied, the observed distributional features reflect *a superposition of two distinct dynamics*. Both include molecular events belonging to the same types but with different rates, corresponding to the distinctive evolutionary modalities of gene-uncorrelated CNEs and of genes (with which gene-proximal CNEs are spatially associated). This superposition of distributions with different features (the slope and the length scale of intervening sequences) is expected to reduce the observed linearity, because of a transformation of part of the superposed linear log-log distributions into a curved shape. Another evidence in support of the divergence between the distributional patterns followed by genes and CNEs stems from the observation reported by several research groups that UCEs and CNEs are abundant in gene deserts, see e.g. [19,20].

A link between CNEs and Segmental Duplication, and an additional insight about the fate of duplicated CNEs is provided in the work of Derti *et al.* [76]. These authors found that segmental duplications (SD) are depleted in UCEs (100% conserved elements). They explained this finding as a result of counter selection of duplications when they contain UCEs, probably due to dosage effects. If their result is valid for constrained elements independently of degree of conservation (denoted herein as CNEs in general), this implies that we should expect a low rate of duplication of CNEs, as is the case for genes under strong dosage dependence. In our proposed evolutionary scenario, we model these rare events that over long evolutionary time may have significantly contributed to the observed distributions. Note that in most of the cases of CNEs studied herein, the time from the divergence of the studied species is sufficient for accumulation of random mutations beyond recognition for a duplicated CNE, which is no longer under purifying selection. The finding of Derti *et al.* about counter selection of duplicated CNEs may drive to the inference that, when a SD containing a CNE is fixed in a population, we may have not only relaxed purifying selection on the second copy, but additionally, due to a deleterious dosage effect a fast rate of accumulation of mutations until all the functions of the duplicated CNE are lost.

In a study of the degree of conservation of distances between UCEs in vertebrates [77], real conservation of distances is found only between closely spaced elements, a range of distances which hardly contribute to the log-log linearity reported herein (see Figure 1 and Plot S1). The absence of considerable interspecies retention of distances between conserved elements is compatible with our claim that an aggregative model (like the one described in the previous section) is suitable for the explanation of the widespread occurrence of power-law type linearities in inter-CNE distance distributions.

Mattick and co-workers, in their article on ultraconserved elements (UCEs) in tetrapod genomes, observed a striking difference in UCE populations between tetrapod genomes, which are rich in UCEs, and fish genomes, where considerably lower UCE numbers have been found [20]. They proposed as the most parsimonious explanation that in the tetrapod lineage a massive exaptation of functional elements occurred, which would probably be required for the more complex morphology and different environmental challenges met by these organisms. On the other hand, the same authors mentioned that this finding might also be explained by the less probable assumption of a massive loss of such elements in the teleost fish genome. The subsequent sequencing of the first cartilaginous fish genome (elephant shark) led to the verification of this latter scenario, as Lee *et al.* found that the jawed vertebrate ancestor had an important number of UCEs which have been eliminated (diverged beyond recognition) at great extent in teleosts, while retained in tetrapods [78]. This finding fits well with our observation that the three globally best scores in linearity extent, as deduced by inspection of Table 1 (datasets **iv**, **v**, **vii**), are all cases of alignments including teleost fish genomes. Additionally, Wang *et al.* directly correlates the observed extended eliminations of UCEs in the teleost fish with the whole genome duplication that had occurred in the ray-fish lineage [79]. Note the significance of duplications (both whole genome and segmental ones) for our proposed model. Their role is twofold, as they make possible the subsequent elimination of duplicated CNEs due to redundancy, and simultaneously provide the necessary sequence extension for the formation of lengthy inter-CNE spacers (see previous subsection).

Evidence for extensive clustering of UCNEs in the vertebrate genome, in groups which are related to the regulation of one gene each, has been presented in a recent work [24]. When loss of duplicated UCNEs occurred, the retention of UCNEs is reported to be far from the expected on the basis of a random retention model. A "winner-takes-all retention pattern" applies, i.e. one gene retains many UCNEs whereas the other paralog loses all of them more often than expected on the grounds of pure chance. An over-representation of survived fish genes, which have lost all their ancestral UCNEs, has been found. This finding is in line with our observation of extended power-law-like linearity observed in the distributional pattern emerging in alignments including teleost fish genomes. Therein, the relatively frequent eliminations of whole UCNE clusters promote the appearance of large inter – UCNEs spacers, which contribute to the long tails and thus to the formation of extended linearity in log-log scale for the corresponding distance distributions [64]. Gene-centered functional clustering described by Bucher and co-workers of CNEs seems to represent one distinct component in the spatial distribution of CNEs. This clustering extends at length scales of the order of single gene neighborhoods. The findings presented herein deals with a broader length scale, the distribution of CNEs at the chromosomal level, resulting in a variety of CNE-rich and CNE-poor domains due to the described aggregative dynamics. These domains follow a fractal-like pattern, as witnessed by the power-law-like inter-CNE distance distributions. The model we propose here consists a better null model than the random positioning of CNEs at a chromosomal level, which then, has to converge with local, gene-specific organization trends. However, the state of our knowledge about the roles of CNEs and their quantitative interactions with genes is currently limited. Further research on evolutionary dynamics and functional roles of CNEs is required in order to better understand the interweaving of the two distributional trends.

The comparison of the extent of power-law-like linearity between Amniotic and Mammalian data sets of Kim and Prichard [19] is also in accordance with the proposed model (Table 1, datasets **iia**, **iib**). We see that the older in evolutionary time Amniotic CNE collection forms the most extended linear segments, as predicted by the hypotheses of the aggregation-elimination model, where, the more the system is exposed (in evolutionary time) to the CNE elimination – sequences insertion dynamics, the more extended the linearity in log-log scale is

expected to get. The same trend is clearly present in four data sets extracted from another study [20]. These datasets (Table 1, **ib** - **ie**) consist of elements exapted in consecutive evolutionary periods: tetrapod, amniote and mammalian speciations. We observe that the mean extents of linearity for the "5 best" chromosomes strictly follow the expected increasing order (from the more recent to the older), while the same holds true when we examine the mean extents for complete chromosomal sets with only one inversion found (between **ic** and **ib**).

In a very interesting work, Martinez-Mekler *et al.* [80] have shown that a broad range of systems consist of elements which, when plotted in rank *vs.* frequency or size diagrams, at semi-logarithmic scale, fit closely to a functional form including two fitting parameters. We have not further elaborated on the relation of the fitness of our data to the rank-ordering distribution approach. The range of application of this approach is quite large, and as these authors suggest, there must be an underlying explanation, possibly of a statistical nature [80]. On the other hand, our principal aim is to focus on the specific events of molecular dynamics origin, which may have caused the linearity in double logarithmic scale, and the evolutionary model proposed herein serves this purpose. A question that remains still open is the molecular dynamics impact of the fitting parameters of the functional form proposed by Martinez-Mekler *et al.*, which however lies beyond the scope of the present work. This task might be more straightforward in the case of the genomic evolution of non-constrained elements [49,50], where the evolutionary modeling is simpler.

In what concerns not *the distances between* functional genomic localizations, but *the sizes of the localizations themselves*, we have to mention again here that a power-law size distribution of "perfectly conserved" sequences between human and mouse (repeat-masked) genomes have been observed [51]. A related finding is that the size distribution of "conserved blocks" between *Drosophila melanogaster* and *Drosophila virilis* genomes fits well to a lognormal distribution [81]. Note that this distribution also presents linear regions in log-log scale [64]. A power-law-like distribution has been reported for 3' untranslated regions earlier [82]. These findings, concerning the sizes of functional genomic sequence stretches, have to be the result of the action of mechanisms entirely different than the evolutionary scenario applied herein, as they extend in very short length scales (tenths to hundreds of nucleotides) while any random-aggregation procedure of the type proposed herein is unsuitable for the formation of functional sequence elements. Aggregative length growth is more suitable for the shaping of large genomic regions (e.g. intervening sequences) with low conservation requirements (see also [47,53]). However, the interweaving of linearities in the form of power-laws at several orders of magnitude and for several functional elements or the distances between them, is probably related to the fractal globule structure reported for the whole human genome when it is in the form of the tightly packed chromatin, which is characterized by extended power-law-type size distributions of the distances between points of the chromosomal thread which come in close mutual contact due to the 3D chromatin folding (see Figure 4A in [83]).

Acknowledgments

The authors wish to thank the academic editor and the two anonymous reviewers for their useful comments. We would like to acknowledge Dr Christoforos Nikolaou and Giannis Tsiagkas for all of their help and support during the preparation of this study and Professors Stavros J. Hamodrakas and George C. Rodakis (both at the Faculty of Biology, University of Athens) for serving as academic advisors to D.P. We are also indebted to Dr John S. Mattick, Dr Slavica Dimitrieva and Dr Philipp Bucher for sharing unpublished datasets of CNEs.

Author Contributions

Conceived and designed the experiments: DP DS YA. Performed the experiments: DP. Analyzed the data: DP DS YA. Contributed reagents/materials/analysis tools: DP DS YA. Wrote the paper: DP DS YA.

References

1. Lindblad-Toh K, Garber M, Zuk O, Lin MF, Parker BJ, et al. (2011) A high-resolution map of human evolutionary constraint using 29 mammals. Nature 478: 476–482. doi:10.1038/nature10530.

2. Bejerano G, Pheasant M, Makunin I, Stephen S, Kent WJ, et al. (2004) Ultraconserved elements in the human genome. Science 304: 1321–1325. doi:10.1126/science.1098119.

3. Elgar G, Vavouri T (2008) Tuning in to the signals: noncoding sequence conservation in vertebrate genomes. Trends Genet 24: 344–352. doi:10.1016/j.tig.2008.04.005.

4. Harmston N, Baresic A, Lenhard B (2013) The mystery of extreme non-coding conservation. Philos Trans R Soc Lond B Biol Sci 368: 20130021. doi:10.1098/rstb.2013.0021.

5. Vavouri T, Walter K, Gilks WR, Lehner B, Elgar G (2007) Parallel evolution of conserved non-coding elements that target a common set of developmental regulatory genes from worms to humans. Genome Biol 8: R15. doi:10.1186/gb-2007-8-2-r15.

6. Glazov EA, Pheasant M, McGraw EA, Bejerano G, Mattick JS (2005) Ultraconserved elements in insect genomes: a highly conserved intronic sequence implicated in the control of homothorax mRNA splicing. Genome Res 15: 800–808. doi:10.1101/gr.3545105.

7. Lockton S, Gaut BS (2005) Plant conserved non-coding sequences and paralogue evolution. Trends Genet 21: 60–65. doi:10.1016/j.tig.2004.11.013.

8. Siepel A, Bejerano G, Pedersen JS, Hinrichs AS, Hou M, et al. (2005) Evolutionarily conserved elements in vertebrate, insect, worm, and yeast genomes. Genome Res 15: 1034–1050. doi:10.1101/gr.3715005.

9. Vavouri T, McEwen GK, Woolfe A, Gilks WR, Elgar G (2006) Defining a genomic radius for long-range enhancer action: duplicated conserved non-coding elements hold the key. Trends Genet 22: 5–10. doi:10.1016/j.tig.2005.10.005.

10. Clarke SL, VanderMeer JE, Wenger AM, Schaar BT, Ahituv N, et al. (2012) Human developmental enhancers conserved between deuterostomes and protostomes. PLoS Genet 8: e1002852. doi:10.1371/journal.pgen.1002852.

11. Retelska D, Beaudoing E, Notredame C, Jongeneel CV, Bucher P (2007) Vertebrate conserved non coding DNA regions have a high persistence length and a short persistence time. BMC Genomics 8: 398. doi:10.1186/1471-2164-8-398.

12. Mikkelsen TS, Wakefield MJ, Aken B, Amemiya CT, Chang JL, et al. (2007) Genome of the marsupial Monodelphis domestica reveals innovation in non-coding sequences. Nature 447: 167–177. doi:10.1038/nature05805.

13. Sandelin A, Bailey P, Bruce S, Engstrom PG, Klos JM, et al. (2004) Arrays of ultraconserved non-coding regions span the loci of key developmental genes in vertebrate genomes. BMC Genomics 5: 99. doi:10.1186/1471-2164-5-99.

14. Sanges R, Kalmar E, Claudiani P, D'Amato M, Muller F, et al. (2006) Shuffling of cis-regulatory elements is a pervasive feature of the vertebrate lineage. Genome Biol 7: R56. doi:10.1186/gb-2006-7-7-r56.

15. Sanges R, Hadzhiev Y, Gueroult-Bellone M, Roure A, Ferg M, et al. (2013) Highly conserved elements discovered in vertebrates are present in non-syntenic loci of tunicates, act as enhancers and can be transcribed during development. Nucleic Acids Res 41: 3600–3618. doi:10.1093/nar/gkt030.

16. Baira E, Greshock J, Coukos G, Zhang L (2008) Ultraconserved elements: genomics, function and disease. RNA Biol 5: 132–134.

17. Calin GA, Liu CG, Ferracin M, Hyslop T, Spizzo R, et al. (2007) Ultraconserved regions encoding ncRNAs are altered in human leukemias and carcinomas. Cancer Cell 12: 215–229. doi:10.1016/j.ccr.2007.07.027.

18. Woolfe A, Elgar G (2008) Chapter 12 Organization of Conserved Elements Near Key Developmental Regulators in Vertebrate Genomes. Adv Genet 61: 307–338. doi:10.1016/s0065-2660(07)00012-0.

19. Kim SY, Pritchard JK (2007) Adaptive evolution of conserved noncoding elements in mammals. PLoS Genet 3: 1572–1586. doi:10.1371/journal.pgen.0030147.

20. Stephen S, Pheasant M, Makunin I V, Mattick JS (2008) Large-scale appearance of ultraconserved elements in tetrapod genomes and slowdown of the molecular clock. Mol Biol Evol 25: 402–408. doi:10.1093/molbev/msm268.

21. Lettice LA, Heaney SJH, Purdie LA, Li L, de Beer P, et al. (2003) A long-range Shh enhancer regulates expression in the developing limb and fin and is associated with preaxial polydactyly. Hum Mol Genet 12: 1725–1735. doi:Doi 10.1093/Hmg/Ddg180.

22. Bishop CE, Whitworth DJ, Qin Y, Agoulnik AI, Agoulnik IU, et al. (2000) A transgenic insertion upstream of sox9 is associated with dominant XX sex reversal in the mouse. Nat Genet 26: 490–494. doi:10.1038/82652.

23. Kikuta H, Laplante M, Navratilova P, Komisarczuk AZ, Engstrom PG, et al. (2007) Genomic regulatory blocks encompass multiple neighboring genes and maintain conserved synteny in vertebrates. Genome Res 17: 545–555. doi:10.1101/gr.6086307.

24. Dimitrieva S, Bucher P (2012) Genomic context analysis reveals dense interaction network between vertebrate ultraconserved non-coding elements. Bioinformatics 28: i395–i401. doi:10.1093/bioinformatics/bts400.

25. Nelson AC, Wardle FC (2013) Conserved non-coding elements and cis regulation: actions speak louder than words. Development 140: 1385–1395. doi:10.1242/dev.084459.

26. Drake JA, Bird C, Nemesh J, Thomas DJ, Newton-Cheh C, et al. (2006) Conserved noncoding sequences are selectively constrained and not mutation cold spots. Nat Genet 38: 223–227. doi:10.1038/ng1710.

27. Sakuraba Y, Kimura T, Masuya H, Noguchi H, Sezutsu H, et al. (2008) Identification and characterization of new long conserved noncoding sequences in vertebrates. Mamm Genome 19: 703–712. doi:10.1007/s00335-008-9152-7.

28. Paparidis Z, Abbasi AA, Malik S, Goode DK, Callaway H, et al. (2007) Ultraconserved non-coding sequence element controls a subset of spatiotemporal GLI3 expression. Dev Growth Differ 49: 543–553. doi:10.1111/j.1440-169X.2007.00954.x.

29. Xie X, Mikkelsen TS, Gnirke A, Lindblad-Toh K, Kellis M, et al. (2007) Systematic discovery of regulatory motifs in conserved regions of the human genome, including thousands of CTCF insulator sites. Proc Natl Acad Sci U S A 104: 7145–7150. doi:10.1073/pnas.0701811104.

30. Ahituv N, Zhu Y, Visel A, Holt A, Afzal V, et al. (2007) Deletion of ultraconserved elements yields viable mice. PLoS Biol 5: e234. doi:10.1371/journal.pbio.0050234.

31. Poulin F, Nobrega MA, Plajzer-Frick I, Holt A, Afzal V, et al. (2005) In vivo characterization of a vertebrate ultraconserved enhancer. Genomics 85: 774–781. doi:10.1016/j.ygeno.2005.03.003.

32. Shin JT, Priest JR, Ovcharenko I, Ronco A, Moore RK, et al. (2005) Human-zebrafish non-coding conserved elements act in vivo to regulate transcription. Nucleic Acids Res 33: 5437–5445. doi:10.1093/nar/gki853.

33. McEwen GK, Goode DK, Parker HJ, Woolfe A, Callaway H, et al. (2009) Early evolution of conserved regulatory sequences associated with development in vertebrates. PLoS Genet 5: e1000762. doi:10.1371/journal.pgen.1000762.

34. Navratilova P, Fredman D, Hawkins TA, Turner K, Lenhard B, et al. (2009) Systematic human/zebrafish comparative identification of cis-regulatory activity around vertebrate developmental transcription factor genes. Dev Biol 327: 526–540. doi:10.1016/j.ydbio.2008.10.044.

35. Ritter DI, Li Q, Kostka D, Pollard KS, Guo S, et al. (2010) The importance of being cis: evolution of orthologous fish and mammalian enhancer activity. Mol Biol Evol 27: 2322–2332. doi:10.1093/molbev/msq128.

36. Sato S, Ikeda K, Shioi G, Nakao K, Yajima H, et al. (2012) Regulation of Six1 expression by evolutionarily conserved enhancers in tetrapods. Dev Biol 368: 95–108. doi:10.1016/j.ydbio.2012.05.023.

37. Matsunami M, Sumiyama K, Saitou N (2010) Evolution of conserved non-coding sequences within the vertebrate Hox clusters through the two-round whole genome duplications revealed by phylogenetic footprinting analysis. J Mol Evol 71: 427–436. doi:10.1007/s00239-010-9396-1.

38. Matsunami M, Saitou N (2013) Vertebrate paralogous conserved noncoding sequences may be related to gene expressions in brain. Genome Biol Evol 5: 140–150. doi:10.1093/gbe/evs128.

39. Hickey DA (2008) Highly similar noncoding genomic DNA sequences: ultraconserved, or merely widespread? Genome 51: 396–397. doi:10.1139/G08-011.

40. Glazko GV, Koonin EV, Rogozin IB, Shabalina SA (2003) A significant fraction of conserved noncoding DNA in human and mouse consists of predicted matrix attachment regions. Trends Genet 19: 119–124.

41. Lettice LA, Horikoshi T, Heaney SJH, van Baren MJ, van der Linde HC, et al. (2002) Disruption of a long-range cis-acting regulator for Shh causes preaxial polydactyly. Proc Natl Acad Sci U S A 99: 7548–7553. doi:DOI 10.1073/pnas.112212199.

42. Loots GG, Kneissel M, Keller H, Baptist M, Chang J, et al. (2005) Genomic deletion of a long-range bone enhancer misregulates sclerostin in Van Buchem disease. Genome Res 15: 928–935.

43. Sagai T, Hosoya M, Mizushina Y, Tamura M, Shiroishi T (2005) Elimination of a long-range cis-regulatory module causes complete loss of limb-specific Shh expression and truncation of the mouse limb. Development 132: 797–803. doi:Doi 10.1242/Dev.01613.

44. Li W, Kaneko K (1992) DNA correlations. Nature 360: 635–636. doi:10.1038/360635b0.

45. Peng CK, Buldyrev S V, Goldberger AL, Havlin S, Sciortino F, et al. (1992) Long-range correlations in nucleotide sequences. Nature 356: 168–170. doi:10.1038/356168a0.

46. Voss R (1992) Evolution of long-range fractal correlations and 1/f noise in DNA base sequences. Phys Rev Lett 68: 3805–3808.

47. Sellis D, Almirantis Y (2009) Power-laws in the genomic distribution of coding segments in several organisms: An evolutionary trace of segmental duplications, possible paleopolyploidy and gene loss. Gene 447: 18–28. doi:10.1016/j.gene.2009.04.028.

48. Athanasopoulou L, Athanasopoulos S, Karamanos K, Almirantis Y (2010) Scaling properties and fractality in the distribution of coding segments in eukaryotic genomes revealed through a block entropy approach. Phys Rev E Stat Nonlin Soft Matter Phys 82: 051917.

49. Klimopoulos A, Sellis D, Almirantis Y (2012) Widespread occurrence of power-law in inter-repeat distances shaped by genome dynamics. Gene. doi:10.1016/j.gene.2012.02.005.

50. Sellis D, Provata A, Almirantis Y (2007) Alu and LINE1 distributions in the human chromosomes: evidence of global genomic organization expressed in the form of power laws. Mol Biol Evol 24: 2385–2399. doi:10.1093/molbev/msm181.

51. Salerno W, Havlak P, Miller J (2006) Scale-invariant structure of strongly conserved sequence in genomic intersections and alignments. Proc Natl Acad Sci U S A 103: 13121–13125. doi:10.1073/pnas.0605735103.

52. Li W (1991) Expansion-modification systems: A model for spatial 1/f spectra. Phys Rev A 43: 5240–5260.

53. Takayasu H, Takayasu M, Provata A, Huber G (1991) Statistical properties of aggregation with injection. J Stat Phys 65: 725–745.

54. Pennacchio LA, Ahituv N, Moses AM, Prabhakar S, Nobrega MA, et al. (2006) In vivo enhancer analysis of human conserved non-coding sequences. Nature 444: 499–502. doi:10.1038/nature05295.

55. Venkatesh B, Kirkness EF, Loh YH, Halpern AL, Lee AP, et al. (2006) Ancient noncoding elements conserved in the human genome. Science 314: 1892. doi:10.1126/science.1130708.

56. Hufton AL, Mathia S, Braun H, Georgi U, Lehrach H, et al. (2009) Deeply conserved chordate noncoding sequences preserve genome synteny but do not drive gene duplicate retention. Genome Res 19: 2036–2051. doi:10.1101/gr.093237.109.

57. Visel A, Prabhakar S, Akiyama JA, Shoukry M, Lewis KD, et al. (2008) Ultraconservation identifies a small subset of extremely constrained developmental enhancers. Nat Genet 40: 158–160. doi:10.1038/ng.2007.55.

58. Quinlan AR, Hall IM (2010) BEDTools: a flexible suite of utilities for comparing genomic features. Bioinformatics 26: 841–842. doi:10.1093/bioinformatics/btq033.

59. Xie X, Kamal M, Lander ES (2006) A family of conserved noncoding elements derived from an ancient transposable element. Proc Natl Acad Sci U S A 103: 11659–11664. doi:10.1073/pnas.0604768103.

60. Clauset A, Shalizi CR, Newman MEJ (2009) Power-Law Distributions in Empirical Data. SIAM Rev 51: 661–703. doi:10.1137/070710111.

61. Adamic LA, Huberman BA (2002) Zipf's law and the Internet. Glottometrics 3: 143–150.

62. Li W (2002) Zipf's law everywhere. Glottometrics 5: 14–21.

63. Stumpf MPH, Porter MA (2012) Critical truths about power laws. Science 335: 665–666. doi:10.1126/science.1216142.

64. Newman MEJ (2005) Power laws, Pareto distributions and Zipf's law. Contemp Phys 46: 323–351. doi:10.1080/00107510500052444.

65. Kumar S, Hedges SB (1998) A molecular timescale for vertebrate evolution. Nature 392: 917–920. doi:10.1038/31927.

66. Viturawong T, Meissner F, Butter F, Mann M (2013) A DNA-Centric Protein Interaction Map of Ultraconserved Elements Reveals Contribution of Transcription Factor Binding Hubs to Conservation. Cell Rep 5: 531–545. doi:10.1016/j.celrep.2013.09.022.

67. De Grassi A, Lanave C, Saccone C (2008) Genome duplication and gene-family evolution: the case of three OXPHOS gene families. Gene 421: 1–6. doi:10.1016/j.gene.2008.05.011.

68. Kehrer-Sawatzki H, Cooper DN (2008) Molecular mechanisms of chromosomal rearrangement during primate evolution. Chromosome Res 16: 41–56. doi:10.1007/s10577-007-1207-1.

69. Kirsch S, Münch C, Jiang Z, Cheng Z, Chen L, et al. (2008) Evolutionary dynamics of segmental duplications from human Y-chromosomal euchromatin/heterochromatin transition regions. Genome Res 18: 1030–1042. doi:10.1101/gr.076711.108.

70. McLysaght A, Enright AJ, Skrabanek L, Wolfe KH (2000) Estimation of synteny conservation and genome compaction between pufferfish (Fugu) and human. Yeast 17: 22–36. doi:10.1002/(SICI)1097-0061(200004)17:1<22::AID-YEA5>3.0.CO;2-S.

71. Bailey JA, Gu Z, Clark RA, Reinert K, Samonte RV, et al. (2002) Recent segmental duplications in the human genome. Science 297: 1003–1007. doi:10.1126/science.1072047.

72. Lynch M (2000) The Evolutionary Fate and Consequences of Duplicate Genes. Science 290: 1151–1155. doi:10.1126/science.290.5494.1151.

73. Gibson TJ, Spring J (2000) Evidence in favour of ancient octaploidy in the vertebrate genome. Biochem Soc Trans 28: 259–264.

74. Sémon M, Wolfe KH (2007) Reciprocal gene loss between Tetraodon and zebrafish after whole genome duplication in their ancestor. Trends Genet 23: 108–112. doi:10.1016/j.tig.2007.01.003.

75. Kasahara M (2007) The 2R hypothesis: an update. Curr Opin Immunol 19: 547–552. doi:10.1016/j.coi.2007.07.009.

76. Derti A, Roth FP, Church GM, Wu C (2006) Mammalian ultraconserved elements are strongly depleted among segmental duplications and copy number variants. Nat Genet 38: 1216–1220. doi:10.1038/ng1888.

77. Sun H, Skogerbo G, Chen R (2006) Conserved distances between vertebrate highly conserved elements. Hum Mol Genet 15: 2911–2922. doi:10.1093/hmg/ddl232.

78. Lee AP, Kerk SY, Tan YY, Brenner S, Venkatesh B (2011) Ancient vertebrate conserved noncoding elements have been evolving rapidly in teleost fishes. Mol Biol Evol 28: 1205–1215. doi:10.1093/molbev/msq304.

79. Wang J, Lee AP, Kodzius R, Brenner S, Venkatesh B (2009) Large number of ultraconserved elements were already present in the jawed vertebrate ancestor. Mol Biol Evol 26: 487–490. doi:10.1093/molbev/msn278.

80. Martínez-Mekler G, Alvarez Martínez R, Beltrán del Río M, Mansilla R, Miramontes P, et al. (2009) Universality of rank-ordering distributions in the arts and sciences. PLoS One 4: e4791. doi:10.1371/journal.pone.0004791.

81. Clark AG (2001) The search for meaning in noncoding DNA. Genome Res 11: 1319–1320. doi:10.1101/gr.201601.

82. Martignetti L, Caselle M (2007) Universal power law behaviors in genomic sequences and evolutionary models. Phys Rev E 76: 021902. doi:10.1103/PhysRevE.76.021902.

83. Lieberman-Aiden E, van Berkum NL, Williams L, Imakaev M, Ragoczy T, et al. (2009) Comprehensive mapping of long-range interactions reveals folding principles of the human genome. Science 326: 289–293. doi:10.1126/science.1181369.

8

FANSe2: A Robust and Cost-Efficient Alignment Tool for Quantitative Next-Generation Sequencing Applications

Chuan-Le Xiao, Zhi-Biao Mai, Xin-Lei Lian, Jia-Yong Zhong, Jing-jie Jin, Qing-Yu He*, Gong Zhang*

Key Laboratory of Functional Protein Research of Guangdong Higher Education Institutes, Institute of Life and Health Engineering, College of Life Science and Technology, Jinan University, Guangzhou, China

Abstract

Correct and bias-free interpretation of the deep sequencing data is inevitably dependent on the complete mapping of all mappable reads to the reference sequence, especially for quantitative RNA-seq applications. Seed-based algorithms are generally slow but robust, while Burrows-Wheeler Transform (BWT) based algorithms are fast but less robust. To have both advantages, we developed an algorithm FANSe2 with iterative mapping strategy based on the statistics of real-world sequencing error distribution to substantially accelerate the mapping without compromising the accuracy. Its sensitivity and accuracy are higher than the BWT-based algorithms in the tests using both prokaryotic and eukaryotic sequencing datasets. The gene identification results of FANSe2 is experimentally validated, while the previous algorithms have false positives and false negatives. FANSe2 showed remarkably better consistency to the microarray than most other algorithms in terms of gene expression quantifications. We implemented a scalable and almost maintenance-free parallelization method that can utilize the computational power of multiple office computers, a novel feature not present in any other mainstream algorithm. With three normal office computers, we demonstrated that FANSe2 mapped an RNA-seq dataset generated from an entire Illunima HiSeq 2000 flowcell (8 lanes, 608 M reads) to masked human genome within 4.1 hours with higher sensitivity than Bowtie/Bowtie2. FANSe2 thus provides robust accuracy, full indel sensitivity, fast speed, versatile compatibility and economical computational utilization, making it a useful and practical tool for deep sequencing applications. FANSe2 is freely available at http://bioinformatics.jnu.edu.cn/software/fanse2/.

Editor: Zhang Zhang, Beijing Institute of Genomics, Chinese Academy of Sciences, China

Funding: This work was collectively supported by the National "973" Projects of China (2011CB910700), National Natural Science Foundation of China (31300649 and 31200612), the Key Project of Chinese Ministry of Education (212207), Guangdong Natural Science Foundation (S2013010013529), Foundation for Distinguished Young Talents in Higher Education of Guangdong, China (2012LYM_0026), the Fundamental Research Funds for the Central Universities (21612202, 21612459, 11610101, 21613343 and 21611201), and the Institutional Grant of Excellence of Jinan University, China (50625072). The funders had no role in study design, data collection and analysis, decision to publish, or preparation of the manuscript.

Competing Interests: The authors have declared that no competing interests exist.

* E-mail: zhanggong@jnu.edu.cn (GZ); tqyhe@jnu.edu.cn (Q-YH)

Introduction

Mapping (aligning) millions of next-generation sequencing (NGS) reads accurately to reference sequences is the basis of all deep sequencing applications that utilize reference genomes or transcriptomes, including variant analysis, gene expression and isoform analysis. Traditional alignment algorithms such as BLAST and BLAT could not process the massive amount of sequencing data in hours (reviewed in [1]). A series of early mapping algorithms such as SSAHA, MAQ and SOAP started to tackle this speed hindrance. These algorithms extended the basic idea of "seeding" (hash table indexing) from BLAST, which is simple in design and easy to implement, bringing the NGS technology into quantitative era (reviewed in [2,3]). The computational time of this type of algorithms is theoretically proportional to the size of reference sequence ([4] and reviewed in [2]). Therefore accurately mapping to large genomes is still time-consuming [5,6]. Another type of algorithms based on Burrows-Wheeler Trasnformation (BWT), e.g. Bowtie and BWA, takes the advantage of the suffix/prefix trie and thus reduces the computational complexity, being typically 5~20x faster than seed-based algorithms (reviewed in [2,7]). Such methods can map tens of millions of reads to human genome within one day on desktop workstations, thus promoting

the blowout of NGS applications. According to a statistics till the end of 2012, two among the top three cited mapping algorithms are of this type (Bowtie and BWA) (reviewed in [6]). In real-world benchmarks, although the sensitivity of earlier BWT-based algorithms like Bowtie and SOAP2 (<80%) is still to be improved when mapping DNA sequencing reads, the sensitivity of the upgraded Bowtie2 is almost the same as the traditional seed-based algorithms while being more than 20x faster [6].

However, deviations between reads and reference sequences set a great challenge of the sensitivity and speed to the mapping algorithms. Origins of the deviation include single nucleotide polymorphisms (reviewed in [8]), PCR amplification [9], base calling (reviewed in [10]) and sequencer errors [11]. When the mismatch rate exceeds 2% or the indel rate exceeds 0.5%, most algorithms lose their accuracy [12]. Due to the principle of BWT, this type of algorithms is less error-tolerant and thus usually less sensitive than seed-based algorithms at higher error rate (reviewed in [1,7]). For RNA-seq, the error rate is higher due to RNA editing [13], modifications [14] and nucleotide misincorporation in reverse transcription [15]. Indeed, in simulated tests, Bowtie and BWA remained 55%~75% accuracy at 4% error rate, while the seed-based algorithms SOAP and Novoalign maintain 80%~90%

accuracy [12]. This result coincides with the real-world test: even when adding the splice-mapped reads, BWT-based algorithms TopHat and SOAPsplice mapped 12~19% of reads less than seed-based algorithms [6]. These algorithms tend to unproportionally lose mappable reads of the medium to low abundance RNA, generating a significant bias in quantification [5]. Moreover, the accuracy of BWT-based algorithms was shown to be highly dependent on the dataset in various comparative tests, from very high [7] to moderate [6,12] to very low [16], in contrast to the seed-based algorithms. The inconsistency of quantitative results given by RNA-seq and microarray may reflect this bias and unrobustness [17–19].

It would be ideal to combine the speed of BWT and the robust accuracy of seed-based algorithms, especially for the cases with higher error rates like RNA-seq. To improve the robustness and indel detection of the BWT-based algorithm Bowtie, the upgraded Bowtie2 partially took the advantage of the seeding principle, and it truly exceeded Bowtie, BWA and SOAP2 [20]. However its accuracy and robustness are difficult to be theoretically estimated. To overcome this problem, we took the advantage of our previously developed FANSe algorithm, which is a seed-based algorithm with theoretical estimation of high accuracy and robustness (miss rate can be as low as 10^{-6}) [5], and further developed FANSe2 algorithm. FANSe2 can map a billion reads to human genome in hours using normal office computers without compromising the high and robust accuracy. We also tested this algorithm using real-world RNA-seq datasets and experimentally validated its results by RT-PCR and microarray.

Materials and Methods

Design of FANSe2

FANSe2 is an iterative and parallel seed-based read mapping algorithm with a simple design to ensure all advantages of FANSe and largely improve the speed and parallelization. The following major steps were implemented: (Figure 1).

Step 1. Segmentation of reference sequences. To reduce the memory consumption, large reference sequences like human genome are split to segments. Two adjacent segments are overlapped with maximum read-length. Each segment will be processed as a task package and assigned to a processor core.

Step 2. Initialize parallel computing environment. To avoid resource competition, FANSe2 parallelizes multiple processes via the industrial standard MPICH2 environment instead of multi-threading. Unlike FANSe that uses the 6- or 8-nt seeds, FANSe2 initially set the seed length as 14-nt.

Step 3. Each CPU core starts to process the assigned task package, mapping all reads to the reference sequence segment using the seed length based on the principle of FANSe. The final refinement of hotspots is performed by calculating Hamming distance (indel detection off) or by using accelerated Smith-Waterman method (indel detection on) [5].

Step 4. After all the task packages are processed, the mapping results are combined and the best mapping location of a read is written to the final result file.

Step 5. FANSe2 decreases the seed length by 2-nt and tries to map the unmapped reads using the shorter seed length (back to step 3). Iterative mapping process stops when the seed length reaches the minimum seed length or all the reads are mapped.

Datasets and reference sequences

To analyze the nucleotide error distribution in the sequencing datasets, we downloaded six datasets from DDBJ Sequence Read Archive (http://trace.ddbj.nig.ac.jp/dra/index_e.shtml), as listed

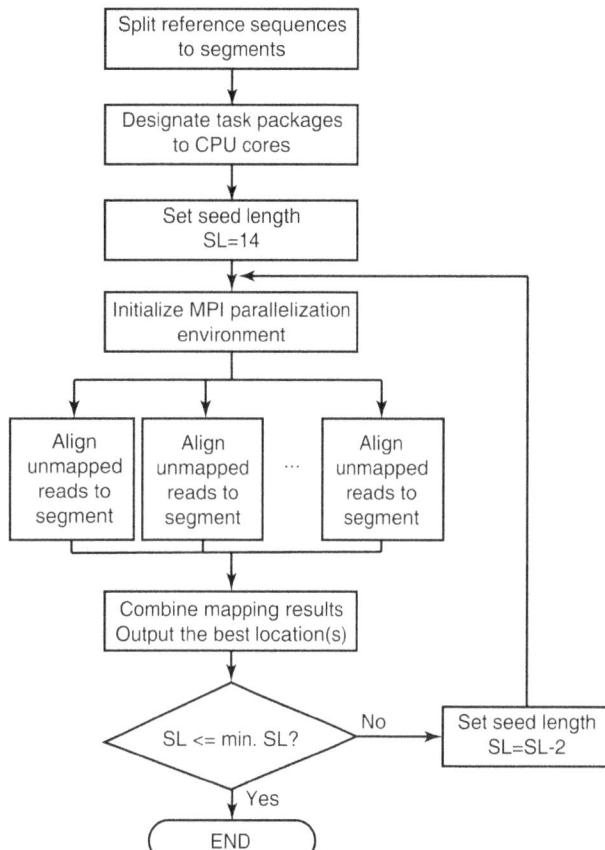

Figure 1. Flowchart of FANSe2. For details please refer to the Materials and Methods section. SL = seed length.

in Table S1 in File S1. Each read was truncated at the nucleotide, whose sequencing quality is lower than 20 in Phred scale. Reads shorter than 18 nt were discarded. The *E. coli* datasets were mapped to *E. coli* K-12 substrain MG1655 genome sequence (NCBI Reference Sequence: NC_000913.2). The yeast datasets were mapped to *S. cerevisiae* genome sequence sacCer3 (downloaded from UCSC genome browser, http://hgdownload.cse.ucsc. edu). FANSe was used to perform these mappings with the errors allowed as listed in Table S1 in File S1 and indel detection on.

The *E. coli* mRNA dataset reported previously was used to test the sensitivity and speed of FANSe2 [5]. The datasets of the whole Flowcell A (FCA) of Human Body Map 2.0 project, containing altogether 608 million 75-nt reads of human polyA$^+$ mRNA sequenced on an Illumina HiSeq-2000 sequencer, were used to test the parallel computing capacity of FANSe2. The human datasets were mapped to human genome sequence hg19/GRCh37 (downloaded from UCSC genome browser).

Simulated datasets with 2% and 4% error rate were generated from human chromosome 1 non-masked and masked genome sequence (hg19/GRCh37). Each datasets contained 500,000 reads, 75-nt long. These reads were generated from the non-masked regions. These datasets were mapped to human chromosome 1 non-masked and masked genome sequence, respectively. Reads with homopolymeric stretch or dinucleotide repeats longer than half of the read length were filtered out to avoid unnecessary and ambiguous alignment [21].

RT-PCR validation of mapping results for RNA-seq

We previously sequenced the total RNA of lung adenocarcinoma A549 cells and sequenced poly-A+ mRNA [22]. The reads were mapped to human mRNA reference sequence (RefSeq) for GRCh37/hg19 (downloaded from UCSC genome browser, accessed on Jan. 21, 2013) using both FANSe2 and Bowtie2. The parameters for FANSe2 was –E7 –I1 –S12, and the parameters for Bowtie2 was —very-sensitive. The mRNAs were quantified using standard rpkM method [3]. Genes with less than 10 reads mapped were considered as unreliable quantified genes and removed [23].

This total RNA sample was reverse transcribed with poly-dT primer using RevertAid Premium reverse transcriptase (Fermentas) and specific genes were amplified using specific primers. PCR was performed using gene-specific primers (Table S2 in File S1) and DreamTaq Green Mix (Fermentas) enzyme. We used the primers in the Whole Transcriptome qPCR Primers Database if available [24], otherwise we used the online tool NCBI PrimerBLAST (http://www.ncbi.nlm.nih.gov/tools/primer-blast/) to design gene-specific primers automatically. The PCR cycle was set as 95°C denaturing for 30 seconds, 59°C annealing for 30 seconds, and 72°C elongation for 30 seconds (amplicon size <500 bp) or 1.2 minutes (amplicon size 500~1200 bp). 35 PCR cycles were conducted for each reaction. The PCR products were resolved on 2.7~3% agarose gels and visualized by SybrGreen staining.

Comparison of NGS and microarray quantifications

The Affymetrix Rat Genome 230 2.0 microarray dataset and Illumina GAIIx RNA-seq dataset of the aristolochic acids treated rat liver sample AA_1 from a previous study [25] was downloaded from Gene Expression Omnibus (GEO) database (accession numbers: GSE5350 and GSE21210). The normalization of the microarray data was performed using RMA method as previously described [25,26]. In case that multiple probe sets were present for a gene, the probe set with the highest signal intensity was used for this gene [27]. The RNA-seq quantification result using Bowtie was downloaded from GSE21210. We mapped the original RNA-seq using FANSe2 to the reference transcriptome sequence RefSeq release 47 as mentioned in that study [25] with the options –L36 –E3 –S10. The splice variants were merged. RNA-seq quantifications were based on rpkM method [3].

Comparison of mapping programs

We compared FANSe2 with FANSe, Bowtie, Bowtie2, BWA, SHRiMP2 and Novoalign (for details please refer to Table S3 in File S1). The performance tests were carried out on quad-core Intel i5-3570K computers with 16 GB RAM. We used –n 3 —tryhard—best for Bowtie and –n 7 –o 1 for BWA. Unless specified, —very-sensitive option was used for all Bowtie2 tests. The memory consumption of these programs was recorded using either Task Manager (Windows) or System Monitor (Linux).

Sensitivity and correctness were defined previously [5]. In brief, a read that is truly originated from the reference sequence can have one of the following three outcomes after being processed by an algorithm: (i) mapped to its correct position (C); (ii) mapped to a wrong position (I); (iii) failed to be mapped to the reference genome (U). Sensitivity is defined as $\frac{C+I}{C+I+U}$, and the correctness is defined as $\frac{C}{C+I}$. Sensitivity can be calculated from a deep-sequencing dataset, which is proportional to the number of mapped reads. Correctness can be only evaluated using simulated random datasets.

Results

Iterative step-down acceleration strategy based on the real-world alignment error distribution

When mapping a read, FANSe takes seeds (6- or 8-nt long) from the read and searches for exact matches in the reference genome with a pre-built look-up table [5]. These exact matches are then merged into hotspots and then refined to determine the best mapping location. An n-nt long seed has in average $N/4^n$ exact matches in the genome (where N is the genome size), a large number for large genomes and $n = 8$, thus creating a heavy workload for the hotspot merging and refinement, especially when indel detection is enabled. Longer seeds decrease the number of exact matches exponentially and thus largely accelerate the mapping: 14-nt seed decreases the number of exact matches $4^{14-8} = 4096$ folds than 8-nt seeds. However, longer seeds are more likely to contain error and may lose the reads with higher number of mismatches, thus impairing the sensitivity. A read containing maximum f errors with a minimal read length of $n(f+1)$ can be reliably mapped to a genome when using n-nt seeds, indicating that a long read with a few errors may be still stably mapped with longer seeds (Figure 2A). For example, up to 5 errors are guaranteed to be detected in 75-nt reads using 14-nt seeds. To achieve theoretical miss rate less than 1%, 12-nt seeds are sufficient for 50-nt reads, whereas 14-nt seeds are more than enough for 100-nt reads (Figure 2B). Decreasing the seed length to 10-nt may reach the theoretical miss rate $10^{-4} \sim 10^{-8}$.

We then analyzed the actual error distribution in real-world datasets. We mapped six datasets including DNA-seq, mRNA-seq and miRNA-seq datasets obtained from various sequencing platforms using FANSe algorithm (Table S1 in File S1). Notably, a large fraction of the mappable reads contained very few errors, regardless in DNA-seq or RNA-seq datasets (Figure 2C). More than half of the mappable reads contain 0 or 1 error in most cases, and they can be reliably mapped with 14-nt seeds in much higher speed. Therefore, we implemented an iterative step-down strategy: long seeds (e.g. 14-nt or 12-nt) are used to map most reads with high speed, and the unmapped reads (a small fraction) are mapped in the next iteration with shorter seed. This iterative process terminates when the seed length reaches the limit set by the user (Figure 1).

We tested this strategy with the E. coli mRNA-seq dataset that was previously used in FANSe test [5]. Stepping down to 8-nt seed length, FANSe2 exported the same mapping result as FANSe at much faster speed using single CPU core when allowing 3 mismatches (Figure 2D and 2E). Indeed, most of the mappable reads were mapped in the initial iteration using 14-nt seeds. When stepped down to 12-nt seeds, FANSe2 mapped 8.26 M reads using 0.28 minutes in total. At this stage, the sensitivity is already higher than the widely-used Bowtie and Bowtie2 (7.93 M and 8.12 M reads, respectively, Figure 2D), while faster than Bowtie2 (1.13 minutes). Stepping further down to 8-nt stage may not be practically necessary, since this significantly increased the running time for three times, however only mapped 0.21 M more reads. Even down to 8-nt stage, the speed of FANSe2 is 3~21x faster than FANSe, Bowtie and BWA, only slightly slower than Bowtie2 (Figure 2E).

Memory consumption, speed and scalability when handling huge datasets

The memory consumption of FANSe2 is tunable by the user, because it is only relevant to the genome segment size: FANSe2 uses 1.2~1.7 GB memory for each activated CPU core when the reference sequences are split to 50~200 Mb segments (Figure 3A).

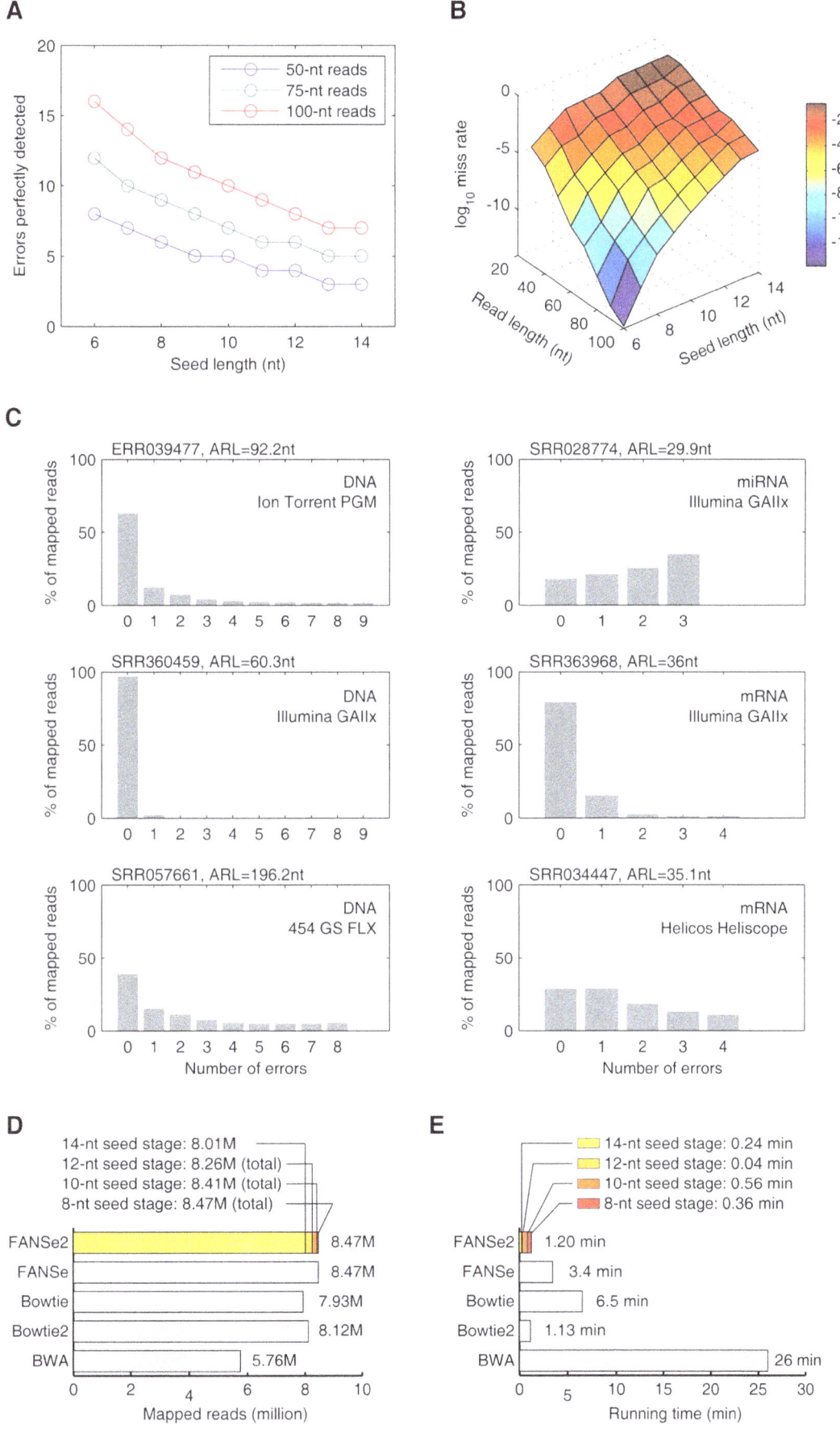

Figure 2. Rational and validation of the iterative strategy of FANSe2. (A) Errors in a read which can be perfectly detected by FANSe algorithm versus the seed length for 50-, 75- and 100-nt reads, respectively. (B) Theoretical miss rate of FANSe algorithm with different seed length for various read length. (C) Error distribution of six sequencing datasets (listed in Table S1 in File S1) sequenced on various types of sequencers, respectively. Reads were mapped with FANSe. ARL = average read length. (D, E) Mapping the *E. coli* mRNA dataset reported in [5]. For FANSe2 algorithm, the reads mapped (D) and the calculation time (E) used using different read length stages were shown in colors. The test was performed in a quad-core Intel i5-3570K computer using one CPU core.

Therefore, FANSe2 can accelerate mapping large datasets by using 2~4 CPU cores on one computer using 4~8 GB RAM. This means that even laptop computers can perform mapping with ease. When mapping reads to human genome, Bowtie2 and Bowtie needs more than 5 GB RAM. BWA and novoalign needs 7.3~7.8 GB RAM, which hardly fits a computer with 8 GB RAM because the operating system usually requires additional 0.7~1.2 GB RAM (Figure 3A). According to the manual, SHRiMP2 needs 48 GB RAM to map reads to human genome, which is already far beyond the capacity of high-end workstations, including our computers (Figure 3A and 3B).

In addition, FANSe can parallelize across multiple normal computers with simple LAN connection, providing a economic and scalable solution for biology labs. This feature is not offered by any other current mainstream mapping tools. We tested the scalability of FANSe2 in our real office environment with three heterologous computers: two Intel i5-3570s and one Intel i5-2500 with 8 GB~16 GB RAM installed, connected with gigabit LAN. One such inexpensive office computer (~$600) mapped an mRNA-seq dataset of Human Body Map 2 generated from an entire Illumina HiSeq-2000 flowcell (8 lanes, 608 M reads, 75-nt) to human reference genome within 10 hours, 3.6x faster than

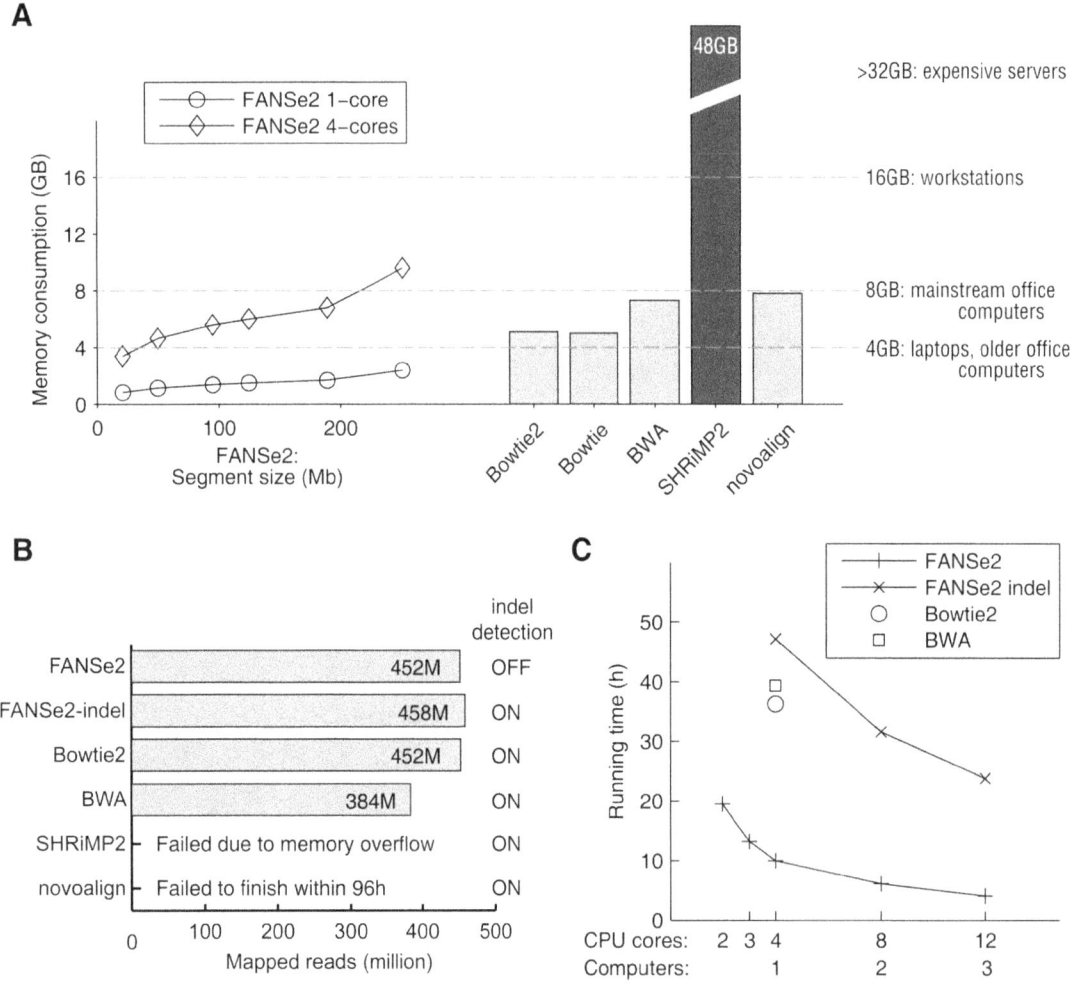

Figure 3. Scalability, sensitivity and speed of FANSe2 compared to other algorithms. (A) Memory consumption of the tested algorithms when mapping 75-nt reads to human genome. The memory consumption of FANSe2 using 1 CPU core and 4 CPU cores are indicated using circles and diamonds, respectively. SHRiMP2 failed to run this test in our 16 GB memory system; thus its memory consumption was taken from its manual. (B, C) Mapping data from an entire Illumina HiSeq-2000 flowcell (608 M 75-nt reads) to masked human genome. (B) The number of reads mapped by the tested algorithms using one computer (4 CPU cores). FANSe2 was tested with indel detection on and off, respectively. SHRiMP2 failed to run in our system due to its high memory consumption. Novoalign failed to finish the task within 96 hours. (C) The time to perform this mapping using different number of CPU cores and computers. Plus sign: FANSe2 without indel detection; cross: FANSe2 with indel detection. Bowtie2 (circle) and BWA (rectangle) do not support automatic parallelization across multiple computers.

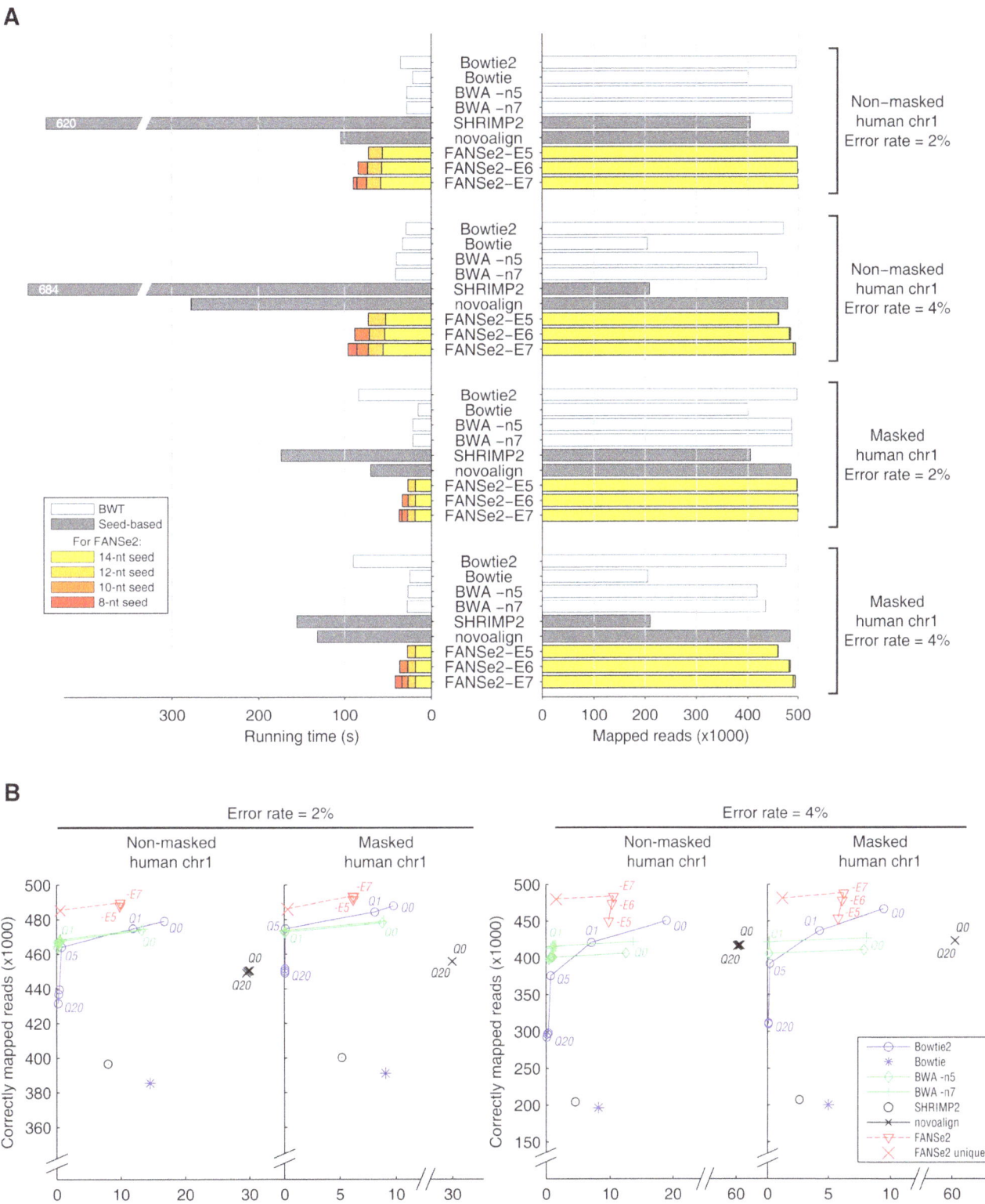

Figure 4. Comparison of FANSe2 and other algorithms on their sensitivity, speed and correctness using simulated datasets from non-masked and masked human chromosome 1 reference sequence (hg19) with 2% and 4% sequencing error rate, respectively. Each dataset contains 500,000 reads with the length of 75-nt. The test parameters are listed in Table S4 in File S1. (A) Comparative test on sensitivity and speed. Reads mapped and the time used at different stages of seed lengths in FANSe2 are shown in colors. The BWT-based algorithms are shown in light blue bars, and the other seed-based algorithms are shown in gray bars. (B) Comparative test on correctness. For Bowtie2, BWA and novoalign, mapped reads were filtered using various mapping quality threshold (Q0~Q20) represented in Phred score scale (black circle). The correctness of FANSe2 results were marked on the same plot when considering all mapping results (red triangle, 5~7 errors allowed) or considering only the reads

that were uniquely mapped (red cross, 7 errors allowed). The results of Bowtie and SHRiMP2 were not filtered according to the mapping quality due to their low mapping sensitivity.

Bowtie2 in very sensitive mode while maintaining the same sensitivity. With three computers and one-click run, the same job finished 4.1 hours by FANSe2 (Figure 3B and 3C). With the indel detection enabled, FANSe2 mapped these reads within 23.8 hours with three computers. FANSe2 with indel search mapped more reads than Bowtie2 and BWA, which were also enabled indel search (Figure 3B). Compared with one computer, three computers accelerated the mapping for 2.43x (Figure 3C). Note that this efficient parallelization was performed with user-friendly graphical user interface. In contrast, SHRiMP2 failed to run because of its high memory demand. Novoalign was unable to finish the task in 4 days (Figure 3B). These results showed that FANSe2, as a seed-based algorithm, is approaching the speed of BWT-based algorithms while maintaining similar or higher sensitivity when handling huge datasets.

Sensitivity and correctness of FANSe2 tested with simulated dataset

Practically, the raw error rates of the current next-generation sequencing platforms were reported as 0.26~13% [11], and the base calling step adds further 2.76~4.86% error rate [10]. Therefore, a mapping algorithm should reliably map reads containing at least such errors. To test the sensitivity and correctness of FANSe2 algorithm, we generated four simulated datasets, each containing 500,000 reads of 75-nt Illumina-like single-end reads, from the non-masked and masked human chromosome 1 genome sequence (hg19) and with substitution rate of 2% and 4%, respectively. For all four cases, the speed of traditional BWT-based algorithms (Bowtie, Bowtie2 and BWA) are generally faster than traditional seed-based algorithms (SHRiMP2 and novoalign). This coincides with the previous comparisons [6]. However, FANSe2 is just slightly slower than BWT-based algorithms in all cases, and is even faster than Bowtie2 when using the masked genome. In all four cases, FANSe2 mapped more reads than all other tested algorithms (Figure 4A). The sensitivity of FANSe2 increased slightly when allowing more errors in a read. When 7 errors were allowed, the sensitivity of FANSe2 reached 99.99% and 99.0% for 2% and 4% error rate, respectively. Again more than 99% of the reads were mapped using 14-nt seeds, exceeding the sensitivity of all other tested algorithms. Stepping down to 12-nt or lower hardly mapped more reads, thus is practically unnecessary. Note that the error allowance for the whole read cannot be explicitly set when using Bowtie2 and novoalign. Some reads mapped with 7 errors were found in their results, showing that this comparison is fair.

Next, we analyzed the correctness of FANSe2 mapping results using the previously described method [20], plotting the number of reads mapped to wrong locations against the number of reads mapped to correct locations (Figure 4B). In all cases, FANSe2 allowing 6~7 mismatches mapped more reads correctly to its original position than all other tested algorithms. For the non-masked genome, FANSe2 allowing 7 mismatches mapped 2.2% and 6.9% more reads to their correct positions than Bowtie2 at 2% and 4% error rate, respectively. Meanwhile, FANSe2 mapped 41.0% and 44.5% less reads than Bowtie2 to their wrong positions. Applying increasing mapping quality threshold, only the uniquely mapped reads were kept. Bowtie2 decreased the wrongly mapped reads in the cost of discarding a considerable fraction of mappable reads. At the threshold of mapping quality of 5, FANSe2 mapped 4.5% and 27.8% more uniquely-mapped reads to its correct place

than Bowtie2 at 2% and 4% error rate, respectively. BWA performed more robust than Bowtie2 in this test, as increasing the mapping quality threshold do not decrease the number of mapped reads dramatically. However it still mapped less reads than FANSe2. Novoalign mapped comparable number of reads as Bowtie2 and BWA, however it mapped 2~3 times more mapped to wrong places than Bowtie2 and BWA, and increasing the mapping quality threshold almost do not increase the correctness. Bowtie and SHRiMP2 mapped considerably less reads than the other algorithms, especially at 4% error rate.

As repetitive sequence creates challenges to correct read mapping, masked genome sequence is widely used in major studies to improve the efficiency of sequence alignment (e.g. NCBI BLAST) [21], polymorphism and mutation discovery [28,29], genome annotation and comparison [30–33], etc. In clinical diagnosis procedures, such as the non-invasive prenatal diagnosis based on next-generation sequencing, mapping reads to masked human genome is also used as a standard [34–37]. Therefore, we also performed read mapping tests using the masked genome sequence provided by UCSC Genome Browser. Compared to the non-masked tests, the sensitivity and correctness of all algorithms increased slightly, because the masked genome sequence is free of repetitive regions. Nevertheless, the scenario remains similar to the non-masked tests: FANSe2 has higher sensitivity while maintaining the correctness.

Experimental validation of the RNA-seq mapping result by FANSe2

The robust sensitivity and correctness of FANSe2 maximizes the usage of data in sequencing datasets. This advantage may be more significant when dealing with RNA-seq data that is more error-prone than DNA-seq. In our previous work, we had shown that BWA and BLAT lose mappable reads in low abundance mRNA unproportionally in a prokaryotic system [5]. We next tested FANSe2 and Bowtie2 with our previously reported mRNA-seq dataset (75 nt single-end reads) of human lung cancer cell line A549 [22]. Aiming at quantitative profiling of known mRNAs, we mapped the reads to RefSeq human RNA reference sequence and the splice variants were merged. Previous study showed that mapping to mature mRNA sequence avoided the error of mapping splice junction reads when using genomic sequence as reference and should be preferentially used for RNA-seq, unless novel splice junctions are to be detected [38,39]. Additionally, protein coding mRNAs consist only a small proportion of the genomic sequence, reducing the computational demand dramatically. Therefore this is an efficient strategy that is widely used by the community [25,38–43]. Genes with less than 10 reads mapped were considered as unreliable quantified genes and removed [23]. We found that the gene expression quantitation of the two algorithms in general coincide for the genes that were identified by both algorithms (Figure 5A).

We next experimentally investigated the genes that were solely identified by FANSe2 (Table 1) or Bowtie2 (Table 2) to check the possible false positives and false negatives. The abundances of the top five RNAs that solely identified by FANSe2 range from 1.47 to 5.77 rpkM (Table 1). They were all validated by RT-PCR with clear bands on the gel at the estimated sizes (Figure 5B). Although the primer specificity of SPIN2A was not high enough so that additional bands appeared in addition to the strongest and expected band, SPIN2A has been detected by microarray in lung

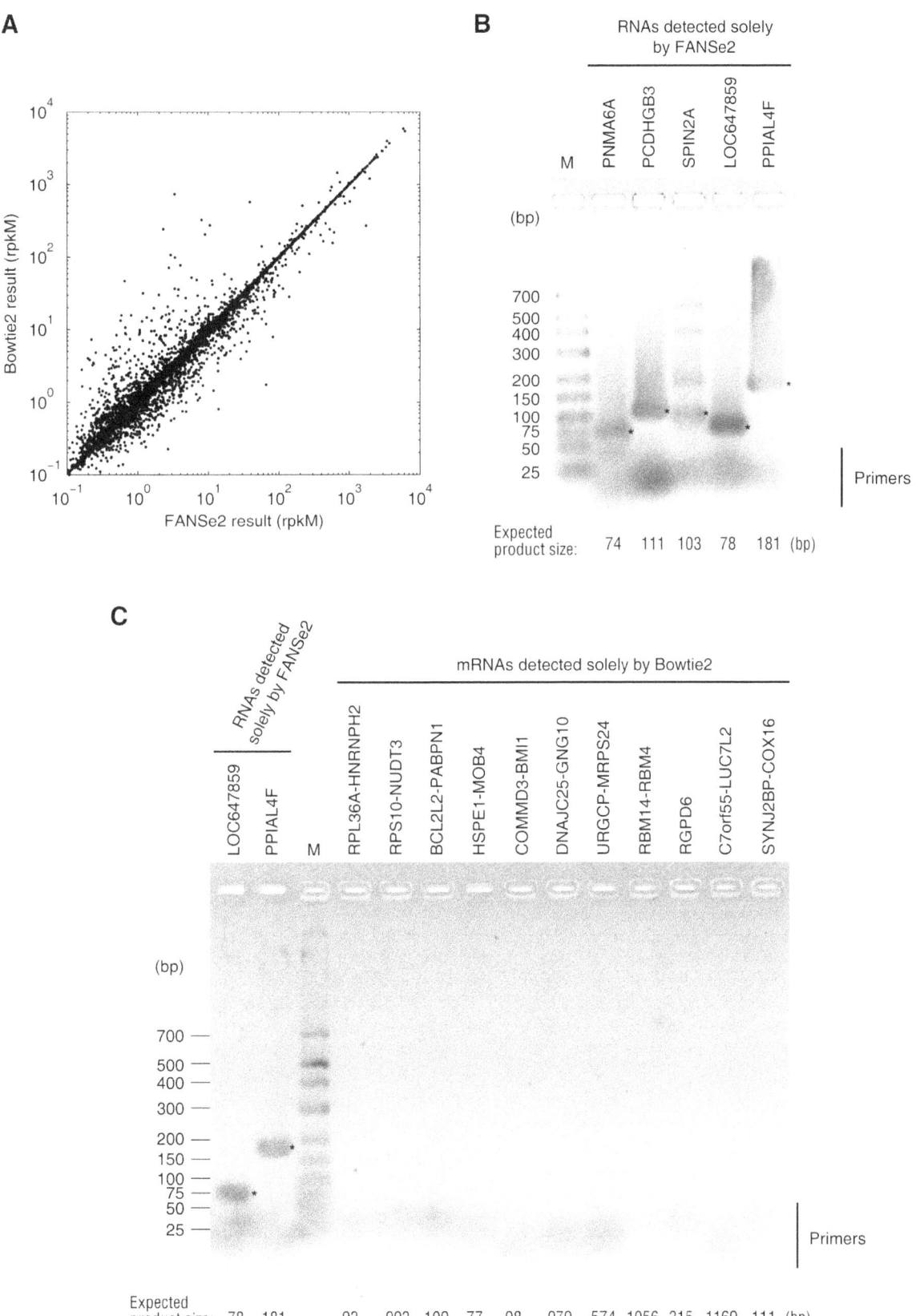

Figure 5. Experimental examination of the results of FANSe2 and Bowtie2. (A) Quantification of mRNA from A549 cells using the mapping results of FANSe2 and Bowtie2. The mRNA sequencing dataset was mapped to the human RefSeq RNA reference sequences and the quantification was performed using the standard rpkM method. (B) RT-PCR validation of mRNAs that were detected by FANSe2 but not by Bowtie2 (see Table 1). 15 μl PCR product were loaded for each lane and resolved on a 3% agarose gel. The bands with the expected product size were marked with stars.

The expected product sizes were noted below. (C) RT-PCR validation of mRNAs that were detected solely by Bowtie2 (See Table 2). Two RNAs detected solely by FANSe2 (LOC647859 and PPIAL4F) were loaded as positive control. 7 μl PCR product were loaded for each lane and resolved on a 2.7% agarose gel. The bands with the expected product size were marked with stars. The expected product sizes were noted below. A faint band appeared at ~200 bp in the lane of BCL2L2-PABPN1 but is quite different than the expected product size.

adenocarcinoma (Expression Atlas) [44]. In contrast, the abundances of the top 20 RNAs that solely identified by Bowtie2 range from 4.58 to 146.46 rpkM (Table 2). Three genes among them, namely LUZP6, PIGY and SNRPN, are identical in sequence to genes MTPN, PYURF and SNURF, respectively. Therefore, they are undistinguishable to algorithms or RT-PCR and thus excluded from our experimental validation. Indeed, FANSe2 identified MTPN, PYURF and SNURF. Fifteen genes among the top 20 "Bowtie2-only" genes were fusion genes with the abundance of 12.66~146.46 rpkM, at least one order of magnitude higher than the "FANSe2-only" genes. We verified 10 protein-coding genes among them using RT-PCR, but none of them can be validated (Figure 5C). The coding gene RGPD6 were also failed in the verification (Figure 5C). This experimental verification showed that Bowtie2 results in both false-negatives and false-positives: it fails to identify genes like PCDHGB3, SPIN2A, while erroneously identified the gene RGPD6 and numerous fusion genes that are actually absent in the sample. Meanwhile, a considerable number of reads were assigned to the false-positive identifications by Bowtie2: 11645 reads were mapped to the 11 "Bowtie2-only" genes that were experimentally determined as absent. This may also influence the quantitation of other genes and may be a source that causes the quantitative deviation from FANSe2 (Figure 5A). In contrast, FANSe2 results can be validated by experiments, showing its reliability.

Gene expression quantifications by FANSe2 coincide to microarray data better than most other algorithms

Microarray is widely used since decades as a reliable approach to quantify gene expression levels. The hybridization nature of microarray do not need read mapping, providing an experimental reference for mapping-based RNA-seq. We downloaded RNA-seq and microarray data from the same sample (the aristolochic acids treated rat liver sample AA_1) from a previous study by Su et al. [25]. We used FANSe2, Bowtie2, BWA and novoalign to map the same RNA-seq dataset to the same transcriptome reference sequence that was used in [25], and the Bowtie result was taken from Su *et al.*'s report [25]. To be comparable to the microarray used in Su et al.'s study, only the reads mapped to the coding genes were considered. Consistent with the tests above, FANSe2 mapped more reads than the other tested algorithms (Figure 6A).

Genes with less than 10 reads mapped were considered as unreliable quantified genes and removed [23]. The FANSe2 result correlates to the microarray data equally good as Bowtie2 (Pearson correlation coefficient $R = 0.81$, Figure 6B and 6D), while the results with other algorithms correlates worse ($R = 0.70$ for Bowtie, $R = 0.74$ for BWA and novoalign, Figure 6C, 6E and 6F). At least in this case on rat, FANSe2 and Bowtie2 provide better consistency to microarray data than other algorithms, facilitating the data integration between different omics platforms. Nevertheless, FANSe2 performed stably also in RNA-seq studies of human cells (Figure 5), showing the advantage of FANSe2 on robustness.

Discussion

In most of the resequencing and RNA-seq applications, mapping is the bottleneck step in the data processing pipeline [20]. BWT-based algorithms such as Bowtie, BWA and SOAP2 have greatly facilitated the sequencing applications since they are fast enough to perform the mapping on desktop workstations instead of supercomputers (reviewed in [2,7]). They performed very well for qualitative applications such as DNA resequencing projects, since the loss of mappable reads can be easily compensated by higher sequencing throughput without affecting the results of sequence variation analysis [6]. However the completeness and robustness of the mapping were compromised [5,6,12,16], leading to unproportional loss of mappable reads [5]. This is inacceptable for quantitative applications like RNA-seq. Seed-based algorithms like FANSe offer very high accuracy, quantitativity and robustness, more suitable for quantitative RNA-seq, but usually with much lower speed [5,6]. To have both advantages, FANSe2 inherited the accuracy of FANSe with largely improved speed due to the iterative step-down strategy based on the statistics of real sequencing datasets. In most sequencing applications, the majority of the reads should be mapped to the reference sequence, and a large fraction of these reads contains very limited number of errors, which can be reliably detected with long seeds at high speed. The accuracy of FANSe2 is ensured by its fully predictable and extremely low theoretical miss rate.

The low correlation between the NGS and microarray platforms in quantitative gene expression studies has been noted in a number of literatures. For miRNA, the Pearson correlation of

Table 1. The top five RefSeq RNAs that are exclusively identified by FANSe2 in A549 mRNA-seq dataset.

| Gene name | RefSeq-ID | FANSe2 | | RT-PCR validation | | Validated (Figure 5) |
		Read count	rpkM	Whole Transcriptome qPCR Primer Database ID	Expected product size (bp)	
PCDHGB3	NM_018924	105	1.47	PCDHGB3_uc003ljw.2_1_2_2	111	Yes
SPIN2A	NM_019003	60	2.92	PB *	103	Yes
LOC647859	NR_026578	52	5.77	OCLN_uc011cru.1_2_1_2	78	Yes
PNMA6A	NM_032882	45	1.95	PB *	74	Yes
PPIAL4F	NM_001164262	44	3.60	PB *	181	Yes

*PB: primer pair not available in whole transcriptome qPCR primer database. The primers are automatically designed using NCBI-PrimerBLAST. Please refer to Table S2 in File S1 for details.

Table 2. The top 20 RefSeq RNAs that are exclusively identified by Bowtie2 in A549 mRNA-seq dataset.

| Gene name | RefSeq-ID | Bowtie2 | | | | RT-PCR validation | | |
		Read count	rpkM	Identical to gene **		Whole Transcriptome qPCR Primer Database ID	Expected product size (bp)	Validated (Figure 5)
SENP3-EIF4A1	NR_037926	9640	146.46					
LUZP6	NM_001128619	3264	54.95	MTPN				
RPL36A-HNRNPH2	NM_001199973	2653	65.24			RPL36A-HNRNPH2_uc022cag.1_3_1_2	93	Failed
RPS10-NUDT3	NM_001202470	2240	57.96			PB *	993	Failed
SLMO2-ATP5E	NR_037929	1490	94.87					
BLOC1S5-TXNDC5	NR_037616	1336	26.06					
BCL2L2-PABPN1	NM_001199864	1297	39.82			BCL2L2-PABPN1_uc001wjh.4_2_2_1	109	Failed
HSPE1-MOB4	NM_001202485	1119	17.10			HSPE1-MOB4_uc021vum.1_1_1_2	77	Failed
HNRNPUL2-BSCL2	NR_037946	1092	17.79					
HIF1A-AS2	NR_045406	985	31.40					
ATP6V1G2-DDX39B	NR_037853	969	27.49					
COMMD3-BMI1	NM_001204062	865	16.69			COMMD3-BMI1_uc009xkg.3_4_2_2	98	Failed
DNAJC25-GNG10	NM_004125	675	29.45			PB *	979	Failed
URGCP-MRPS24	NM_001204871	644	50.25			PB *	574	Failed
PIGY	NM_001042616	607	29.27	PYURF				
RBM14-RBM4	NM_001198845	559	22.65			PB *	1056	Failed
SNRPN	NM_022807	559	20.88	SNURF				
RGPD6	NM_001123363	535	4.58			PB *	315	Failed
C7orf55-LUC7L2	NM_001244584	532	12.66			PB *	1169	Failed
SYNJ2BP-COX16	NM_001202547	526	17.59			SYNJ2BP-COX16_uc021rv2m.1_2_1_1	111	Failed

*PB: primer pair not available in whole transcriptome qPCR primer database. The primers are automatically designed using NCBI-PrimerBLAST. Please refer to Table S2 in File S1 for details.
**Identical to gene: two genes have identical sequence and thus are non-distinguishable by the mapping algorithm or RT-PCR.

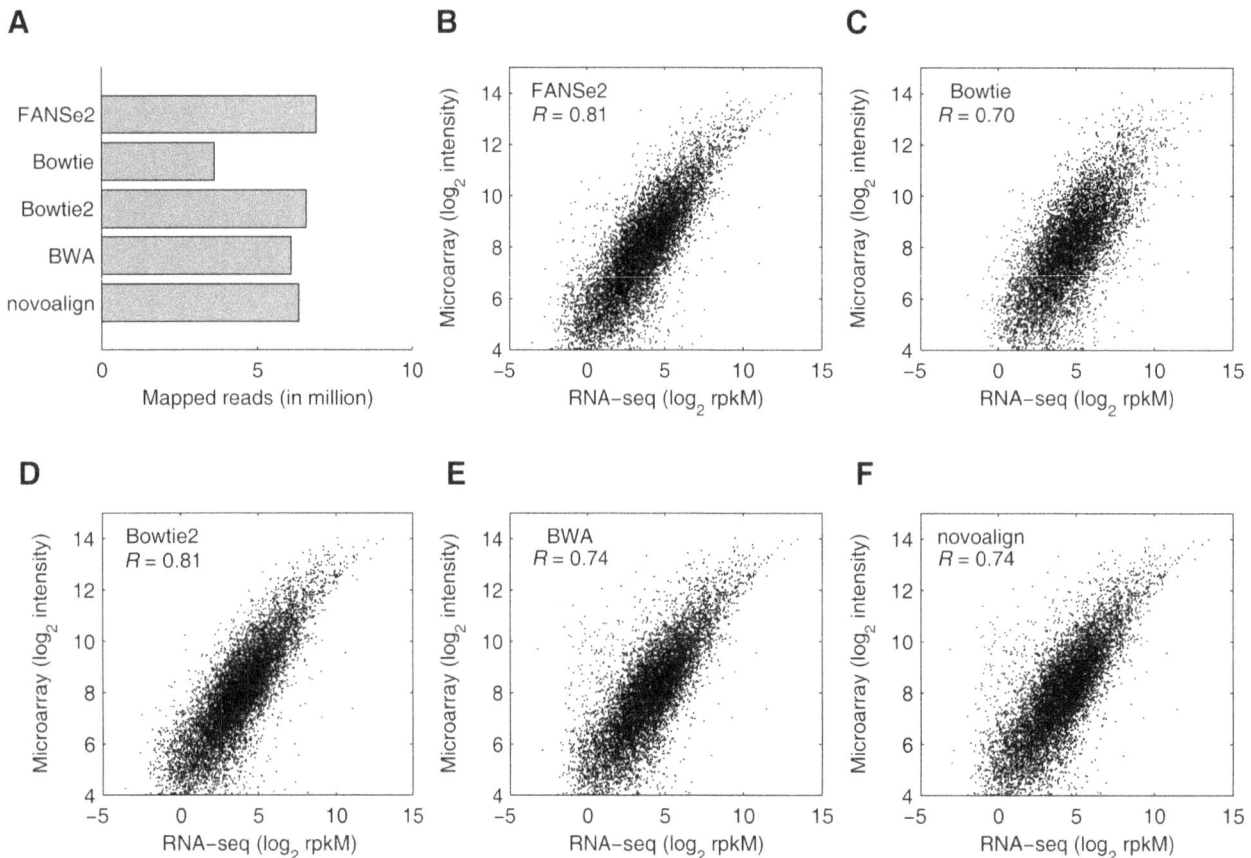

Figure 6. Comparison of the gene expression level calculated by RNA-seq and microarray. (A) The mapped reads to coding genes (NM_*) by FANSe2, Bowtie, Bowtie2, BWA and novoalign. Three mismatches were allowed in the mapping by FANSe2 and BWA. The results of Bowtie were obtained from GEO database GSE21210 and described by Su et al. [25]. (B–F) Correlations of RNA-seq and microarray. Only the reads that mapped to coding genes were taken into consideration to be consistent to the microarray data. The rpkM values of RNA-seq were calculated from the mapping results by FANSe2 (B), Bowtie (C), Bowtie2 (D), BWA (E) and novoalign (F).

both techniques reaches only $R = 0.52 \sim 0.66$ [18,19]. The false negative rates of Illumina NGS platform was as high as 12%, much higher than the microarray platforms (0.97%~3.1%) [17]. This is also consistent with other studies [18]. Considering the enormous throughput and dynamic range of NGS, this low correlation and high false negative rates is not likely to be caused by the throughput, but by the data processing. As experimentally shown in this study, mapping to RNA reference sequences especially requires the sensitivity and correctness of the mapping algorithm. Algorithms lacking robustness can result in numerous false positives and false negatives in gene identifications and may affect the gene quantification (Figure 5). Also, FANSe2 showed remarkable better consistency to the microarray data than most other algorithms, bridging the gap between the NGS and microarray and leading to better reproducibility and confidence, which is in great demand for the NGS-based studies (reviewed in [45]).

Furthermore, almost all mapping algorithms offer numerous parameters, and small alteration of parameters may lead to significant change of result. This already leads to low reproducibility and low robustness of many next-generation sequencing studies (reviewed in [45]). In contrast, the parameter settings of FANSe2 almost did not affect the sensitivity and correctness (Figure 4B), providing a remarkable simplicity and robustness of usage.

Previous algorithms require more memory for larger reference genomes. For human genome, $3 \sim 14$ GB memory is usually required [46]. To reduce the memory consumption when parallelized, some common data, e.g. the reference sequences and the index, need to be shared by multiple CPU cores, increasing the risk of access contention, i.e. simultaneous access of the same data by different CPU cores. This may trigger an unpredictable error or needs additional handling, leading to reduced stability or speed. This problem remains as an open challenge in computational science [47–49]. In contrast, FANSe2's memory consumption is almost independent of the reference genome size, since it splits the large reference genome into user-specified segments (Figure 3A). Reducing the segment size can significantly reduce the memory demand without impairing the result, facilitating the parallelization especially in normal office computers. The small and user-adjustable memory consumption also allows parallelization of multiple processes instead of threads, since there is no need to share any common data in the memory, thus eliminating the instability caused by access contention.

Importantly, this merit makes FANSe2 the first algorithm with the feature of flexible, scalable and almost maintenance-free parallelization across multiple computers, efficiently utilizing the computational power of inexpensive office computers and even laptop computers. With just three computers, FANSe2 mapped

608 million reads to human genome within 4.1 hours. This might be the first mapping algorithm that matches the speed of the coming generation of sequencers like Ion Torrent Proton P2 (660 M reads in several hours) running in normal computers. There is no need for expensive, exclusive and maintenance-intensive clusters or workstations.

FANSe2 runs under various operating systems including Windows, providing user-friendly graphical user interfaces, bringing convenience to the biological researchers who are not familiar with computational issues. With simple online video tutorials, everyone knows how to install and use it in 15 minutes. The ability of mapping billions of reads in hours using normal office computers with robust accuracy makes FANSe2 a good candidate to remove the bottleneck of data processing pipeline, leading to much faster, more reliable and quantitative analysis, able to handle the future sequencing applications.

Supporting Information

File S1   File S1 includes the following: Table S1. Six datasets downloaded from DDBJ Sequence Read Archive to analyze the error distribution. **Table S2.** The gene-specific PCR primers for validation of gene identifications. The genes identified solely by Bowtie2 are shaded as gray, and the genes identified solely by FANSe2 are not shaded. **Table S3.** Mapping programs tested in this study. **Table S4.** Test parameters for Figure 4.

Acknowledgments

We thank Dr. Hanqi Yin (Shanghai Biotechnology Corporation) for his help on analyzing the microarray data.

Author Contributions

Conceived and designed the experiments: GZ CLX. Performed the experiments: XLL JYZ JJJ. Analyzed the data: GZ CLX ZBM. Wrote the paper: GZ QYH.

References

1. Trapnell C, Salzberg SL (2009) How to map billions of short reads onto genomes. Nat Biotechnol 27: 455–457.
2. Li H, Homer N (2010) A survey of sequence alignment algorithms for next-generation sequencing. Brief Bioinform 11: 473–483.
3. Mortazavi A, Williams BA, Mccue K, Schaeffer L, Wold B (2008) Mapping and quantifying mammalian transcriptomes by RNA-Seq. Nat Methods 5: 621–628.
4. Myers E (1986) AnO(ND) difference algorithm and its variations. Algorithmica 1: 251–266.
5. Zhang G, Fedyunin I, Kirchner S, Xiao C, Valleriani A, et al. (2012) FANSe: an accurate algorithm for quantitative mapping of large scale sequencing reads. Nucleic Acids Res 40: e83.
6. Fonseca NA, Rung J, Brazma A, Marioni JC (2012) Tools for mapping high-throughput sequencing data. Bioinformatics 28: 3169–3177.
7. Schbath S, Martin V, Zytnicki M, Fayolle J, Loux V, et al. (2012) Mapping reads on a genomic sequence: an algorithmic overview and a practical comparative analysis. J Comput Biol 19: 796–813.
8. Gilad Y, Pritchard JK, Thornton K (2009) Characterizing natural variation using next-generation sequencing technologies. Trends Genet 25: 463–471.
9. Makridakis NM, Phipps T, Srivastav S, Reichardt JK (2009) PCR-free method detects high frequency of genomic instability in prostate cancer. Nucleic Acids Res 37: 7441–7446.
10. Ledergerber C, Dessimoz C (2011) Base-calling for next-generation sequencing platforms. Brief Bioinform 12: 489–497.
11. Quail MA, Smith M, Coupland P, Otto TD, Harris SR, et al. (2012) A tale of three next generation sequencing platforms: comparison of Ion Torrent, Pacific Biosciences and Illumina MiSeq sequencers. BMC Genomics 13: 341.
12. Ruffalo M, LaFramboise T, Koyuturk M (2011) Comparative analysis of algorithms for next-generation sequencing read alignment. Bioinformatics 27: 2790–2796.
13. Peng Z, Cheng Y, Tan BC, Kang L, Tian Z, et al. (2012) Comprehensive analysis of RNA-Seq data reveals extensive RNA editing in a human transcriptome. Nat Biotechnol 30: 253–260.
14. Iida K, Jin H, Zhu JK (2009) Bioinformatics analysis suggests base modifications of tRNAs and miRNAs in Arabidopsis thaliana. BMC Genomics 10: 155.
15. Alvarez M, Barrioluengo V, Afonso-Lehmann RN, Menendez-Arias L (2013) Altered error specificity of RNase H-deficient HIV-1 reverse transcriptases during DNA-dependent DNA synthesis. Nucleic Acids Res 41: 4601–4612.
16. Homer N, Merriman B, Nelson SF (2009) BFAST: an alignment tool for large scale genome resequencing. PLoS One 4: e7767.
17. Willenbrock H, Salomon J, Sokilde R, Barken KB, Hansen TN, et al. (2009) Quantitative miRNA expression analysis: comparing microarrays with next-generation sequencing. Rna 15: 2028–2034.
18. Git A, Dvinge H, Salmon-Divon M, Osborne M, Kutter C, et al. (2010) Systematic comparison of microarray profiling, real-time PCR, and next-generation sequencing technologies for measuring differential microRNA expression. Rna 16: 991–1006.
19. Kelly AD, Hill KE, Correll M, Hu L, Wang YE, et al. (2013) Next-generation sequencing and microarray-based interrogation of microRNAs from formalin-fixed, paraffin-embedded tissue: preliminary assessment of cross-platform concordance. Genomics 102: 8–14.
20. Langmead B, Salzberg SL (2012) Fast gapped-read alignment with Bowtie 2. Nat Methods 9: 357–359.
21. Morgulis A, Gertz EM, Schaffer AA, Agarwala R (2006) WindowMasker: window-based masker for sequenced genomes. Bioinformatics 22: 134–141.
22. Wang T, Cui Y, Jin J, Guo J, Wang G, et al. (2013) Translating mRNAs strongly correlate to proteins in a multivariate manner and their translation ratios are phenotype specific. Nucleic Acids Res.
23. Bloom JS, Khan Z, Kruglyak L, Singh M, Caudy AA (2009) Measuring differential gene expression by short read sequencing: quantitative comparison to 2-channel gene expression microarrays. BMC Genomics 10: 221.
24. Zeisel A, Yitzhaky A, Bossel Ben-Moshe N, Domany E (2013) An accessible database for mouse and human whole transcriptome qPCR primers. Bioinformatics 29: 1355–1356.
25. Su Z, Li Z, Chen T, Li Q-Z, Fang H, et al. (2011) Comparing next-generation sequencing and microarray technologies in a toxicological study of the effects of aristolochic acid on rat kidneys. Chemical research in toxicology 24: 1486–1493.
26. Irizarry RA, Bolstad BM, Collin F, Cope LM, Hobbs B, et al. (2003) Summaries of Affymetrix GeneChip probe level data. Nucleic Acids Res 31: e15.
27. Prasad A, Kumar SS, Dessimoz C, Jaquet V, Bleuler S, et al. (2013) Global regulatory architecture of human, mouse and rat tissue transcriptomes. BMC genomics 14: 716.
28. Evrony GD, Cai X, Lee E, Hills LB, Elhosary PC, et al. (2012) Single-neuron sequencing analysis of L1 retrotransposition and somatic mutation in the human brain. Cell 151: 483–496.
29. Hillier LW, Marth GT, Quinlan AR, Dooling D, Fewell G, et al. (2008) Whole-genome sequencing and variant discovery in C. elegans. Nat Methods 5: 183–188.
30. Jex AR, Liu S, Li B, Young ND, Hall RS, et al. (2011) Ascaris suum draft genome. Nature 479: 529–533.
31. Lewis NE, Liu X, Li Y, Nagarajan H, Yerganian G, et al. (2013) Genomic landscapes of Chinese hamster ovary cell lines as revealed by the Cricetulus griseus draft genome. Nat Biotechnol 31: 759–765.
32. Zhang G, Liu X, Quan Z, Cheng S, Xu X, et al. (2012) Genome sequence of foxtail millet (Setaria italica) provides insights into grass evolution and biofuel potential. Nat Biotechnol 30: 549–554.
33. Li Z, Zhang Z, Yan P, Huang S, Fei Z, et al. (2011) RNA-Seq improves annotation of protein-coding genes in the cucumber genome. BMC Genomics 12: 540.
34. Chiu RW, Chan KC, Gao Y, Lau VY, Zheng W, et al. (2008) Noninvasive prenatal diagnosis of fetal chromosomal aneuploidy by massively parallel genomic sequencing of DNA in maternal plasma. Proc Natl Acad Sci U S A 105: 20458–20463.
35. Canick JA, Kloza EM, Lambert-Messerlian GM, Haddow JE, Ehrich M, et al. (2012) DNA sequencing of maternal plasma to identify Down syndrome and other trisomies in multiple gestations. Prenat Diagn 32: 730–734.
36. Palomaki GE, Deciu C, Kloza EM, Lambert-Messerlian GM, Haddow JE, et al. (2012) DNA sequencing of maternal plasma reliably identifies trisomy 18 and trisomy 13 as well as Down syndrome: an international collaborative study. Genet Med 14: 296–305.
37. Dames S, Chou LS, Xiao Y, Wayman T, Stocks J, et al. (2013) The development of next-generation sequencing assays for the mitochondrial genome and 108 nuclear genes associated with mitochondrial disorders. J Mol Diagn 15: 526–534.
38. Ju YS, Kim J-I, Kim S, Hong D, Park H, et al. (2011) Extensive genomic and transcriptional diversity identified through massively parallel DNA and RNA sequencing of eighteen Korean individuals. Nat Genet 43: 745–752.
39. Kinsella M, Harismendy O, Nakano M, Frazer KA, Bafna V (2011) Sensitive gene fusion detection using ambiguously mapping RNA-Seq read pairs. Bioinformatics 27: 1068–1075.

40. Rowley JW, Oler AJ, Tolley ND, Hunter BN, Low EN, et al. (2011) Genome-wide RNA-seq analysis of human and mouse platelet transcriptomes. Blood 118: e101–e111.

41. McDaneld TG, Smith TPL, Harhay GP, Wiedmann RT (2012) Next-Generation Sequencing of the Porcine Skeletal Muscle Transcriptome for Computational Prediction of MicroRNA Gene Targets. PLoS ONE 7: e42039.

42. Ramskold D, Luo S, Wang Y-C, Li R, Deng Q, et al. (2012) Full-length mRNA-Seq from single-cell levels of RNA and individual circulating tumor cells. Nat Biotech 30: 777–782.

43. Xiao W, Tran B, Staudt LM, Schmitz R (2013) High-Throughput RNA Sequencing in B-Cell Lymphomas. Lymphoma: Springer. pp. 295–312.

44. Petryszak R, Burdett T, Fiorelli B, Fonseca NA, Gonzalez-Porta M, et al. (2014) Expression Atlas update—a database of gene and transcript expression from microarray- and sequencing-based functional genomics experiments. Nucleic Acids Res 42: D926–932.

45. Nekrutenko A, Taylor J (2012) Next-generation sequencing data interpretation: enhancing reproducibility and accessibility. Nat Rev Genet 13: 667–672.

46. Li R, Yu C, Li Y, Lam TW, Yiu SM, et al. (2009) SOAP2: an improved ultrafast tool for short read alignment. Bioinformatics 25: 1966–1967.

47. Chang DY-c, Lawrie D (1982) Performance of multiprocessor systems with space and access contention. Urbana, Ill.: Dept. of Computer Science, University of Illinois at Urbana-Champaign. 80 p. p.

48. Zhuravlev S, Blagodurov S, Fedorova A (2010) Addressing Shared Resource Contention in Multicore Processors via Scheduling. Asplos Xv: Fifteenth International Conference on Architectural Support for Programming Languages and Operating Systems: 129–141.

49. Chen KY, Chang JM, Hou TW (2011) Multithreading in Java: Performance and Scalability on Multicore Systems. Ieee Transactions on Computers 60: 1521–1534.

9

Purifying Selection in Deeply Conserved Human Enhancers Is More Consistent than in Coding Sequences

Dilrini R. De Silva[1,2], Richard Nichols[2], Greg Elgar[1]*

1 Systems Biology, MRC National Institute for Medical Research, Mill Hill, London, United Kingdom, **2** School of Biological and Chemical Sciences, Queen Mary University of London, London, United Kingdom

Abstract

Comparison of polymorphism at synonymous and non-synonymous sites in protein-coding DNA can provide evidence for selective constraint. Non-coding DNA that forms part of the regulatory landscape presents more of a challenge since there is not such a clear-cut distinction between sites under stronger and weaker selective constraint. Here, we consider putative regulatory elements termed Conserved Non-coding Elements (CNEs) defined by their high level of sequence identity across all vertebrates. Some mutations in these regions have been implicated in developmental disorders; we analyse CNE polymorphism data to investigate whether such deleterious effects are widespread in humans. Single nucleotide variants from the HapMap and 1000 Genomes Projects were mapped across nearly 2000 CNEs. In the 1000 Genomes data we find a significant excess of rare derived alleles in CNEs relative to coding sequences; this pattern is absent in HapMap data, apparently obscured by ascertainment bias. The distribution of polymorphism within CNEs is not uniform; we could identify two categories of sites by exploiting deep vertebrate alignments: stretches that are non-variant, and those that have at least one substitution. The conserved category has fewer polymorphic sites and a greater excess of rare derived alleles, which can be explained by a large proportion of sites under strong purifying selection within humans – higher than that for non-synonymous sites in most protein coding regions, and comparable to that at the strongly conserved trans-dev genes. Conversely, the more evolutionarily labile CNE sites have an allele frequency distribution not significantly different from non-synonymous sites. Future studies should exploit genome-wide re-sequencing to obtain better coverage in selected non-coding regions, given the likelihood that mutations in evolutionarily conserved enhancer sequences are deleterious. Discovery pipelines should validate non-coding variants to aid in identifying causal and risk-enhancing variants in complex disorders, in contrast to the current focus on exome sequencing.

Editor: Arnar Palsson, University of Iceland, Iceland

Funding: This work was funded by MRC core funding (U117597141) to GE and a Queen Mary University of London PhD studentship to DRD. The funders had no role in study design, data collection and analysis, decision to publish, or preparation of the manuscript.

Competing Interests: The authors have declared that no competing interests exist.

* Email: gelgar@nimr.mrc.ac.uk

Introduction

The effect of mutations in coding sequence is a well-characterised phenomenon whereby non-synonymous changes alter the resulting protein product. Therefore, these non-synonymous changes tend to be under stronger purifying selection than synonymous mutations. There are only rare instances where synonymous changes have been demonstrated to have phenotypic effects [1]. This difference can be exploited in the analysis of genomes; for example, since synonymous changes are assumed to be largely neutral in effect, their genetic diversity can be used as a null distribution against which to test for evidence to identify loci at which non-synonymous changes have been established by positive selection (e.g. Macdonald-Kreitman test, [2]). Conversely, non-synonymous changes are treated as being more likely to be the cause of an altered phenotype. Indeed, genome-wide scans for causative mutations often concentrate on exons (the 'exome') and have been successful in detecting non-synonymous changes responsible for numerous genetic diseases, [3]. By contrast, there is scant data on the effect of mutations in regulatory sequences; in part because regulatory regions can be difficult to identify. Nevertheless, a few studies have been able to identify *cis*-regulatory

mutations implicated in human diseases. For example, beta thalassemia and haemophilia B are both instances of human diseases caused by mutations affecting transcription-factor binding sites in regulatory sequences (reviewed in [4]). However, not all transcription-factor binding motifs are well characterised, and other types of sequence may also have regulatory activity. In these cases evolutionary conservation provides a means to identify sequences of functional importance to the organism. If a non-coding sequence has been conserved across large evolutionary periods so that it is found in a number of diverse species, then this is most likely to be a consequence of selection against mutations that would modify it. Occasionally such mutations have been identified within a species and are indeed found to be deleterious. For example Lettice et al. found several heterozygous dominant point mutations in a highly conserved enhancer 1 Mb away from the Shh locus in humans that segregate with pre-axial polydactyly [5]. Similarly, Benko et al, found a heterozygous point mutation in a highly conserved non-coding element flanking the SOX9 locus that alters binding of the transcription factor MSX1 associated with the Pierre Robin syndrome [6]. More recently, two rare variants in the 5′ UTR and an intron of the *RBM8A* gene were found to segregate in individuals with Thrombocytopenia (reduced platelet count) with Absent Radii (TAR) syndrome when no exonic mutations were found in affected individuals at that locus [7].

Generally it has only been when exome sequencing fails to identify any causative mutations that non-coding DNA is surveyed. Now that whole genomes are being sequenced in greater numbers, such as in the 1000 Genomes Project [8], there is the opportunity to identify more non-coding variants. However, there are still hurdles to assessing their importance. Firstly large re-sequencing projects have tended to neglect non-exonic regions resulting in much lower coverage. Consequently, the stringent quality control measures used in SNP calling from next-generation sequencing data mean that rare variants in non-coding DNA may be filtered out as putative sequencing errors. Secondly, it is much harder to predict the consequence of mutation/variation in regulatory sequences because their grammar is poorly understood. An effective way of identifying putative regulatory sequences, in particular those that are under strong selection, has been to use cross species comparisons, often across large evolutionary distances - a method referred to as phylogenetic footprinting [9]. The various methods used to predict *cis*-regulatory modules often depend on a signal of evolutionary conservation [10], [11], [12]. In contrast, phylogenetic shadowing [13] using species that are more closely related helps identify regions that have diverged recently, and may have acquired a lineage-specific role.

Here, we focus our efforts on sequences that are conserved across all jawed vertebrates and that likely define a set of developmental regulatory elements (57 of the 1809 CNEs in this dataset have been tested functionally for enhancer activity in zebrafish and the data is available on the CONDOR website [14]). This core set of conserved non-coding elements (CNEs) was identified through multiple alignments of mammalian and Fugu genomes [14]. CNEs differ from other sets of conserved non-coding sequence that have been identified by comparative analyses in not overlapping known exons (e.g. Ultra-Conserved Elements identified by human-mouse-rat genomic comparisons [15] and Highly Conserved Elements [16]). About 80% of CNEs tested experimentally in zebrafish embryos show tissue-specific enhancer activity at various stages of embryonic development. 23 out of 25 CNEs tested from four different loci drive expression in zebrafish embryos [17] including SOX21 and PAX6-associated CNEs directing GFP-expression in the developing eye. A second study found 8 out of 10 elements drive reproducible reporter expression in the developing embryo [18]. The enrichment of vertebrate CNEs for conserved binding site motifs such as the Pbx-Hox heterodimer [19] and the over-representation of several transcription factor position weight matrices in mammalian conserved non-coding sequences [20] suggest that conservation of non-coding sequences is likely to be due, at least in part, to the presence of common transcription factor binding sites. Some transcription factors are highly sequence-specific and only bind to genomic regions with the exact transcription factor binding sequence [21]. In such highly specific interactions, any variation in the transcription factor binding sequence will have an effect on the transcription factor binding and subsequent gene expression, as seen with allele-specific binding of CTCF [22]. Conversely, those positions that are less strongly conserved may be less important for the functioning of the sequences. This logic is used in the construction of the Position Weight Matrix (PWM), which reflects the affinity of transcription factors to their preferred binding sites [23].

On average CNEs are about 200 bp in length (the maximum being *c*. 800 bp), yet their conservation cannot be explained by our current knowledge of transcription-factor binding sites, since most known binding targets are only 4-10 bp long. In fact the rate of evolution of known binding sites is faster than that of CNEs. One possibility would be if the binding sites overlap each other and the order of overlap is necessary to retain the proper function of the *cis*-regulatory module (as discussed in [24]). Another hypothesis is that conserved non-coding sequences (CNSs) represent mutational 'coldspots', however this explanation has been rejected [25] because of the excess of rare derived alleles observed within CNSs relative to polymorphisms outside CNSs that cannot be explained by population bottleneck effects or background selection. CNEs can be defined by a large number of completely (evolutionarily) conserved sites (Non Variable Regions), as well as a number of more variable sites (Restricted Variable Regions) based on their conservation across seven divergent vertebrate species. We evaluate the hypothesis that restricted variable regions in CNEs have been accumulating substitutions in the human lineage due to relaxed evolutionary constraint, resulting in more within-species polymorphism than non-variable regions. Using the occurrence and allele frequencies of SNPs from both the HapMap and 1000 Genomes Projects in CNEs we show evidence that a) non-variable regions within CNEs are under stronger selective constraint than restricted variable regions, b) the distribution of selective effects in CNEs is comparable to that in non-synonymous sites and c) there are discrepancies between the results obtained from HapMap and 1000 Genomes Project datasets.

Results and Discussion

We compared the proportion of polymorphic sites and the derived allele-frequency spectra of single nucleotide polymorphisms (SNPs) in CNEs with other regions of the human genome, for the purpose of exploring selective constraint in CNEs, and in a wider context understanding the evolutionary forces that define large *cis*-regulatory modules in vertebrate genomes. Comparisons were made with three categories of region; non-synonymous sites (those resulting in amino acid changes), synonymous sites from coding regions, and non-exonic sequences that do not overlap any CNEs or other annotated regulatory features. The non-synonymous sites act as a positive control, since it is known that a subset of non-synonymous changes is counteracted by relatively strong purifying selection [26]. Conversely, on the assumption that they are under negligible selection, we have used synonymous and non-coding sites as our negative controls. We obtained SNPs that map to CNEs, coding and non-coding regions from the public databases of both the HapMap Project and the 1000 Genomes Project and used the derived allele frequencies of SNPs to compare selective constraint in these sequences. We built alignments of CNEs from the human, macaque, mouse, chicken, frog, zebrafish and fugu genomes and defined two categories of sites in CNEs after masking the human sequence from the alignment. The human sequences were masked to avoid an ascertainment bias, since the second step of the analysis uses human data to identify polymorphic sites. These species last shared a common ancestor over 450 million years ago and in total, the divergence between these species represents approximately 1.9 billion years of evolution during which CNEs have retained a high degree of sequence similarity (divergence times to common ancestor [27] are given in Table 1). CNE sites that are invariant across all six species (i.e. excluding human) in our alignments are defined as Non-Variable Regions (hereafter referred to as NVRs) and sites where at least one substitution is present in any of the species are called Restricted Variable Regions (hereafter referred to as RVRs). By excluding the human reference sequence in our definition of NVRs and RVRs, we avoided mis-classifying rapidly evolving human sites that are represented by the derived allele in the human reference genome. In reality, given that the human reference genome is generated from a consensus of a number of individuals, that we have also included another primate (macaque) in the alignments and that CNEs in general appear to be under strong and deep evolutionary constraint, the number of instances where a human site alone would change the definition from NVR to RVR is very small (see Methods). Conservation of sequences across such large divergence times indicates strong historical evolutionary constraint, most likely reflecting the importance of such sequences as functional elements. The proportion of polymorphic sites found in 1,809 CNEs spanning a length of 318,286 bp together with coding and non-coding regions are given in Table 2.

Reduced diversity in CNEs is not due to a bias in variant calling

From Table 2 we observe a reduced frequency of variants in CNEs relative to synonymous sites suggesting that mutations in CNEs are subject to continuing purifying selection in the human genome. Although the 1000 Genomes Project uses a whole genome approach, the Variant Quality Score Recalibrator algorithm used by GATK to call SNPs relies on HapMap 3 sites to build the Gaussian Mixture model used to classify a variant as a "true site" [28]. This could mean that fewer SNPs in non-coding regions are being reported, resulting in the low levels of variation we observe in CNEs. If this was indeed the case, then the same bias should extend to non-conserved non-coding sites. However, the proportion of CNE sites that are polymorphic is significantly lower than that at non-conserved non-coding sites demonstrating that the observed low levels of variation at CNE sites is not an artefact of the variant calling procedure. Polymorphism is also significantly lower at synonymous sites compared with non-conserved non-coding sites. This trend is expected under a model of neutral evolution, since the synonymous sites are closely linked to non-synonymous sites that can experience either positive or purifying selection. The consequent effects of hitchhiking and background selection depress effective population size and hence polymorphism in coding regions [29], along with other influences such as the effects of epigenetic modification [30].

Sites within CNEs are subject to different levels of constraint

There are two broad explanations for the excess of substitutions that have occurred in Restricted Variable Regions of CNEs (compared with neighbouring NVRs): they could be neutral (or nearly neutral) changes persisting through genetic drift, or they could be adaptive changes fixed by selection. The first possibility would imply relaxed purifying selection acting on these sites, which should be apparent in greater within-species polymorphism. On the other hand, in the second case, purifying selection could be of comparable strength in the NVRs and RVRs. In order to investigate whether selective constraint in both classes of CNE sites is comparable, we therefore looked at both the proportion of polymorphic sites and spectra of derived allele frequencies. The imputation accuracy for low coverage imputed SNPs in the 1000 Genomes Project was highest for SNPs with an allele count of at least six [31]. Any biases introduced by imputation should affect both classes of CNE sites equally. Therefore, we chose to use this cut-off in our derived allele frequency spectra analyses of CNEs in comparisons with coding sequences. In fact the conclusions of the analysis are unaltered if alleles with lower counts are included: the derived allele frequency spectra obtained taking into account SNPs where the derived allele count is at least one is given in Figure S1. When we consider all observed SNPs in the two CNE categories, there are significantly fewer polymorphisms in NVRs relative to RVRs (Table 2). This observation indicates stronger selective constraint at sites that have been conserved across all lineages suggesting mutations in these regions may have functional consequences. In Figure 1 we observe a significant difference between the derived allele frequency spectra between the two classes of CNE sites, whereby NVRs have an excess of rare derived alleles compared to RVRs. This distinction between the two classes of sites within a CNE indicate that CNEs are composed of sites that are subject to different levels of evolutionary constraint and may have different roles in a regulatory context.

We note that the G+C content varies considerably in the different classes of constrained sites (NVR 35.8%, RVR 41.9%; non-synonymous 48.9% and synonymous 50.7%), and that an increase in G+C content is associated with an increase in heterozygosity in the human genome [32]. This increased heterozygosity results, at least in part, from an increase in the mutation rate brought about by the deamination of C to T at CpG sites where the cytosine is methylated [33]. In agreement with this, the proportion of SNPs occurring at the cytosine in ancestral CpG sites also varies across the different classes of site (NVR 6.7%, RVR 14.7%; non-synonymous 12.3% and synonymous 21.1%). Even after normalising for G+C content, purifying selection at CpG sites appears strongest at NVRs and considerably stronger than at RVRs (normalised values assuming 50% G+C are; NVR 9.35%, RVR 17.5%; non-synonymous 12.6%, synonymous 20.8%) in agreement with overall levels of constraint across the different classes of sites. In addition, all else being equal, the frequency of non-synonymous segregating sites would be slightly depressed, since the frequency of mutations appears to be additionally affected by neighbouring bases, in a manner that reduces non-synonymous mutations [34]; a pattern which might have been shaped by selection to reduce mutational load. Not all conserved non-coding sites show exceptionally low heterozygosity: an elevation in heterozygosity in HapMap and Environmental Genome Project data at a different set of conserved non-coding sites (CNSs), has been attributed to weaker selective effects on CNSs [35]. Similarly, we argue that the higher level of heterozygosity in RVRs suggests a relatively smaller proportion of deleterious mutations than at NVRs and non-synonymous sites.

Purifying selection in CNEs is more consistent than in coding sequences

Previous studies have shown that UltraConserved Elements (UCEs) defined by human-mouse-rat comparisons experience stronger purifying selection than coding regions [36]. UCEs are

Table 1. Divergence time since last common ancestor.

Organism	Divergence time since last common ancestor (Mya)
Macaque	29.2
Mouse	92.4
Chicken	301.7
Frog	371.2
Zebrafish	400.1
Fugu	400.1

The divergence time since last common ancestor with human obtained from Time Tree (Hedges et al., 2006).

Table 2. Variants from 1000 Genomes Low Coverage Data across the different categories.

Type of site	No. of sites surveyed	DAC >= 1		DAC >= 6	
		No. of SNPs	% sites with SNPs	No. of SNPs	% sites with SNPs
CNEs (Total)	318286	3182	1.00	1119	0.35
a) NVR	161361	1259	0.78	400	0.25
b) RVR	156925	1923	1.22	719	0.46
Coding sequences (Total)	680811	6118	0.9	2340	0.34
c) Non-synonymous (64%)	437283	3492	0.79	1161	0.26
d) Synonymous (36%)	243528	2626	1.07	1179	0.48
Non-coding	398240	5259	1.32	2429	0.6

The proportion of sites that is polymorphic in each category using SNPs where at least one derived allele is reported (Derived Allele Count (DAC) > = 1) and those SNPs where at least six derived alleles are reported (DAC > = 6).

(by definition) 100% conserved across at least 200 bp in the three mammalian genomes used to detect them. However, they often overlap exons where non-synonymous mutations can contribute to the excess of rare derived alleles. By contrast, CNEs do not overlap any known exons and represent a larger set of sequences conserved across all jawed vertebrates and with a broader range of sequence identity. Nevertheless,

NVRs within CNEs exhibit a similar proportion of sites that are polymorphic relative to non-synonymous sites (Table 2). The low diversity at NVRs in CNEs is accompanied by a derived allele frequency spectrum that shows a significant excess of rare derived

alleles relative to non-synonymous sites (Figure 1, Table 3), indicating stronger purifying selection at NVRs. A comparable pattern of polymorphism in mouse ultraconserved elements was interpreted in a similar manner [37]. In a separate study involving human-mouse Conserved Non-Coding sequences (CNCs) an excess of rare derived alleles relative to non-synonymous sites was observed in only those CNCs spanning 5′ and 3′ UTRs (UnTranslated Regions), indicating that conserved sequences in UTRs have an excess of weakly deleterious alleles compared to non-synonymous sites [38].

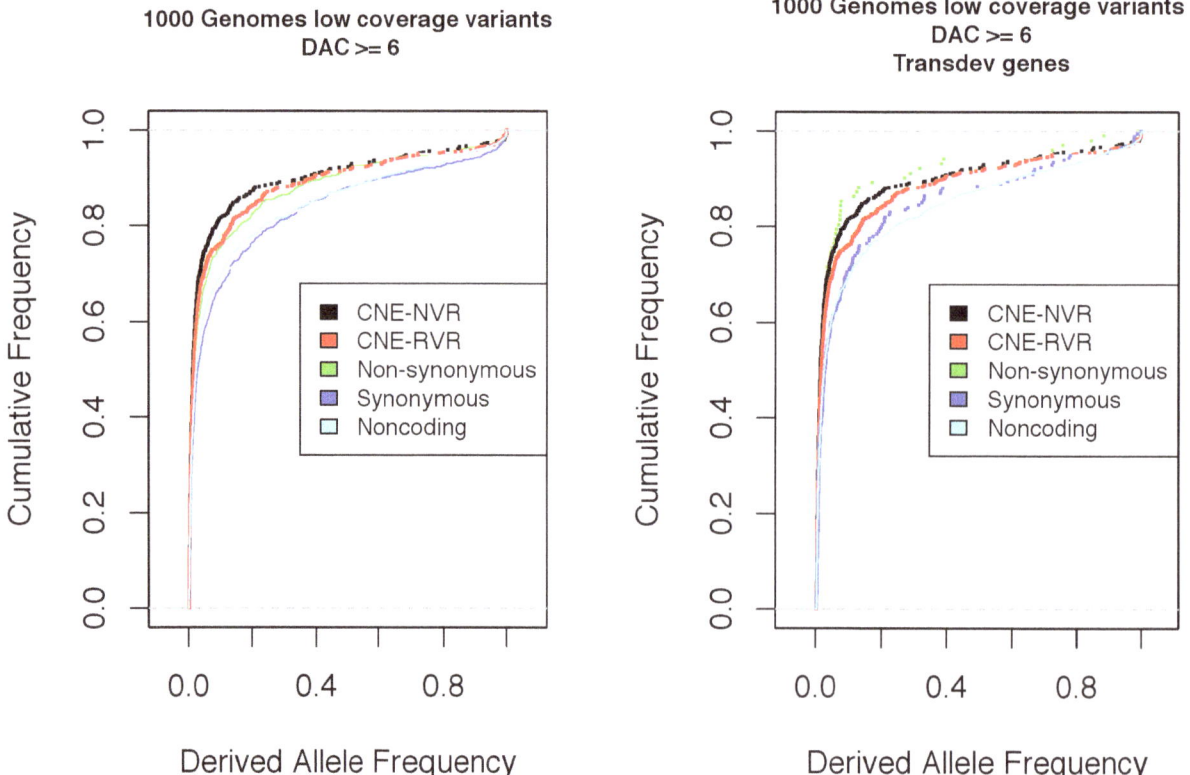

Figure 1. Cumulative derived allele-frequency in CNEs and control regions from 1000 Genomes Project. An excess of rare derived alleles is observed in CNE-NVRs, CNE-RVRs and Non-synonymous sites relative to Synonymous and Non-coding controls (using polymorphic sites with Derived Allele Count (DAC) > =6). The spectra at CNEs is comparable to the spectra at highly conserved trans-dev genes associated with CNEs.

In contrast, our dataset comprised sequences conserved over much longer evolutionary periods (fugu:mammals) and contained approximately 5% UTR sequences. Nevertheless we found evidence for strong purifying selection: there was no significant difference between the observed derived allele frequency spectra in CNE-NVRs (KS test, p-value $= 0.9045$) and the non-synonymous spectrum of derived alleles at the transcriptional and developmental genes (trans-dev genes) the CNEs are associated with. Strong purifying selection is expected at these trans-dev genes, since they are fundamentally important to vertebrate development and amongst the most highly conserved of all vertebrate genes (Figure 1).

We employed the approach of interpreting unfolded site-frequency spectra to infer the distribution of selective effects, as developed for comparisons between synonymous and non-synonymous sites in protein coding regions ([39], [40], [41]). The underlying logic is that mutations under weaker purifying selection have a higher probability of segregating in the population and drifting to higher frequencies than mutations with non-neutral effects. We explored the site-frequency spectra of different classes of sites in the Yoruba population of the 1000 Genomes Project to explain the observed levels of heterozygosity, combined with lower frequency of common alleles found at NVRs, relative to non-synonymous sites. Torgeson et al. [38] report both single estimates and distributions of gamma on the fitness effects of mutations in CNCs. The site-frequency spectrum at non-synonymous sites in our data is best explained by a gamma distribution of selective effects (shape $= 0.1$, rate $= 6.25$), consistent with previous findings using human polymorphism data [41]. The spectrum for NVRs is best described by a gamma distribution of selective effects with higher mean (shape $= 0.18$, rate $= 7.8$), with selection sufficient to keep mutations predominantly at lower frequencies (Figure 2A). The probability densities of the fitted gamma distributions (Figure 2B) indicate a larger proportion of lethal sites (selection coefficient $|s| > 1\%$) in CNE-NVRs (32%) compared to non-synonymous sites (21%), which suggests that mutations in CNE-NVRs are consistently deleterious.

Since the shape parameter of the gamma distribution of selective effects at non-synonymous sites from the 1000 Genomes data in this study is different from that inferred from previous datasets (e.g. [40]), we used the method of Eyre-Walker et al., 2006 [36] on the allele-frequency spectrum obtained from the 1000 Genomes data for comparison (Table 4). The shape parameter for

non-synonymous sites estimated from the method of Boyko et al., 2008 [41] falls within the 95% credibility intervals estimated using the method of Eyre-Walker et al. [40] and is consistent with a larger proportion of effectively neutral mutations in non-synonymous sites relative to CNE-NVRs.

The differences in the estimates of the distribution of selective effects could be attributable to differences in ascertainment in the datasets used to estimate its parameters. For example, Eyre-Walker et al. [40] used Environmental Genome Project (EGP) data from 320 genes involved in biologically important pathways (in 90 individuals) while Boyko et al [41] used genome-wide non-synonymous sites (in 19 individuals), both in US populations. We reason that the more varied SNP discovery panel and whole genome sequencing approach used by the 1000 Genomes Project discovered a larger proportion of rare variation, which contributes to greater diversity at non-synonymous sites than observed in previous datasets.

The different patterns of polymorphism observed at CNEs and protein coding loci can be interpreted in terms of their consequences for the molecular biology of the organism carrying the mutations. Non-synonymous changes can have major effects on protein structure and function, particularly through truncation. One might expect that there will be a specific number of non-synonymous changes in any coding sequence that might render the resultant protein completely non-functional, whereas other more conservative changes might have little or no effect on protein function. This would be reflected in a wide spectrum of selective pressures at a limited number of non-synonymous sites, with some being essentially immutable (thus no variant alleles) and others having relatively high derived-allele frequencies. CNEs represent an entirely different form of functional unit, possibly mediating their action through the binding of large numbers of transcription factors. In general, transcription factor binding sites are highly redundant with a rapid turnover rate (reviewed in [42]) but it has been proposed that large *cis*-regulatory modules such as CNEs might be composed of overlapping sets of binding sites thereby imposing a greater evolutionary constraint at each nucleotide position (reviewed in [24]). Mutations in regulatory motifs can cause tangible changes without being lethal. For example, polymorphisms in binding regions have been associated with allele-specific differential binding of RNA polymerase II and nuclear factor κB [43], and functional differences in transcriptional activity [44]. In *Drosophila*, reduced levels of polymorphism

Table 3. P-values from Chi-square and K-S tests.

Categories of sites (Derived Allele Count (DAC) >=6)		χ^2 P value	K-S P value
CNE - NVR	CNE - RVR	<2.20E-016	0
CNE - NVR	Non-synonymous	0.2364	0
CNE - NVR	Synonymous	<2.20E-016	7.07E-011
CNE - NVR	Non-coding	<2.20E-016	3.57E-008
CNE - RVR	Non-synonymous	<2.20E-016	0.11
CNE - RVR	Synonymous	0.24	1.65E-008
CNE - RVR	Non-coding	1.37E-011	1.55E-005
Non-synonymous	Synonymous	<2.20E-016	2.06E-005
Non-synonymous	Non-coding	<2.20E-016	0
Synonymous	Non-coding	<2.20E-016	0.01

P-values from the χ^2 test (df = 1) to detect differences in proportion of observed polymorphic sites between the different categories and the Kolmogorov-Smirnov test to detect differences in the derived allele-frequency spectra between categories.

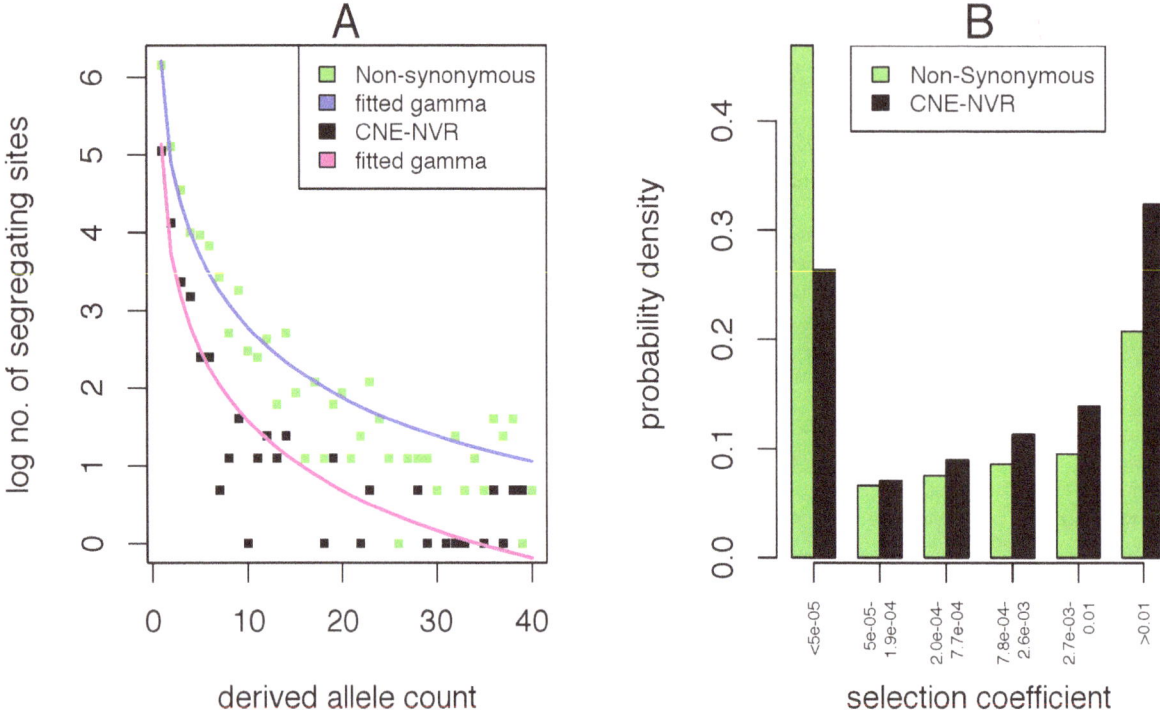

Figure 2. Site-frequency spectra in CNE-NVRs and non-synonymous sites. A) Site-frequency spectra in YRI binned into 3 units. The observed non-synonymous site-frequency spectrum fits a gamma distribution of selective effects (shape = 0.1, rate = 6.25). The observed site-frequency spectrum in CNE NVRs fits a gamma distribution of selective effects with a higher mean (shape = 0.18, rate = 7.8). B) The cumulative probability densities of the fitted gamma distributions indicate a larger proportion of lethal sites (>1%) in CNE-NVRs (32%) compared to non-synonymous sites (21%).

have been observed at functional transcription factor binding sites (i.e. transcription factor bound motifs) relative to instances of the same motif outside of the bound region that are not deemed functional [23]. Bound transcription factor-motifs have also been shown to be under stronger purifying selection than unbound motifs [45]. Consequently, some stretches of CNEs are likely to be under strong purifying selection than others.

The site frequency spectrum in CNE RVRs is consistent with a selective effect that is weaker than at CNE NVRs, reflected by the higher level of heterozygosity in RVRs and the higher proportion of derived alleles that drift to high frequencies. The relative relaxation in the selective effect at RVRs implies a difference in functionality. There is evidence that the length of sequence separating functional binding sites is more important than the composition of sequences in some DNA-protein interactions. For example, transcription factor p73 interacts with its half-sites differently in the presence of spacers of various lengths [46]. Therefore, it is possible that a proportion of some RVRs function as spacers that maintain the functional binding sites in NVRs. Alternatively, the more degenerate positions within a binding site

Table 4. Distribution of selective effects as inferred by the method of Eyre-Walker et. al, 2006.

	Non-synonymous sites (Eyre-Walker et al., 2006)	Non-synonymous sites	CNE-Nonvariable regions	CNE-Restricted variable regions
Shape	**0.23** (0.19, 0.27)	**0.079** (0.041, 0.17)	**0.19** (0.11, 0.30)	**0.26** (0.13, 0.46)
Mean $4N_e s$	**425** (225, 766)	**65900** (80.8, 475000)	**11840** (387, 92579)	**28.2** (10.7, 71.4)
$N_e s$	**Proportion of mutations**			
<1 (effectively neutral)	0.19	0.48	0.21	0.42
1–10 (slightly deleterious)	0.14	0.095	0.11	0.31
10–100 (strongly deleterious)	0.23	0.11	0.18	0.25
100–1000 (strongly deleterious)	0.31	0.31	0.5	0.026
1000–10000 (strongly deleterious)	0.13	0.005	0	0

The parameters estimates are accompanied by 95% credibility intervals (in brackets). The shape parameter for non-synonymous sites from the 1000 Genomes data in this study is different from that inferred from previous datasets (e.g. Eyre-Walker et al., 2006). Note that the categories of $N_e s$ are different to those from the method of Boyko et al., 2009 and the proportion of mutations in the various categories of $N_e s$ from the two methods are not directly comparable.

could be concentrated in RVRs resulting in a higher tolerance to mutations.

Discrepancies between the HapMap and 1000 Genomes datasets

Derived allele frequency spectra from HapMap [47] genotype data have previously been used to compare selective constraint between different types of sequences. For example, Conserved Non-Coding sequences defined by human-mouse and human-dog comparisons, which reflect much smaller divergence times than across the CNEs in this study, have been shown to be under stronger selective constraint than non-conserved regions and under similar constraint to non-synonymous mutations [25]. Before the 1000 Genomes data was publicly available we also looked at the derived allele-frequency spectra from HapMap Release #27. The derived allele frequency spectra of both categories of CNE SNPs obtained from the HapMap Project (Figure 3) are not significantly different to the spectrum at synonymous sites.

These discrepancies reflect a bias in the data toward coding SNPs due to the ascertainment procedure in the HapMap Project, given that it was designed to capture common variants, particularly in or near coding regions. In the HapMap Project, SNPs with a minor allele frequency of >0.05 in a panel of individuals with African, European and Asian ancestry were given preference. Much rare variation that is private to a specific population is lost in this way resulting in a large number of rare variants in non-exonic regions being excluded (reviewed in [48]). Furthermore, selected ENCODE regions were re-sequenced, hence rare variation in these regions are more likely to be captured. Notably, CNEs do not overlap any of the 15 ENCODE regions resequenced in Phases I and III of the HapMap Project. Preference was also given to validated SNPs in the HapMap Pilot Project [47]. Most non-coding variants are unlikely to be validated by other studies given the focus on mutation screening in exomes and there may still be a great number of singletons - i.e. present only in a single copy in the sampled population - being discarded as false positives by rigorous filtering.

With the HapMap dataset it is still possible to observe that non-synonymous sites have an excess of rare derived alleles relative to synonymous sites. Similarly we observe that NVRs in CNEs have an excess of rare derived alleles relative to RVRs. However it is impossible to determine that NVRs in CNEs are under stronger purifying selection than non-synonymous sites because the derived allele frequency spectra at both types of non-coding sequence (conserved and non-conserved) is biased downward relative to the spectra at both types of coding sequence. The results are clearer with the 1000 Genomes dataset because, albeit at low coverage, the whole genome sequencing approach captures a greater number of rare variants outside of the exome.

Conclusions

Unlike earlier studies, we have focused specifically on deeply conserved, pan-vertebrate, non-coding elements that have high regulatory potential during early development. Given their likely functional roles, we wanted to examine the strength of selective constraint at base pair resolution within these elements. We find different levels of purifying selection, as measured by both evolutionary conservation and human variation, across these large elements and show that some NVR sites are under consistently stronger purifying selection than non-synonymous sites in coding sequences, critically suggesting that mutations in NVRs within CNEs are likely to be deleterious.

The focus on exome-wide sequencing may benefit identifying causal variants in diseases/disorders that follow Mendelian patterns of inheritance [49] [50] [51]. However, the study of complex genetic diseases/disorders, for example developmental disorders determined by perturbation of regulatory networks, warrants either whole genome sequencing or targeted re-sequencing of putative regulatory regions to identify alleles that contribute to an increased risk of occurrence. Variant discovery pipelines in many genome-wide re-sequencing projects discard non-coding variants altogether resulting in potentially important data being lost. Because evolutionarily conserved non-coding DNA represents a small fraction of the vast non-coding landscape, the addition of loci spanning such regions to existing exome selection strategies may be particularly valuable.

We have combined deep, historical phylogenetic footprinting with the occurrence of SNPs and their derived allele frequencies in human populations to identify two classes of sites (Non Variable Regions and Restricted Variable Regions) in CNEs that experience different effects of selective pressure. Increasingly, approaches are combining phylogenetic footprinting with population genomics to effectively identify evolutionarily conserved non-coding sites that are likely to be functional and thus contribute to phenotypic variability and genetic disease [31] [52] [53]. We also found in our analyses that estimates for the distribution of selective effects at non-synonymous sites is different to those inferred from previous datasets, which is possibly attributable to the ascertainment strategy of variants where a whole genome approach across a geographically diverse variant discovery panel results in an unprecedented number of variants being captured.Future work will focus on mapping transcription factor binding sites from ChIP-Seq and other data to CNEs to further explore the relationship between deep evolutionary conservation and binding site degeneracy, paving the way for a better understanding of the role of mutations in regulatory regions in genetic disease. This type of approach has recently proven successful in an analysis of type 2 diabetes risk loci [54]. We might also get a clearer picture of whether overlapping known transcription factor binding sites might explain the signature of unexpectedly strong purifying selection acting on invariant sites in CNEs.

Methods

Generating multiple sequence alignments

CNE sequences in fasta format were downloaded from the CONDOR database [14] for the following species: *Homo sapiens, Macaca mulatta, Mus musculus, Gallus gallus, Xenopus tropicalis, Danio rerio* and *Takifugu rubripes*. Out of ~7000 human-fugu CNEs spanning a combined length of ~ 800,000 bp, we identified a subset of 1809 CNEs spanning 318286 bp that could be aligned across all seven species. Clustalw with default parameters was used to align the fasta sequences. Fasta sequences for the seven vertebrate species can be downloaded from the CONDOR website using the list of CNE IDs provided in Text S1.

Classifying NVRs and RVRs in CNEs

We defined two classes of sites within CNEs based on their conservation across the six vertebrate species by masking the human sequence. Custom Perl scripts were used to distinguish between sites that are identical in all six species (NVRs) and those that vary in at least one species (RVRs). In instances where the presence of the alternate allele in the human reference sequence made a non-variable region a variable region, the site was reclassified as a Non-Variable Region that is polymorphic. There were 27 polymorphic sites reclassified in this way. 161361 bp

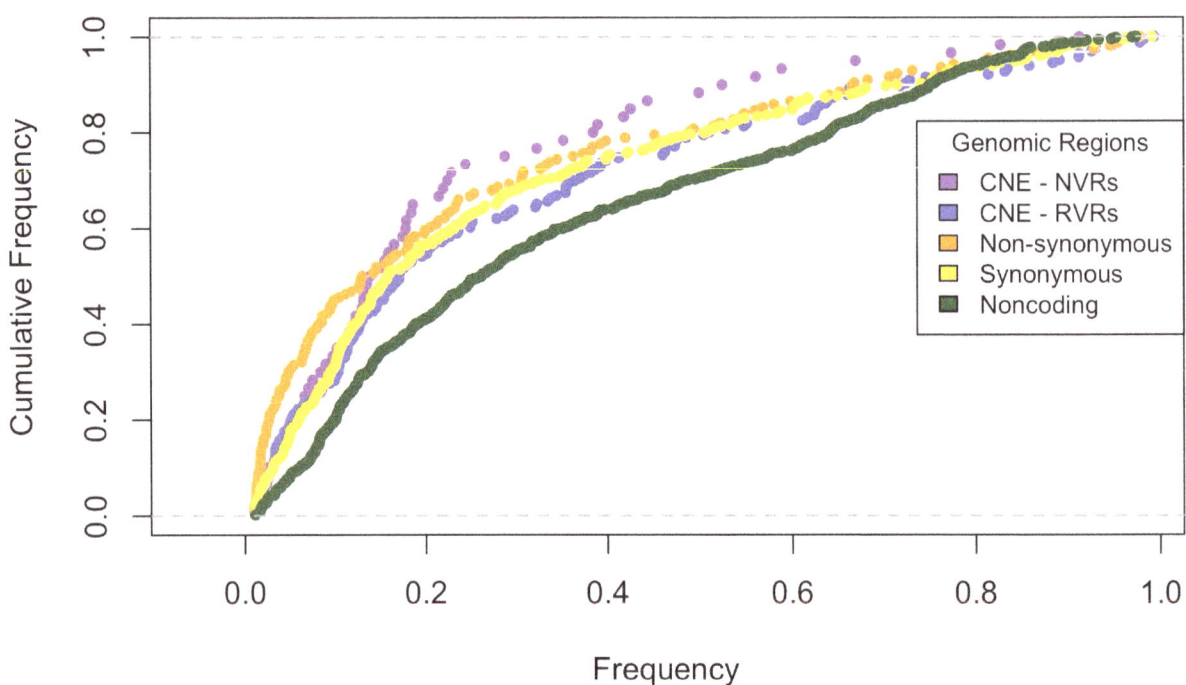

Figure 3. Cumulative derived allele-frequency in CNEs and control regions from the HapMap Project. An excess of rare derived alleles is observed only in Non-synonymous sites relative to Synonymous and Non-coding controls. CNE-NVRs have an excess of rare derived alleles compared to CNE-RVRs however, the derived allele-frequency spectra in CNEs resemble that at synonymous sites as a result of ascertainment bias in the HapMap dataset.

across 1809 CNEs were non-variable in all seven species while 156925 bp were variable in at least one species.

Control regions

a) Coding. The Biomart [55] tool on Ensembl 71 [56] was used to retrieve the exon coordinates of all transcripts on forward strand genes within 500 kbp of the CNEs in the analysis and the trans-dev genes associated with CNEs according to the CONDOR database. The transcript with the longest coding sequence was retained. The number of 0-fold, 2-fold and 4-fold degenerate sites in the full transcript with the longest coding sequence was obtained by using the software MEGA 5.1 [57]. The relative proportion of non-synonymous sites (0-fold degenerate) in the genic sequences was thus determined to be 64% whereas the proportion of synonymous sites (2 and 4-fold degenerate) was determined to be 36%. The genomic coordinates of the coding sequence for each transcript was used to extract variants. The consequence of the SNP to the transcript was determined from Ensembl VEP (Variant Effect Predictor) web tool [58].

b) Non-coding. The non-coding regions for the HapMap analysis only were randomly chosen from five different chromosomes. In the analysis of the 1000 Genomes Project data, the main dataset in this study, genomic regions within 5–6 kb away from the CNEs in the analysis that do not overlap known exons and CNEs constitute the non-coding control. The distance of 5–6 kb was small (corresponding to 0.006 cM in the human genome), so the sites would be exposed to the local reduction in effective population size due to any background selection. Previous analyses

(e.g. Hernandez et al., 2011, [59] have shown that these effects extend over ten times that range (>0.08 cM) with a twofold reduction in the levels of diversity being observed beyond genetic distances of 0.043 cM. Any bases in the non-coding controls that overlapped annotated GERP elements were excluded from the analysis.

Extracting allele frequencies from the HapMap Project

We used the marker IDs of SNPs reported from Biomart to extract the allele frequencies of SNPs from the HapMap Release #27 dataset with a customised XML query. 721 SNPs in ~800,000 bp of CNE regions were reported from HapMap Release #27. 182 non-synonymous, 400 synonymous and 982 non-conserved non-coding SNPs were obtained from length-matched control regions. The number of SNPs that mapped to 318286 bp is 70 SNPs in CNE NVRs and 176 SNPs in CNE RVRs. The allele frequencies were averaged across all the populations to obtain a global allele frequency for a given variant.

Extracting variants from 1000 Genomes Project

Variants were extracted from the file; "ftp://ftp.1000genomes.ebi. ac.uk/vol1/ftp/release/20110521/supporting/ALL.wgs.project_ consensus_vqsr2b.20101123.snps.low_coverage.sites.vcf.gz" using tabix [60]. The resulting vcf file was parsed using custom Perl scripts to obtain derived allele frequencies of variants. Variants with an alternate allele count of zero in the vcf file were excluded.

The ancestral state of all variants in the analysis was obtained from the vcf files located in; "ftp://ftp.1000genomes.ebi.ac.uk/

vol1/ftp/technical/working/20120213_phase1_integrated_release_version1/". All genomic coordinates were based on Hg19 Feb 2009 assembly of the human genome, Genome Reference Consortium GRCh37.

Significance testing

Kolmogorov-Smirnov and Chi-squared significance tests were carried out in R [61].

Base composition

The number of each type of nucleotide and the number of 'CG' occurrences were calculated from the fasta sequences of CNEs and coding controls using a combination of Perl and Biopython scripts.

Site-frequency spectra analyses

We used the full synonymous site-frequency spectra from the YRI population in the 1000 Genomes project as a neutral standard to fit a gamma distribution of selective effects using the population expansion model in the prfreq software [41]. The demographic parameters of tau (scaled time since non-stationary population dynamics) and omega (ratio of ancestral to current population size) derived from African-American data in [41] were not good predictors of the non-synonymous site-frequency spectrum in YRI, reflecting admixture in the African-American data. We found that using a smaller effective population size of Ne = 12818 predicted the observed site-frequency spectra better, reflecting a relatively milder population expansion in YRI. We then used the estimated demographic parameters under a gamma distribution of selective effects (shape = 0.1, rate = 6.25) to fit the observed non-synonymous site-frequency spectrum in YRI. Since there was a significant difference in the observed derived allele-frequency spectra between synonymous sites and non-coding regions in the vicinity of the CNEs, the site frequency spectrum at

non-coding sites was used to infer the demographic parameters (tau = 0.45, omega = 0.75, Ne = 16000) for assessing the distribution of selective effects in CNE-NVRs. A gamma distribution of selective effects with a higher mean (shape = 0.18, rate = 7.8) fitted the observed spectra at CNE-NVRs.

Supporting Information

Figure S1   Cumulative derived allele-frequency in CNEs and control regions from 1000 Genomes Project (full spectrum). An excess of rare derived alleles is observed in CNE-NVRs, CNE-RVRs and Non-synonymous sites relative to Synonymous and Non-coding controls when the full spectrum (Derived Allele Count (DAC) > = 1) of variants is used.

Acknowledgments

We thank Adam Boyko for instructions and guidance using the prfreq software in the site-frequency spectra analyses, Adam Eyre-Walker for his back of the envelope calculations and help with the analysis using the model of Eyre-Walker et al, 2006 [41], and once again Adam Eyre-Walker and two further anonymous reviewers for their comments and suggestions. We also thank Riwa Meshaka for her help with the design in extracting data from Ensembl using Biomart and from the HapMap Project using HapMart.

Author Contributions

Conceived and designed the experiments: GE RAN DRD. Performed the experiments: DRD. Analyzed the data: DRD GE RAN. Wrote the paper: DRD GE RAN.

References

1. Todorova A, Halliger-Keller B, Walter MC, Dabauvalle MC, Lochmller H, et al. (2003) A synonymous codon change in the LMNA gene alters mRNA splicing and causes limb girdle muscular dystrophy type 1B. J Med Genet 40: e115.
2. Cai JJ, Macpherson JM, Sella G, Petrov DA (2009) Pervasive hitchhiking at coding and regulatory sites in humans. PLoS Genet 5: e1000336.
3. Singleton AB (2011) Exome sequencing: a transformative technology. Lancet Neurol 10: 942–946.
4. Epstein DJ (2009) Cis-regulatory mutations in human disease. Brief Funct Genomic Proteomic 8: 310–316.
5. Lettice LA, Heaney SJ, Purdie LA, Li L, de Beer P, et al. (2003) A long-range Shh enhancer regulates expression in the developing limb and fin and is associated with preaxial polydactyly. Hum Mol Genet 12: 1725–1735.
6. Benko S, Fantes JA, Amiel J, Kleinjan DJ, Thomas S, et al. (2009) Highly conserved non-coding elements on either side of SOX9 associated with Pierre Robin sequence. Nat Genet 41: 359–364.
7. Albers CA, Newbury-Ecob R, Ouwehand WH, Ghevaert C (2013) New insights into the genetic basis of TAR (thrombocytopenia-absent radii) syndrome. Curr Opin Genet Dev.
8. Abecasis GR, Altshuler D, Auton A, Brooks LD, Durbin RM, et al. (2010) A map of human genome variation from population-scale sequencing. Nature 467: 1061–1073.
9. Tagle DA, Koop BF, Goodman M, Slightom JL, Hess DL, et al. (1988) Embryonic epsilon and gamma globin genes of a prosimian primate (Galago crassicaudatus). Nucleotide and amino acid sequences, developmental regulation and phylogenetic footprints. J Mol Biol 203: 439–455.
10. Berezikov E, Guryev V, Plasterk RH, Cuppen E (2004) CONREAL: conserved regulatory elements anchored alignment algorithm for identification of transcription factor binding sites by phylogenetic footprinting. Genome Res 14: 170–178.
11. Dubchak I, Ryaboy DV (2006) VISTA family of computational tools for comparative analysis of DNA sequences and whole genomes. Methods Mol Biol 338: 69–89.
12. Philippakis AA, He FS, Bulyk ML (2005) Modulefinder: a tool for computational discovery of cis regulatory modules. Pac Symp Biocomput: 519–530.
13. Boffelli D, McAuliffe J, Ovcharenko D, Lewis KD, Ovcharenko I, et al. (2003) Phylogenetic shadowing of primate sequences to find functional regions of the human genome. Science 299: 1391–1394.
14. Woolfe A, Goode DK, Cooke J, Callaway H, Smith S, et al. (2007) CONDOR: a database resource of developmentally associated conserved non-coding elements. BMC Dev Biol 7: 100.
15. Bejerano G, Pheasant M, Makunin I, Stephen S, Kent WJ, et al. (2004) Ultraconserved elements in the human genome. Science 304: 1321–1325.
16. Siepel A, Bejerano G, Pedersen JS, Hinrichs AS, Hou M, et al. (2005) Evolutionarily conserved elements in vertebrate, insect, worm, and yeast genomes. Genome Res 15: 1034–1050.
17. Woolfe A, Goodson M, Goode DK, Snell P, McEwen GK, et al. (2005) Highly conserved non-coding sequences are associated with vertebrate development. PLoS Biol 3: e7.
18. McEwen GK, Woolfe A, Goode D, Vavouri T, Callaway H, et al. (2006) Ancient duplicated conserved noncoding elements in vertebrates: a genomic and functional analysis. Genome Res 16: 451–465.
19. Parker HJ, Piccinelli P, Sauka-Spengler T, Bronner M, Elgar G (2011) Ancient Pbx-Hox signatures define hundreds of vertebrate developmental enhancers. BMC Genomics 12: 637.
20. Minovitsky S, Stegmaier P, Kel A, Kondrashov AS, Dubchak I (2007) Short sequence motifs, overrepresented in mammalian conserved non-coding sequences. BMC Genomics 8: 378.
21. Stormo GD, Zhao Y (2010) Determining the specificity of protein-DNA interactions. Nat Rev Genet 11: 751–760.
22. McDaniell R, Lee BK, Song L, Liu Z, Boyle AP, et al. (2010) Heritable individual-specific and allele-specific chromatin signatures in humans. Science 328: 235–239.
23. Spivakov M, Akhtar J, Kheradpour P, Beal K, Girardot C, et al. (2012) Analysis of variation at transcription factor binding sites in Drosophila and humans. Genome Biol 13: R49.
24. Elgar G, Vavouri T (2008) Tuning in to the signals: noncoding sequence conservation in vertebrate genomes. Trends Genet 24: 344–352.
25. Drake JA, Bird C, Nemesh J, Thomas DJ, Newton-Cheh C, et al. (2006) Conserved noncoding sequences are selectively constrained and not mutation cold spots. Nat Genet 38: 223–227.

26. Hughes AL, Packer B, Welch R, Bergen AW, Chanock SJ, et al. (2003) Widespread purifying selection at polymorphic sites in human protein-coding loci. Proc Natl Acad Sci U S A 100: 15754–15757.

27. Hedges SB, Dudley J, Kumar S (2006) TimeTree: a public knowledge-base of divergence times among organisms. Bioinformatics 22: 2971 2972.

28. DePristo MA, Banks E, Poplin R, Garimella KV, Maguire JR, et al. (2011) A framework for variation discovery and genotyping using next-generation DNA sequencing data. Nat Genet 43: 491–498.

29. Stephan W (2010) Genetic hitchhiking versus background selection: the controversy and its implications. Philos Trans R Soc Lond B Biol Sci 365: 1245–1253.

30. Keller I, Bensasson D, Nichols RA (2007) Transition-transversion bias is not universal: a counter example from grasshopper pseudogenes. PLoS Genet 3: e22.

31. Abecasis GR, Auton A, Brooks LD, DePristo MA, Durbin RM, et al. (2012) An integrated map of genetic variation from 1,092 human genomes. Nature 491: 56–65.

32. Sachidanandam R, Weissman D, Schmidt SC, Kakol JM, Stein LD, et al. (2001) A map of human genome sequence variation containing 1.42 million single nucleotide polymorphisms. Nature 409: 928–933.

33. Piganeau G, Mouchiroud D, Duret L, Gautier C (2002) Expected relationship between the silent substitution rate and the GC content: implications for the evolution of isochores. J Mol Evol 54: 129–133.

34. Antezana MA, Jordan IK (2008) Highly conserved regimes of neighbor-base-dependent mutation generated the background primary-structural heterogeneities along vertebrate chromosomes. PLoS One 3: e2145.

35. Asthana S, Noble WS, Kryukov G, Grant CE, Sunyaev S, et al. (2007) Widely distributed non-coding purifying selection in the human genome. Proc Natl Acad Sci U S A 104: 12410–12415.

36. Katzman S, Kern AD, Bejerano G, Fewell G, Fulton L, et al. (2007) Human genome ultraconserved elements are ultraselected. Science 317: 915.

37. Halligan DL, Oliver F, Guthrie J, Stemshorn KC, Harr B, et al. (2011) Positive and negative selection in murine ultraconserved noncoding elements. Mol Biol Evol 28: 2651–2660.

38. Torgerson DG, Boyko AR, Hernandez RD, Indap A, Hu X, et al. (2009) Evolutionary processes acting on candidate cis-regulatory regions in humans inferred from patterns of polymorphism and divergence. PLoS Genet 5: e1000592.

39. Piganeau G, Eyre-Walker A (2003) Estimating the distribution of fitness effects from DNA sequence data: implications for the molecular clock. Proc Natl Acad Sci U S A 100: 10335–10340.

40. Eyre-Walker A, Woolfit M, Phelps T (2006) The distribution of fitness effects of new deleterious amino acid mutations in humans. Genetics 173: 891–900.

41. Boyko AR, Williamson SH, Indap AR, Degenhardt JD, Hernandez RD, et al. (2008) Assessing the evolutionary impact of amino acid mutations in the human genome. PLoS Genet 4: e1000083.

42. Dowell RD (2010) Transcription factor binding variation in the evolution of gene regulation. Trends Genet 26: 468–475.

43. Kasowski M, Grubert F, Heffelfinger C, Hariharan M, Asabere A, et al. (2010) Variation in transcription factor binding among humans. Science 328: 232–235.

44. Butter F, Davison L, Viturawong T, Scheibe M, Vermeulen M, et al. (2012) Proteome-wide analysis of disease-associated SNPs that show allele-specific transcription factor binding. PLoS Genet 8: e1002982.

45. Mu XJ, Lu ZJ, Kong Y, Lam HY, Gerstein MB (2011) Analysis of genomic variation in non-coding elements using population-scale sequencing data from the 1000 Genomes Project. Nucleic Acids Res 39: 7058–7076.

46. Ethayathulla AS, Tse PW, Monti P, Nguyen S, Inga A, et al. (2012) Structure of p73 DNA-binding domain tetramer modulates p73 transactivation. Proc Natl Acad Sci U S A 109: 6066–6071.

47. Consortium IH (2005) A haplotype map of the human genome. Nature 437: 1299–1320.

48. Teo YY, Small KS, Kwiatkowski DP (2010) Methodological challenges of genome-wide association analysis in Africa. Nat Rev Genet 11: 149–160.

49. Guerreiro RJ, Lohmann E, Kinsella E, Br·s JM, Luu N, et al. (2012) Exome sequencing reveals an unexpected genetic cause of disease: NOTCH3 mutation in a Turkish family with Alzheimer's disease. Neurobiol Aging 33: 1008.e1017–1023.

50. Hammer MB, Eleuch-Fayache G, Gibbs JR, Arepalli SK, Chong SB, et al. (2013) Exome sequencing: an efficient diagnostic tool for complex neurodegenerative disorders. Eur J Neurol 20: 486–492.

51. Bras JM, Singleton AB (2011) Exome sequencing in Parkinson's disease. Clin Genet 80: 104–109.

52. Loots GG, Ovcharenko I (2010) Human variation in short regions predisposed to deep evolutionary conservation. Mol Biol Evol. 27: 1279–1288.

53. Ritchie GR, Dunham I, Zeggini E, Flicek P. (2014) Functional annotation of non-coding sequence variants. Nat Methods 11: 294–296.

54. Claussnitzer M, Dankel SN, Klocke B, Grallert H, Glunk V, et al. (2014) Leveraging cross-species transcription factor binding site patterns: from diabetes risk loci to disease mechanisms. Cell 156: 343–358.

55. Kasprzyk A (2011) BioMart: driving a paradigm change in biological data management. Database (Oxford) 2011: bar049.

56. Flicek P, Ahmed I, Amode MR, Barrell D, Beal K, et al. (2013) Ensembl 2013. Nucleic Acids Res 41: D48–55.

57. Tamura K, Peterson D, Peterson N, Stecher G, Nei M, et al. (2011) MEGA5: molecular evolutionary genetics analysis using maximum likelihood, evolutionary distance, and maximum parsimony methods. Mol Biol Evol 28: 2731–2739.

58. McLaren W, Pritchard B, Rios D, Chen Y, Flicek P, et al. (2010) Deriving the consequences of genomic variants with the Ensembl API and SNP Effect Predictor. Bioinformatics 26: 2069–2070.

59. Hernandez RD, Kelley JL, Elyashiv E, Melton SC, Auton A, et al. (2011) Classic selective sweeps were rare in recent human evolution. Science 331: 920–924.

60. Li H (2011) Tabix: fast retrieval of sequence features from generic TAB-delimited files. Bioinformatics 27: 718–719.

61. R Development Core Team (2012) R: A language and environment for statistical computing. R Foundation for Statistical Computing, Vienna, Austria. ISBN 3-900051-07-0, URL http://www.R-project.org/.

10

A Comparison of Peak Callers Used for DNase-Seq Data

Hashem Koohy[1,2]*, **Thomas A. Down**[1], **Mikhail Spivakov**[2], **Tim Hubbard**[1]*

1 The Babraham Institute, Babraham Research Campus, Cambridge, United Kingdom, **2** Wellcome Trust Sanger Institute, Wellcome Trust Genome Campus, Cambridge, United Kingdom

Abstract

Genome-wide profiling of open chromatin regions using DNase I and high-throughput sequencing (DNase-seq) is an increasingly popular approach for finding and studying regulatory elements. A variety of algorithms have been developed to identify regions of open chromatin from raw sequence-tag data, which has motivated us to assess and compare their performance. In this study, four published, publicly available peak calling algorithms used for DNase-seq data analysis (F-seq, Hotspot, MACS and ZINBA) are assessed at a range of signal thresholds on two published DNase-seq datasets for three cell types. The results were benchmarked against an independent dataset of regulatory regions derived from ENCODE in vivo transcription factor binding data for each particular cell type. The level of overlap between peak regions reported by each algorithm and this ENCODE-derived reference set was used to assess sensitivity and specificity of the algorithms. Our study suggests that F-seq has a slightly higher sensitivity than the next best algorithms. Hotspot and the ChIP-seq oriented method, MACS, both perform competitively when used with their default parameters. However the generic peak finder ZINBA appears to be less sensitive than the other three. We also assess accuracy of each algorithm over a range of signal thresholds. In particular, we show that the accuracy of F-Seq can be considerably improved by using a threshold setting that is different from the default value.

Editor: Manuela Helmer-Citterich, University of Rome Tor Vergata, Italy

Funding: This work was supported by a Wellcome Trust grant [098051]. The funders had no role in study design, data collection and analysis, decision to publish, or preparation of the manuscript.

* E-mail: hashem.koohy@babraham.ac.uk (HK); th@sanger.ac.uk (TH)

Introduction

Over the past decade, our ability to interrogate the features of the chromatin state has benefitted greatly from high-throughput sequencing (HTS) technologies. Genome-wide profiling of protein-DNA interactions has been made possible by Chromatin Immunoprecipitation coupled with high throughput sequencing (ChIP-seq) for a remarkable number of protein targets [1–3]. Similarly, HTS can be combined with the established DNase I hypersensitivity assay (DNase-seq) to profile open chromatin regions [4–7]. This approach has led to the detection of a total of nearly three million DNase I Hypersensitive Sites (DHS) across the human genome in about 140 different cell type [3,8].

Probing the chromatin state using ChIP-seq and DNase-seq requires sophisticated data analysis pipelines once the sequence reads have been collected, but at their core, all analysis approaches involve gauging the significance of enrichment of short read tags in a given region relative to an expected background distribution. Algorithms used for this purpose are generally known as peak callers [6,9].

Analysis of ChIP-seq data has received a great deal of attention and an enormous range of peak callers have been implemented [6,10–14], benchmarked and extensively reviewed [2,9–11,15,16]. However, DNase-seq has thus far received less attention and to the best of our knowledge there has been no systematic comparison of the performance of algorithms for calling DHSs from DNase-seq data. This places the end user in an uncertain situation, with little

evidence to base decisions on as to which tools to use and with what parameter settings.

The properties of enriched regions vary greatly between different HTS-based chromatin interrogation technologies. For example, TF-ChIP experiments typically yield very sharp and punctate signals, while histone-ChIP for modifications such as H3K36me3 are much more broadly distributed. Signals from DNase-seq data, in turn, appear neither as sharp as those in TFBS ChIP-seq, nor as broad as in a typical histone modification ChIP [17,18]. Therefore, peak callers that have been originally developed with ChIP-seq data in mind are usually not recommended for DNase-seq data, at least without additional parameter tuning [19].

To address this problem, a number of approaches have been presented. The Hotspot [7,18] and F-Seq tools [20] have been implemented specifically for use with DNase-seq data (although F-Seq has also been used for ChIP-seq and FAIRE-seq data [21]). In contrast, Zero-Inflated Negative Bionomial Algorithm (ZINBA) [17] has been proposed as a generic tool for handling a variety of HTS data types including DNase-seq, FAIRE-seq, ChIP-seq and RNA-seq. Finally, several published studies have used the Model-based Analysis of ChIP-seq (MACS) peak caller [13] for the analysis of DNase-seq data [22]. As we will see, these tools are based on a diverse range of mathematical models, have different parameter spaces, and deal differently with the problem of background estimation.

In this paper, we compare the performance of the aforementioned four tools (all of which are open-sources and publicly available) on several DNase-seq datasets from the ENCODE project. The analysis has been performed on the chromosome 22 of the human genome GRCh37 assembly. The key aim of our analysis is to present a framework within which the user can decide which peak caller is more applicable to their data and whether or not the default signal threshold is appropriate in their case. In what follows, we first provide the reader with a brief overview of each of these peak callers and then present the results of our analyses.

Results

An Overview of Peak Callers

In this section we provide the reader with a brief description of each of the four tools used for benchmarking. More specifics about these algorithms including the version number, run time, the language in which they have been implemented and their original references are summarized in Table 1.

Hotspot. The Hotspot [7,8] algorithm is the underlying algorithm used for the discovery of DHSs in the ENCODE project. The idea behind Hotspot is to gauge the enrichment of sequence tags in a region compared to the background distribution. Enrichment is measured as a $Z-$ score, taking the binomial distribution of tag frequencies as the null model. Considering a small window of length 250 bp centred in a larger window of length 50 kb, the probability of each tag in the larger window hitting the small window is denoted as p which is defined as the ratio of the number of uniquely mappable tags in the smaller window to those in the larger window. (Note that p may differ in different regions because not all $k-$ mers in a window can be aligned uniquely to the reference genome).

Assuming n tags hitting the smaller window and N tags hitting the larger window, the expected number can be calculated as $\mu = Np$, the standard deviation as $\sigma = \sqrt{Np(1-p)}$, and the $Z-$ score (that is then assigned to the small window) as $z = \frac{n-\mu}{\sigma}$. Using this method, each tag is assigned a $Z-$ score which is equal to the $Z-$ score of a small window centred at that tag position. Then a "hotspot" region is defined as a succession of tags having a $Z-$ score above a specific threshold (assumed to equal two by default). Hotspot infers its final hotspots after two phases. Some highly enriched regions are detected as the first phase hotspots and the corresponding tags are filtered out from the set of short read tags. In the second phase, Hotspot tries to discover weaker but reproducible peaks that might have been overshadowed by the most enriched regions. Finally, the results of these two phases are combined and subjected to false discovery rate analysis. For this, Hotspot generates a set of random tags that is uniformly distributed over the mappable region of the genome. For a given $Z-$ score threshold T, the *FDR* for the observed peaks centered at each tag with a threshold greater than or equal to T is defined as a ratio of the number of random tags with $Z-$ scores greater than or equal to T to the number of observed tags falling within the same score range.

Hotspot is mainly programmed in C^{++}, but the statistical analyses have been implemented in R. Some parts of the algorithm are also written in Python and as Unix shell scripts. The package depends on BEDOPS [23] and BEDTools [24].

A new implementation of Hotspot named "Dnase2hotspots" has been reported by Baek et al. [18]. The key difference between the two versions seems to be the merging of the two-pass detection in the original Hotspot algorithm into a single pass. At the time of our analyses, Dnase2hotspots required MATLAB for running, and was therefore excluded from the benchmarking. However, as this manuscript was at a late stage of revision, we learned that an updated version of Dnase2hotspots became available that no

Table 1. DNase I Peak Callers Benchmarked in This Study.

Algorithm	Version	CPU Time(sec.)	Max Memory(Processes)	Control Data	Mappablity Data	Language	Refs
Hotspot	V3	6824	2288 MB(10)	Not Required	Required	C^{++}	[7]
F-Seq	1.84	1296	4304 MB(3)	Not Required	Not Required	JAVA	[20]
MACS	macs2(2.0.10)	562	1452 MB(3)	Optional	Not Required	Python	[13]
ZINBA	zinba_2.01	412093	10318 MB(10)	Optional	Required	R	[17]

This table shows a number of properties of the four peak calling algorithms used in this study, highlighting some of the differences (as evaluated on an Intel(R) Xeon(R) CPU E5440 @ 2.83GHz machine). The CPU Time in seconds was obtained when running each algorithm on the K562 data (for the other two cell type comparable figures are obtained, data not shown). The Max Memory column shows the maximum memory used by the algorithm. The numbers in parentheses show the maximum number of processes used when generating this data. For MACS and ZINBA control data is optional and is believed to improve the accuracy of the algorithm, but in this study all algorithms have been run without control data sets. Hotspot and ZINBA require mappability data and both algorithms provide it for a set of specific lengths including 36 bp. The core Hotspot algorithm is implemented in C^{++}, however the Hotspot pipeline involves R, python and bash scripts. ZINBA comes as an R package having its core implemented in C.

A

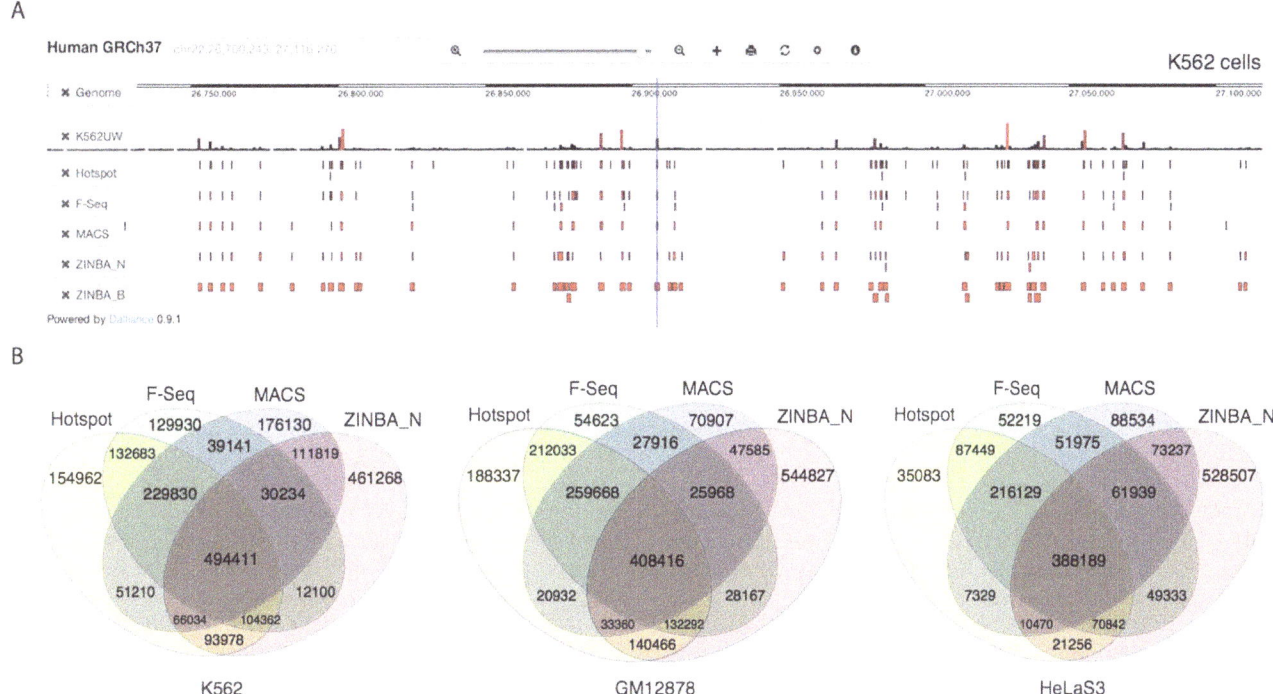

Figure 1. Comparison of the Four Peak Callers in a Representitive Genomic Region. (A) A screenshot from Dalliance [29] showing peaks called by the four peak callers in about 400 kb of chromosome 22 in K562 cells. The first row in this figure labelled as 'K562UW' illustrates the distribution of short read tags of K562 (replicate 1) from University of Washington (see Methods for full details). The following rows show the statistically significant regions (peaks) according to each of the algorithms with their default signal thresholds. (B) Overlap between peaks called by each algorithm. Venn diagrams showing the overlap between peaks called by each of the four algorithms using their default parameters in K562 cells (left), GM12878 cells (middle) and HeLaS3 cells (right). The numbers correspond to the number of basepairs called.

longer requires MATLAB (http://sourceforge.net/projects/dnase2hotspots/).

F-Seq. F-Seq [20] was developed with the aim of summarising DNase-seq data over genomic regions. The authors identified problems with histogram based-peak calling algorithms, in which the enrichment of tags is measured across equal-sized bins. Such algorithms suffer from boundary effects and difficulties in selecting bin widths.

In F-Seq, it is assumed that n short tags $\{x_i\}$ are independently and identically distributed along the chromosome i.e. $x_i \cong p(x)$ such that the probability density function is inferred as: $\hat{p}(x) = \frac{1}{nb}\sum_{i=1}^{n} K(\frac{x-x_i}{b})$ in which b is the bandwidth parameter to control the smoothness and $K()$ is a Gaussian kernel function. Although this algorithm was initially developed for DNase-seq data, it has also been used for ChIP-seq peak detection [20].

ZINBA. ZINBA [17] is a generic algorithm for genome-wide detection of enrichment in short-read data that was proposed for the analysis of a broad range of genomic enrichment datasets. ZINBA first divides each chromosome into small non-overlapping windows (250 bp by default) based on the number of reads. These read count values, alongside other covariates including G/C content, mappablility scores, copy number variation and an estimation of background distribution make up the parameters of a mixture regression model. This model then assigns each region into one of three classes: enriched, background, or zero (windows for which no read is assigned due to insufficient sequencing coverage). The relationship between the covariates and the signal for various experimental data is then inferred through an Expectation Maximisation-based implementation of a mixture regression model. ZINBA is supplied as an R package.

MACS. MACS [13] is one of the most popular peak callers for ChIP-seq data [14] that has recently been used for DNase-seq [22]. As a ChIP-seq tool, MACS has been reviewed and benchmarked in a number of studies [9,10,12]. The key advantage of MACS compared to previous peak callers is that it models the shift size of tags and can also allow for local biases in sequencability and mappability through a dynamic Poisson background model. MACS is written in Python and can be run with or without an input control dataset. The only required input for this model is a set of short read tag alignments.

The Sensitivity and Specificity of the Peak Callers

To systematically evaluate the performance of Hotspot, F-Seq, MACS and ZINBA, we ran them on the publicly available DNase-seq data sets for K562, GM12878 and HelaS3 cell type over human chromosome 22 [8] (see Methods for the availability of these data sets). A visual inspection of peaks generated by these peak callers (Figure 1A) at their default signal threshold showed that their were not fully consistent. In particular, it can be seen that while some regions of strong enrichment were consistently detected, there was a significant variation in the detection of weaker regions, as well as in the sizes of the recovered DHS peaks. To our surprise, only ~11.5% of the reference set (at the base pair level) were consistently detected by all four tools (8% in K562, 13% in GM12878 and 14% in HeLaS3, respectively). Overall, peaks detected by at least one tool spanned on average 41% of the reference set (30% in K562, 48% in GM12878 and 46% in HeLaS3, respectively). This is likely due to a combination of factors, including the genuine mapping of some TF binding sites in the reference set outside of regions of increased chromatin accessibility, some "true" DHSs missed by the DNase-seq protocol and the false-negative rates of the peak detection tools themselves. The base-pair overlap of the peak regions detected by each algorithm in the three cell lines is shown in Figure 1B. Significant differences were also observed in the running times of the algorithms, with ZINBA taking on average $370\times$ longer and using $4.5\times$ more memory than the rest (Table 1).

We then ran each of these four tools over a range of signal thresholds and compared the peaks detected by each algorithm at each threshold level to the "reference sets" of regulatory regions. These sets were generated by pooling the ChIP binding profiles of multiple transcription factors (TFs) in each of the three cell types (the ChIP data was produced by ENCODE [15], see also Supplementary Data S1 (Files GM12878, K562 and HeLaS3) for the list of TFs used). Using TF-binding profiles to produce the reference set has been motivated by the fact that the majority of TF binding sites map to regions of increased chromatin accessibility that are detectable as DNase hypersensitive sites [5,8]. Although our reference set is inevitably incomplete, since the ChIP data is only available for a subset of TFs, it still allows us to robustly assess the relative performance of the DHS-calling algorithms (as used previously in [25]).

For each of the four algorithms, we estimated the sensitivity (expressed in the terms of the True Positive Rate, TPR) and specificity (expressed as $1 - FDR$, False Discovery Rate) from the degree of the overlap (at base pair level) of their respective DNase I peaks at each signal threshold with each of the reference sets. This approach is presented in more detail in the Methods section. The sensitivity-specificity analysis revealed further substantial differences between the peak finders (Figure 2). In particular, we found ZINBA to underperform all other tested tools in terms of both TPR and FDR. Its "narrow peaks" output (ZINBA_N) showed the lowest FDR among all algorithms, but also the lowest TPR, meaning that ZINBA_N may miss many true DHSs. On the other hand, ZINBA's "broad peaks" output (ZINBA_B) still had a relatively low TPR but also showed the highest FDR, meaning that its broad peaks showed a poorer overlap with the reference set compared to the other three peak callers.

Both the TPR and FDR of the other three peak callers (Hotspot, F-Seq and MACS) segregated by nearly 10% on the data from the GM12878 cell type. As we can see from Figure 2, in this cell type, F-Seq showed the highest TPR and Hotspot showed the best (lowest) FDR. More similar FDR and TPR values were observed in the other two cell types, with both F-Seq and Hotspot having only slightly lower TPR and higher FDR compared to MACS (Figure 2).

We asked if the relative performance of the algorithms is affected by the choice of a specific DNase-seq protocol. Currently, there are two DNase-seq protocols commonly used by the community: the "end capture" protocol [26] and the "double hit" protocol [27]. While this study so far focused on the "double hit" protocol, we also evaluated the performance of the algorithms with the "end capture" protocol using the ENCODE data for the K562 cell type [8]. However, we found the relative performance of the algorithms to remain generally consistent across the two protocols (Figure S1).

Overall these results suggest that F-Seq, Hotspot and MACS generally outperform ZINBA with DNase-seq data in terms of both specificity and sensitivity, with the F-Seq algorithm showing the best performance of all four algorithms tested.

$$z = 1,2,3,4$$

Comparison of the Summary Statistics of the Detected Peaks

We next sought to evaluate how the differences in the performance of the four algorithms are reflected in the summary statistics of the respective peaks. As shown in Figures 3 and 4, peaks detected by the four algorithms vary both in the total number and their length distributions. In particular, MACS produced the smallest number of peaks compared to the other three algorithms, followed by ZINBA (for which the numbers of broad and narrow peaks were equal). The peaks from F-Seq and Hotspot outnumbered both MACS and ZINBA peaks, with either F-Seq or Hotspot yielding the highest number depending on the cell type.

ZINBA's broad peaks were on average the longest compared to all other datasets, ranging from 1 kb to 10 kb (Figure 4). These were followed, sequentially, by MACS peaks (with a median of around 2700 bp over all three cell type), ZINBA narrow peaks and Hotspot peaks (median length 2.5 kb). F-Seq peaks were on average the shortest, with a median of 2 kb but notably, they showed a considerably higher variance of peak lengths (Figure 4).

These differences prompted us to look at the overall peak coverage produced by each algorithm, which we defined as the ratio of the number of base pairs covered by the peaks to the length of the chromosome. Note that chromosome 22 has an active arm of about 35 Mb. It can be seen from Figure 5, with the exception of ZINBA.B (broad) peaks showing an appreciably higher coverage than the rest, the peaks from all four algorithms (including ZINBA's narrow peaks) showed a comparable coverage. On average, MACS showed the lowest coverage and ZINBA.N showed the greatest coverage among the narrow peaks of algorithms. The highest spread of coverage (1.35%) was observed in GM12878 cells, between ZINBA.N (3.88%) and MACS (2.53%). The lowest spread of 0.6% was observed in K562 cells. The similarity in the peak coverage produced by the four algorithms at their respective default parameter settings suggests that these settings were generally appropriate for a relative evaluation of the tools' performance.

Effects of Algorithm-specific Parameters

So far, we have compared the algorithms' performance across the range of a single parameter that was common to all four peak callers: the overall signal threshold for making a peak call. Although a number of additional, mostly algorithm-specific, parameters exist, we kept them at their default values. A comprehensive evaluation of the peak callers over their full parameter spaces is challenging due to the algorithm-specificity of some parameters and also to the extensive number of parameter combinations. Some of these parameters, however, are unlikely to affect the sensitivity or specificity of the algorithms, as they are concerned either with other data types (eg ChIP-seq) and/or file formats. For example, in MACS one may see "–broad" and "–call-summits" for data type, "–g" for genome size and "–f" for file format. However, a number of tunable parameters, in particular in Hotspot and ZINBA seemed to affect the key parts of the respective algorithms, prompting us to ask whether they have a significant effect on the results.

For Hotspot, we evaluated the effects of the $z-$ score and the merging size threshold. As shown in Figure S3, the distribution of peaks' lengths is nearly indistinguishable when merging peaks closer than 150 bp (default) or not merging them at all. Similarly, we found that the performance of the Hotspot remains almost invariable at a range of $z-$ scores ($z = 1,2,3,4$; Figure S4).

For ZINBA, we assessed the effect of the number of hits per read allowed during mapping process ("athreshold"), and of

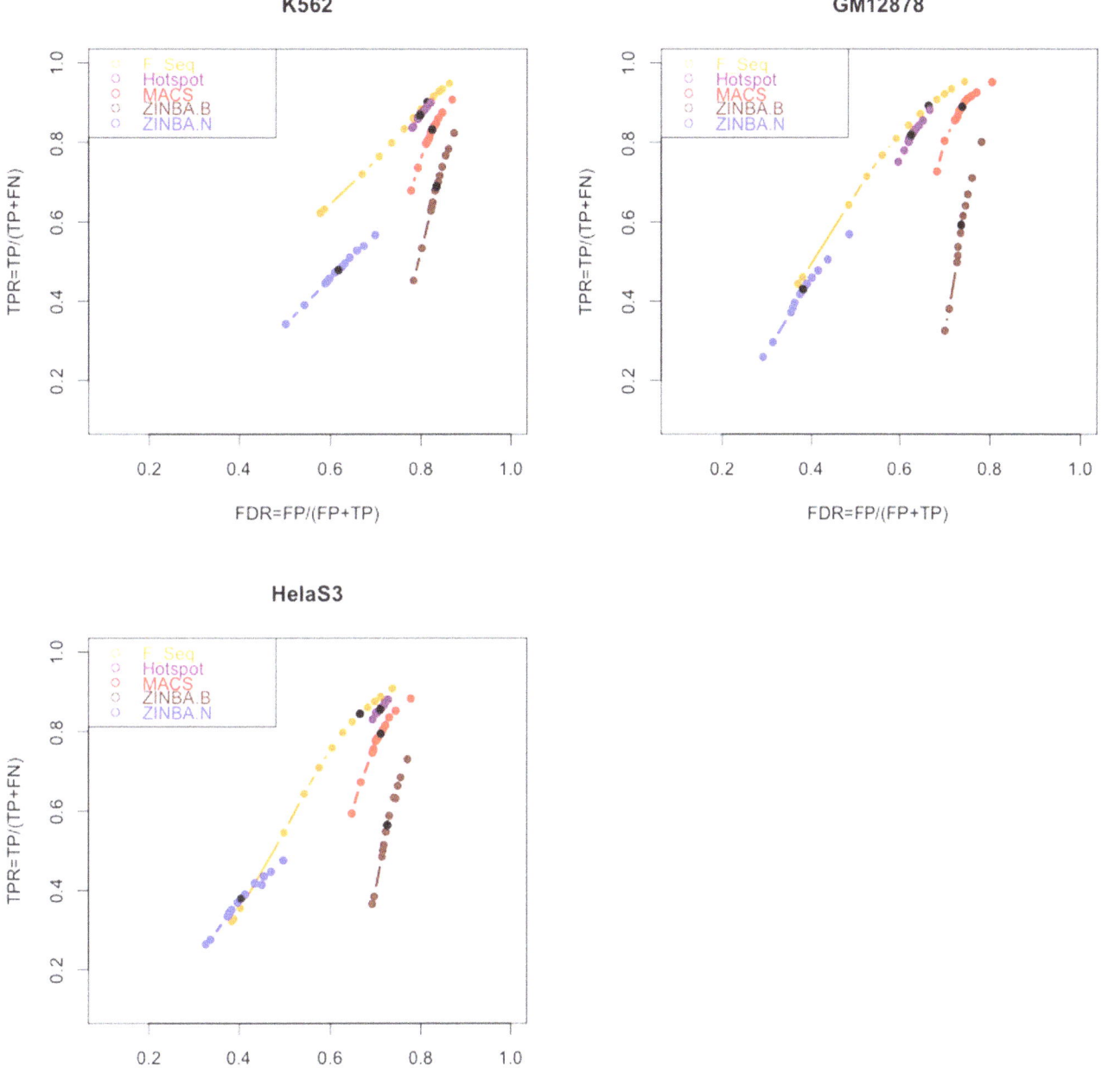

Figure 2. Comparison of the Peak Calling Algorithms Based on Estimated True Positive and False Discovery Rates. Each algorithm was run over 13 values of a parameter that controls the false discovery. These values for Hotspot, MACS and ZINBA range from 0.001 to 0.2 and for F-Seq it ranges from 0.001 up to 6 (see methods for more details). For each value the overlap between the calls and the "reference set of regulatory regions" for that cell type was measured. The black dots show the default value for each algorithm.

average fragment library length ("extension") on its performance. As can be seen from Figure S5, peak coverage remained insensitive to varying the "athreshold" parameter. In contrast, increasing the "extension" parameter from the default resulted in the peak coverage increasing beyond the range observed for all other peak callers.

In conclusion, we found no evidence that adjusting the algorithm-specific parameters of Hotspot and ZINBA leads to improved performance compared to their default parameter settings.

Adjusting the Default Signal Threshold Setting Improves the Performance of F-Seq

As a final step in our analyses, we set out to determine the peak signal threshold settings that ensure an optimal tradeoff between sensitivity and specificity. To this end, we expressed the sensitivity and specificity data for each peak caller generated over a range of signal thresholds (described above and shown in Figure 2) in terms of the $F-$ score metric which is commonly used in information retrieval. The $F-$ score combines both the sensitivity and specificity such that the higher $F-$ score values indicate a more

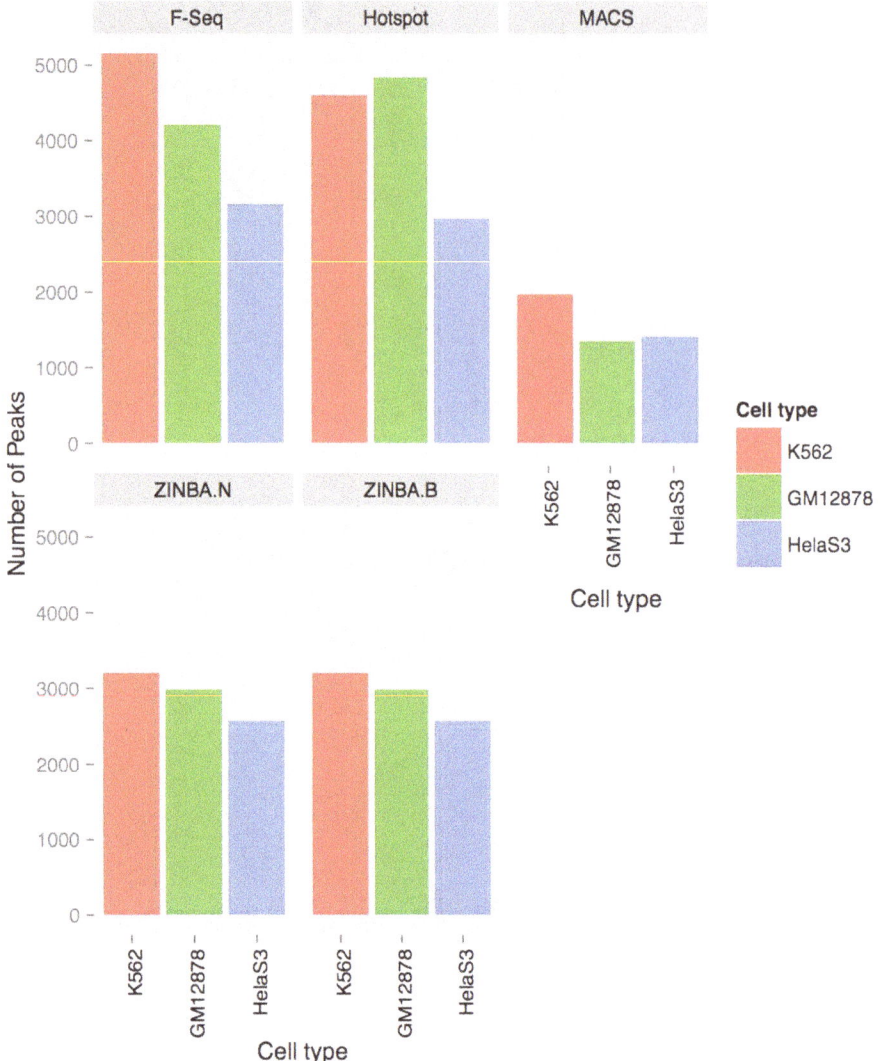

Figure 3. Number of Peaks Detected by Each Peak Caller Using Their Default Parameters. The number of peaks obtained by each algorithm at their default signal threshold.

optimal performance (see Methods and also [25]). The relative contribution of sensitivity and specificity is weighted by the β parameter that we assumed to be 0.5 to place a higher emphasis on specificity over sensitivity (see Methods for more detail).

In Figure 6, we plotted the $F-$ scores corresponding to a range of peak thresholds for each of the tools. As can be seen, F-Seq showed an improved performance when its signal threshold (defined by the "standard deviation threshold" parameter) was reduced from the default value of 4 to a value between 2 and 3. In contrast, Hotspot performance remained largely unchanged over the range of its threshold parameter. For MACS, the default threshold settings seemed optimal. ZINBA on the other hand, showed continuously decreasing $F-$ scores with increasing threshold, suggesting no clear-cut optimal threshold setting.

In conclusion, while Hotspot and MACS showed a near-optimal performance at the default signal threshold settings, the performance of F-Seq can be further improved by reducing the threshold parameter.

Discussion

In this study, four open-source peak callers proposed for the analysis of DNase-seq data were benchmarked and briefly reviewed. Our results showed that there is, in fact, a considerable discrepancy in the tools' performance. Of the four peak callers, F-Seq showed the best performance with DNase-seq data, particularly when run with a signal threshold level slightly lower than default. Both Hotspot and MACS also showed appreciable performance, only slightly lagging behind F-Seq in both sensitivity and specificity. In contrast, and despite its reported performance with RNA-seq, ChIP-seq and FAIRE-seq data [17], ZINBA showed to be less suitable for DNase-seq data analysis, both in terms of specificity, sensitivity and the computational time. To the best of our knowledge, this peak caller has not been used with DNase-seq in any published studies.

Although both ChIP-seq and DNase-seq experiments generate short-read tags, there exist a number of differences between these data types that caution against the application of ChIP-seq peak callers to DNase-seq data, at least without re-tuning their

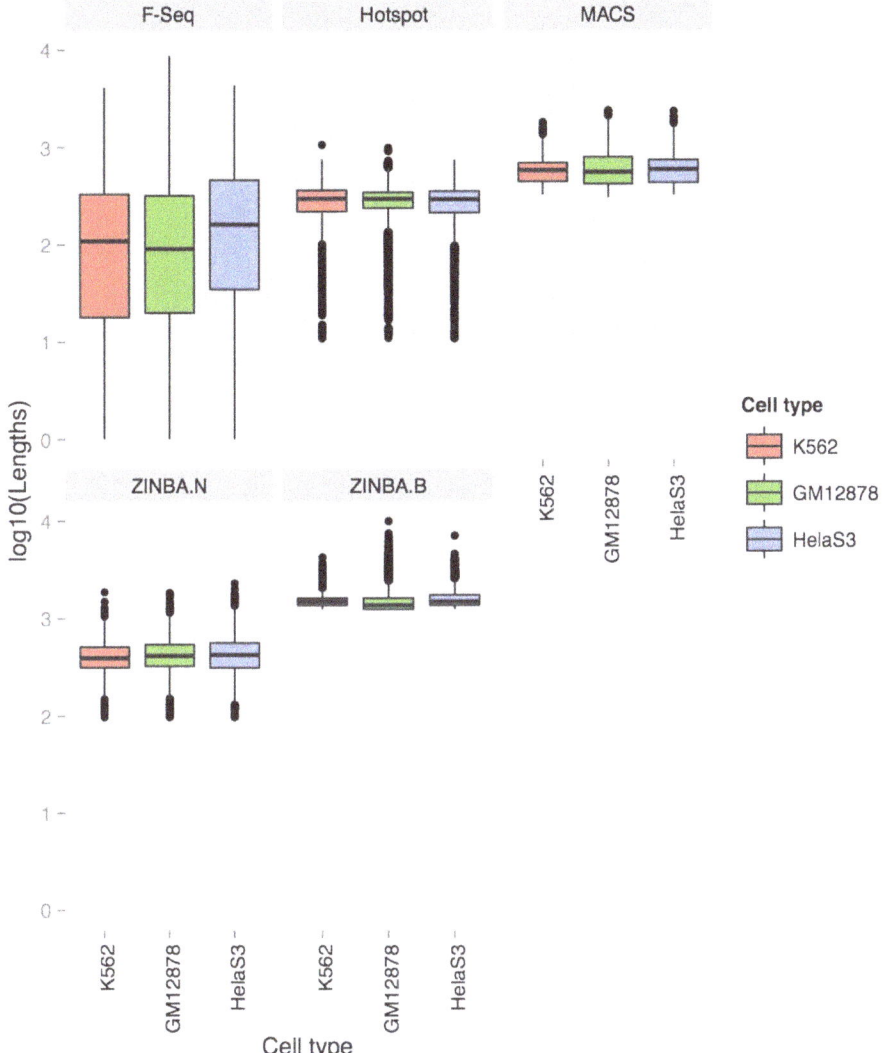

Figure 4. Distribution of Lengths Depending on Peak Callers and Their Parameter Settings. Distribution of peak lengths found by each of the algorithms, when ran with their default parameters, are compared between cell types. ZINBA.N and ZINBA.B represent narrow peaks and broad peaks (respectively) obtained from ZINBA.

parameters. The key differences include: a) ChIP-seq data usually shows a higher signal-to-noise ratio compared to DNase-seq, making ChIP-seq peaks easier to detect; b) ChIP-seq data, unlike DNase-seq data, are strand-specific with a shift in the signal between strands; c) as the general hallmarks of open chromatin regions, DHSs may cover wider regions, spanning the binding positions of different regulators and differentially modified histones; therefore DHSs vary more broadly in length compared to typical ChIP-seq peaks [9,19]. Taking these differences into account, one may conclude that the ChIP-seq-oriented peak caller MACS performs relatively well for DNase-seq data.

In our analyses we benchmarked the performance of each algorithm against a "reference set" of regulatory regions, generated from the union of multiple TF-binding profiles from ENCODE. This allowed us to compare the results of the peak callers with a "standard" that is based on a different type of experimental data and that is analysed using a different set of tools. It must be noted that, despite the large number of TFs used, our "reference set" is necessarily incomplete and may have its own

inherent biases. It seems unlikely that these biases would selectively favour the performance of some DNase-seq algorithms over others. The continued expansion of the range of TFs profiled by ChIP will make it possible to further improve the precision of such reference sets in the future.

Furthermore, we recently showed that DNase I has DNA binding preferences [25] that potentially present a source of bias in DHS detection. This largely unexpected property of the DNase I enzyme is currently unaccounted for by any peak caller. There may therefore be scope for a new generation of DHS peak calling algorithms taking this factor into account.

Primarily due to ZINBA's extended run time (see Table 1), benchmarking was limited to chromosome 22. To the best our knowledge, chromosome 22 is a representative part of the human genome, at least with respect to the density and distribution of TF ChIP peaks and DHSs. It is therefore expected that the benchmarking results obtained on chromosome 22 are applicable genome-wide.

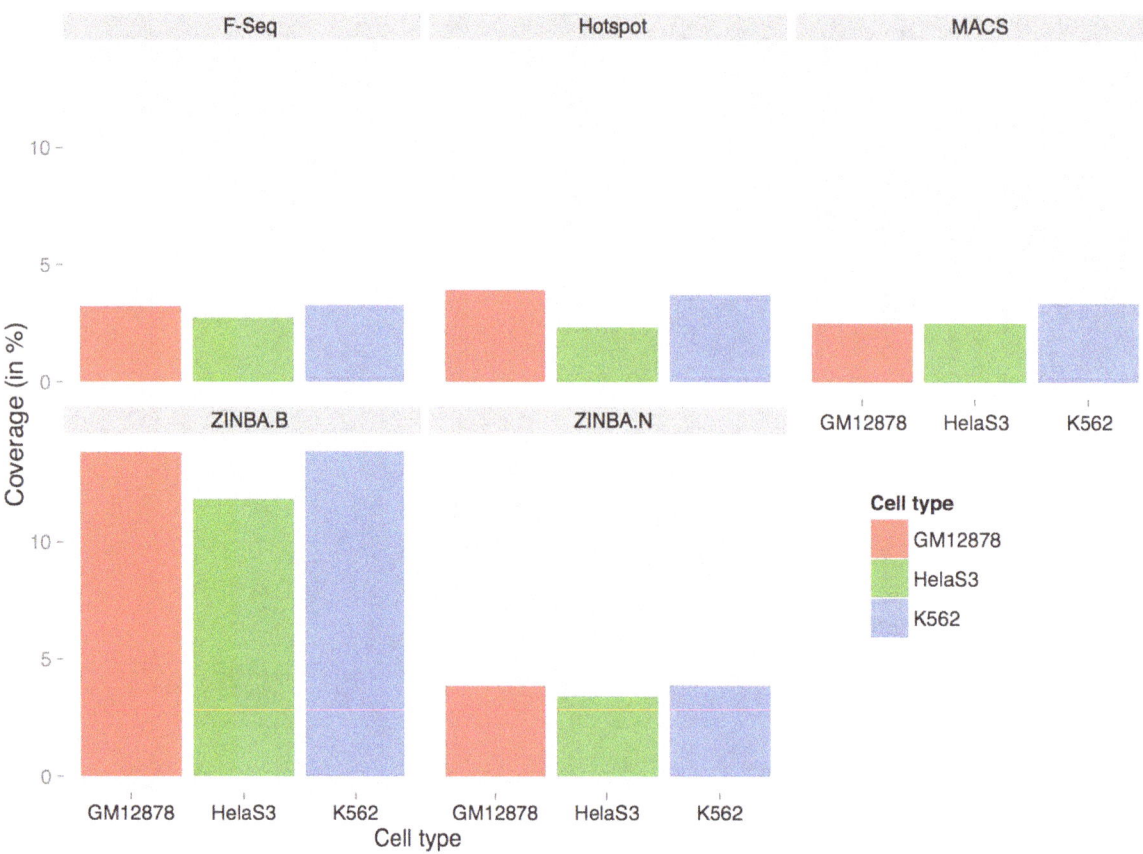

Figure 5. Coverage of Peaks Detected by Each Peak Caller Using Their Default Parameters. Illustrated here is the percentage of chromosome 22 covered by peaks from each peak caller over three cell type.

Finally, it is worth mentioning that in addition to the quality of peak calling per se, factors such as documentation and the overall user friendliness may play a role in the choice of DNase-seq analysis software, particularly by experimental biologists. To this end, F-Seq, MACS and ZINBA are published and well-documented (see [20], [13] and [17]). Hotspot has been partly described in [7], but its source code and some more documentation are available at http://www.uwencode.org/proj/hotspot-ptih/.

DNase-seq is gaining popularity as a genome-wide chromatin accessibility analysis method, and its applications have led to new insights into genome function and variation [8,28]. Robust peak detection on these data is therefore instrumental to the research community, particularly when it is provided by publicly available, well-documented and user-friendly software that can be easily used in any lab.

Materials and Methods

The performance of four peak calling algorithms was compared over a range of the false discovery rate thresholds for Hotspot, MACS and ZINBA and a range of the standard deviation threshold for F-Seq. Each of the methods was used on the DNase-seq short-read data from three cell type (K562, GM12878 and HelaS3) that was obtained from the ENCODE project [8,26]. We assessed the performance of these methods by comparing the peaks reported from each of these algorithms to the "reference sets of regulatory regions" generated from a union of peaks from a set of transcription-factor binding ChIP experiments for each of the

three cell type. Our analyses were restricted to chromosome 22, primarily due to the very significant compute times taken by ZINBA. All data in this study was mapped to the GRCh37 (hg19) human genome assembly. All computations were run on an Intel(R) Xeon(R) CPU $E5440$ @ $2.83 GHz$, with 6GiB of RAM.

Our experimental design was as follows:

Step 1: Input files

We downloaded University of Washington DNase I short read tags for K562, GM12878 and HelaS3 from http://hgdownload.cse.ucsc.edu/goldenPath/hg19/encodeDCC/wgEncodeUwDnase/ and for Duke University from http://hgdownload.cse.ucsc.edu/goldenPath/hg19/encodeDCC/wgEncodeOpenChromDnase/ as BAM files which are labeled as wgEncodeUwDnaseK562AlnRep1.bam, wgEncodeUwDnaseGm12878AlnRep1.bam and wgEncodeUwDnaseHelas3AlnRep1.bam. The number of short read tags mapped to chromosome 22 were 434301, 426770 and 255489 respectively for K562, GM12878 and HelaS3.

Step 2: Running peak callers at different thresholds

We ran Hotspot, F-Seq, ZINBA and MACS with the aligned datasets listed above (either directly from the BAM files or converted to BED format if required) with the following thresholds:

Hotspot. Keeping all other parameters in Hotspot as their defaults, we tried the *FDR* threshold with values equal to *0:001, 0:005, 0:01, 0:02, 0:03, 0:04, 0:05, 0:06, 0:07, 0:08, 0:09, 0:1, 0:2, 0:3.*

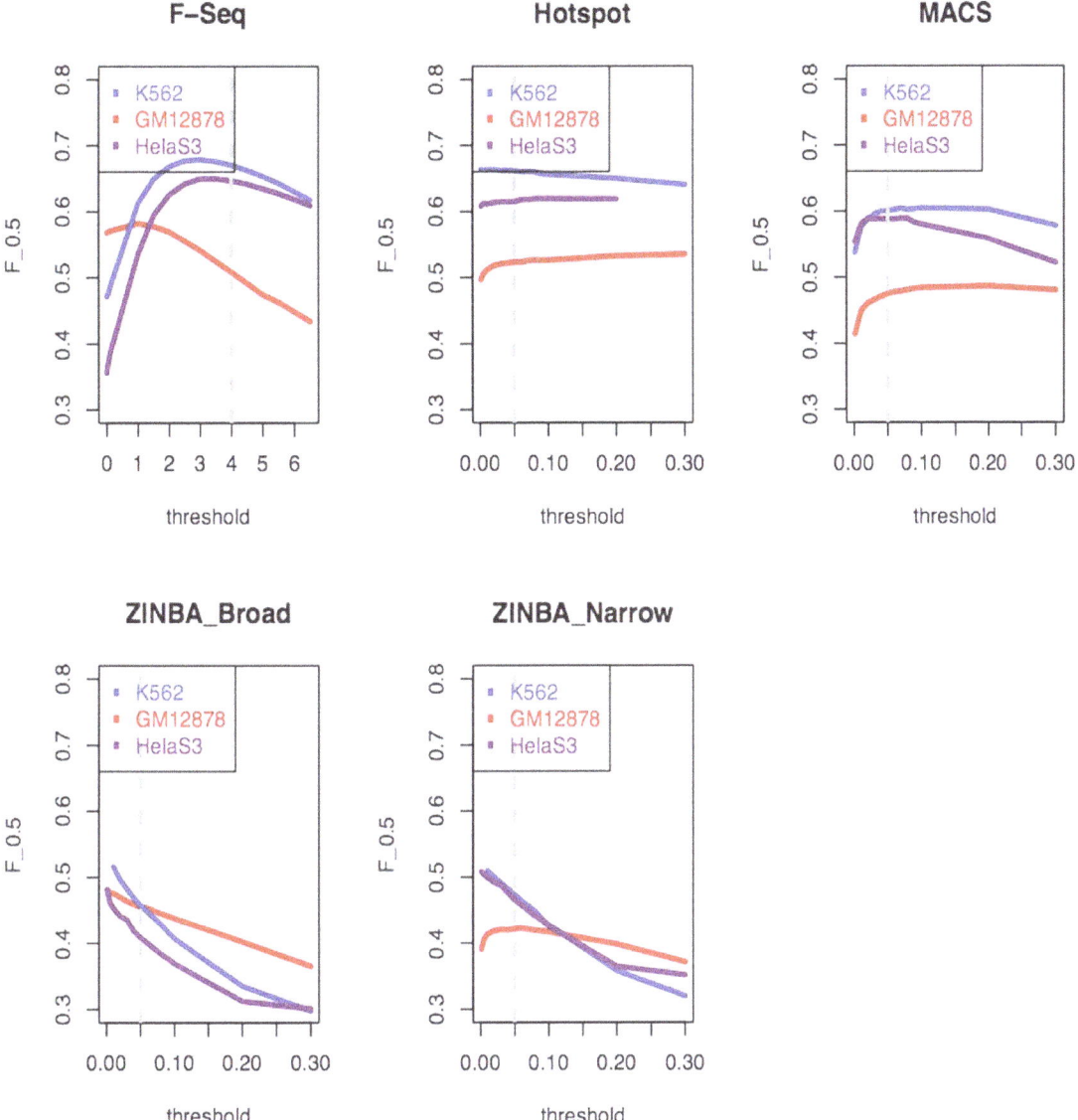

Figure 6. F Scores of Algorithms Over Three Cell Types from the "Double Hit" Protocol. Each algorithm was evaluated to gauge the enrichment of short read tags in each of the three cell types obtained from University of Washington "double hit" protocol [27]. The overlap of peaks from each of the cell types was measured against the cell type's "reference set of regulatory regions". The accuracy of each algorithm was defined as the value of the F score (see Methods for more details) by running it over a range of thresholds. The dashed vertical grey line depicts the value of F score when the algorithm is run with its default parameter. Note that Hotspot failed when ran with $FDR = 0.3$ for HelaS3 cell type and therefore its corresponding curve is shorter by one data point.

F-Seq. Although there isn't a parameter defined in F-Seq to directly control FDR, the standard deviation threshold t defined in F-Seq has an inverse correlation with FDR [20]. The default t in F-Seq is equal to 4. In this analysis we therefore ran it with an t equal to *0:001, 0:005, 0:05, 0:1, 0:5, 1, 1:5, 2, 2:5, 3, 3:5, 4, 4:5, 6.*

The feature length parameter (representing the bandwidth) was equal to 600 bp by default.

MACS. The parameter controlling the FDR in MACS is called q-value and its default is 0.05. In our analysis we ran it with a q equal to *0:001, 0:005, 0:01, 0:02, 0:03, 0:04, 0:05, 0:06, 0:07 0:08 0:09 0:1, 0:2, 0:3.*

ZINBA. In ZINBA the signal threshold controlling the FDR is

called "threshold", with a default value of 0.05. In this study we ran it with thresholds of *0:001, 0:005, 0:01, 0:02, 0:03, 0:04, 0:05, 0:06, 0:07, 0:08, 0:09, 0:1, 0:1, 0:2, 0:3* Inspired by the developers' demonstration for the FAIRE-seq data, we set $numProc = 5$ and $extension = 150$.

Step 3: Making a reference set of regulatory regions

For each of the cell types K562, GM12878 and HelaS3, we downloaded the narrow peaks of 99, 53 and 56 TFBSs respectively from the ENCODE project repository at http://hgdownload.cse. ucsc.edu/goldenPath/hg19/encodeDCC/wgEncodeSydhTfbs/ (these were all the available TFBSs as SydhTfbs for these three cell type) See Files S1, S2 and S3). Then we computed the union of TFBSs (using [23]) at each cell type and took it as our reference set of regulatory regions specific for that cell type.

Step 4: Measuring the performance of the algorithms

We defined the overlap (at base pair level) between peak calls of each algorithm at each threshold and our reference set of regulatory regions as a metric for measuring the performance of each of the algorithms. More precisely, for each algorithm and for each threshold, the True Positive Rate (also known as sensitivity) was defined as $TPR = \frac{TP}{TP+FN}$ which is in fact the ratio of the number of correctly predicted base pairs to the number of base pairs in the union of TF set. Similarly, the False Discovery Rate was defined as $FDR = \frac{FP}{TP+FP}$, which is the ratio of the number of falsely found bases as peaks to the whole set of peaks found. The reader should take care to distinguish between the FDR that we have defined here and the false discovery threshold parameter defined in each of Hotspot, MACS and ZINBA algorithms.

The specificity (or precision) in this context was defined as $Spec = \frac{TP}{TP+FP}$ and the sensitivity was defined as $Sen = \frac{TP}{TP+FN}$, which is sometimes called "recall". For each experiment the TPR was plotted against FDR.

Common to information retrieval, the overall performance of algorithms was defined as an $F-$ measure:

$$F_\beta = (1+\beta^2)\frac{Spec.Sen}{\beta^2 Spec + Sen} \qquad (1)$$

As can be seen from this equation, F_β assigns β times as much weight (or importance) to sensitivity as specificity. Normally, in situations where both specificity and sensitivity are of equal importance, β is set to 1, and the score is known as F_1 or as the "harmonic mean". In our analysis, however, because of incompleteness of our reference data set (TFs), we used $F_{0.5}$ to put more emphasis on specificity than sensitivity. Our choice of β reflects our prior belief about the incompleteness of the reference set. Using other reasonable values of β does not significantly affect our conclusions about the relative performance of the algorithms. For example, Figure S6 shows the results from Figure 6, but assuming $\beta = 1$ (i.e. an equal emphasis on specificity and sensitivity), instead of $\beta = 0.5$.

Supporting Information

Figure S1 Performance of Algorithms Over One Cell Type From the "End Capture" Protocol. Similar to Figure 6, the performance of each algorithm was evaluated using GM12878 cell type obtained from Duke University "end capture" protocol [26].

Figure S2 Comparison of TPR and FDR of Peak Callers with "End Capture" Data. Depicted here is the result of our $TPR - FDR$ comparison of four algorithms over data obtained from Duke University end capture protocol.

Figure S3 Effect of Hotspot "merge" Parameter on the Distribution of Peak Lengths. Distribution of Hotspot peak length merged (default: peaks closer than 150 bp are merged) versus not merged in UW K562 cells.

Figure S4 Effect of Hotspot "zscore" Parameter on its Performance. Hotspot was run at a range of z-score threshold ranging from 0.5 to 4 and all other parameters were kept as default. The other three algorithms were also run at a range of signal threshold (as described in main text).

Figure S5 Effect of the Number of Hits and Extension on ZINBA Coverage. Depicted here is the coverage (as defined in main text) of ZINBA when run at various combinations of number of hits per read known as "athreshold"(run at values equal to 1, 2, 3, 4) and the average of fragment lengths known as "extension"(run at values equal to 135, 200 and 300 bp).

Figure S6 The F-scores of the Algorithms Across the Three Cell Types Assuming $\beta = 1$. Illustrated here is the data shown in Figure 6, but computed assuming the β parameter equal to 1, which corresponds to same weight associated with both sensitivity and specificity. The vertical dash lines show the default threshold values in each algorithm.

Acknowledgments

We thank Alistair Rust, Ignacio Vazques Garcia and Daniel Bolland for their helpful comments. Advice from Naim Rashid and Paul Giresi on generating mappability data is appreciated.

Author Contributions

Conceived and designed the experiments: HK TH. Performed the experiments: HK TD MS. Analyzed the data: HK TD MS. Contributed reagents/materials/analysis tools: HK MS. Wrote the paper: HK TD MS TH.

References

1. Kim TH, Ren B (2006) Genome-wide analysis of protein-DNA interactions. Annual review of genomics and human genetics 7: 81–102.
2. Tong Y, Falk J (2009) Genome-wide analysis for protein-DNA interaction: ChIP-chip. Methods in molecular biology (Clifton, NJ) 590: 235–251.
3. ENCODE Project Consortium, Bernstein BE, Birney E, Dunham I, Green ED, et al. (2012) An integrated encyclopedia of DNA elements in the human genome. Nature 488: 57–74.
4. Crawford GE, Holt IE, Whittle J, Webb BD, Tai D, et al. (2006) Genome-wide mapping of DNase hypersensitive sites using massively parallel signature sequencing (MPSS). Genome research 16: 123–131.
5. Song L, Zhang Z, Grasfeder LL, Boyle AP, Giresi PG, et al. (2011) Open chromatin defined by DNaseI and FAIRE identifies regulatory elements that shape cell-type identity. Genome research 21: 1757–1767.
6. Zeng W, Mortazavi A (2012) Technical considerations for functional sequencing assays. Nature Immunology 13: 802–807.
7. John S, Sabo PJ, Thurman RE, Sung MH, Biddie SC, et al. (2011) Chromatin accessibility predetermines glucocorticoid receptor binding patterns. Nature genetics: 1–7.

8. Thurman RE, Rynes E, Humbert R, Vierstra J, Maurano MT, et al. (2012) The accessible chromatin landscape of the human genome. Nature 489: 75–82.
9. Pepke S, Wold B, Mortazavi A (2009) Computation for ChIP-seq and RNA-seq studies. Nature Methods 6: S22–32.
10. Kim H, Kim J, Selby H, Gao D, Tong T, et al. (2011) A short survey of computational analysis methods in analysing ChIP-seq data. Human genomics 5: 117–123.
11. Szalkowski AM, Schmid CD (2011) Rapid innovation in ChIP-seq peak-calling algorithms is out-distancing benchmarking efforts. Briefings in Bioinformatics 12: 626–633.
12. Rye MB, Sætrom P, Drabløs F (2011) A manually curated ChIP-seq benchmark demonstrates room for improvement in current peak-finder programs. Nucleic Acids Research 39: e25.
13. Zhang Y, Liu T, Meyer CA, Eeckhoute J, Johnson DS, et al. (2008) Model-based analysis of ChIP-Seq (MACS). Genome Biology 9: R137.
14. Ramagopalan SV, Heger A, Berlanga AJ, Maugeri NJ, Lincoln MR, et al. (2010) A ChIP-seq defined genome-wide map of vitamin D receptor binding: associations with disease and evolution. Genome research 20: 1352–1360.

15. Landt SG, Marinov GK, Kundaje A, Kheradpour P, Pauli F, et al. (2012) ChIP-seq guidelines and practices of the ENCODE and modENCODE consortia. Genome research 22: 1813–1831.

16. Kharchenko PV, Tolstorukov MY, Park PJ (2008) Design and analysis of ChIP-seq experiments for DNA-binding proteins. Nature biotechnology 26: 1351–1359.

17. Rashid NU, Giresi PG, Ibrahim JG, Sun W, Lieb JD (2011) ZINBA integrates local covariates with DNA-seq data to identify broad and narrow regions of enrichment, even within amplified genomic regions. Genome Biology 12: R67.

18. Baek S, Sung MH, Hager GL (2011) Quantitative analysis of genome-wide chromatin remodeling. Methods in molecular biology (Clifton, NJ) 833: 433–441.

19. Madrigal P, Krajewski P (2012) Current bioinformatic approaches to identify DNase I hypersensitive sites and genomic footprints from DNase-seq data. Frontiers in genetics 3: 230.

20. Boyle AP, Guinney J, Crawford GE, Furey TS (2008) F-Seq: a feature density estimator for high-throughput sequence tags. Bioinformatics (Oxford, England) 24: 2537–2538.

21. Gaulton KJ, Nammo T, Pasquali L, Simon JM, Giresi PG, et al. (2010) A map of open chromatin in human pancreatic islets. Nature genetics 42: 255–259.

22. Wang YM, Zhou P, Wang LY, Li ZH, Zhang YN, et al. (2012) Correlation between DNase I hypersensitive site distribution and gene expression in HeLa S3 cells. PLOS ONE 7: e42414.

23. Neph S, Kuehn MS, Reynolds AP, Haugen E, Thurman RE, et al. (2012) BEDOPS: high-performance genomic feature operations. Bioinformatics (Oxford, England) 28: 1919–1920.

24. Quinlan AR, Hall IM (2010) BEDTools: a exible suite of utilities for comparing genomic features. Bioinformatics (Oxford, England) 26: 841–842.

25. Koohy H, Down TA, Hubbard TJ (2013) Chromatin Accessibility Data Sets Show Bias Due to Sequence Specificity of the DNase I Enzyme. PloS one 8: e69853.

26. Song L, Crawford GE (2010) DNase-seq: a high-resolution technique for mapping active gene regulatory elements across the genome from mammalian cells. Cold Spring Harbor protocols 2010: pdb.prot5384.

27. Sabo PJ, Kuehn MS, Thurman R, Johnson BE, Johnson EM, et al. (2006) Genome-scale mapping of DNase I sensitivity in vivo using tiling DNA microarrays. Nature methods 3: 511–518.

28. Degner JF, Pai AA, Pique-Regi R, Veyrieras JB, Gaffney DJ, et al. (2012) DNaseI sensitivity QTLs are a major determinant of human expression variation. Nature 482: 390–394.

29. Down TA, Piipari M, Hubbard TJP (2011) Dalliance: interactive genome viewing on the web. Bioinformatics 27: 889–890.

11

Mapping the Structure and Dynamics of Genomics-Related MeSH Terms Complex Networks

Jesús M. Siqueiros-García[1]*, **Enrique Hernández-Lemus**[2,3], **Rodrigo García-Herrera**[2], **Andrea Robina-Galatas**[1]

1 Ethical, Legal and Social Studies Department, National Institute of Genomic Medicine, Mexico City, D.F., Mexico, **2** Computational Genomics Department, National Institute of Genomic Medicine, Mexico City, D.F., Mexico, **3** Complexity in Systems Biology, Center for Complexity Sciences, National Autonomous University of Mexico, Mexico City, D.F., Mexico

Abstract

It has been proposed that the history and evolution of scientific ideas may reflect certain aspects of the underlying socio-cognitive frameworks in which science itself is developing. Systematic analyses of the development of scientific knowledge may help us to construct models of the collective dynamics of science. Aiming at scientific rigor, these models should be built upon solid empirical evidence, analyzed with formal tools leading to ever-improving results that support the related conclusions. Along these lines we studied the dynamics and structure of the development of research in genomics as represented by the entire collection of genomics-related scientific papers contained in the PubMed database. The analyzed corpus consisted in more than 49,000 articles published in the years 1987 (first appearance of the term *Genomics*) to 2011, categorized by means of the Medical Subheadings (MeSH) content-descriptors. Complex networks were built where two MeSH terms were connected if they are descriptors of the same article(s). The analysis of such networks revealed a complex structure and dynamics that to certain extent resembled small-world networks. The evolution of such networks in time reflected interesting phenomena in the historical development of genomic research, including what seems to be a phase-transition in a period marked by the completion of the first draft of the Human Genome Project. We also found that different disciplinary areas have different dynamic evolution patterns in their MeSH connectivity networks. In the case of areas related to science, changes in topology were somewhat fast while retaining a certain core-stucture, whereas in the humanities, the evolution was pretty slow and the structure resulted highly redundant and in the case of technology related issues, the evolution was very fast and the structure remained tree-like with almost no overlapping terms.

Editor: Jesus Gomez-Gardenes, Universidad de Zarazoga, Spain

Funding: The authors have no support or funding to report.

* E-mail: jsiqueiros@inmegen.gob.mx

Introduction

Complex networks theory is taking over where other *theoretical* approaches to complexity –such as synergetics, chaos theory and self-organized criticality– have had limited success. This is what Lászlo Barabási recently published in a paper called the *The Network Takeover* [1]. Complex networks theory has become the most recent attempt to tackle complexity, it has developed a great variety of methods and techniques with a wide scope of generality. So far, it has been possible to find properties such as power-law behavior in node degree distributions that are common among different kinds of systems, from regulatory transcription networks, to friendship networks, to epidemic networks [2–4]. But what seems to have been the difference between complex networks theory and the former approaches to complexity is its ability to incorporate (or build upon) empirical data: massive amounts of data. So, it has become an effective way to deal with the –so to speak– real complexity in which we are embedded, in a simple and beautiful way.

Science itself has been the subject of complex networks analysis [5–10]. Most studies are close to or framed inside scientometrics, the discipline that studies science by measuring and analysing its products. Network-based scientometrics rely on collaborations, coauthorship and citation networks. These kinds of networks are interesting for complex network analysis for reasons that can be found in current literature. Nevertheless, scientometrics has also produced many valuable findings that help us understand the sociology of science: the flows of knowledge, its cultural and disciplinary differences, as well as its political and economical aspects [11–13]. On those efforts we build upon. We are very much interested in how scientific knowledge is generated by a descentralized collectivity, and how it is organized. Our view is a little bit different from scientometrics in that we are mainly concerned on how to frame these data, patterns, processes and networks from the perspective of complex systems analysis, as an epistemological issue and as a subject of philosophy of science. We believe this is something that as far as we know, has not been systematically studied yet. For these matters, our methods are close to those of scientometrics, however, our interests are even closer to sociology and philosophy of science.

In this paper we explore the networks formed by the terms that describe the content of an article. These terms are known as MeSH, an acronym that stands for *Medical Subject Headings*. We are interested in the structure and dynamics of a network of the set of MeSH terms related to the term *Genomics*; we believe that to some extent, this particular MeSH terms network represents the image of a part of the biomedical human knowledge evolution and its current state in a specific time in history.

To be clear, MeSH is the controlled vocabulary defined by the National Library of Medicine of the United States that is used for indexing, cataloging, and searching for biomedical and health-related information and documents. It is composed of terms that name the descriptor of an object that can be a scientific paper, a film, a book or any other format in which medical knowledge is recorded and transmitted [14]. MeSH vocabulary constitutes what is called a MeSH tree, and it is structured alphabetically and hierarchically. By 2013, the tree included 17 hierarchies. Each hierarchy is a branch that goes from general to specific and every hierarchy is independent from the others, this means there is no dependance relationship between terms belonging to different hierarchies. For example: Hierarchy number 2, tagged [B], corresponds to *Organisms*; *Eukaryota* is the first subdivision of *Organisms*, that places *Eukaryota* in the classification system in *B01*. *Drosophila melanogaster* belongs to hierarchy [B01] but it goes all the way down to [B01.050.500.131.617.289.310.250.500]. The *Organisms* hierarchy is independent from the rest of hierarchies such as hierarchy [K], that makes reference to the *Humanities*.

From the 17 branches, 16 define the existing headings and 1 is a branch for all the existing subheadings to date. The 16 branches start with the root [A] that stands for the branch *Anatomy*, followed by [B] for *Organisms*, [C] being the root for *Diseases* and it goes all the way to the letter [N] that is for *Health Care*, then, it skips to letters [V] and [Z] for *Publication Characteristics* and *Geographicals* respectively. The remaining root is [Y] which includes all subheadings and is not part of the main MeSH tree. In order to define a precise meaning, a MeSH term starts with a heading and it is refined using subheadings. Every heading begins with uppercase letters and all subheadings begin in lowercase. When a composed MeSH term is formed, headings are followed by subheadings separated by slashes, for example: *Breast Neoplasms/ diagnosis/genetics/psychology*.

MeSH is not a static vocabulary. The first list of terms was published in the 1950's, and the first catalog of Medical Subject Headings appeared in the 1960's. The 1960 MeSH edition included 4,400 descriptors and the second edition of 1963 got up to 5,700. The 2013 MeSH contains as many as 26,853 descriptors [15]. The MeSH catalog is updated every year and it is the work of a staff specialized in the subjects. The National Library of Medicine also receives recommendations regarding new terms and indexing. The whole enterprise is supported by many external professionals who are consulted for their expertise. The MeSH is used to index articles from the most important biomedical journals worldwide for the MEDLINE/PubMED database [14].

In this paper we explore the organization of knowledge that has been developed along with genomics research. Our networks display the emergence and decay of subjects, topics and disciplines as research in genomics has changed and we wanted to know how such behaviors have been taking place. Analyzing these networks we pretend to have a better understanding on how the topologies have changed during 25 years, how close are subjects from each other –what is the average shortest paths between nodes–, how different subjects and topics cluster, and how different areas of knowledge behave. Particularly we explore the structure and dynamics of different subnetworks defined by MeSH terms related to the areas of *science*, the *humanities* and *technology*.

The main findings in this work are that MeSH networks present a topology similar to that of small-world networks which is maintained along the years, independently of the rate of growth of such networks. Also we perceived differences in the connectivity patterns between different disciplines. Such patterns seem to be characteristic of each type of discipline. With regards to the dynamics of network growth we found evidence that seem to point out to the presence of three different regimes, roughly corresponding to the pre-Human Genome Project, the completion of the Human Genome Project (from now HGP) –whose behavior resembles a dynamic phase transition– and the post-genomic eras.

The rest of the manuscript is organized as follows, a section on Results and Discussion dealing with general aspects on MeSH networks topology, with network dynamics as well as a detailed analysis of some theme-specific subnetworks. Those networks have been chosen to represent different fields of inquiry. Representing a science-based theme are the subnetworks around the MeSH *Neoplasms*, for the humanities the networks chosen were encompassed by the MesH terms *Ethics* as well as *History*; and for the case of more technical/technological issues we chose the subnetworks related to the MeSH terms *Computational Biology* and *PCR*. After this discussion, a brief section dealing with the Materials and Methods used is included.

Results and Discussion

Networks Topology

We would like to report three main topological results. First, global networks (GNs) displayed an intricate and complex topology as can be seen in their respective network structure parameters [See Table S1]. Interstingly enough global networks always consist of only one connected component. This seems to imply that the whole corpus of biomedical knowledge related to genomic research is somehow integrated. This non-trivial structure may become evident by analyzing the values of quantities such as network density $\langle p \rangle$ and clustering coefficient $\langle C \rangle$. It becomes clear that the structure of such networks cannot be the result of random generated connections. In MeSH GNs, density decreased as the networks grow bigger (in 1991 $\langle p \rangle < 0.1$ and since 2001 $\langle p \rangle < 0.01$). However $\langle C \rangle$ for all networks remained constant and high-valued for the whole history of genomics research ($\langle C \rangle > 0.8$). The topology of GNs, and specially their clustering coefficient values would be very unlikely in an Erdös-Rényi network with such density. It is also important to mention that $\langle C \rangle > 0.8$ is indicative of non-tree-like networks.

The shortest average distance between any nodes i, j remained low for all GNs: $\langle l \rangle < 2.3$. We believe such average distances would not be possible without a particular topology, such as in the Watts-Strogatz model, in which long distant links bring closer every node in the network that otherwise would be quite apart from each other. We also believe that there are communities that may lead to a small-world-like topology [16]. We may recall that Erdös-Rényi networks display low values of average shortest paths, and their clustering coefficient is also very low.

All global networks displayed are sparse networks [Figures 1 and 2]. According to Watts and Strogatz, small-world topologies might be common to large, sparse or low density networks found in nature [17]. It has been pointed out by others that many real-world complex networks have a small-world effect, but they are different from a real small-world network in that their ...*average path length increases slower than any polynomial function of the system size...* [16,18]. If we look up at Figure 3 and Table S1 we may see that there is a strong resemblance of our GNs to small-world networks. Many of the small-world properties networks seem to be modular [16] and this also has implications for our work. If our GNs have a small-world effect topology then it means that genomics, as it has been growing, is becoming a vast sea that is quite navigable with islands of knowledge and lanes between them to be traveled. If GNs have such modular structure is something that remains to be elucidated and is part of our future agenda. Particularly, we would like to see if there are any emergent modules, how are they composed and how do they connect and affect each other. Most of all, emergent modularity might indicate non-trivial ways of organizing collectively produced knowledge, which would be interesting to be studied from a sociological, historical and philosophical point of view.

Networks Dynamics

Following the parameters of the whole sequence of networks, from 1987 to 2011, it can be appreciated that they seem to grow in a structured manner. The first GN corresponds to 1987 (our initial time-point). During the first couple of years, the number of terms (nodes) in the network increased by addition of approximately ten new terms each year, starting with 25 nodes and 216 edges in 1987. From 1990 to 1999 the number of terms were added by hundreds and in the last decade it increased by thousands, so that by the year 2011, the number of terms summed up to 13,169 with 213655 edges. It is noteworthy that for the years that passed from the beginning of the Human Genome Project to the announcement of a working draft of the Human Genome in the year 2000,

the networks seem to be limited in adding new terms. For instance, from 1990 to 1999 new terms were added at a rate of 58.2 per year, a very conservative number compared to a rate of 1035.8 terms per year from 2000 to 2011, in average. Specially interesting is the fact that in the year 2000 the curves for the number of nodes and edges start growing much faster than before. The sudden change in slope for the year 2000 curve suggests the beginning of a substantial increase in the exploration of genomic-related issues, contrary to the very limited scope before 2000 which might have been the result of the somehow narrow objective of sequencing the human genome in a limited time. It is in the year 2000 where there is the only significant change in the average shortest path length, from $\langle l \rangle < 2$ to $2 < \langle l \rangle < 2.3$.

From the 25 nodes in 1987, five nodes were the most connected ones. Among these, the term *Human Genome Project* was present in every GN. Another node that was one of the most connected terms over time is *Humans*, but in this first network it was not relevant ($k = 12$). From 1989 to 1998 these two terms centralized the networks without rivals. In 1999 *Humans* had a higher degree of connectivity than *Human Genome Project*, and also, for the first time in ten years, a new term that will be important in the networks for the years to come emerges, such term was *Animals*. The year 2001 was the year in which the term *Genomics* became visible altogether with the term *Humans*, *Animals* and *Human Genome Project*. From this

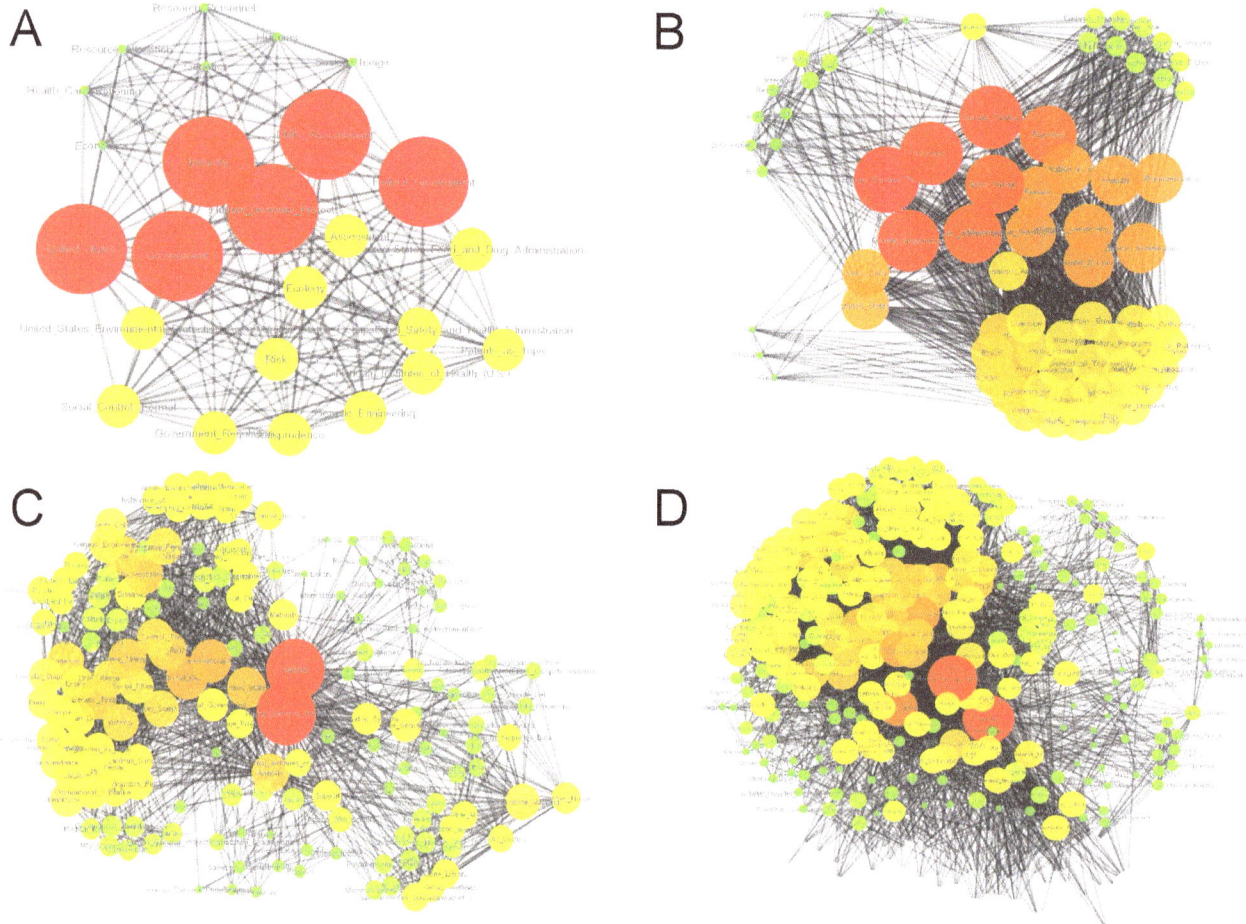

Figure 1. Global MeSH networks for the period (1987–1990). It is noticeable that there is a progressively growth of the network that induces greater variability in the connection patterns. New terms arise that lead to the generation of a more complex connectivity structure that reduces the relative importance of terms that were initially dominant. Nodes are size and color-coded according with their respective connectivity degree, i.e. big red nodes present a high connectivity, whereas small green nodes have lower values. This is an example of only four years.

Global Networks topological dynamics

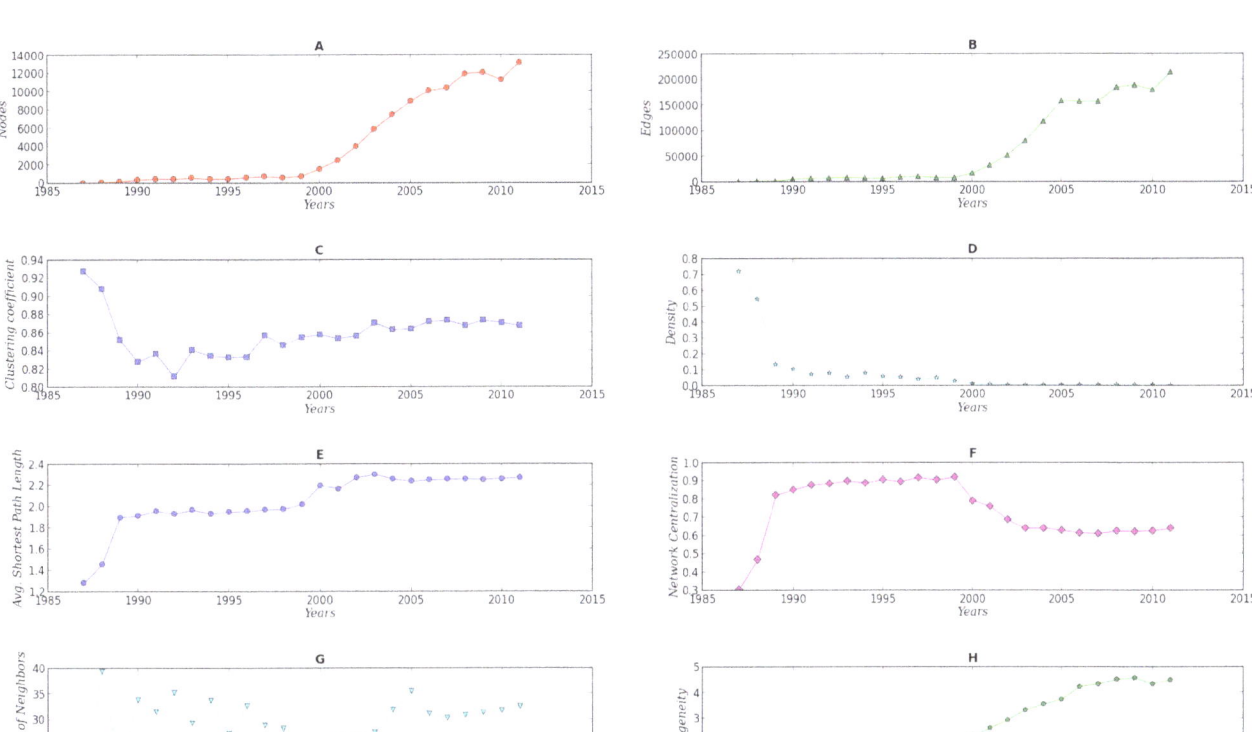

Figure 2. Dynamics of the topological parameters for the global networks. Panels A–H.

year onwards, this last term declined as an important, well-connected node, since its connectivity started to decrease reaching a k around the 300 in 2005, and remained close to this value for the following networks. Another moment that seems to be important in the history of genomics was the year 2003. During this year the term *Proteomics* gained in connectivity, comparable to the big nodes already mentioned. And for the year 2005, *Proteomics* along with *Proteomics, Methods*, were more connected than *Genomics* and it stayed like that for the following years. In our last network (*i. e.*, 2011), the most connected node was *Humans*, followed by *Animals*, *Proteomics*, and *Genomics*.

The complex connectivity structure within different MeSH terms lies behind the links among abstract terms, responsible for the conceptual coherence of the biomedical publication corpus represented in the PubMed database. In Figure 4 we can see a Circos plot [19] showing the interconnectivity between all different headings whose MeSH terms categorize PubMed-indexed publications related to *Genomics*, corresponding to the year 2011. Labels in every section correspond to the key in Table S2. We can notice that a blue histogram (shown in the inner circle) represents the corresponding depth of specific MeSH tree levels for a given term. Higher bars thus refer to more specific terms. The outer layer displays an orange histogram showing the base-10 logarithm of the number of published papers categorized for every MeSH term. Higher bars are then *hot topics*, generating a very large number of related publications.

As one can see in Figure 4, a multitude of different areas of knowledge converge to conform a highly multidisciplinary corpus in the research related to genomics. There is a dense, non-random

interconnection pattern spanning across traditional fields of research (for instance, areas so-apparently disparate as pathology, history, molecular biology, sociology, and computer science are conceptually connected in this corpus). This may be indicative of current phenomena leading to an increase in adequacy and a reshape of the boundaries between contemporary research subjects in the evolution of afore-mentioned traditional fields of research. Very likely, this multidisciplinarity may be the driving force (or at least one of the forces) behind the establishment of new relations between already structured research areas, as well as the evolution of trending research topics. Further evidence in this regard may be observed when looking at the outer circle histograms in Figure 4 that present the number of articles published (on a logarithmic scale) that contain the connected MeSH terms. By looking at the blue histograms in the inner circle of Figure 4 we may notice that there is no apparent trivial relationship between the specificity of research topics (as represented by the number of hierarchical levels in the MeSH tree structure spanned by such topic) and the impact that such research may have in the overall biomedical community (as represented by the number of papers in the area given by the height of the orange histograms in the outer circle).

Since every heading belongs to a branch (or category), we also wanted to know the proportion of every branch in each GN by counting and plotting the normalized frequency of every root of each heading branch. The heading [V] never appeared in our networks, so we did not plot it. Semantic relationships between these terms are built and are behind the conceptual structure intrinsic to research in genomics. Figure 5 shows that in the first

Clustering coefficient dynamics

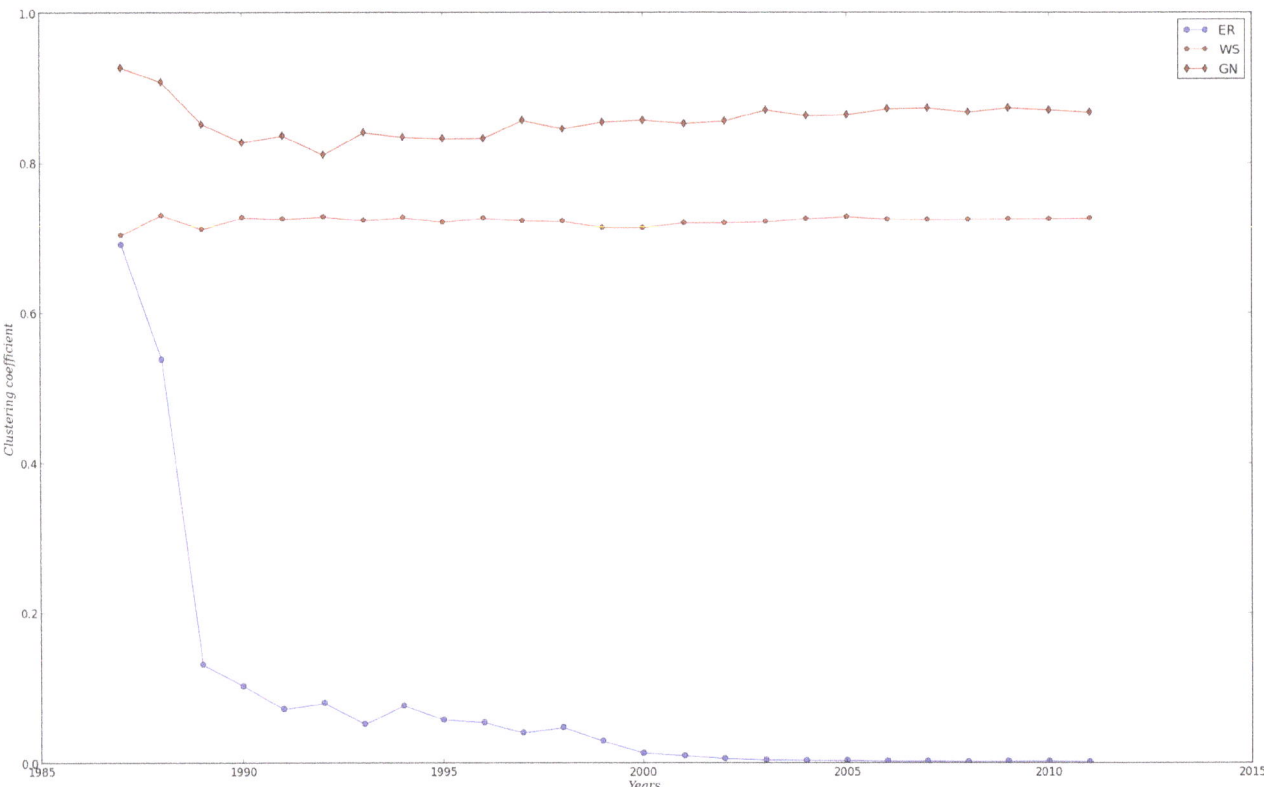

Figure 3. Average clustering coefficient versus time. We may see the average clustering coefficient for MeSH global (GN) networks, compared to random Erdös-Rényi (ER) and to small-world Watts-Strogatz (WS) networks with the same parameters as these. We can notice that MeSH global networks are not similar to tree-like networks (clustering coefficient = 0), neither to random networks (with clustering coefficient rapidly decaying) but seem to be close to small-world networks with large and persistent values of clustering. In the other hand, MeSH networks are also different from lattices since the average shortest path lenght values are not proportional to the size of the network but rather to the logaritm of such size.

two-year period the proportions of headings get the positions that will keep for the following 7 years. This means that some headings abruptly increased while some others decreased their proportions in the 1987 and 1988 networks. From 1989 to 1995 most headings had a steady position. During this period the most represented headings were H, and N, the first one standing for *Disciplines and Occupations* and the second one standing for *Health care*. In the middle region (second period), one may find headings such as L for *Information science*, E for *Analytical, Diagnostic and Therapeutic Techniques and Equipment* and I for *Anthropology, Education, Sociology and Social Phenomena*. The least represented headings were A for *Anatomy*, C for *Diseases*, D for *Chemicals and Drugs*, F for *Psychiatry and Psychology*, J for *Technology, Industry, Agriculture*, K for *Humanities*, M for *Named Groups*, and Z for *Geographicals*.

From 1995 to 2004 things changed again: H and L went upwards and then downwards between 1996 to 1999. In 1999, the proportions of headings E, D and G grew rapidly. Contrary to these categories, category I abruptly started to decrease in 1997. The proportions for the remaining headings grew or diminished slightly. 2004 is the year when heading proportions stabilized and reached the positions that they have maintained since then. Headings D, E, G, H are placed at the top and moved all together as a group, although in 2009 E, G begun to split away. Heading B is alone, underneath the group of headings just mentioned. Under heading B, there is another well packed group of four headings. This group is composed by A, C, L and N. There is one last group

of headings for which since 2004 their representation in the networks is close to zero, this group includes F, I, J, K, M and Z. It is noteworthy that I begun as an important heading during the first years, but its importance decreased, becoming almost unrepresented in later networks.

Figure 5 shows the dynamics in the proportion of branches (i.e. main categories in Table S2) for every GN. Analysis of this graph led us to suggest that there are three somewhat distinguishable periods in the evolution of genomics and related issues in the biomedical literature. The plot, as well as the corresponding networks, show that the benchmark for the identification of these periods is located between years 1999 and 2003. It is noteworthy that 2001 (the middle point in that interval) was the year that the draft of the sequence of the Human Genome was published in *Nature* and Celera's paper on the methods used in the sequence draft was published in *Science* [20,21]. Analysis of this graph in conjunction with the parameters obtained from the GNs lead us to hypothesize that in the period that ranged between 1999 and 2003, something similar to a *phase transition* took place. There were rearrangements in the proportions of each category that ended up in the actual configuration. Such transition may have been triggered by an important change in the networks parameters that occured during the 1999–2000 years, most surely due to the completion and publication of the first draft of the human genome. The number of nodes and edges increased substantially (from hundreds to thousands) and network centralization dropped from

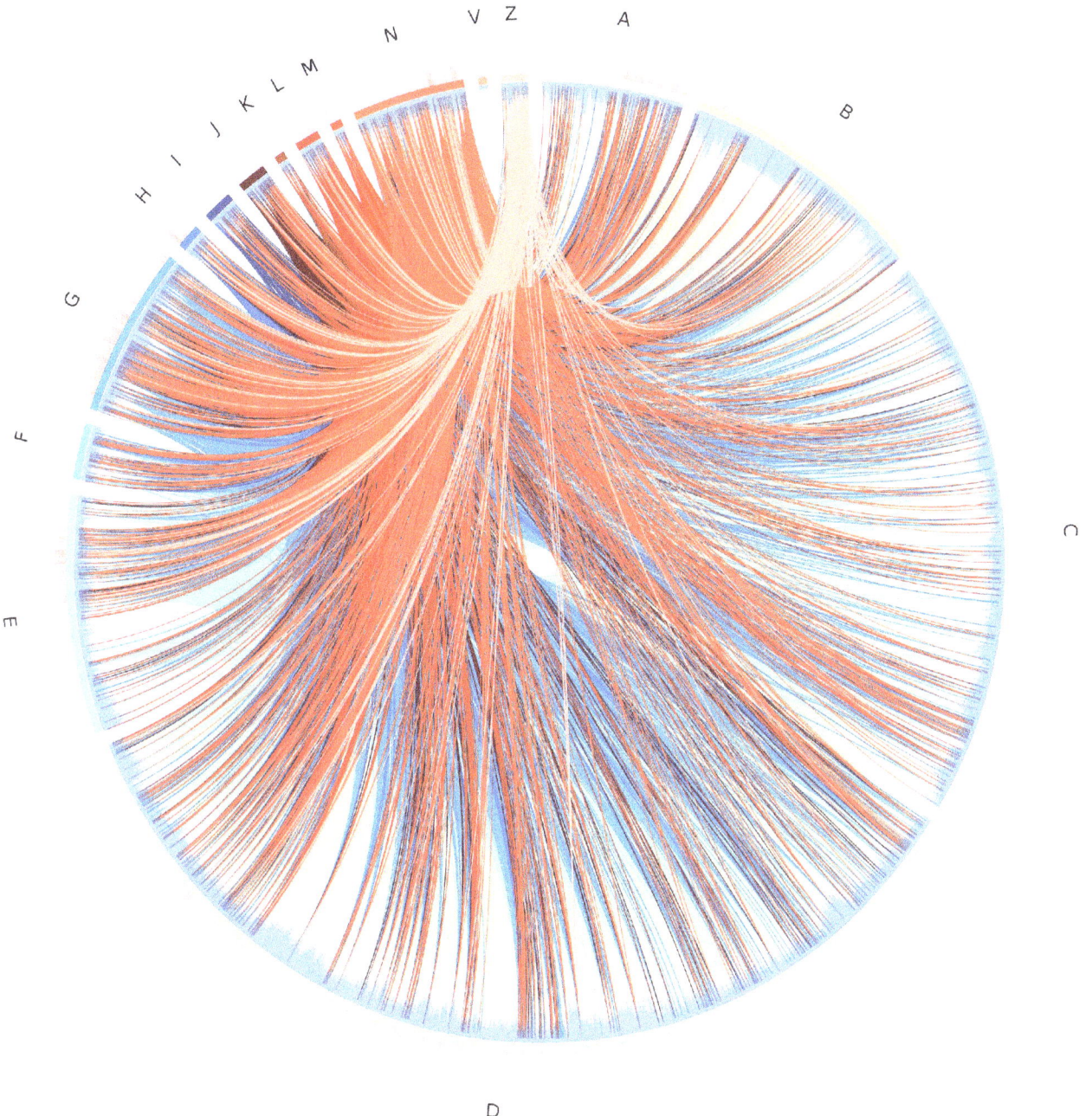

Figure 4. Circos plot of MeSH-tree headings. Circos plot displaying the interconnectivity among different headings belonging to genomics-related PubMed-indexed articles published in 2011. Letters correspond to the key in Table S2. The blue histogram in the inner circle represents the corresponding depth of MeSH tree levels for a given term. Higher bars are more specific terms. The orange histogram in the outer circle displays the base-10 logarithm of the number of published papers categorized for every MeSH term.

0.91 in 1999 to 0.78 in 2000 and to 0.63 in 2003, due to the emergence of new terms, such as *Animals* and *Genomics* as highly centralized nodes. These two nodes rivaled with *Humans* and *Human Genome Project* as the two main hubs during the pre-genomic era. In summary, what we see in this apparent phase transition is an explosion in the variety of terms related to genomic research. So to speak, the number of nodes is at the same time, the number of different terms and different topics that might be seen as proxies for new grounds for genomics-framed exploration. In these years a new regime shown in Figure 2, came to be and its growth rate has been sustained so far. This must mean somehow that what sciences

produce is sufficient at least to create the same or more diverse knowledge, a knowledge that is well connected to the major component and is well integrated as the clustering coefficient seems to suggest.

Our data about GNs also shows how in the postgenomic era, different headings (a proxy for general subjects) cluster together, and we assume that, at least in some cases, this is due to their conceptual affinity. For example we find that heading *H*, that stands for disciplines (in our networks mostly represented by genomics and proteomics) is highly correlated with the understanding of biological processes and chemicals, and with technol-

Dynamic evolution of MeSH-tree headings

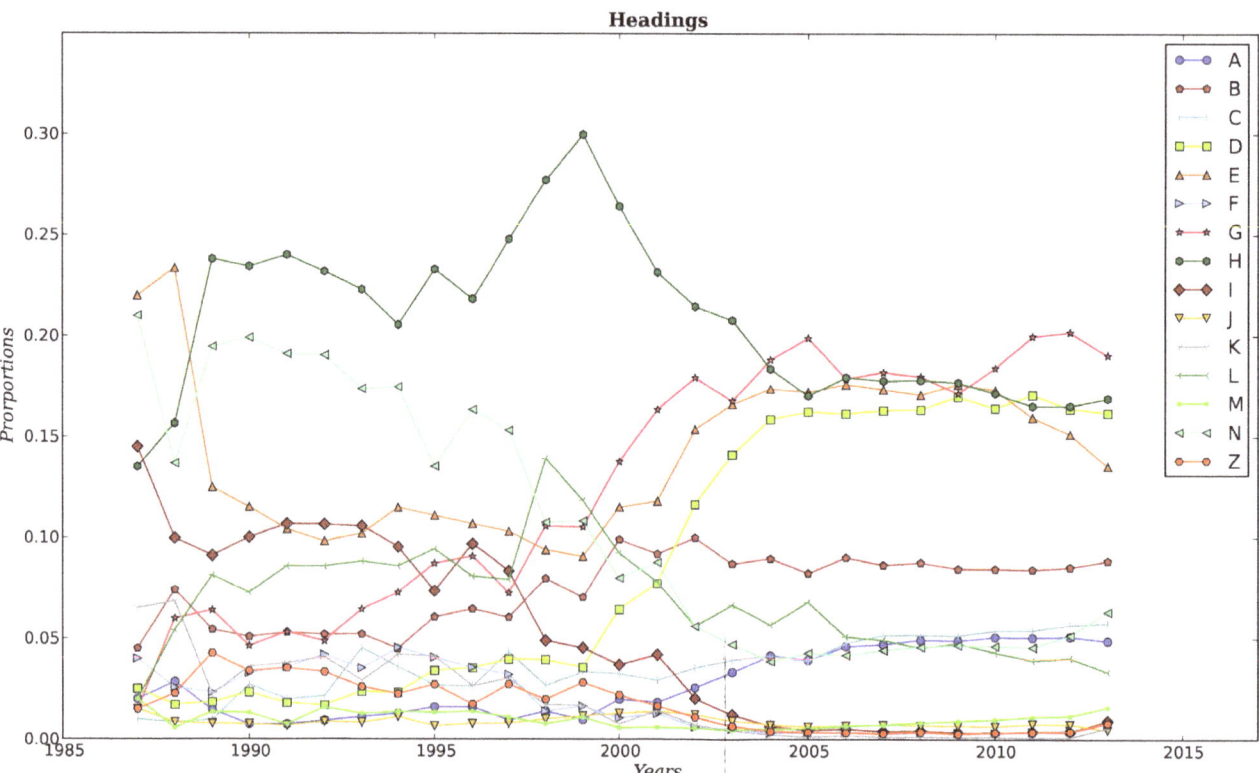

Figure 5. Dynamic evolution of MeSH-tree headings. Dynamics of the proportion of participation for every heading in terms of the number of PubMed indexed publications each year. There are noticeable changes in the trends for different issues that are consistent with different periods of time within the *genomics age*.

ogy and methods (such as PCR) [see Figure 5]. It is interesting that the areas related to *Ethical, Legal and Social Issues of Genomics* (ELSI), lost a great part of their representation share once the Human Genome Project was accomplished. Since the graph shows proportions, it might be the case that the production of ELSI research is the same now as it was 20 years ago, but at the end the message is that it did not grow accordingly to the amount of subjects investigated, new areas of research, and new technologies –all this represented by the ever increasing size of the networks. ELSI research simply became less relevant once the project was over. Something similar happens to *Health care*, or heading *N*. Personalized, preventive and predictive medicine was the promise of genomics. All these important concepts were part of the vision of the proximal future in *Health care*. But once again, as other issues of public interest (or at least, closer to the public), its relevance in the share of proportions, faded away in the postgenomic era [Figure 5].

Along with the history of genetics and genomics [22], the dynamics of the proportions tells us a story of how genomics has departed from a set of preconceptions regarding human nature (as beings with rights and dignity) inherited from a bioethical and science policy and society agenda, in order to set the limits to human research. Nevertheless, the development of genomics has now created an image of what a living being is and how we fit into that description. It also tells us about what are the tools that have been applied in order to create that image. Such tools might be technological or conceptual. Furthermore, the very idea of genomics, reflected in the first networks and closely related to

the HGP, must have changed considerably after 2003 and 2004. Today genomics may not be as much as the study of the whole genome but more of a generic name for a systemic view of the biology of living things. This drive, if there is such a thing, might be the focal point leading to a better understanding of how different levels of biological organization interact among each other, as well as to how different biological systems interact in an ecological fashion, as it is currently studied by metagenomics.

A closer analysis of the hot-topics was made by selecting the top-10 most connected MeSH terms for each year. It reveals interesting facts and trends while supporting our previous discussion. In figure 6 we can see that during the former years of genomic research (1987–2001) there were a lot of issues under discussion, dominated by health care and policy matters, as well as social phenomena, in what may be called an ELSI stage of genomics. In the other hand, most recent years show a completely different trend by having fewer issues, most of them related to more technical and scientific aspects of genome research.

By examining the color code in figure 6 (which corresponds with the log-2 of the degree of each node) we can see that the discussion has become more rich in recent years with a much larger number of edges for the Top-10 topics that in those corresponding to the former years.

Subnetworks

Since the GNs include the whole set of MeSH terms related to the term *Genomics* we were able to substract smaller networks based on particular terms such as *Neoplasms, Ethics, History, Computational*

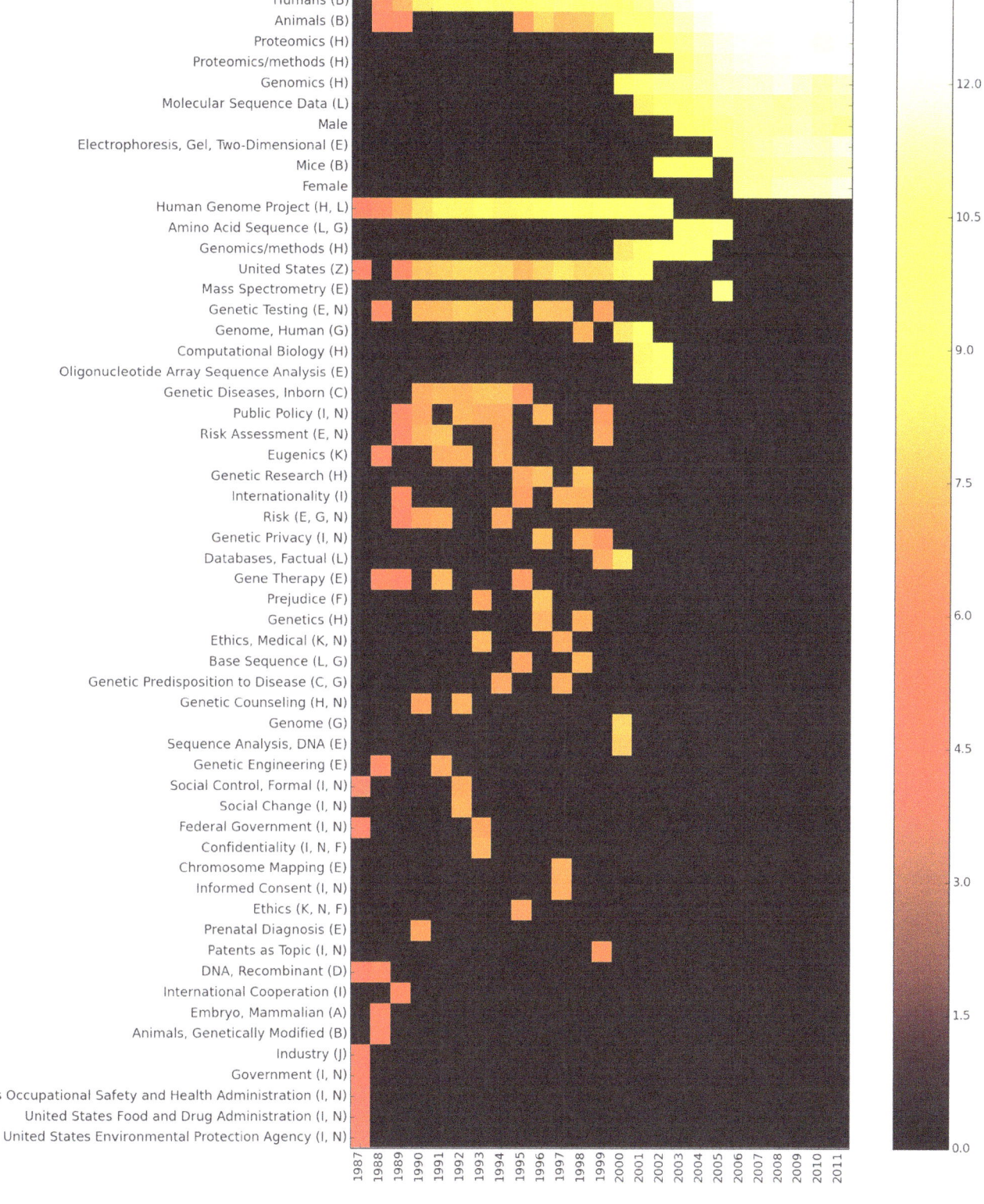

Figure 6. Dynamics of the Top-10 of the most connected MeSH terms. The Y-axis corresponds to a list of the 57 most connected MeSH terms in the global networks from 1987 to 2011. X-axis is the corresponding year. We may notice that some terms remain in the Top-10 list for several years, and even appear and dissappear from the list. Color-intensity is given by the $\log_2$ of the degree. The relative importance of the top-10 nodes reflects the fact that in the former years there was a lower number of potential connections in the global networks.

Ethics subnetworks topological dynamics

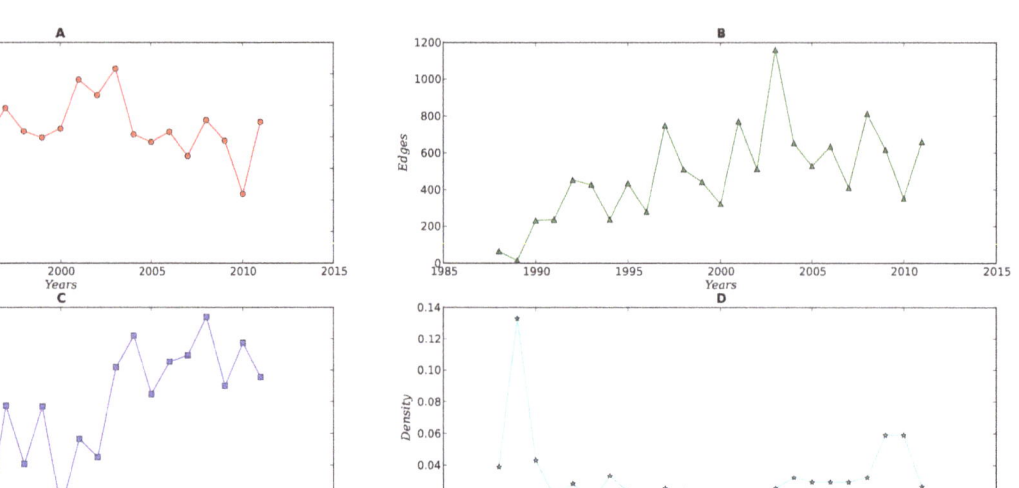

Figure 7. Dynamics of the topological parameters for the Ethics networks. Panels A–F.

biology, and *Polymerase Chain Reaction* (commonly known as *PCR*). The networks reflect the use of the specific term in such a context for every year explored. Some terms like *Ethics* appeared in the earliest networks (1988), some others like *Computational biology* come to be part of the GN in the year of 1996. We chose these subjects because we were very much interested in the approach of different "disciplines" or perspectives towards a particular topic (in our case that topic is genomics), how they behave over the years and if in that regard there are substantial differences between the humanities (*e.g. Ethics and History*), science (*e.g. Neoplasms* research) and technological development (*e.g. Computational biology and PCR*). For these set of SNs we recorded values for the number of nodes *n* and edges *m*, clustering coefficient $\langle C \rangle$, network centralization *NC*, shortest path length $\langle l \rangle$ and density $\langle p \rangle$ [see Tables S3–S7].

Every set of subnetworks (SNs) display a different structure and dynamics [see Figures 7, 8, 9, 10, 11]. When compared against each other, we noticed that despite the low density of each network, there were important differences in the clustering coefficient values. While the set of networks for the humanities had high clustering coefficient values [Figures 7, 8 Panel C], the science subnetwork (represented by the SN *Neoplasms*), were slightly higher than those presented by a random network [Figure 9 Panel C]. Contrary to the results of the humanities and science SNs clustering coefficients, the technology SNs were below the results for a random network [Figures 10, 11 Panel C]. We believe that these results somehow mirror the nature of the different areas. The results of these thematic SNs suggest that not all areas of inquiry behave in the same way. The humanities appear to move at a slower pace as compared to the other areas. The humanities seem

to be quite redundant in their subjects and concepts. For instance, in the case of *Ethics*, concepts that are central to debates are words like *justice*, *dignity*, *equity*, words that have been in the ethics vocabulary for hundreds of years. Interestingly enough, in an article recently published in the *New York Times*, Nicholas Christakis makes reference to an apparent state of stagnation in the Social Sciences [23]. From what we see in our *Ethics* and *History* SNs, it seems that what is to be for the Social Sciences it might be also true for the Humanities –although the pupose of the formers is different from the latters, since the Social Sciences purport themselves as *sciences*. We also explored the content of the triangles (responsible for the clustering coefficient) of a fraction of all SNs for each of these subjects, nevertheless, we were able to see that for *History*, a network with very high clustering coefficient and network centralization [see Tables S3–S7], the terms with the highest connectivity were the terms present as two of the nodes in the triangle. Quite different from this, the *Ethics* SNs had a high clustering coefficient and network centralization, still triangles were not dominated by highly connected nodes, on the contrary, clusters were more diverse.

In the case of science, we think that the low but constant clustering is a sign of the fact that in science there must be some conserved knowledge as a scaffold or dynamic supportive structure on which novelty and new ideas are built upon. We wonder what is the nature of such conserved knowledge and we will try to identify it as part of our future work. However we think that perhaps these scaffolds rest on the form of models, and model organisms. Finally, technology seems to be an area that is always *on a rush*. These SNs have average clustering coefficients $\langle C \rangle = 0$

History subnetworks topological dynamics

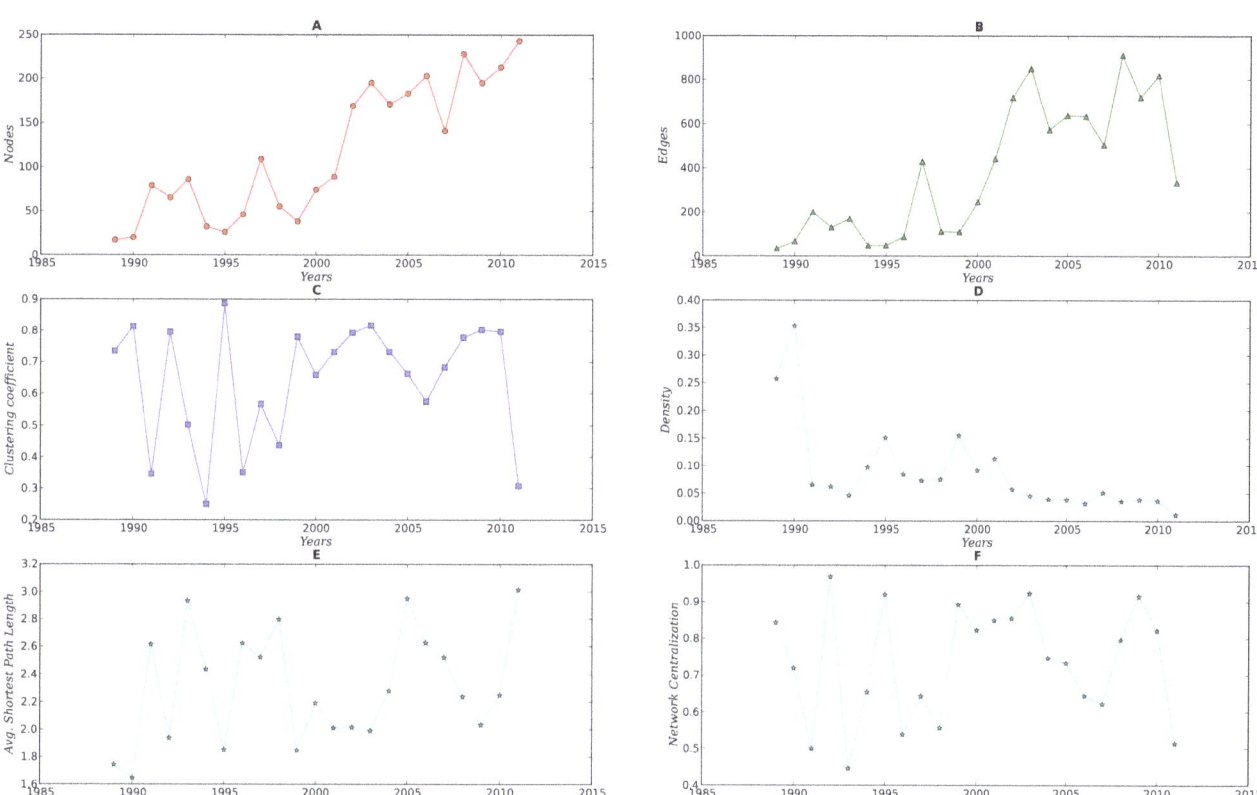

Figure 8. Dynamics of the topological parameters for the History subnetworks. Panels A–F.

for most of the years. We suggest that this is the case because technology, although built on previous technologies, it might not need to make reference to them because, possibly, there are different technical solutions to the same problem. Even if this can be the most common case, we found that for the *PCR* case there are some years in which there was a clustering coefficient over zero. In these *PCR* networks, one of the nodes in the triangle was the new technology, the other the precedent technology and another node that could change but probably was relevant for the new technology, such as gene expression and in a lesser degree, sequencing, proteomics, methods, and organisms. In summary, the humanities change very slowly compared to science and technology. Science develops fast but needs to rely on previous knowledge, and technology always goes ahead, almost with no explicit reference to previous works. In order to test the validity of these ideas and as part of our future work, we will perform studies to properly characterize these areas as communities or neighborhoods and study their local dynamics compared to those of the GNs [16,24].

In the networks we studied, there were terms that were used both as headings and subheadings. The terms *Neoplasms*, *Computational Biology* and *Polymerase Chain Reaction* exist in the MeSH database as headings only, which means that they only appear at the beginning of the MeSH term. The terms *Ethics* and *History* exist as headings and subheadings, that is, they can be at the beginning or in the middle of a MeSH term. Headings and subheadings can be distinguished from each other because headings always start with a capital letter and subheadings always begin in lowercase, *e. g.*, the term *Ethics*, can be found as **Ethics** or as **ethics**, and can

be the heading of a MeSH term such as: *Ethics*, *Medical*, or as a subheading like: *Informed Consent/ethics/legislation & jurisprudence*. We believe that this is interesting because ethics was an important area of research that was promoted since the beginning of genomics and the Human Genome Project. We think that the fact that ethics moved from a heading into a subheading as years went by, is telling us that there is a *cultural* progressive shift regarding the place of moral values in science. At the beginning, the concept of ethics was about a discipline and linked necessarily to the Human Genome Project and as things were developing, ethics became part of many other, more specific areas related to genomics, biomedical research and its clinical and societal consequences. Ethics changed from being the big word for socially legitimating the Human Genome Project to the ethics of many things. Still, it is important to mention that even if the results regarding the shift from ethics as a heading to ethics as a subheading might suggest some ideas related to the role of moral values and the dynamics of science, we recognize that due to the lack of more evidence (that we will explore in the future) we cannot generalize any of these results.

It is no news to say that the Human Genome Project was primarily a State project with the Department of Energy and the National Institutes of Health of the United States as the main actors behind one of the biggest scientific, technological and economical enterprises in the course of research history. It is evident that genomics research somehow was born with the HGP. Even if the HGP was launched officially in 1990, there is a record of publications regarding it since 1987. Such early publications clearly reflect the nature of the project. All articles published from

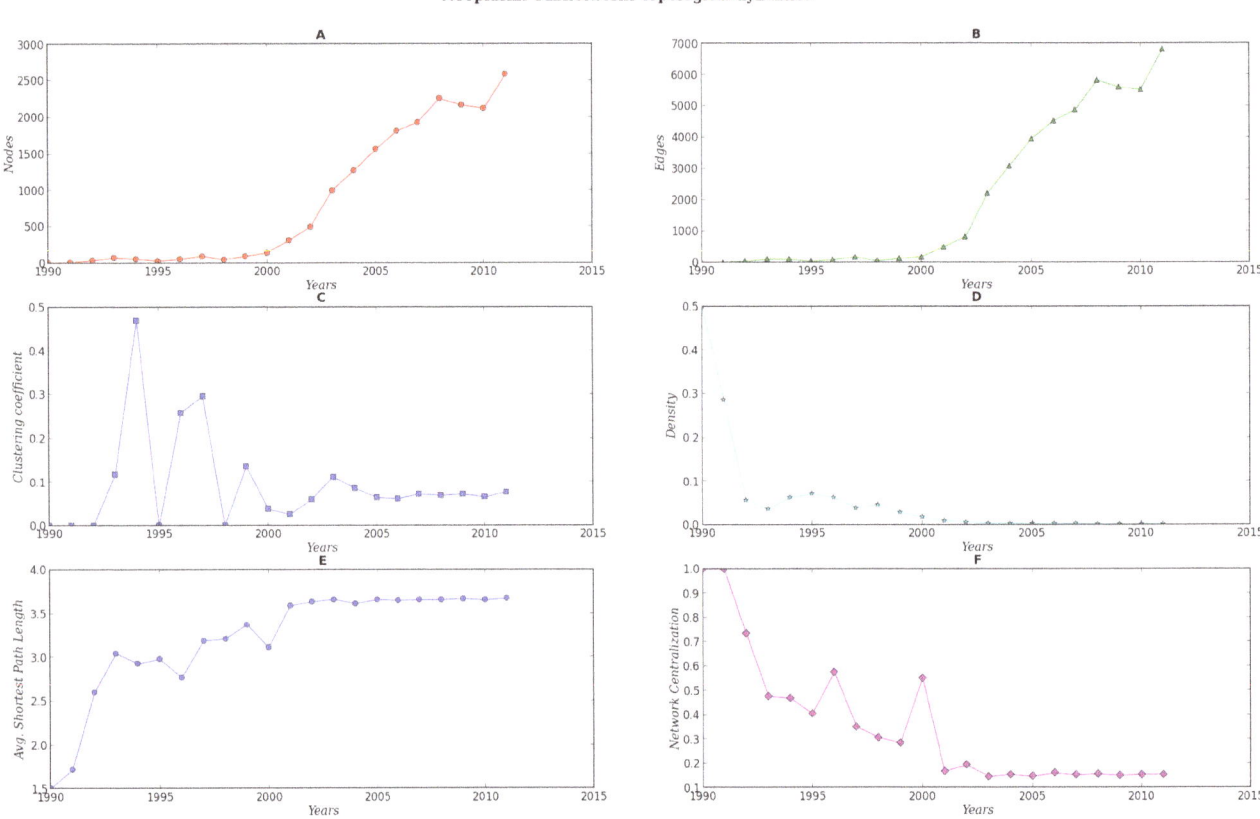

Figure 9. Dynamics of the topological parameters for the Neoplasms networks Panels A–F.

1987 to 1989 are by no means technical in terms of science but technical regarding what was needed economically, socially and politically for the proper development of the HGP. Some of the terms that appear in the first three years networks were *United States*, *Risk Assessment*, *Resources Allocation*, *Government Regulation*, and *Ecology*, as well as others of more ethical concern, like *Eugenics*, *Abortion*, and *Christianity*.

This is interesting in several ways: it is so for the history of science, technology, and society studies because the HGP is probably one of the first big scientific projects that it was planned ahead and looked for a formal way to gain social legitimacy; this strategy was an antecedent for another big project, that of Nanotechnology. It was not only planned in terms of budget, it somehow implied political planning (in order to get the budget approved), but it also looked for social approval. It is no coincidence that James Watson –being the head of the National Center for Human Genome Research– suggested (even before the HGP was formally launched) to include a research area for the study of the Ethical, Social and Legal implications of Genomics.

From the history of the HGP and the networks we built, it seems that genomics was the outcome of a *top-down* style of planning science. Genomics might have appeard in the long run as a consequence of technological and scientific interests as it has happened to many other scientific areas. For example, the Sanger sequencer was first developed in 1977, which means that there were some antecedents heading towards the emergence of genomics as a discipline, and of course, it played an important part in its future appearence. But it was in the context of the HGP that genomics really fluorished. In the particular case of genomics, we think that the HGP played a central role in accelerating the

process. The behavior we just described for the first 10 years of the HGP might have be the reflection of a centralized, *top-down*, control over the development of the project, with a very precise objective in mind. Once the draft of the Human Genome was released and published, –results along with the technology developed in the former years– opened the door to a *bottom-up* style science, in which scientific communities were relatively free to set their particular problems and follow their own interests. Such change might be related to the change in the rate of new terms added per year, which somehow can be seen as a proxy of new emerging research subjects and areas.

Finally, it is also interesting the fact that methodologies like those of complex network theory can be useful for historians. Data mining and complex networks visualization and analysis can be a way for supporting ideas and intuitions with data, regarding how and why science changes and how does it interact with technology and society. In this regard, the whole set of terms related to the Ethical, Legal and Social issues (from now on ELSI) found in the first 10 years of the HGP is the most informative corpus of data on the social perceptions, conceptions and concerns about the human genome. The human genome as the blueprint of life, as the deepest repository of our identity as individuals and as a species. What is even more interesting is that the HGP shaped and materialized such concerns that were slowly framed over the 20th century as has been noted by other authors [25].

At the end, *what are our networks?* MeSH terms that form the networks are a simplified representation of the content of scientific papers. Scientific papers are the result of many people's work, a work that goes beyond authors. A paper is the outcome of many individuals, such as the peers that review the proposal in order to

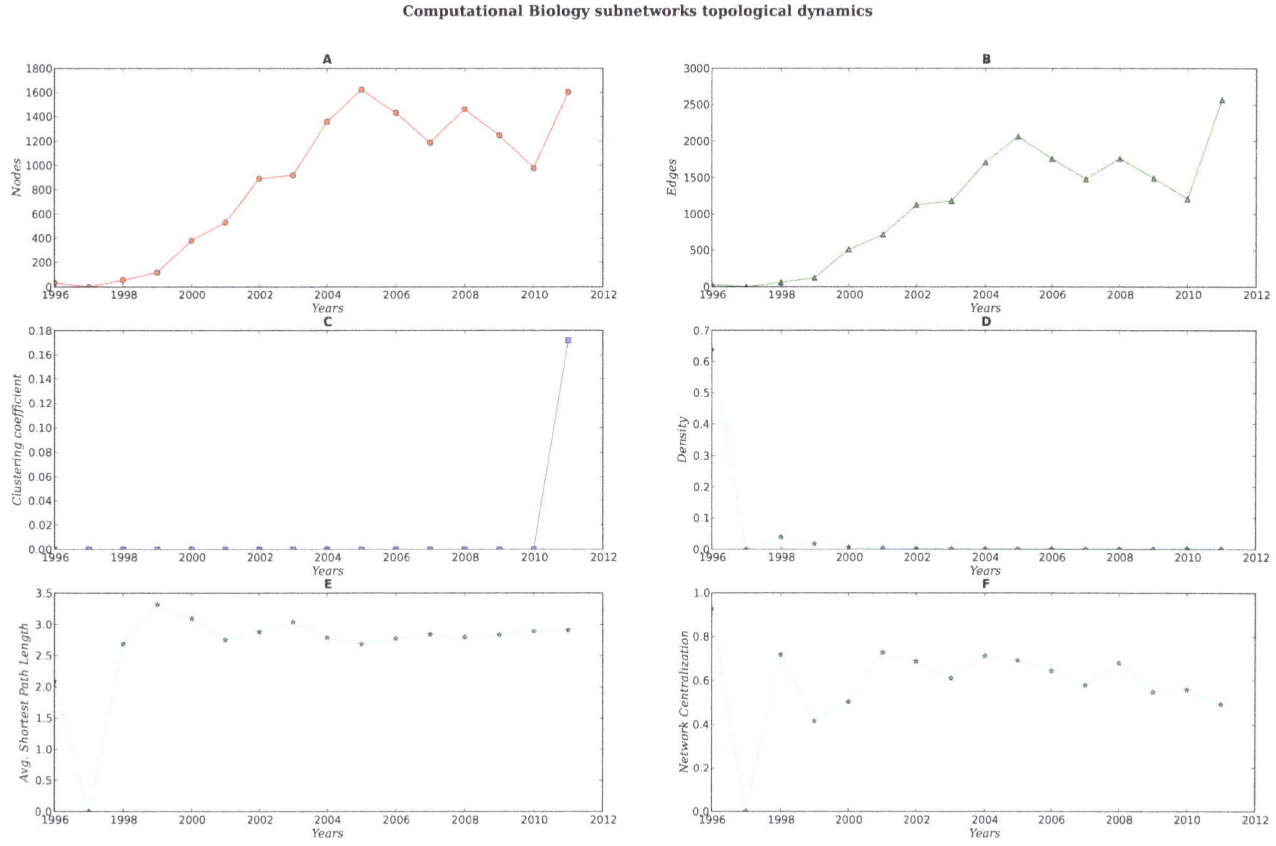

Computational Biology subnetworks topological dynamics

Figure 10. Dynamics of the topological parameters for the Computational Biology subnetworks. Panels A–F.

get funding, research ethics committees, and the paper reviewers as well (just to mention a few). It is also part of an institutional enterprise. So to speak, every paper is a piece of knowledge that is socially agreed (at least in terms of the community that believe that the work should be published). In our specific case, the pieces of knowledge are not only technical and scientific in their nature, but also include knowledge from the humanities and the social sciences. The content of every paper includes directly or indirectly a diversity of topics, from a diversity of sources. By this, we do not imply that, for example, to be part of a paper, ethical issues need to have a MeSH term associated to the subject Ethics in it. What we mean is that there are papers that in order to be published, need many people to have agreed upon its content, including those of the ethics committee that had to approve and follow up the project. Therefore, our networks are the concise description of these pieces of knowledge interacting massively among each other. They are the image and the collective evolution of a world-wide community around *Genomics*, arguing and agreeing on *what* and *how* ideas are or may become *knowledge*. At the end, this is the substance of our networks [10] and the main object of study in this work.

A plethora of questions arise from these studies. The implications of the highly structured connectivity patterns in these networks for the evolution of scientific knowledge (or at least for the case of biomedical research) are taunting and yet to be discovered. Also intriguing is whether such structure emerges from the particular classification approach in the case of MeSH terms or is interwoven in other, more general approaches to knowledge

classification, perhaps even with ontological and epistemological implications.

There is more work to be done. In particular, there are some issues that we want to address in the near future. Some of these issues are related to community-identification and how these communicate among each other; what are the implications of a small-world topology for knowledge organization, and how the diversity of terms and edges impacts the topology and dynamics of the networks. We would also like to study in more detail, from a mathematical as well as from a historical point of view, the suggested phase transition.

Finally, quantitative studies in the evolution of knowledge are now arising. These studies may help us gain in understanding of the structure and evolution of ideas behind academic publications, not only (as is the present case) in the biomedical sciences, but in every documented field of human inquiry. Computational studies, as well as data and text mining techniques are now being supplemented with analysis and visualization tools that will allow researchers in this nascent discipline to construct more insightful models of cognition in a wide variety of fields. These changes may lead to an eventual development of data-driven approaches to epistemological studies of science.

Materials and Methods

We extracted the MeSH terms that describe all the articles indexed in PUBMED that included the MeSH term *Genomics*. Data-mining was done using custom scripts written in Python. To create and analyze the networks we used Cytoscape 3.0.0 [26] and

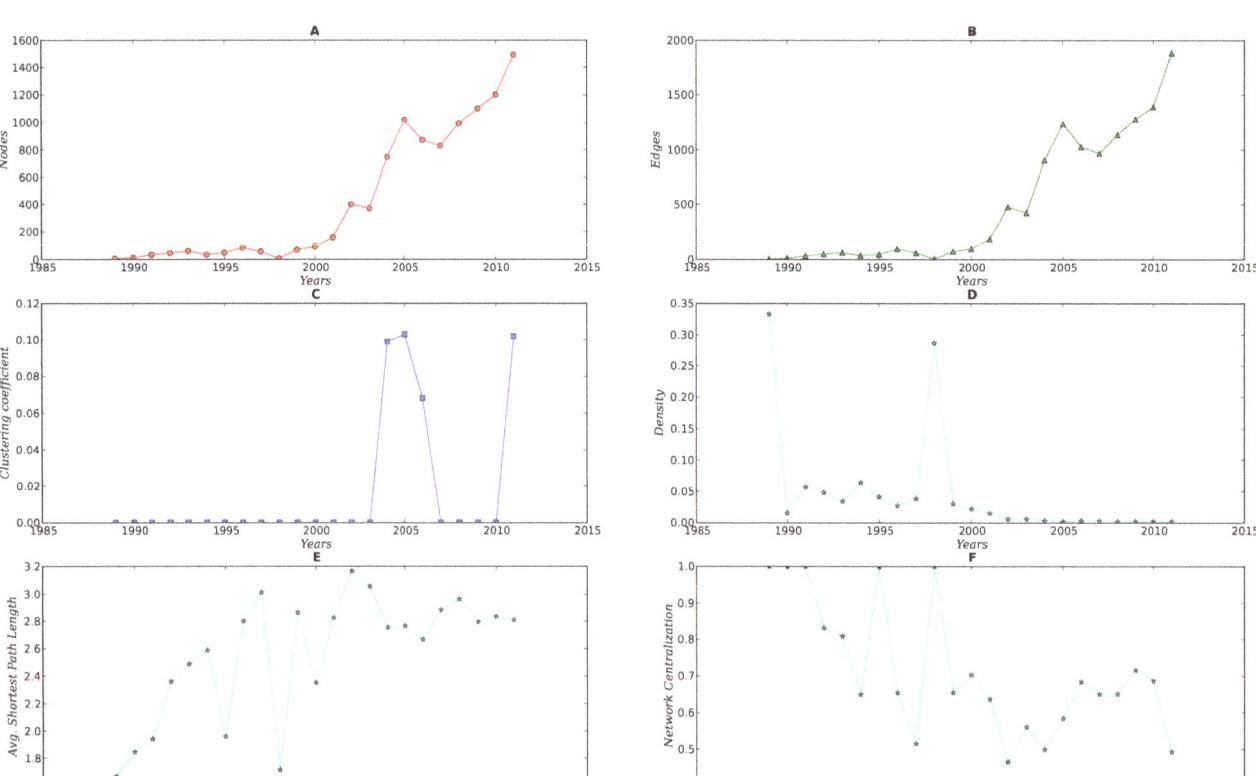

Figure 11. Dynamics of the topological parameters for the PCR subnetworks. Panels A–F.

NetworkX [27], a Python library for the study of complex networks.

We searched for all the articles that included the MeSH term *Genomics* to December 2011, and retrieved their information in the MEDLINE format, as it is offered by the PUBMED webpage. We implemented BioPython's API in order to parse our MEDLINE files and generate a connectivity map in which source and target nodes are the MeSH terms that describe the articles including the MeSH term *Genomics* and the link between the nodes are the PUBMED IDs, or PMIDs.

Major topics, or the main topic of an article is a MeSH term marked with an asterisk. Since we were not interested at the moment in separating major from minor topics, we eliminated the asterisk from the terms in the connectivity map. Once the connectivity map was clean, we created a connectivity map for every year, starting from 1987 and ending in 2011. To every connectivity map a bash script was run to count and give order to the pairs of MeSH terms. By doing this we replaced the PMID that linked every pair of terms with a "weight" according to the number of occurrences of that pair. For example if a pair of terms like *animals* and *humans* comes to occur 100 times, then, the weight of that pair is 100. For this paper we did not analyze the role of the weight of edges in the structure and dynamics of the networks, but that will be part of our future work.

We used Cytoscape 3.0.0 to generate the MeSH terms network for every year; we named these networks *global networks* or GN. For each year we counted the frequencies of every main category heading [see Table S2], that is, recorded how many *A's*, *B's*, *C's* an so forth. The results were normalized and plotted using Python. We also counted the number of levels for each MeSH term from

the table distributed by the National Library of Medicine. Given this, we encoded them to the Circos plot histogram format (blue histogram) 4. For the orange histogram, for each term we counted the number of articles containing it for every year from our data base and encoded it in the Circos plot histogram format 4. For the links connecting terms in the central area of the Circos plot, we used the information already obtained for our networks. In order to do this, we encoded the information in the Circos plot link format 4.

We identified the nodes and edges that are in the intersection of the global networks (GN) for the first network in 1987, to the 1990, 2001 and 2011 GNs. The intersections helped us to identify a small set of nodes such as: *Humans, Animals, Genomics, Human Genome Project, Risk* in order to analyze the dynamics of their degree, clustering, centrality, betweenness centrality and specially, to see who were their neighbors along this history of genomics throughout the years. For this purpose we used NetworkX. By using Cytoscape 3.0.0 [28], we identified the most centralized terms and the emergence of new central terms at different moments of history.

We extracted a group of subnetworks, for terms that are related to different areas of knowledge that have become part of genomic research. These areas are the study of *Neoplasms*, for a scientific object of research, for the humanities we generated a subnetwork for *Ethics* and *History*, and finally, for the technological areas we did it for *Computational biology* and *Polymerase Chain Reaction*. In order to extract these SNs, we searched the connectivity map of each year and chose every pair of nodes that included the term of interest. In the case of *Computational biology* we looked for the root *comput*, for the others we searched for the whole word, that is *Neoplasms, Ethics,*

History and *Polymerase Chain Reaction.* We also extracted the triangles of the *Polymerase Chain Reaction,* and *Computational Biology* SNs to those few years with a clustering coefficient above zero; we also extracted the triangles to the *History* SNs because we wanted to see what was behind the high network centralization and clustering coefficient. Subnetworks were extracted using a bash script and Cytoscape 3.0.0 was used to visualize and analyze them; triangles were extracted with a bash script as well.

Supporting Information

Table S1 Data for the set of MeSH terms **Global Networks** for 25 years. First column contains the year, second column (n) is the number of nodes, third column (m) is the number of edges, fourth column $\langle C \rangle$ is the average clustering coefficient, fifth column $\langle p \rangle$ is the networks' density, sixth column $\langle l \rangle$ is the shortest average path length, and seventh column (NC) is the network centralization.

Table S2 MeSH terms are organized according to a hierarchical multilayered structure of main categories, subcategories and so on. This table displays the two principal layers and the issues spanned by them. Additional layers provide specificity to the conceptual ontology.

Table S3 Data for the MeSH term **Neoplasms** networks for 22 years. First column contains the year, second column (n) is the number of nodes, third column (m) is the number of edges, fourth column $\langle C \rangle$ is the average clustering coefficient, fifth column $\langle p \rangle$ is the networks' density, sixth column $\langle l \rangle$ is the shortest average path length, and seventh column (NC) is the network centralization.

Table S4 Data for the MeSH term **Ethics** networks for 24 years. First column contains the year, second column (n) is the number of nodes, third column (m) is the number of edges, fourth column $\langle C \rangle$ is the average clustering coefficient, fifth column $\langle p \rangle$ is the networks' density, sixth column $\langle l \rangle$ is the shortest

average path length, and seventh column (NC) is the network centralization.

Table S5 Data for the MeSH term **History** networks for 23 years. First column contains the year, second column (n) is the number of nodes, third column (m) is the number of edges, fourth column $\langle C \rangle$ is the average clustering coefficient, fifth column $\langle p \rangle$ is the networks' density, sixth column $\langle l \rangle$ is the shortest average path length, and seventh column (NC) is the network centralization.

Table S6 Data for the of MeSH term **Computational Biology** networks for 16 years. First column contains the year, second column (n) is the number of nodes, third column (m) is the number of edges, fourth column $\langle C \rangle$ is the average clustering coefficient, fifth column $\langle p \rangle$ is the networks' density, sixth column $\langle l \rangle$ is the shortest average path length, and seventh column (NC) is the network centralization.

Table S7 Data for the MeSH term **Polymerase Chain Reaction** networks for 23 years. First column contains the year, second column (n) is the number of nodes, third column (m) is the number of edges, fourth column $\langle C \rangle$ is the average clustering coefficient, fifth column $\langle p \rangle$ is the networks' density, sixth column $\langle l \rangle$ is the shortest average path length, and seventh column (NC) is the network centralization.

Acknowledgments

We gratefully acknowledge Dr. Marcela Varela for the time she spent sharing her knowledge on the PCR history.

Author Contributions

Conceived and designed the experiments: JMSG EHL RGH. Performed the experiments: JMSG EHL RGH ARG. Analyzed the data: JMSG EHL RGH ARG. Contributed reagents/materials/analysis tools: JMSG EHL RGH. Wrote the paper: JMSG EHL RGH.

References

1. Barabási AL (2012) The Network Takeover. Nature Physics 8: 1416.
2. Barabási AL, Albert R (1999) Emergence of scaling in random networks. Science 286: 509512.
3. Newman MEJ (2003) The structure and function of complex networks, SIAM Review 45: 167–256.
4. Newman MEJ (2010) *Networks. An Introduction.* New York: Oxford University Press.
5. Barabási AL, Jeong H, Ravasz R, Néda Z, Vicsek T, et al. (2002) On the topology of the scientific collaboration networks. Physica A 311: 590–614.
6. Newman MEJ (2004) Coauthorship networks and patterns of scientific collaboration. PNAS (suppl. 1): 5200–5205.
7. Cardillo A, Scellato S, Latora V (2006) A topological analysis of scientific coauthorship networks. Physica A: Statistical Mechanics and its Applications 372(2): 333–339.
8. Pan RK, Sinha S, Kaski K, Saramaki J (2012) The evolution of interdisciplinarity in physics research. Scientific Reports 2, 551 doi:10.1038/srep00551.
9. Perc M (2013) Self-organization of progress across the century of physics. Scientific Reports 3, 1720 doi:10.1038/srep01720.
10. Sun X, Kaur J, Milojevic S, Flammini A, Menczer F (2013) Social Dynamics of Science. Scientific Reports 3, 1069 doi:10.1038/srep01069.
11. Yousefi-Nooraie R, Akbari-Kamrani M, Hanneman RA, Etemadi A (2008) Association between co-authorship network and scientific productivity and impact indicators in academic medical research centers: A case study in Iran. Health Research Policy and Systems 6, 9 doi:10.1186/1478-4505-6-9.
12. Pan RK, Kaski K, Fortunato S (2012) World citation and collaboration networks: uncovering the role of geography in science. Scientific Reports 2, 902 doi:10.1038/srep00902.
13. Erjia Y, Ying D (2012) Scholarly network similarities: How bibliographic coupling networks, citation networks, cocitation networks, topical networks,

coauthorship networks, and coword networks relate to each other. Journal of the American Society for Information Science and Technology 63(7): 1313–1326.
14. *U.S. National Library of Medicine.* National Institutes of Health. http://www.nlm.nih.gov/pubs/factsheets/mesh.html (Accessed 2013 July 23).
15. *Introduction to MeSH - 2014.* http://www.nlm.nih.gov/mesh/introduction.html (Accessed 2013 October 23).
16. Pan KR, Sinha S (2009) Modularity produces small-world networks with dynamic time-scale separation. Europhys. Lett. 85: 68006.
17. Watts DJ, Strogatz SH (1998) Collective dynamics of 'small-world' networks. Nature 393: 440–442.
18. Dorogovtsev SN, Mendes JFF (2004) The shortest path to complex networks. Am J Med Genet 71: 47–53.
19. Krzywinski M, Schein J, Birol I, Connors J, Gascoyne R, et al. (2009) Circos: an Information Aesthetic for Comparative Genomics. Genome Res 19: 1639–1645.
20. International Human Genome Sequencing Consortium (2001) Initial sequencing and analysis of the human genome. Nature 409(6822): 860–921.
21. Venter JC, Adams MD, Myers EW, Li PW, Mural RJ, et al. (2001) The sequence of the human genome. Science 291(5507): 1304–1351.
22. Barnes B, Dupré J (2008) *Genomes and What to Make of Them.* Chicago: University Of Chicago Press.
23. Christakis NA. (2013) Lets Shake Up the Social Sciences. New York Times, 2013 July 21. http://www.nytimes.com/2013/07/21/opinion/sunday/lets-shake-up-the-social-sciences.html?smid=pl-share (Accessed 2013 October 23).
24. Lancichinetti A, Kivelä M, Saramäki J, Fortunato S (2010) Characterizing the Community Structure of Complex Networks. PLoS ONE 5(8): e11976, doi:10.1371/journal.pone.0011976.
25. Keller EF (2000) *The Century of the Gene.* Cambridge: Harvard University Press.

26. Shannon P, Markiel A, Ozier O, Baliga NS, Wang JT, et al. (2003) Cytoscape: a software environment for integrated models of biomolecular interaction networks. Genome research, 13(11): 2498–504.

27. Hagberg AA, Schult DA, Swart PJ. Exploring network structure, dynamics, and function using NetworkX. In: *Proceedings of the 7th Python in Science Conference (SciPy2008)*, Gäel Varoquaux, Travis Vaught, Jarrod Millman (Eds), (Pasadena, CA USA), 1115, Aug 2008.

28. Saito R, Smoot ME, Ono K, Ruscheinski J, Wang PL, et al. (2012) A travel guide to Cytoscape plugins. Nature methods 9(11): 1069–76.

12

Benchmarking Undedicated Cloud Computing Providers for Analysis of Genomic Datasets

Seyhan Yazar[1], George E. C. Gooden[1], David A. Mackey[1,2], Alex W. Hewitt[1,2,3]*

1 Centre for Ophthalmology and Visual Science, University of Western Australia, Lions Eye Institute, Perth, Western Australia, Australia, **2** School of Medicine, Menzies Research Institute Tasmania, University of Tasmania, Hobart, Tasmania, Australia, **3** Centre for Eye Research Australia, University of Melbourne, Department of Ophthalmology, Royal Victorian Eye and Ear Hospital, Melbourne, Victoria, Australia

Abstract

A major bottleneck in biological discovery is now emerging at the computational level. Cloud computing offers a dynamic means whereby small and medium-sized laboratories can rapidly adjust their computational capacity. We benchmarked two established cloud computing services, Amazon Web Services Elastic MapReduce (EMR) on Amazon EC2 instances and Google Compute Engine (GCE), using publicly available genomic datasets (*E.coli* CC102 strain and a Han Chinese male genome) and a standard bioinformatic pipeline on a Hadoop-based platform. Wall-clock time for complete assembly differed by 52.9% (95% CI: 27.5–78.2) for *E.coli* and 53.5% (95% CI: 34.4–72.6) for human genome, with GCE being more efficient than EMR. The cost of running this experiment on EMR and GCE differed significantly, with the costs on EMR being 257.3% (95% CI: 211.5–303.1) and 173.9% (95% CI: 134.6–213.1) more expensive for *E.coli* and human assemblies respectively. Thus, GCE was found to outperform EMR both in terms of cost and wall-clock time. Our findings confirm that cloud computing is an efficient and potentially cost-effective alternative for analysis of large genomic datasets. In addition to releasing our cost-effectiveness comparison, we present available ready-to-use scripts for establishing Hadoop instances with Ganglia monitoring on EC2 or GCE.

Editor: Maureen J. Donlin, Saint Louis University, United States of America

Funding: This work was supported by funding from the BrightFocus Foundation, the Ophthalmic Research Institute of Australia and a Ramaciotti Establishment Grant. CERA receives operational infrastructure support from the Victorian government. The funders had no role in study design, data collection and analysis, decision to publish, or preparation of the manuscript.

Competing Interests: The authors have declared that no competing interests exist.

* Email: hewitt.alex@gmail.com

Introduction

Through the application of high-throughput sequencing, there has been a dramatic increase in the availability of large-scale genomic datasets [1]. With reducing sequencing costs, small and medium-sized laboratories can now easily amass many gigabytes of data. Given this dramatic increase in the volume of data generated, researchers are being forced to seek efficient and cost-effective measures for computational analysis [2]. Cloud computing offers a dynamic means whereby small and medium-sized laboratories can rapidly adjust their computational capacity, without concern about its physical structure or ongoing maintenance [3–6]. However, transitioning to a cloud environment presents with unique strategic decisions [7], and although a number of general benchmarking results are available (http://serverbear.com/benchmarks/cloud; https://cloudharmony.com/; Accessed 2014 Aug 7), there has been a paucity of comparisons of cloud computing services specifically for genomic research.

We undertook a performance comparison on two established cloud computing services: Amazon Web Services EMR on Amazon EC2 instances and GCE. Paired-end sequence reads of publicly available genomic datasets (*Escherichia coli* CC102 strain and a Han Chinese male genome) were analysed using Crossbow, a genetic annotation tool, on Hadoop-based platforms with

equivalent system specifications [8–10]. A standard analytical pipeline was run simultaneously on both platforms multiple times (Figure 1 and 2). The performance metrics of both platforms were recorded using Ganglia, an open-source high performance computing monitoring system [11].

Results

Wall-clock time for complete mapping and SNP calling differed by 52.9% (95% CI: 27.5–78.2) and 53.5% (95% CI: 34.4–72.6) for *E.coli* and human genome alignment and variant calling, respectively, with GCE being more efficient than EMR. Table 1 displays the key metrics for data analysis using both services. The proportion of central processing unit (CPU) usage by Crossbow differed between platforms when aligning and SNP calling each genome, with GCE having better utilisation as the genome size increased. There was considerably more free memory on GCE for the smaller *E.coli* dataset and on EMR for larger human genome runs. The CPU idle percentage, the percentage of time where the CPU was idle without waiting for disk input/output (I/O), was greater on EMR for the human genome while CPU waiting for I/O (WIO) was considerably lower on the same platform. The CPU idle and CPU WIO percentages were both significantly higher on EMR for the *E.coli* genome. The cost of running this Crossbow

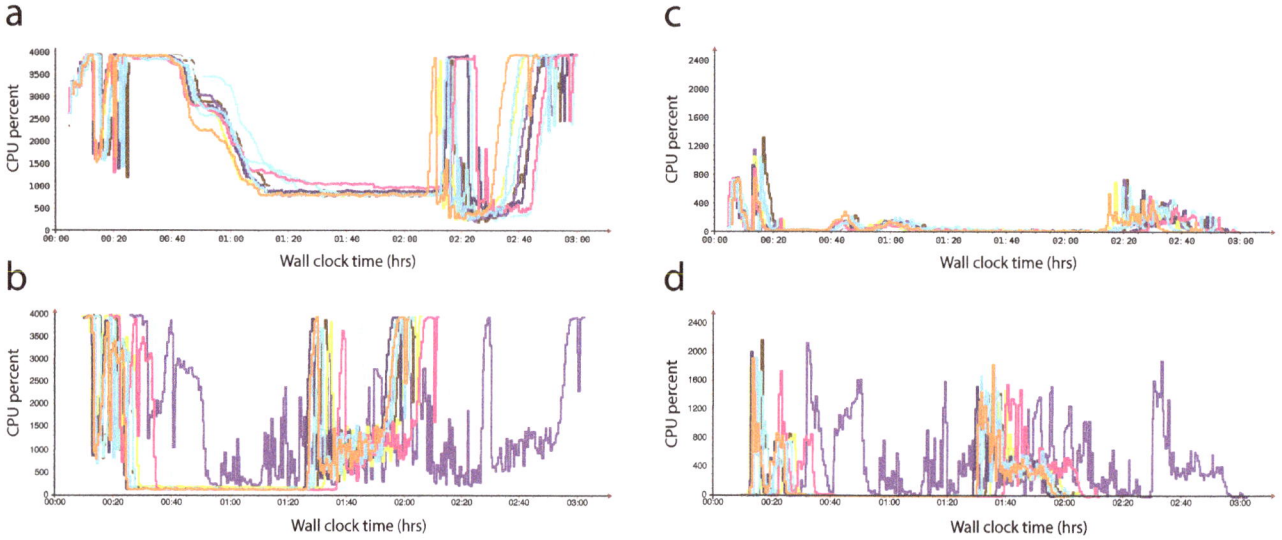

Figure 1. Comparison of undedicated cloud computing performances. The panel includes results of Amazon Web Services Elastic MapReduce (EMR) on Amazon EC2 instances (panels a & c) versus Google Compute Engine (GCE) (panels b & d) for human genome alignment and variant calling. In this 40 node cluster the total CPU percent for CPU idle (a and b) and waiting for disk input/output (c and d) is displayed. Note the greater consistency in performance of Crossbow, though generally longer wall clock times for complete analysis, on EMR compared to GCE.

pipeline on EMR and GCE also differed significantly (p<0.001), with the costs on EMR being 257.3% (95% CI: 211.5–303.1) and 173.9% (95% CI: 134.6–213.1) more expensive than GCE for *E.coli* and human assemblies, respectively. For ∼36x coverage of a human genome, at a current sequencing cost of ∼US$1000, the median cost for computation on GCE was US$29.81 (range: US$28.86 to US$45.99), whilst on EMR with a fixed hourly rate it was US$69.60 (range: US$69.60 to US$92.80).

Although runtime variability was inevitable and present in both platforms when assembling each genome, GCE had a considerably greater variability with the larger human genome compared to EMR (coefficient of variation $(COV)_{EMR} = 4.48\%$ vs $COV_{GCE} = 16.72\%$). We identified a single outlier in run time

on GCE during the human genome analysis. This occurred due to the virtual cluster having a slower average network connection (1.55 MB/s compared to the average of the other GCE clusters of 2.02 MB/s) and a higher CPU WIO percentage than the average for the other GCE runs (9.56% versus 3.52%). The variation in cluster performance likely reflects an increase in network congestion amongst GCE servers.

Runtime predictably is an important issue in undedicated cloud computing. The existing workload of the cloud at the time of service usage is one of the main determinants of variability in runtime of undedicated services [12]. In our benchmarking, EMR was more consistent, though slower, in overall wall-clock time compared to GCE. This may suggest that GCE is more susceptible

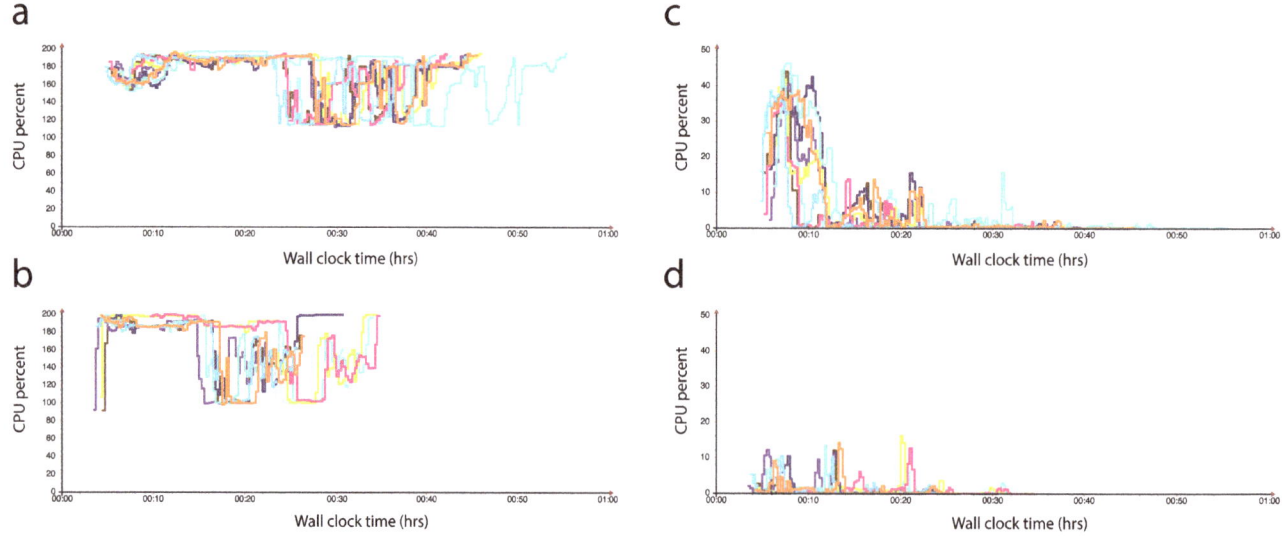

Figure 2. Comparison of undedicated cloud performance of Amazon Web Services Elastic MapReduce (EMR) on Amazon EC2 instances (panels a & c) versus Google Compute Engine (GCE) (panels b & d) for *E.coli* genome alignment and variant calling. In this two node cluster the total CPU percent for CPU idle (a and b) and waiting for disk input/output (c and d) is displayed. Note the shorter wall clock times for complete analysis on GCE compared to EMR.

Table 1. Comparison of performance metrics for genomic alignment and SNP calling.

Metric	*E.coli* Genome			Human Genome		
	EMR (n = 10)	GCE (n = 10)	p-value*	EMR (n = 10)	GCE (n = 10)	p-value*
Wall clock time (mean)	0:46:30	0:31:50	<0.001	2:58:24	2:14:12	<0.001
Pre-processing short reads time (mean)	0:14:37	0:12:46	0.109	0:07:29	0:06:23	0.116
Alignment with Bowtie time (mean)	0:07:04	0:05:03	<0.001	1:51:06	1:15:07	0.003
Calling SNPs with SOAPsnp time (mean)	0:05:05	0:02:51	<0.001	0:35:31	0:29:31	0.033
Post-processing time (mean)	0:04:51	0:00:57	<0.001	0:01:23	0:01:03	<0.001
CPU user (mean %)	17.44±1.30	22.31±3.14	<0.001	43.80±1.87	58.05±6.20	<0.001
CPU idle (mean %)	72.75±1.23	65.76±4.63	<0.001	47.48±2.30	22.17±3.14	<0.001
CPU wio (mean %)	3.88±1.06	0.70±0.16	<0.001	1.86±0.19	4.54±1.82	0.001
Bytes in (MB/sec)	1.15±0.09	2.12±0.42	<0.001	1.58±0.07	2.00±0.19	<0.001
Memory free (GB)	2.19±0.13	6.17±0.42	<0.001	0.91±0.07	0.70±0.03	<0.001

All times are presented as hr:min:sec and remaining metrics are shown as mean ± standard deviation.
*Calculated by paired *t*-test.

to server congestion than EMR; though service usage data is difficult to obtain.

Discussion

Our findings confirm that cloud computing is an efficient and potentially cost-effective alternative for analysis of large genomic datasets. Cloud computing offers a dynamic, economical and versatile solution for large-scale computational analysis. There have been a number of recent advances in bioinformatic methods utilising cloud resources [4,9,13], and our results suggest that a standard genomic alignment is generally faster in GCE compared to EMR. The time differences identified could be attributed to the hardware used by the Google and Amazon for their cloud services. Amazon offers a 2.0 GHz Intel Xeon Sandy Bridge CPU, whilst Google uses a 2.6 GHz Intel Xeon Sandy Bridge CPU. This clock speed variability is considered the main contributing factor to the difference between the two undedicated platforms. It must also be noted that the resource requirements of Ganglia may have had a small impact on completion times [11].

There are a number of technical differences between GCE and EMR, which are important to consider when running standard bioinformatic pipelines. Running Crossbow on Amazon Web Services was simplified by an established support service, which provides an interface for establishing and running Hadoop clusters (Text S1). In contrast, there is currently no built-in support for GCE in Crossbow (Text S2). The current process to run a Crossbow job on GCE requires users to complete various steps such as installing and configuring the required software on each node in the cluster, transferring input data onto the Hadoop Distributed File System (HDFS), downloading results from the HDFS and terminating the cluster on completion. All of these steps are automatically performed by Crossbow on EMR. Python scripts offering similar functionality for GCE that Crossbow provides for EMR were created and are available (https://github.com/hewittlab/Crossbow-GCE-Hadoop).

While our findings confirm that cloud computing is an attractive alternative to the limitations imposed by the local environment, it is noteworthy that better performance metrics and lower cost were found with GCE compared to its established counterpart,

Amazon's EMR. Currently, a major limitation of these services remains at the initial transfer of large datasets onto the hosted cloud platform [14]. To circumvent this in the future, sequencing service providers are likely to directly deposit data to a designated cloud service provider, thereby eliminating the need for the user to double handle the data transfer [15]. Once this issue is resolved, it is foreseen that demand for these services is likely to increase considerably, given the low cost, broad flexibility and good customer support for cloud services [15]. The development of

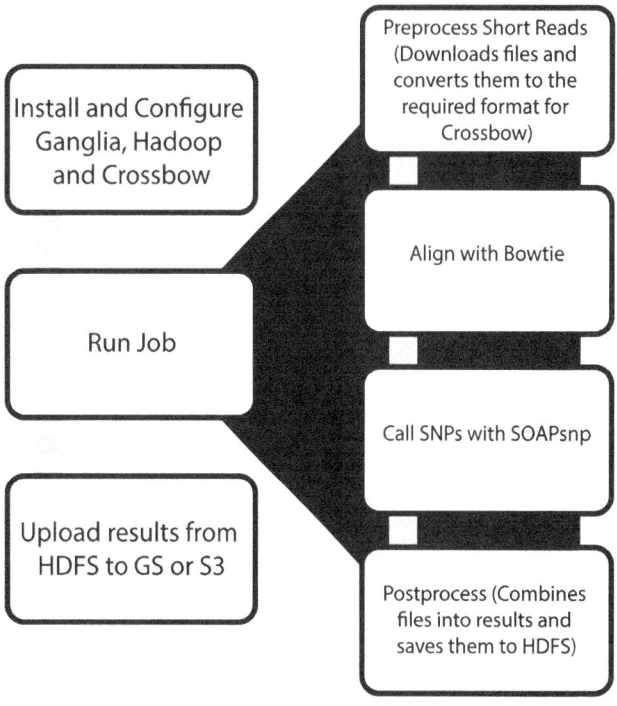

Figure 3. Analytical pipeline demarcating each step required to complete the Crossbow job in the cloud.

Table 2. Specification of used computational nodes for each system.

	Virtual Cores	Memory (GB)	Included Storage (GB)	Price (USD/Hour)^
Amazon Elastic Compute Cloud (EC2) + Elastic MapReduce (EMR) [c1.xlarge]	8	7	4×420	$0.640
Google Compute Engine [n1-highcpu-8]	8	7.2	0#	$0.352

^Date accessed: April to June 2014; prior to this period, pricing was $0.700 and $0.520 in Amazon and Google respectively.
#for each instance we added the minimum storage quota of 128 GB.

additional tools specific to genomic analysis in the cloud, which offer flexibility in choice of providers, is clearly required.

Methods

Datasets and Analytical Pipeline

We benchmarked two platforms by a single job that completed read alignment and variant calling stages of next generation sequencing analysis simultaneously on two independent cloud platforms. To investigate the impact of data size on undedicated cluster performance, one small (*Escherichia coli* CC102 strain (3 GB SRA file; Accession: SRX003267) and one large (a Han Chinese male genome (142 GB Fastq files; Accession: ERA000005) publicly available genomic dataset was selected for analysis [8,10]. For each job in this experiment, a parallel workflow was designed using Crossbow. This workflow included the following four steps: (1) Download and conversion of files; (2) Short read alignment with Bowtie; (3) SNP call with SOAPsnp; and (4) Combination of the results (Figure 3). Crossbow was the preferred genetic annotation tool in this experiment, as it has built in support for running via Amazon's EMR and Hadoop clusters [16].

Cluster construction and architecture

Instances were simultaneously established on Amazon's EMR (http://aws.amazon.com/ec2/; Accessed 2014 Aug 7) and GCE (http://cloud.google.com/products/compute-engine.html; Accessed 2014 Aug 7). Undedicated clusters were optimized by selection of computational nodes as suggested for Crossbow [9]. Nodes with equivalent specifications were selected for each system (Table 2), these being c1.xlarge node in EMR and the closest specification node n1-highcpu-8 in GCE. For the *E.coli* genome, two nodes (one master and one slave) were used on each platform. On the other hand, for the human genome, the cluster was built with 40 nodes (one master and 39 slaves). As GCE did not provide any included storage for each instance, a 128 GB drive (the default storage quota provided by GCE) was added for each node. This was at the additional cost of $0.04/GB/Month or $0.000056/GB/Hour (Jan to June 2014).

Each cluster was run using Apache Hadoop, an open-source implementation of the MapReduce algorithm [17]. MapReduce was used to organise distributed servers, manage the communication between servers and provide fault tolerance allowing tasks to be performed in parallel [18].

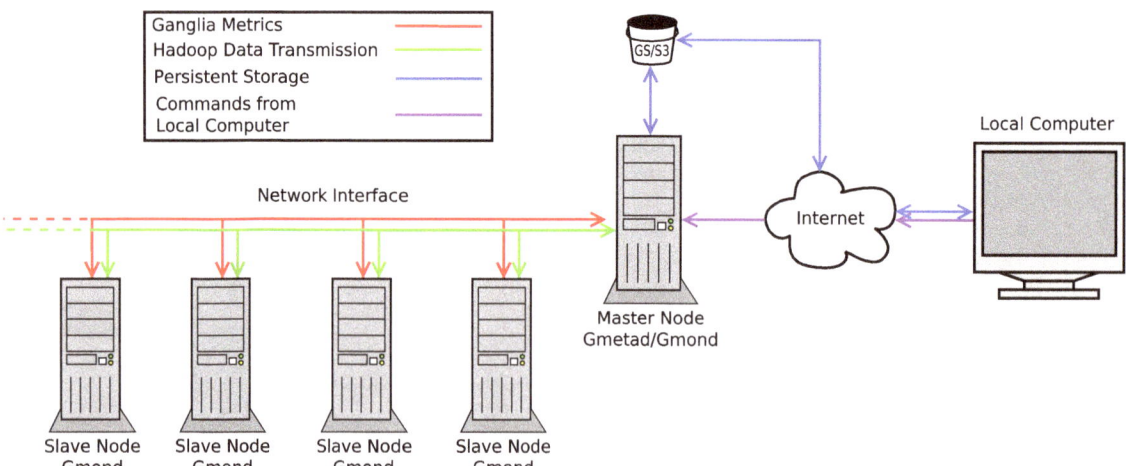

Figure 4. Directions and types of network transfers in our cloud-computing model. There are a variety of different network transfers between the nodes for each of the services in use in our model. Hadoop requires a bidirectional transmission of data between the master node and the slave nodes. This is required to coordinate the parallel processing of the cluster, and to allow for data transfer between nodes. Ganglia uses a unidirectional connection from the slave nodes to the master node to transfer the recorded metrics for storage and visualization. The persistent storage (provided by Amazon S3 (Simple Storage Service) or Google Storage, or an alternative method such as an FTP server) is accessed via the master node. The master node uses it to download input files for Crossbow, such as the manifest file and the reference Jar, and to use for persistent storage of the results of the Crossbow job as the instances destroy their storage on termination. Our local computer can also access the persistent storage via the Internet to allow access to upload the input files, or to download the results. The local computer needs to access the master node to initiate Crossbow. In EMR, this is replaced by a web interface and a JavaScript Object Notation Application Programming Interface (JSON API). In GCE, the user is required to remotely log in via Secure Shell (SSH) to commence the job.

To explore the effect of network activity differences between the platforms, each job was run simultaneously; same day (including weekdays and weekends) and same time. Detailed description of the set up and scripts to run the jobs can be found in Text S1 and Text S2.

Cluster Monitoring

In both EMR and GCE, multiple components of cloud infrastructure including CPU utilisation, memory usage and network speeds were monitored and recorded for each node using Ganglia. The default setting of Ganglia for distributing incoming requests is multicast mode; however, since EMR and GCE environments do not currently support multicast Ganglia, it was configured in unicast mode (Figure 4). The metric output files constructed in.rrd format were converted into.csv format with a Perl script (Text S3). For comparison between performance and costs between platforms, the Student t-test was undertaken using the statistical software R (R Foundation for Statistical Computing version 3.0.2; http://www.r-project.org/). In the analysis, cost of each run was calculated using current pricing (June 10th 2014); however, all *E.coli* runs and one human genome run were performed prior to a recent decrease in price on both platforms. The COV for runtime variability was calculated as the ratio of the standard deviation to the mean time (mins) for each system.

Author Contributions

Conceived and designed the experiments: SY AWH. Performed the experiments: SY GECG. Analyzed the data: SY GECG. Contributed reagents/materials/analysis tools: DAM AWH. Contributed to the writing of the manuscript: SY GECG DAM AWH.

References

1. Marx V (2013) Biology: The big challenges of big data. Nature 498: 255–260.
2. Patro R, Mount SM, Kingsford C (2014) Sailfish enables alignment-free isoform quantification from RNA-seq reads using lightweight algorithms. Nat Biotechnol 32: 462–464.
3. Schatz MC, Langmead B, Salzberg SL (2010) Cloud computing and the DNA data race. Nat Biotechnol 28: 691–693.
4. Angiuoli SV, White JR, Matalka M, White O, Fricke WF (2011) Resources and Costs for Microbial Sequence Analysis Evaluated Using Virtual Machines and Cloud Computing. PLoS ONE 6: e26624.
5. Fusaro VA, Patil P, Gafni E, Wall DP, Tonellato PJ (2011) Biomedical Cloud Computing With Amazon Web Services. PLoS Comput Biol 7: e1002147.
6. Drake N (2014) Cloud computing beckons scientists. Nature 509: 543–544.
7. Marx V (2013) Genomics in the clouds. Nat Meth 10: 941–945.
8. Parkhomchuk D, Amstislavskiy V, Soldatov A, Ogryzko V (2009) Use of high throughput sequencing to observe genome dynamics at a single cell level. Proc Natl Acad Sci USA 106: 20830–20835.
9. Langmead B, Schatz MC, Lin J, Pop M, Salzberg SL (2009) Searching for SNPs with cloud computing. Genome Biol 10: R134.
10. Wang J, Wang W, Li R, Li Y, Tian G, et al. (2008) The diploid genome sequence of an Asian individual. Nature 456: 60–65.
11. Massie ML, Chun BN, Culler DE (2004) The ganglia distributed monitoring system: design, implementation, and experience. Parallel Comput 30: 817–840.
12. Schad J, Dittrich J, Quiané-Ruiz J-A (2010) Runtime measurements in the cloud: observing, analyzing, and reducing variance. Proceedings VLDB Endowment 3: 460–471.
13. Onsongo G, Erdmann J, Spears MD, Chilton J, Beckman KB, et al. (2014) Implementation of Cloud based Next Generation Sequencing data analysis in a clinical laboratory. BMC Res Notes 7: 314.
14. Schadt EE, Linderman MD, Sorenson J, Lee L, Nolan GP (2010) Computational solutions to large-scale data management and analysis. Nat Rev Genet 11: 647–657.
15. Stein LD (2010) The case for cloud computing in genome informatics. Genome Biol 11: 207.
16. Crossbow project homepage. Available: http://bowtie-bio.sourceforge.net/crossbow/index.shtml. Accessed 2014 Aug 7.
17. Hadoop - Apache Software Foundation project homepage. Available: http://hadoop.apache.org/. Accessed 2014 Aug 7.
18. Dean J, Ghemawat S (2008) MapReduce. Commun ACM 51: 107–113.

13

Partition Enrichment of Nucleotide Sequences (PINS) - A Generally Applicable, Sequence Based Method for Enrichment of Complex DNA Samples

Thomas Kvist, Line Sondt-Marcussen, Marie Just Mikkelsen*

Samplix ApS, Ballerup, Denmark

Abstract

The dwindling cost of DNA sequencing is driving transformative changes in various biological disciplines including medicine, thus resulting in an increased need for routine sequencing. Preparation of samples suitable for sequencing is the starting point of any practical application, but enrichment of the target sequence over background DNA is often laborious and of limited sensitivity thereby limiting the usefulness of sequencing. The present paper describes a new method, Probability directed Isolation of Nucleic acid Sequences (PINS), for enrichment of DNA, enabling the sequencing of a large DNA region surrounding a small known sequence. A 275,000 fold enrichment of a target DNA sample containing integrated human papilloma virus is demonstrated. Specifically, a sample containing 0.0028 copies of target sequence per ng of total DNA was enriched to 786 copies per ng. The starting concentration of 0.0028 target copies per ng corresponds to one copy of target in a background of 100,000 complete human genomes. The enriched sample was subsequently amplified using rapid genome walking and the resulting DNA sequence revealed not only the sequence of a the truncated virus, but also 1026 base pairs 5′ and 50 base pairs 3′ to the integration site in chromosome 8. The demonstrated enrichment method is extremely sensitive and selective and requires only minimal knowledge of the sequence to be enriched and will therefore enable sequencing where the target concentration relative to background is too low to allow the use of other sample preparation methods or where significant parts of the target sequence is unknown.

Editor: Ruslan Kalendar, University of Helsinki, Finland

Funding: The authors received no specific funding for this work. All authors are employed by Samplix ApS. Samplix ApS provided support in the form of salaries for authors TK, LS-M and MJM but did not have any additional role in the study design, data collection analysis, decision to publish, or preparation of the manuscript.

Competing Interests: All authors are affiliated with the company Samplix ApS, Denmark, which holds rights to commercial use of the method described in patent application WO2014096421 (Probability-directed isolation of nucleotide sequences (PINS)). Samplix ApS seeks to enable the use and broaden the knowledge of the PINS method broadly within the scientific community. There are no further patents or products to declare.

* Email: mjm@samplix.com

Introduction

Molecular biology methods are increasingly used for diagnosis, prognostication and prediction of disease and efficiency of therapy. While PCR is now widely used, sequence analysis on relevant samples such as swabs, blood or feces needs preparation as the collection of sequence information is complicated by the high level of background DNA from e.g. blood cells or microbial cells [1]. As the target cell or molecule may only be present in a few copies in the complex sample the selectivity of a sample preparation method must be exceptionally high. Next generation sequencing offers extremely high throughput and low cost per sequenced base pair. Samples are typically prepared by generating PCR fragments of a few hundred base pairs, containing adapter sequences at both ends. The fragments are then clonally amplified before the actual sequencing. The small sequence fragments are aligned and a full sequence is constructed. In next generation sequencing, the error rate is typically at least 0.1%, even after stringent filtering based on quality scores and a given mutation must therefore be present in at least 1% of the sequences of a given region for the investigator to

be fairly sure that the purported mutation is not a falsely interpreted sequencing error [2]. When the target sequence constitutes less than 1%, enrichment needs to be included in the sample preparation. This can be done using a procedure such as ICE-COLD PCR, but will only retrieve a small fragment containing the mutation [3]. Enrichment of specified regions of a genome can be performed using hybrid capture techniques, typically using biotin-linked capture probes [4], however the selectivity of these techniques is limited and the preparation of capture probe libraries is laborious and expensive.

PINS is a deceptively simple and powerful technology for DNA enrichment, only relying on PCR detection of a short specific sequence. It combines terminal dilution, target detection and whole genome amplification (Figure 1). The prerequisite for using the technology is that the target molecule contains a known sequence and that this sequence can be detected in the original sample. The first step of enrichment consists of repeated dilutions of the sample to the point, where the target is no longer present in all wells when the final dilution is partitioned into a number of replicate samples. The distribution of target DNA molecules among these partitions follows Poisson statistics, and at the limiting dilution, most reactions contain either one or zero target DNA molecules [5]. All replicate diluted samples are now amplified, using multiple displacement amplification (MDA), and the presence/absence of the target is detected in the individually amplified wells. The MDA reaction will produce fragments of more than 10,000 base pairs on average [6]. Wells containing a target DNA (positive wells) will have a higher concentration of target fragments relative to background, a principle that is also known from e.g. digital droplet PCR [7,8]. The ratio of positive to negative wells will determine the degree of target enrichment. This procedure of end point dilution and amplification is now repeated until the desired abundance of positive fragment is achieved, and the sample is ready for downstream applications such as sequencing.

The sensitivity of PINS enrichment is determined only by the sensitivity of detection and is therefore high compared to other means of enrichment such as hybridization-based techniques. In the results section below, it is shown that enrichment can be performed from a starting point of 0.001% integrated human papilloma virus (HPV) copies relative to human background genomes corresponding to less than three target molecules per µg of total DNA. HPV is the principal cause of virtually all cervical cancers [9] and subsets of head and neck cancers [10]. HPV18, which is used in the current study, is the second most prevalent carcinogenic type of HPV, causing around 16% of all cervical cancer cases [11]. PINS enrichment of HPV containing sequences from e.g. a blood sample provides not only the sequence of the virus, but also sequence of the integration site and nearby region.

The PINS enrichment method successfully demonstrated in this study enables sequencing of samples with far lower target to background ratio than previously seen and can be applied to viruses as well as a broad range of other applications through changing only one set of detection primers.

Materials and Methods

The data were analyzed anonymously and consent was therefore not needed. The leukocytes used for extraction of negative background DNA were kindly donated by Quantibact A/S, Denmark. They were originally provided by Department of Clinical Immunology - Blood Bank (Rigshospitalet, Denmark) where they were discarded for clinical use. The leukocyte sample was part of a sample used in a previous study [12], the sample was fully anonymized and it could not be traced back to the original donor.

Nucleic acid templates

HeLa DNA containing HPV18 was purchased from New England Biolabs in a concentration of 100 ng/µL (NEB-N4006S). Background DNA was extracted from leukocytes from an HPV18 negative blood sample and DNA was extracted using a DNA Extraction kit (Fermentas – GeneJET^{TM}Genomic DNA Purification Kit).

Both templates were pre-amplified using Phi29 amplification (described below) and visually inspected by gel electrophoresis prior to quantification using a Quantous Fluorometer (Promega BioSystems) according to the instructions provided by the manufacturer (Instructions for use of product E6150).

Phi29 amplification

Phi29 MDA reactions were carried out using the Sampliphi kit as described by the manufacturer (Samplix ApS, Denmark): Two solutions, Mix 1 and Mix 2 were prepared separately.

Mix1: 1 µL DNA template, 1 µL "Solution A", 2.5 µL freshly prepared "Solution B", and 0.5 µL nuclease free (NF) water (Sigma-Aldrich) was thoroughly mixed using a vortexer. Then, the mixture was heated to 94°C for 3 minutes and immediately transferred to ice.

Mix2: 4 µL "Solution C", 2 µL "Solution D", 8.5 µL NF water, and 0.5 µL (10 U/µL) Phi29 polymerase was added to the reaction.

After combining Mix 1 and 2, the reaction was mixed thoroughly and incubated at 30°C for 16 hours. Finally, the reaction was terminated by heat inactivation at 65°C for 10 minutes. No reactions were observed in any of the included negative controls using NF water (Sigma) as "non-template".

Setting up the initial template

A Phi29 amplified sample of HeLa DNA (228 ng/µL) was diluted into Phi29 amplified HPV18-negative background DNA (245 ng/µL) to create an initial sample containing 0.6 target copies per µL in a total DNA concentration of 211 ng/µL.

Primer design

HPV18 PCR primers were designed using sequence information (GenBank accession number: GQ180790) from GenBank in the primer design application of CLC Main workbench v6 (CLCbio, a Qiagen company) targeting a desired product size of 100-130 bp. Numbers in HPV primer IDs corresponds to the

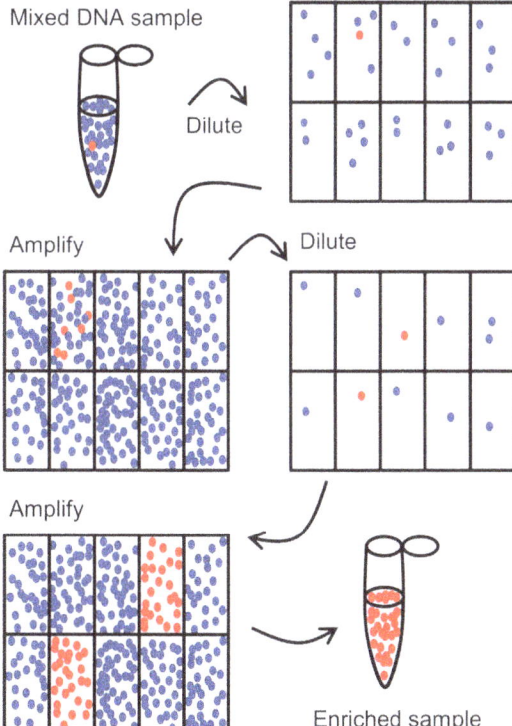

Mixed DNA sample

Dilute

Amplify

Dilute

Amplify

Enriched sample

Figure 1. The principle behind the PINS enrichment is a repetition of the steps of end-point dilution and general amplification. The mixed DNA sample is diluted until less than one out of 5–10 wells contain the target DNA molecule to be enriched and the DNA in all wells is then generally amplified using Phi29 polymerase. A sample containing target DNA is selected for next round of dilution and general amplification based on the result of qPCR directed towards the target.

Table 1. PCR Primers and oligos used for target detection, sequencing and genome walking.

Primer ID	Sequence (5′ – 3′)	Target/Description
HPV5901f	GTTTAGTGTGGGCCTGTGC	Target: HPV18
HPV5994r	GGCATGGGAACTTTCAGTGT	Target: HPV18
HPV6640r	GAGACTGTGTAGAAGCACATATTG	Target: HPV18
HPV6552f	CATAAGGCACAGGGTCATAAC	Target: HPV18
HPV7459f	CCGATTTCGGTTGCCTTTG	Target: HPV18
HPV7819r	CCCAACCTATTTCGGTTGC	Target: HPV18
WalkID	GCGCTGCAGGCATGCGAGCTC	Primer for oligo cassette [14]
WalkOL	GCGCTGCAGGCATGCGAGCTCCCAAGCTTGATCG	Part of oligo cassette
RGW-HindIII	AGCTCGATCAAGCTTGGGAGCTCGCATGCCTGCAGCGC	Part of oligo cassette
RGW-BglII	GATCCGATCAAGCTTGGGAGCTCGCATGCCTGCAGCGC	Part of oligo cassette

location in nucleotide sequence (GenBank accession number: GQ180790). A total overview of all primers used in this study is listed in Table 1.

Quantitative PCR (qPCR) conditions

qPCR reactions were performed using BioRad Sso ADV SYBR Green Supermix in the following mixture: 10 µl SSO, 7.4 µL NF water, 0.8 µL (5 µM) HPV5901f, 0.8 µL (6 µM) HPV-5994r, using PCR conditions of: (94°C/65.8°C/72°C) at time intervals of (15sec/15sec/15sec) for 50 cycles.

Most Probable Number (MPN) Calculations

Determination of the HPV18 target copy number in the samples was based on MPN quantification as described in [13], and all quantifications were carried out as five-tube assays (MPN5). Serial dilutions were carried out as either 5- or 10-fold dilutions and calculated according to the relevant MPN tables. Due to the high content of DNA in the first two series of analysis, evaluation of the PCR product formation was established by agarose gel electrophoresis. Correct size target amplification was 113 bp, and was easily recognizable in a 2% agarose gel. After completion of the two first rounds of amplification, 10 fold dilutions were performed prior to qPCR, thereby minimizing the background fluorescence from the reaction template and agarose gel analysis was only performed as occasional controls in enrichment round 3–6.

Abundance

The abundance of target relative to total human genomes in the sample was calculated based on the concentration of DNA. Assuming an average molecular weight per base pair of 660 g*mol^{-1}*bp^{-1}, a human genome size of 3.2 billion base pairs, and using Avogadros constant of 6.022+10^{23} an average of 285 human genomes per ng human DNA is reached. Consequently, the target abundance for the initial sample is:

$$\text{Target abundance (initial sample)} = \frac{0.6 \frac{copies(MPN)}{sample}}{(211 \frac{ng}{\mu L} * 285 \frac{genomes}{ng})}$$

$$= 0.00001 = 0.001\%$$

where 0.6 and 211 is the target number (copies/µL) and DNA concentration (ng/µL) respectively (Table 2).

Rapid Genome Walking (RGW)

Sequencing of the PINS enriched sample was carried out as described by Kilstrup and Kristiansen [14]. In brief, 3 µL Phi amplified sample (fifth enrichment) was digested by fast digest enzyme BglII (Thermo Scientific) in fast digest buffer (FD buffer) provided by the manufacturer. Digestion was done at 37°C for 30 minutes in a total digest volume of 30 µL. After digestion the enzyme was inactivated at 80°C for 10 minutes. WalkOL-BglII cassette ("WalkOL" +"RGW-BglII" oligos) was ligated to the digested sample in the following mixture: 15 µL BglII digested sample (fifth enrichment), 0.4 µL (0.5 pmol/µL) WalkOL-BglII, 3 µL (10x) T4 Ligase buffer, 1 µL T4 Ligase (5 U/µL) and 10.6 µL NF water (Sigma). The ligation process was set to run at 16°C overnight and was thereafter kept at 4°C. PCR was performed using (1 pmol/µL) WalkID primer and (10 pmol/µL) HPV5994r using DreamTaq polymerase (Thermo Scientific). PCR conditions were 35 cycles at temperatures of 94°C/64.4°C/72°C and at time intervals of 15sec/15sec/90sec. A specific PCR product was excised from a 1% agarose gel and was purified using GeneJet Gel Extraction Kit (Thermo Scientific). The eluted product was re-amplified using PCR conditions identical to those described above with the only change being that 10 pmol/µL WalkID primer was added instead of 1 pmol/µL. The sample was purified to remove excess primers and nucleotides and was sequenced by Eurofins MWG (Germany) using HPV5994r and WalkID primers. The RGW procedure was repeated using HindIII for digestion, a "WalkOL + RGW-HindIII" cassette and primer HPV7459f instead of HPV5994r.

PCR for sequence assembly

PCR was carried out by mixing the following: 2.5 µL (10x) reaction buffer, 2.5 µL 2 mM dNTP, 2.5 µL 25 mM magnesium, 1 µL forward primer, 1 µL reverse primer, 1 µL BSA, 0.2 µL DreamTaq polymerase (Thermo Fisher Scientific) and NF water to a total volume of 25 µL. Amplification was carried out using 35 cycles at temperatures of 94°C/60°C/72°C and time intervals of 15 sec/15 sec/90 sec. Two µL PCR product was loaded to 1% agarose gels.

Table 2. Quantification of results from rounds 0–6 of the PINS enrichment of HPV18 containing DNA.

Round of enrichment	DNA conc.	Target number[1]	Target number[2]	Increase[3]	Total Increase[4]	Increase[3]	Total increase[4]	Abundance[5]
	ng/ μL	per μL	per ng	per μL	per μL	per ng	per ng	%
0	211	0.6	0.0028	n.a.	n.a.	n.a.	n.a.	0.001
1	234	30	0.1	50	50	45	45	0.04
2	219	170	0.8	5.7	283	6.1	273	0.27
3	165	540	3.3	3.2	900	4.2	1,151	1.15
4	186	2,800	15.1	5.2	4,667	4.6	5,294	5.28
5	112	14,000	125	5.0	23,333	8.3	43,958	43.8
6	42	33,000	786	2.4	55,000	6.3	276,310	275

[1]Number of targets per μL of total DNA in the sample as determined by MPN,
[2]Number of targets per ng of total DNA in the sample as determined by MPN,
[3]Increase in target number relative to the previous round of enrichment,
[4]Increase in target number relative to initial sample (round 0),
[5]Number of targets per human genome (3.2 billion base pairs).

Sequence Assembly, Blast and alignment

Assembly of both RGW-sequences and PCR from forward and reverse sequencing was done using the assembly function of "CLC Main Workbench" with a minimum requirement of 30 bp overlap. The assembled sequence was blasted against the GenBank database [15], and most related sequences from the database were downloaded for alignment analysis.

Results

PINS Enrichment

In order to demonstrate enrichment of an extremely rare target using PINS, HPV18 containing DNA was mixed into a target-negative human background at a ratio of 1:100,000 or 0.001% (w/w). 10 individual 1 μL samples of the mix were amplified using Phi29 amplification. The Phi29 amplified products were easily recognized as smears on a 0.7% agarose gel resembling those described using Exo-resistant primers in [16]. To identify amplified samples containing the HPV18 target, PCR was performed using primers HPV5901f and HPV5994r detecting a 113 bp fragment in the L1 gene of HPV18.

One sample out of ten was found to have an increased amount of HPV18 target, as analyzed by MPN5. Further analysis showed that the positive sample contained 30 HPV18 targets per μL corresponding to 0.1 targets/ng total DNA or approximately 50-fold more than the original sample. The results from all six rounds of PINS enrichment can be found in Table 2.

The procedure was repeated six times with the only modification being that the amplified sample, with progressively higher target copy number, was diluted more for each repetition. Thus, the second round of enrichment was initiated from the sample selected from the first round of enrichment by dilution until on average one to two Phi29 amplified products resulted in detection of the target HPV DNA after Phi29 amplification. 1 μL diluted template was added to each of the 10 Phi29 amplifications. The second round of amplification resulted in 3 positive samples from where the HPV18 DNA could be detected by PCR. Of these three, one sample was markedly higher in copy number than the two others. Consequently, only the sample with the highest quantity of targets was selected for further analysis. MPN5 from this sample was determined to be 170 targets/μL, corresponding to 0.8 targets/ng total DNA or approximately 280 fold higher than the original sample. Subsequently, the procedure was repeated four additional times resulting in a final target concentration of 33000 copies/μL or 786 targets/ng. In total, the HPV18 containing fragment was enriched more than 275,000 fold as compared to the original sample.

Rapid genome walking (RGW) and sequence analysis of the enriched sample

At an abundance of 786 targets/ng, the final sample could have been sequenced directly using next generation sequencing technology. However, even the PINS round 5 sample could be sequenced using the rapid genome walking method [14]. In RGW, the DNA is fragmented using a restriction enzyme and linkers are ligated to the ends of the fragment. PCR products are then produced using an internal primer and a linker primer and the fragments are sequenced by Sanger sequencing.

Two PCR products spanning the 5′ and 3′ integration site of HPV18 respectively were produced from the sample selected in PINS enrichment round 5 and sequenced using RGW ("WalkID + HPV5994r" and "HPV7459f + WalkID", Figure 2). Two additional PCR products "HPV5901f + HPV6640r" and "HPV6552f + HPV7819r" were produced from the same sample

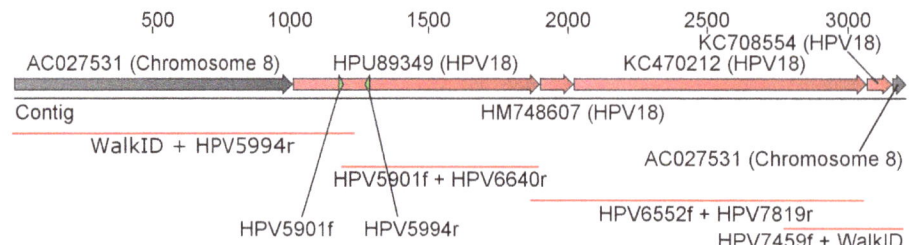

Figure 2. Alignment of RGW sequence and HPV specific sequencing from this study to published sequences on human chromosome 8 and HPV18 sequences. Positions of the primers used for detection of the HPV18 target sequence in the enrichment (HPV5901f and HPV5994r) are illustrated by small triangles in HPU89349.

to cover the remaining HPV18 sequence. The PCR products were sequenced in both directions. The eight sequences were assembled into one continuous sequence of 3208 base pairs (Table S1).

Compared to the data available in GenBank, position 1–1011 of the assembled sequence aligned with 99.1% identity to chromosome 8 of the human genome sequence (GenBank accession number: AC027531). Position 3158–3208 aligned with 100% identity to the same region of chromosome 8, 3441 base pairs downstream of the 5′ integration point (GenBank accession number: AC027531). The less than 100% identity (99.1%) in the upstream sequence was due to an 8 bp deletion. Sequencing of a pure HeLa DNA sample as provided from NEB revealed the same deletion and the deletion was therefore not introduced during PINS enrichment.

The 2147 base pair fragment from position 1012 to 3158 aligned with 100% identity to published HPV18 sequences (Genbank accession numbers: HPU89349, HM748607, KC470212, and KC708554). A graphical representation of the assembly and associated GenBank sequences can be seen in Figure 2.

Discussion

In this study we present a novel method for enrichment of DNA fragments based on minimal sequence information. The enrichment can be performed on very low abundance target DNA, in this case down to 0.0028 copies of target per ng of total DNA, and is generally applicable for enrichment of DNA fragments from complex samples. It is shown that an HPV18 fragment can be enriched more than 275,000 fold through six rounds of enrichment thereby enabling sequencing of a fragment of more than 3000 bp fragment containing a short known sequence used to direct the enrichment.

The enrichment in each round of PINS ranged from a 4.2-fold to a 45-fold increase (targets per ng total DNA) with an average of 12 fold, and the degree of enrichment in each round correlated to the ratio of positive to negative wells as expected. After six rounds of amplification, the sample contained 2.8 targets per human genome (abundance of 275%) and sequencing was thereby enabled. Rapid genome walking and PCR employing approximately 1000 bp fragments could be performed from samples with at least 0.44 targets per genome, i.e. from PINS rounds 5 and 6. At target concentrations below this level, no larger PCR fragments were obtained and sequencing could therefore not be performed (data not shown).

Using the relative simple RGW method it was shown that, although the DNA region of interest had been amplified through five consecutive rounds of PINS, the sequences from the pure HeLa sample and the enriched fragment were 100% identical and the HPV18 sequence was identical to previously published

sequences. This observation correlates well with the low error rate previously reported for Phi29 [17], and is of high importance for the final evaluation of the sequences.

Although the enrichment was based on PCR detection of a short specific PCR fragment of only 113 base pairs it resulted in enrichment of a much larger genome fragment as determined by sequencing. The data analysis of the present study describes the exact integration site of HPV18 into human chromosome 8 including 1026 base pairs of upstream sequence and correlating to identical findings in previous studies [18]. To further analyze the enriched sequence, an additional DNA sequence of 2182 base pairs was amplified, sequenced and analyzed confirming the integration site in chromosome 8 and showing 2147 bp of the integrated HPV18 virus. Phi29 is known to produce fragments of an average length of approximately 70 kb [19] and it is therefore likely that the enriched fragment is significantly longer than the 3208 bp reported here. The exact length of the fragment could possibly be deduced from next generation sequencing.

Earlier publications have pointed out, that amplification bias is an inherent feature of Phi29 amplification, and most reactions mixtures are known to produce DNA solely from the kit components [20]. The phenomenon is generally described as non-template-reactions (NTR) or template-independent products (TIPs), and various attempts have been introduced to circumvent the problem [21]. Phi29 kits producing no NTR are commercially available, as exemplified by the Replig-g Ultra-Fast mini kit (Qiagen), but the requirement for template quantity in such kits is typically high (>10 ng), leaving amplification of very dilute samples impossible. In the current study we have shown that the applied reaction mixtures provided an efficient amplification with templates as dilute as 8 pg corresponding to less than 2.5 human genomes, without observing any reaction in the negative control (NTR) and without requirement of additional UV-treatment [20] or addition of inhibitory compounds [21]. Moreover, no amplification errors were observed after six rounds of amplification.

Regardless of application, PINS provides a powerful method to gain sequence information from a minute sub-fraction of a complex sample with a minimum of prior sequence information. In the present example, at least 3208 base pair of sequence was enriched from a very complex background based on just 39 base pairs of sequence information (the two detection primers).

The current example is based on an integrated virus in a complex background and may be used to determine other virus sequences and integration sites. However, we also see an obvious possibility to broaden the usage to include applications comprising other partially unknown sequences in complex samples such as cells or DNA in blood samples, swabs, biopsies, feces samples or complex environmental samples, thereby enabling sequencing or dramatically reducing the volume of sequencing reads.

Acknowledgments

We gratefully acknowledge Anders Weber and Michael Obermayer for careful review of the manuscript.

Author Contributions

Conceived and designed the experiments: TK MJM LS-M. Performed the experiments: TK LS-M. Analyzed the data: TK LS-M MJM. Contributed to the writing of the manuscript: MJM TK LS-M.

References

1. Bidard FC, Weigelt B, Reis-Filho JS (2013) Going with the flow: from circulating tumor cells to DNA. Sci Transl Med 5: 207ps214.
2. Gibbons JG, Janson EM, Hittinger CT, Johnston M, Abbot P, et al. (2009) Benchmarking next-generation transcriptome sequencing for functional and evolutionary genomics. Mol Biol Evol 26: 2731–2744.
3. Milbury CA, Li J, Makrigiorgos GM (2011) Ice-COLD-PCR enables rapid amplification and robust enrichment for low-abundance unknown DNA mutations. Nucleic Acids Res 39: e2.
4. Koboldt DC, Steinberg KM, Larson DE, Wilson RK, Mardis ER (2013) The next-generation sequencing revolution and its impact on genomics. Cell 155: 27–38.
5. Sykes PJ, Neoh SH, Brisco MJ, Hughes E, Condon J, et al. (1992) Quantitation of targets for PCR by use of limiting dilution. Biotechniques 13: 444–449.
6. Dean FB, Hosono S, Fang L, Wu X, Faruqi AF, et al. (2002) Comprehensive human genome amplification using multiple displacement amplification. PNAS 99: 5261–5266.
7. Myers LE, McQuay LJ, Hollinger FB (1994) Dilution assay statistics. J Clin Microbiol 32: 732–739.
8. Vogelstein B, Kinzler KW (1999) Digital PCR. PNAS 96: 9236–9241.
9. Bosch FX, Lorincz A, Munoz N, Meijer CJ, Shah KV (2002) The causal relation between human papillomavirus and cervical cancer. J Clin Pathol 55: 244–265.
10. Gillison ML, Koch WM, Capone RB, Spafford M, Westra WH, et al. (2000) Evidence for a causal association between human papillomavirus and a subset of head and neck cancers. J Natl Cancer Inst 92: 709–720.
11. Li N, Franceschi S, Howell-Jones R, Snijders PJ, Clifford GM (2011) Human papillomavirus type distribution in 30,848 invasive cervical cancers worldwide: Variation by geographical region, histological type and year of publication. Int J Cancer 128: 927–935.
12. Schneider UV, Mikkelsen ND, Lindqvist A, Okkels LM, Johnk N, et al. (2012) Improved efficiency and robustness in qPCR and multiplex end-point PCR by twisted intercalating nucleic acid modified primers. PLoS One 7: e38451.
13. Oblinger J, Koburger J (1975) Understanding and teaching the most probable number technique. J Milk Food Technol 38: 540–545.
14. Kilstrup M, Kristiansen KN (2000) Rapid genome walking: a simplified oligo-cassette mediated polymerase chain reaction using a single genome-specific primer. Nucleic Acids Res 28: e55.
15. Benson DA, Karsch-Mizrachi I, Lipman DJ, Ostell J, Wheeler DL (2005) GenBank. Nucleic Acids Res 33 Database Issue: D34–D38.
16. Dean FB, Nelson JR, Giesler TL, Lasken RS (2001) Rapid amplification of plasmid and phage DNA using Phi 29 DNA polymerase and multiply-primed rolling circle amplification. Genome Res 11: 1095–1099.
17. Han T, Chang CW, Kwekel JC, Chen Y, Ge Y, et al. (2012) Characterization of whole genome amplified (WGA) DNA for use in genotyping assay development. BMC Genomics 13: 217.
18. Landry JJ, Pyl PT, Rausch T, Zichner T, Tekkedil MM, et al. (2013) The genomic and transcriptomic landscape of a HeLa cell line. G3 (Bethesda) 3: 1213–1224.
19. Lasken RS, Huges S (2005) Multiple displacement amplification of genomic DNA. In: Huges S, editor. Whole Genome Amplification. Oxfordshire: Scion Publishing Ltd. 2005. pp.99–118.
20. Woyke T, Sczyrba A, Lee J, Rinke C, Tighe D, et al. (2011) Decontamination of MDA reagents for single cell whole genome amplification. PLoS ONE 6: e26161.
21. Pan X, Urban AE, Palejev D, Schulz V, Grubert F, et al. (2008) A procedure for highly specific, sensitive, and unbiased whole-genome amplification. PNAS 105: 15499–15504.

14

Analysis of Genome-Wide Copy Number Variations in Chinese Indigenous and Western Pig Breeds by 60 K SNP Genotyping Arrays

Yanan Wang[1❞], Zhonglin Tang[2❞], Yaqi Sun[1], Hongyang Wang[1], Chao Wang[1], Shaobo Yu[1], Jing Liu[1], Yu Zhang[1], Bin Fan[1], Kui Li[2]*, Bang Liu[1]*

1 Lab of Molecular Biology and Animal Breeding, Key Laboratory of Agricultural Animal Genetics, Breeding and Reproduction of Ministry of Education, Huazhong Agricultural University, Wuhan, PR China, 2 Key Laboratory of Farm Animal Genetic Resources and Germplasm Innovation of Ministry of Agriculture, Institute of Animal Science, Chinese Academy of Agricultural Sciences, Beijing, PR China

Abstract

Copy number variations (CNVs) represent a substantial source of structural variants in mammals and contribute to both normal phenotypic variability and disease susceptibility. Although low-resolution CNV maps are produced in many domestic animals, and several reports have been published about the CNVs of porcine genome, the differences between Chinese and western pigs still remain to be elucidated. In this study, we used Porcine SNP60 BeadChip and PennCNV algorithm to perform a genome-wide CNV detection in 302 individuals from six Chinese indigenous breeds (Tongcheng, Laiwu, Luchuan, Bama, Wuzhishan and Ningxiang pigs), three western breeds (Yorkshire, Landrace and Duroc) and one hybrid (Tongcheng×Duroc). A total of 348 CNV Regions (CNVRs) across genome were identified, covering 150.49 Mb of the pig genome or 6.14% of the autosomal genome sequence. In these CNVRs, 213 CNVRs were found to exist only in the six Chinese indigenous breeds, and 60 CNVRs only in the three western breeds. The characters of CNVs in four Chinese normal size breeds (Luchuan, Tongcheng and Laiwu pigs) and two minipig breeds (Bama and Wuzhishan pigs) were also analyzed in this study. Functional annotation suggested that these CNVRs possess a great variety of molecular function and may play important roles in phenotypic and production traits between Chinese and western breeds. Our results are important complementary to the CNV map in pig genome, which provide new information about the diversity of Chinese and western pig breeds, and facilitate further research on porcine genome CNVs.

Editor: Marinus F.W. te Pas, Wageningen UR Livestock Research, Netherlands

Funding: This work was supported by NSFC (31072012) (http://www.nsfc.gov.cn/), Major International Cooperation NSFC (31210103917) (http://www.nsfc.gov.cn/), National 863 programs (2011AA100302, 2011AA100304) (http://program.most.gov.cn/) and National Key Project (2013ZX08009-001) (http://program.most.gov.cn/). The funders had no role in study design, data collection and analysis, decision to publish, or preparation of the manuscript.

Competing Interests: The authors have declared that no competing interests exist.

* Email: kuili@caas.cn (KL); liubang@mail.hzau.edu.cn (BL)

❞ These authors contributed equally to this work.

Introduction

Copy number variations (CNVs) refer to the structurally genomic variations from hundreds of bases to several kilo-bases and the relevant complex mutations in the construction of chromosomes. Since the duplication of *Bar* gene in Drosophila melanogaster was first reported by Bridge to cause the Bar eye phenotype, more and more scientists focused on such DNA structural duplication [1]. In 2004, Iafrate firstly illustrated numerous structural variations in human genome, and then Redon *et al*. defined the copy number variations in human genome [2,3].

Compared with the most frequent SNP marker, CNVs cover wider genomic regions and have potentially larger effects to change gene structure and dosage, exposing recessive alleles, and alternating gene regulation and other mechanisms. In humans, most studies on CNVs are shown to associate with Mendelian diseases and complex genetic disorders, such as major depressive disorder [4], schizophrenia [5], cancer [6,7], body mass index

[8,9], and various congenital defects [10]. Besides disorders, CNVs are also important to maintain the normal phenotypic variability. Many studies were conducted on the influence of CNVs on the phenotype of domestic animals, such as copy number variation in intron 1 of *SOX5* leading to the Pea-comb phenotype in chickens [11], a 4.6-kb intronic duplication in *STX17* (Syntaxin 17) causing hair greying and melanoma in horse [12,13], and the duplication of *FGF3*, *FGF4*, *FGF19* and *ORAOV1* resulting in hair ridge and predisposition to dermoid sinus in Ridgeback dogs [14]. A 110 kb microdeletion in the maternally imprinted *PEG3* domain was found, which results in a loss of paternal *MIMT1* expression and causes late term abortion and stillbirth in cattle [15], and ectopic *KIT* copy number variation may be associated with gonadal hypoplasia in Northern Finncattle and Swedish Mountain cattle [16]. George. Liu and his team detected CNVRs of cattle genomes by using different methods in diverse cattle breeds, and they also found evidence of CNVs relating with residual feed intake and resistance to gastrointestinal nematodes [17–21]. In pigs, only a few such studies are reported. For example, for the color of pig

coat, the duplication and the exon-17-skipping mutation of *KIT* are responsible for the dominant white phenotype and peripheral blood cell [22–24].

Currently, CNVs can be identified using several technologies based on either ultra-dense genotyping with SNP chips, the hybridization of DNA in BAC/PAC/oligonu-cleotide arrays or high-throughput sequencing. The comparative genomic hybridization (CGH) based approach and high-throughput sequencing have excellent performance in refined resolution and relative signal intensities, while the SNP genotyping array has the advantage in both genome-wide association studies (GWAS) and CNV detection. The SNP arrays can collect normalized total signal intensity (Log R ratio-LRR) and allelic intensity ratios (B allele frequency-BAF) which represent overall copy numbers and allelic contrasts [25]. Besides, the SNP arrays need fewer samples than CGH arrays in an experiment, thus being more cost-effective and allowing users to increase the number of tested samples on a limited budget [26]. Nowadays, SNP arrays have been routinely used for CNV detection in humans and other organisms, and manufacturers of SNP genotyping arrays have incorporated non-polymorphic markers into their arrays to improve the coverage of SNP arrays for CNV analysis [27].

Since the accomplishment of the first human genome CNV map, many reports have been published on the characterization of the genomic architecture of CNVs in domestic species. Low-resolution CNV maps were produced for cattle, dog, pig, goat, sheep, chicken, duck, turkey and horse, showing that these structural polymorphisms comprise a significant part of these genomes [17,28–32]. Based on porcine SNP60 Beadchip and aCGH, Fadista *et al.* [33], Ramayo-Caldas *et al.* [29], Wang *et al.* [34,35], Chen *et al.* [36], and Li *et al.* [37] have identified a large amount of CNVs in pig genome among different breeds, including several Chinese breeds.

Chinese indigenous breeds have larger genetic diversity than European breeds, leading to the tremendous phenotypic differences among them. In the present study, we analyzed the difference in CNVs between Chinese indigenous breeds and western breeds by using Porcine SNP60 BeadChip and PennCNV algorithm, and performed a genome-wide CNV detection in 302 pigs from six Chinese indigenous breeds, three European breeds and one hybrid. This study produced a comprehensive map of CNVs in the pig genome, which could give new insight to the interspecific diversity of different breeds and facilitate further research on porcine genome CNVs.

Materials and Methods

Ethics Statement

The whole blood samples were collected in strict accordance with the protocol approved by the Biological Studies Animal Care and Use Committee of Hubei Province, PR China. All efforts were made to minimize any discomfort during blood collection.

Animal samples

The animals were composed of 302 pigs from nine pure breeds and one hybrid, including six Chinese indigenous breeds (45 Tongcheng pigs-TC, 23 Laiwu pigs-LW, 40 Luchuan pigs-LC, 23 Bama pigs-BM, 26 Wuzhishan pigs-WZS, 24 Ningxiang pigs-NX) and three western breeds (33 Yorkshire pigs-YS, 33 Landrace pigs-LD, 32 Duroc pigs-Dur) and one hybrid (23 Tongcheng × Duroc crossbred pigs-BC).

Genomic DNA samples were extracted from whole blood of all pigs using a standard phenol/chloroform method. All DNA samples were analyzed by spectrophotometry and agarose gel electrophoresis.

SNP array genotyping and quality control

All 302 pigs were genotyped with the Porcine SNP60 Genotyping BeadChip (Illumina Inc., USA) using the Infinium II Multisample assay (Illumina Inc.). SNP arrays were scanned using iScan (Illumina Inc.) and analyzed using GenomeStudio (Version 3.2.2, Illumina, Inc.).

In order to exclude poor-quality DNA samples and decrease potential false-positive CNVs, only the samples at a call rate > 98% and call frequency >90% were reserved. After quality control, 286 of the 302 samples were retained for CNV detection (43 Tongcheng pigs, 22 Laiwu pigs, 39 Luchuan pigs, 21 Bama pigs, 23 Wuzhishan pigs, 23 Ningxiang pigs, 31 Yorkshire pigs, 31 Landrace pigs, 30 Duroc pigs and 23 Tongcheng ×Duroc crossbred pigs). For subsequent data analysis, a subset of 52,089 SNPs was selected by removing the SNPs located in sex chromosomes and those not mapped in the Sscrofa10.2 assembly.

Identification of pig CNVs

PennCNV was used for CNV identification by integrating a Hidden Markov Model (HMM) for high resolution copy number variation detection with whole-genome SNP genotyping data [38]. This algorithm incorporates multiple sources of information, including total signal intensity data of log R Ratio (LRR) and B allele frequency (BAF) at each SNP marker, the distance between neighboring SNPs, the population frequency of B allele (PFB) of SNPs, and the pedigree information where available. The LRR and BAF were exported using Illumina BeadStudio software. There were three arguments in PennCNV including -test, -trio and -joint. Individual-based CNV calling was performed using the -test with default parameters of the HMM model by integrating Log R Ratio, BAF, population allele frequency and the SNP distance. To reduce the false discovery rate in CNV calling, we adopted the calling criteria that the standard deviation (SD) of LRR was under or less than 0.35, and the CNV contained three or more consecutive SNPs. All putative CNVs identified in this study were pooled across breeds. Finally the CNV regions (CNVRs) were determined by aggregating the overlapping CNVs identified across all samples according to the previously published protocols [3].

Gene contents and functional annotation

Gene contents in the identified CNVRs were retrieved from the Ensembl Genes 70 Database using the BioMart (http://asia. ensembl.org/biomart/martview/) data management system. Functional annotation of these genes was performed with the DAVID bioinformatics resources 6.7 (http://david.abcc.ncifcrf. gov/) for Gene Ontology (GO) terms and Kyoto Encyclopedia of Genes and Genomes (KEGG) pathway analysis. Considering the limited number of genes in the pig genome have been annotated, we first converted the pig Ensembl gene IDs to homologous human Ensembl gene IDs by BioMart, and then carried out the GO and pathway analysis. Statistical significance was assessed by using P value of a modified Fisher's exact test and Benjamini correction for multiple testing.

Validation of CNVRs by qPCR

CNVRs were confirmed by qPCR using the Roche Light-CyclerW 480 Detection System and the $2^{-\Delta\Delta Ct}$ method which compares the ΔCt (cycle threshold (Ct) of the target region minus Ct of the control region) value of samples with CNV to the ΔCt of

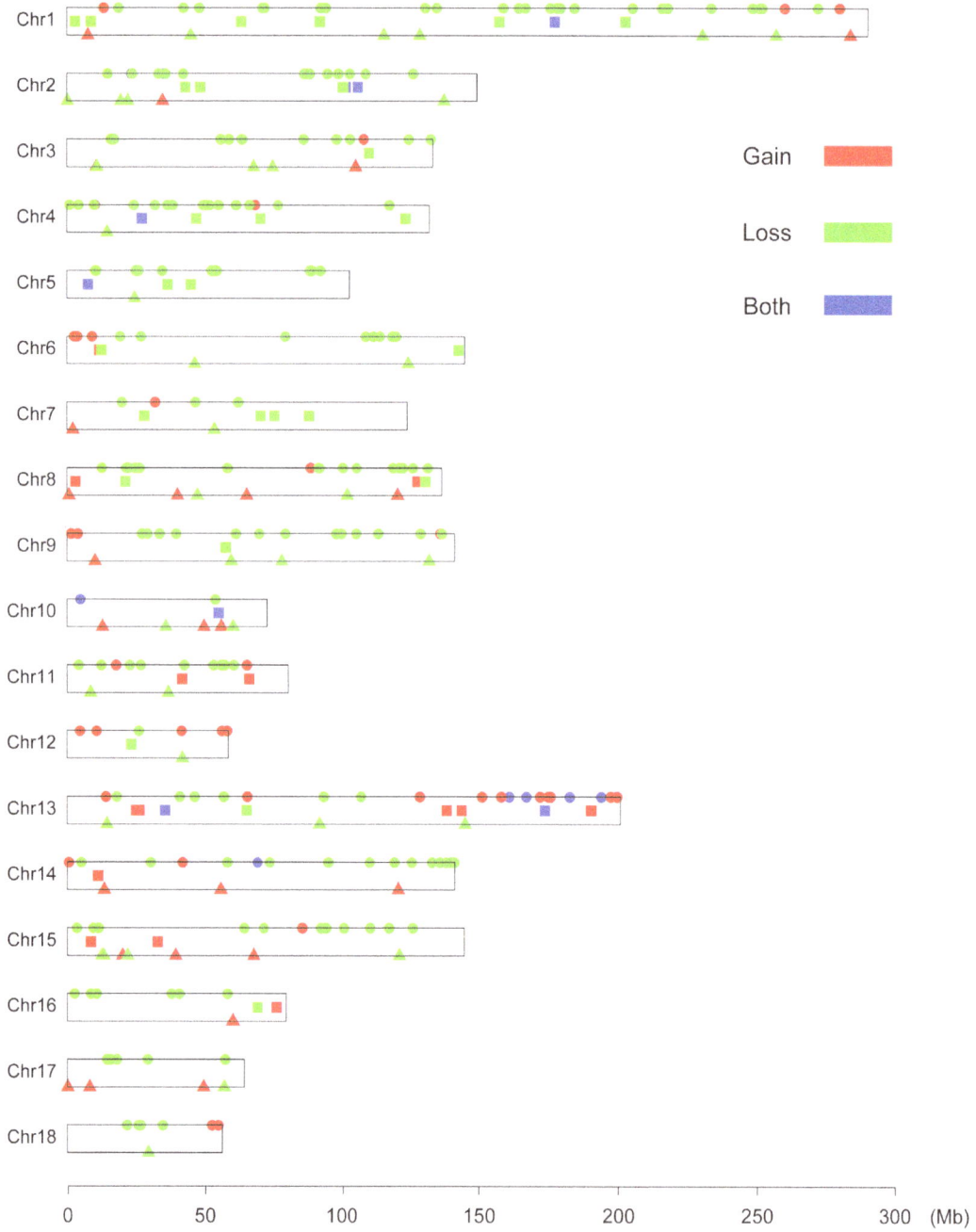

Figure 1. Genomic distribution of CNVRs in 18 pairs of autosomal chromosomes of pigs. The chromosomal locations of 348 CNVRs are indicated by lines. Y-axis values are chromosome names, and X-axis values are chromosome positions in Mb, which are proportional to the real size of swine genome sequence assembly (10.2). Round represents CNVRs identified only in Chinese indigenous breeds; triangle represents CNVRs identified only in western breeds; and quadrate represents those identified both in Chinese and western breeds.

a calibrator without CNV [39]. The primers were designed using the Primer Premier 5 software (Table S10 in File S1). As previously reported, the copy number of each CNVR was normalized against the *GCG* gene, a control region in the genome with no variation in the copy numbers between the pigs [29]. Triplicate wells of reactions (10 μL) contained 5 μL SYBR Green Real-time PCR Master Mix, 1 μL of 50 ng/μL gDNA, 0.3 μL 10 μM of each primer and 3.4 μL ddH₂O. The cycling conditions consisted of 95°C for 10 min, 40 cycles (at 94°C for 30 sec, 60°C for 30 sec, and 72°C for 10 sec), and fluorescence acquisition at

72°C in single mode. The specific PCR products were confirmed by the results of melting curve analysis and agarose gel electrophoresis.

Results and Discussion

Genome-wide CNVs detection

A total of 1272 CNVs were assessed by PennCNV on 18 pairs of autosomal chromosomes and 348 CNVRs were acquired by aggregating overlapping CNVs (Table S1a in File S1), covering

Table 1. Chromosome distribution of CNVRs in pigs.

Chr	No. of CNVRs	No. of genes	Length of CNVRs (bp)	Length of chromosomes (bp)	Percentage (%)
1	43	73	23217110	315321322	7.36
2	26	59	11921008	162569375	7.33
3	18	23	2788734	144787322	1.93
4	22	51	8591089	143465943	5.99
5	18	35	5889799	111506441	5.28
6	16	45	3648860	157765593	2.31
7	12	40	2619625	134764511	1.94
8	28	20	8873483	148491826	5.98
9	22	53	5312989	153670197	3.46
10	10	14	1565092	79102373	1.98
11	17	20	5477336	87690581	6.25
12	8	15	741280	63588571	1.17
13	32	230	45721621	218635234	20.91
14	23	59	11437761	153851969	7.43
15	23	18	5627033	157681621	3.57
16	11	18	2673814	86898991	3.08
17	12	17	2291744	69701581	3.29
18	7	8	2087925	61220071	3.41
Total	348	798	150486303	2450713522	6.14

150.49 Mb of the pig genome and about 6.14% of the autosomal genome sequence, with an average number of 4.45 CNVs per individual. Among the 348 CNVR events, 88 were found to be gain, 243, loss and 17, both (loss and gain within the same region) events, which were dispersed all over the 18 autosomes and ranged from 4.93 Kb to 12.41 Mb in length with a mean of 443.24 Kb and median of 170.77 Kb. The location and characteristics of all CNVRs on autosomal chromosomes are presented in Figure 1, showing that these CNVRs are not uniformly distributed among different chromosomes. The proportion of CNVRs on the 18 pairs of autosomal chromosomes varies from 1.17–20.91% and chromosome 1 harbors the greatest number (43) of CNVRs (Table 1).

Excluding the CNVR event detected only in one individual, we obtained 166 CNVRs, with 107 loss, 42 gain and 17 both events. These CNVRs covered 103.01 Mb, and ranged from 5.75 Kb to

12.41 Mb with a mean of 620.57 Kb and median of 217.76 Kb (Table S1b in File S1). Previous studies always analyzed their results after eliminating the only individual event, but when compared with their results, more than 36% of the only individual events in our results overlapped with their data, suggesting the loss of much useful information by eliminating the only individual events. Thus, we analyzed the whole 348 CNVRs during the following research.

In our study, the 6.14% CNVR coverage in the autosomal genome sequence was consistent with the 0.31% to 5.84% coverage of analyzed genome on pigs reported previously [29,34,36]. In humans, CNVR coverage was reported to be as high as 12% of the genome when Redon et al first identified 1447 CNVRs in human genome [3]. The CNVs were anticipated to cover up to 13% of the human genome [40]. In bovines, the

Table 2. Sample size and CNVs number detected in nine breeds.

Breed	sample size	CNVs number	CNVs per sample	CNVRs number	unique CNVRs	Frequency (%)
Tongcheng	43	194	4.51	84	33	22.70
Laiwu	22	160	7.27	76	41	20.54
Luchuan	39	279	7.15	125	64	33.78
Bama	21	90	4.29	36	6	9.73
Wuzhishan	23	67	2.91	35	15	9.46
Ningxiang	23	120	5.22	61	25	16.49
Yorkshire	31	113	3.65	54	25	14.59
Landrace	31	107	3.45	48	19	12.97
Duroc	30	79	2.63	37	12	10.00

CNVR coverage was reported to be from 0.68% to 4.6% of the genome [17,41,42].

A great difference was found in the CNVR numbers among the nine breeds. Among the six Chinese indigenous breeds, the maximum number of CNVRs was detected in Luchuan pigs, accounting for 125 CNVRs (33.78%), followed by Tongcheng 84(22.7%) and Laiwu 76 (20.54%) pigs. The minimum number of CNVRs was only 35 (9.46%) in Wuzhishan pigs. With respect to European breeds, 54 (14.59%), 48 (12.97%) and 37 (10%) CNVRs were found in Yorkshire, Landrace and Duroc pigs, respectively. Altogether 240 unique CNVRs were detectable in the nine breeds, with Luchuan pigs harboring the maximum number (Table 2).

The number of CNVs among individuals was variable. 39 individuals were only found one CNV event, 49 individuals were found two CNV events, and most individuals were three CNV events (Table S2 in File S1). With an increase in copy number variation, the detected individuals became fewer and fewer, indicating that most animals can survive through only a few CNVs. CNV numbers differed greatly among different pig populations too. The average number of CNVs per population was 127.2, ranging from 67 (Wuzhishan) to 279 (Luchuan). The maximum number of CNVs per sample was detected in Laiwu pigs (7.32 CNVs per sample on average) against the minimum number of 2.63 CNVs per animal in Duroc pigs. Similar to the finding in human [43], most CNVRs (68.77%) were restricted to one population, probably due to sampling variances or recent evolution events.

Gene content of pig CNVRs

Totally, 798 genes within the identified CNVRs were retrieved from the Ensembl Genes 70 Database using the BioMart data management system, including 651 protein-coding genes, 10 pseudogenes, 30 miRNA, 49 snRNA, 35 snoRNA, 13 miscRNA, 7 rRNA, and 3 processed-transcripts. In CNVRs, 455 of the 798 genes were identified to be loss events, 303 and 39 genes, gain and both events, respectively (Table S3 in File S1). The average number of genes per Mb of 348 CNVRs was 5.29, which was less than that on the whole genome(8.62) according to the *Sus crofa* 10.2 assembly in Ensembl (http://a sia.ensembl.org/), suggesting that CNVs are located preferably in gene-poor regions, probably because changes in copy number for genes that perform essential functions are subject to strong purifying selection [44,45].

In order to provide insight into the functional enrichment of the CNVs, Gene Ontology (GO) and Kyoto Encyclopedia of Genes and Genomes (KEGG) pathway analyses were performed with the DAVID bioinformatics resources. The Gene Ontology (GO) analysis revealed that CNV genes mainly participated in cell adhesion, phosphorus metabolic process, cell projection, phosphorylation, cellular component morphogenesis, cell differentiation, muscle cell development and other basic metabolic processes. The KEGG pathway analysis indicated that genes in CNVRs were involved in seven pathways including phosphatidylinositol signaling system, inositol phosphate metabolism, oocyte meiosis, cell cycle, leukocyte transendothelial migration, Alzheimer's disease and Huntington's disease (Table S4 in File S1 and S5 in File S1).

Additionally, 3258 QTLs out of 8000 which affect a wide range of traits, such as immune capacity, disease resistance, meat quality, growth and litter size, were found in 240 CNVRs by comparing the overlapping of CNVRs with QTLs in the pig QTLdb (http://www.animalgenome.org/cgi-bin/QTLdb/SS/index) (Table S6 in File S1).

Differences between Chinese normal size pig and minipig breeds

In our population, four of the six Chinese breeds are normal size (Tongcheng, Laiwu, Luchuan and Ningxiang), the other two are minipig breeds. There are many differences of CNVs between these two type pigs. The four normal size breeds harbored 179 CNVRs, covering 80.59 Mb of pig genome sequence and 366 ensemble genes (table S7a and 7b in File S1). These genes mainly participate in cell adhesion, regulation of cell cycle, detection of stimulus, phosphate metabolic process, ATP biosynthetic process, muscle cell differentiation, purine nucleotide metabolic process, and regulation of growth. Many CNV-associate genes in these regions appear to be certain gene clusters or families, such as ubiquitin-conjugating enzyme family (*UBE2B*, *UBE2G2*), ATPase family (*ATP2A1*, *ATP13A5*, *ATP5J*), trefoil factor family (*TFF1.TFF2*, *TFF3*), and claudin family (*CLDN8*, *CLDN17*). These genes mainly involved in immune system and some human diseases [46–53]. *VCAN* (versican), *ADAM17* (ADAM metallopeptidase domain 17), *ITGB1BP1* (integrin beta 1 binding protein), *CDH19*, *ITGAD* and *PCDH15* were also playing important role in inflammation and other diseases [54–59]. Moreover, *CDH19* was evidenced as a copy number alterations (CNAs) target gene to impact central nervous system. *ZWINT* (ZW10 interacting protein PIK3C3), *PIK3C3* (phosphoinositide-3-kinase class 3) and *PTGS2* (prostaglandin-endoperoxide synthase 2) were important for cell proliferation [60–62], and *PIK3C3* and *PTGS2* were proved as candidate marker for production and reproductive traits in pigs [63–65]. These findings indicated that CNVs may have potential effect on immune response, production and reproductive traits of these pigs.

Bama pigs and Wuzhishan pigs are two famous Chinese minipig breeds, whose body weight are less than one third of modern commercial breeds. 21 CNVRs that only detected in these two breeds were picked out to investigate whether CNVs were contribute to their phenotypes. 49 ensemble genes were retrieved overlapped with these CNVRs, including 42 protein-coding genes, 4 snoRNAs, 2 snRNAs and 1 pseudogene (Table S7c and 7d in File S1). Among these genes, *STX17* was evidenced that a 4.6-kb intronic duplication of it would cause hair greying and melanoma in horse [12]; *INPP5A* (inositol polyphosphate-5-phosphatase), *TCERG1L* (transcription elongation regulator 1-like), *FOXL1* (forkhead box L1) and *POLR1D* (polymerase (RNA) I polypeptide D) mainly participated in human cancer and some other diseases [66–69]. Some genes such as *MTCH2* (mitochondrial carrier 2), *BNIP3* (BCL2/adenovirus E1B 19 kDa interacting protein 3) and *DPYSL4* (dihydropyrimidinase-like 4) were playing essential role in cell apoptosis [70–72], some as *CDK13* (cyclin-dependent kinase 13) and *SGSM1* (small G protein signaling modulator 1) participated in regulation of cell circle, differentiation and proliferation [73,74]. According to the function of CNV-associated genes in the six breeds, we assumed that CNVs may contribute to disease resistance and stress resistance of all these breeds, but unfortunately, we didn't find sound evidence for CNVs impacting the growth of these two minipig breeds.

An average of 6.04 CNVs was obtained for each sample in the four Chinese normal size breeds, while only 3.6 was in the two minipig populations (table 2). This means more variations occurred in four nomal size breeds than minipigs. Despite the population size of these breeds, the most convincing reason for such small number of CNVs in minipigs is domestic methods and artificial selection. These two breeds were both raised in mountain area, which means highly inbreeding was inevitable because of terrible traffic condition, leading to less variation in these breeds.

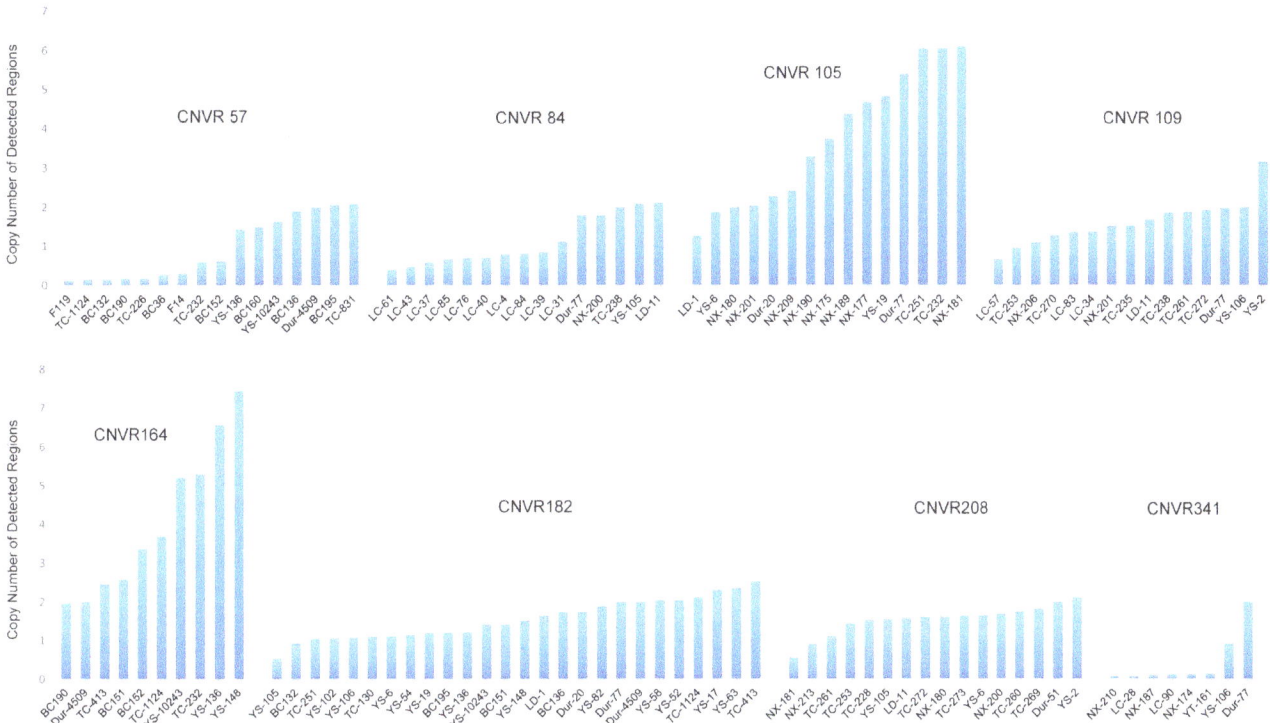

Figure 2. QPCR validation of 8 identified CNVRs. The x-axis represents the animals and the y-axis shows the relative quantification value.

After long period of natural and artificial selection, these two breeds became more and more conservative and steady.

Differences of CNVRs in Chinese indigenous breeds and western breeds

A total of 213 CNVRs were identified to exist only in the six Chinese native breed populations, 60, in three western breeds and 49, in all. Fewer CNVRs were detected in western breeds, with only one CNVR detected on chromosomes 4, 5, 12 and 18 (Figure 1, Table S8a in File S1). Two potential reasons may explain such a small number of CNVRs. Firstly, the population size of the western breeds of pigs in this study is smaller than that of Chinese native breeds. Secondly, strongly artificial selection of the western commercial breeds tended to make them purified, thus decreasing variations in the population. However, Chinese native pigs were subjected to lower selection intensities and had fewer selection signatures, which may conserve most variations after evolution.

A total of 109 ensemble genes overlapped with CNVRs of western breeds, including 99 protein-coding genes, 2 pseudogenes, 1 miRNA, 5 snRNA and 2 snoRNA. GO analysis revealed that these genes were mainly involved in cell adhesion, regulation of phosphorylation and cell proliferation. Differ from western breeds, 499 ensemble genes were retrieved overlapping CNVRs of Chinese indigenous breeds, including 389 protein-coding genes, 6 pseudogenes, 26 miRNA, 35 snRNA, 28 snoRNA, 8 miscRNA, 5 rRNA, and 2 processed-transcripts. By GO analysis, we found that these genes were involved in not only cell adhesion and regulation of phosphorylation, but also behavior, neuron differentiation and regulation of T cell receptor signaling pathway (Table S8b in File S1). For further pathway analysis, we found Chinese breeds specific CNV-associated genes such as *ITGB2*, *CLDN8/14/17*, *CDK13*, *JAM2*, *ALCAM* and *VCAN* played

important role in immune system, Leukocyte transendothelial migration, T cell receptor signaling pathway and other pathways. Some of the important genes were mentioned in proceeding part. These results illustrated that CNVs may play important role in immune system among Chinese breeds and growth among western breeds. As we all know that Chinese indigenous breeds and western breeds show obvious differences in many aspects, our results may provide a genetic explanation for the difference in disease resistance and growth rate.

Comparison with previous reports

Our results were compared with previous reports on porcine genomic CNVs (Table S9 in File S1). The first research on CNVs of pig genome by using Porcine SNP60 Beadchip was reported by Ramayo-Caldas [29], who detected 49 CNVRs from 55 pigs of Iberian x Landrace cross. Thirty out of the 49 CNVRs were overlapped with our results. Wang *et al.* detected 382 CNVRs based on the Porcine SNP 60 genotyping data of 474 individuals from three pure breed populations and one Duroc × Erhualian crossbred population [34]. Among the 382 CNVRs, 97 regions were found to overlap with our results. Using the same method, Chen *et al.* detected 565 CNVRs in 1693 pigs from 18 diverse populations, and 179 of them were overlapped with our results. Using 720 K array CGH, Li *et al.* identified 259 CNVRs in 12 animals including three Chinese native pigs, 5 European pigs, 2 synthetic pigs and 2 crossbred pigs (Landrace × DIV pigs) [37], and only 15 regions were found to be identical or overlapped with our data. Only 3 regions were found in all reports, containing 9 genes, with 7 of them being protein coding genes (table S9e in File S1).

The potential reasons for the differences between our results and other studies are as follows. First, there was a difference in population size and genetic background between our study and others. In our study, the CNVs of the four Chinese indigenous breeds including Luchuan, Laiwu, Ningxiang and Wuzhishan pigs were reported for the first time, involving lots of new breed-specific regions. Second, platforms (SNP genotyping array and CGH array) and calling algorithms, which are different in the calling technique, resolution and genome coverage, might contribute to the discrepancy of the CNVs detected. Third, our results were based on genome assembly *sus scrofa* 10.2, and except for Chen's research, the aforementioned reports were all based on genome assembly *sus scrofa* 9.0 version. As our data needed to be converted from 10.2 to 9.0, great differences might arise during the transformation, causing the deviation between our results and others. This is also occurred in CNV studies of other mammals [17,75,76].

Validation of CNVR by real-time quantitative PCR (qPCR)

Quantitative real time PCR (qPCR) was used to validate 12 CNVRs chosen from the 348 CNVRs detected in the study. These 12 CNVRs represent different predicted status of copy numbers (i.e., loss, gain and both) and different CNVR frequencies. 9 (75%) of them were in agreement with the prediction by PennCNV (Figure 2).

Except for CNVR187, 11 CNVRs contained important functional genes. The CNVR164 contained Mast/stem cell growth factor receptor gene, also known as *KIT* gene. *KIT* gene was considered to affect colors and their distribution in pigs with a 450 kb long duplication and the exon-17-skipping mutation. The copy number of *KIT* in dominant white color was higher than any other patterns [77,78]. In our data, the copy numbers of three detected Yorkshire pigs were from 4 to 6, while the other detected samples including Tongcheng (two-black-end) and Duroc pigs (red) had approximately 2 copies. This result is consistent with the previous reports and prediction of PennCNV analysis based on SNPs chip. The CNVR182 locus contained *MAPK10* gene, which is a member of mitogenactivated protein kinase (MAPK) superfamily, and plays an important role in cancer and some other diseases [79]. The *MAPK10* is also associated with Hirschsprung disease as a candidate CNV gene [80]. In our data, the copy numbers of 11 samples out of 26 were identified as loss events, corresponding with PennCNV prediction. There has been no report so far about the function of *MAPK*10 in pigs, and our results indicate that *MAPK*10 may play an important role in pigs. Another 7 genes were detected in our validated CNVRs, such as *VCAN*, *CIB4*, *VCPIP1*, *ALG14*, *FAM5C*, *ZFPM2* and *DOK5*. Copy number variations were identified in all these seven genes, and these variations may affect their function in immunity, development, and growth.

Conclusions

In this study, we described a map of porcine CNVRs between six Chinese indigenous breeds and three western breeds based on Porcine SNP60 genotyping data of 302 pigs. The results revealed that 213 CNVRs belong to the Chinese native breed populations, and 60 CNVRs, to the western breeds. We also discussed the CNV characters of four Chinese normal size breeds (Luchuan, Tongcheng and Laiwu pigs) and two minipig breeds (Bama and Wuzhishan pigs). Functional annotation suggested that these CNVRs and CNV-associate genes are involved in variety of molecular function and may play important roles in phenotypic and production traits difference between Chinese and western pig breeds.

Supporting Information

File S1 Supporting tables. Table S1, Information of identified CNVRs. The description and information of all 348 identified CNVRs were showed in Table S1a. The information of 166 CNVRs which were identified by excluding the CNVR event detected only in one individual were showed in Table S1b. **Table S2, The CNVs numbers detected in 286 individuals. Table S3, Information of genes in the identified CNVRs. Table S4, Gene ontology (GO) analysis of genes in identified CNVRs. Table S5, Pathway analysis of genes in identified CNVRs. Table S6, Previously reported QTLs overlapping with identified CNVRs. Table S7, Information of CNVRs in Chinese normal size and minipig breeds.** Table S7a showed the information of CNVRs in four Chinese normal size pig breeds. Table S7b showed the information of genes in the CNVRs of four Chinese normal size pig breeds. Table S7c showed the information of CNVRs in two Chinese minipig breeds: Bama and Wuzhishan pigs. Table S7d showed the information of genes in the CNVRs of Bama and Wuzhishan pigs. **Table S8, Information of CNVRs and CNV-involved genes in Chinese indigenous and western breeds.** Table S8a showed information of CNVRs in Chinese indigenous and western breeds. Table S8b showed information of genes in the CNVRs of Chinese indigenous and western breeds. **Table S9, Comparison between identified CNVRs and those reported in other pig CNVR papers. Table S10, Information and the primers used in qPCR analysis of the 12 CNVRs chosen to be validated.**

Acknowledgments

The authors are grateful to Sanping Xu in Tongcheng Animal Husbandry Bureau in Hubei Province for samples collection.

Author Contributions

Conceived and designed the experiments: BL KL YNW. Performed the experiments: YNW ZLT YQS HYW SBY JL YZ. Analyzed the data: YNW ZLT CW BF. Contributed reagents/materials/analysis tools: YNW ZLT YQS HYW SBY JL YZ. Wrote the paper: YNW.

References

1. Bridges CB (1936) The Bar "Gene" a Duplication. Science 83: 210–211.
2. Iafrate AJ, Feuk L, Rivera MN, Listewnik ML, Donahoe PK, et al. (2004) Detection of large-scale variation in the human genome. Nat Genet 36: 949–951.
3. Redon R, Ishikawa S, Fitch KR, Feuk L, Perry GH, et al. (2006) Global variation in copy number in the human genome. Nature 444: 444–454.
4. Perlis RH, Ruderfer D, Hamilton SP, Ernst C (2012) Copy number variation in subjects with major depressive disorder who attempted suicide. PLoS One 7: e46315.
5. Vacic V, McCarthy S, Malhotra D, Murray F, Chou HH, et al. (2011) Duplications of the neuropeptide receptor gene VIPR2 confer significant risk for schizophrenia. Nature 471: 499–503.
6. Liu W, Sun J, Li G, Zhu Y, Zhang S, et al. (2009) Association of a germ-line copy number variation at 2p24.3 and risk for aggressive prostate cancer. Cancer Res 69: 2176–2179.
7. Jin G, Sun J, Liu W, Zhang Z, Chu LW, et al. (2011) Genome-wide copy-number variation analysis identifies common genetic variants at 20p13 associated with aggressiveness of prostate cancer. Carcinogenesis 32: 1057–1062.
8. Sha BY, Yang TL, Zhao LJ, Chen XD, Guo Y, et al. (2009) Genome-wide association study suggested copy number variation may be associated with body mass index in the Chinese population. J Hum Genet 54: 199–202.
9. Jarick I, Vogel CI, Scherag S, Schafer H, Hebebrand J, et al. (2011) Novel common copy number variation for early onset extreme obesity on chromosome 11q11 identified by a genome-wide analysis. Hum Mol Genet 20: 840–852.

10. Sailani MR, Makrythanasis P, Valsesia A, Santoni FA, Deutsch S, et al. (2013) The complex SNP and CNV genetic architecture of the increased risk of congenital heart defects in Down syndrome. Genome Res 23: 1410–1421.

11. Wright D, Boije H, Meadows JR, Bed'hom B, Gourichon D, et al. (2009) Copy number variation in intron 1 of SOX5 causes the Pea-comb phenotype in chickens. PLoS Genet 5: e1000512.

12. Sundstrom E, Komisarczuk AZ, Jiang L, Golovko A, Navratilova P, et al. (2012) Identification of a melanocyte-specific, microphthalmia-associated transcription factor-dependent regulatory element in the intronic duplication causing hair greying and melanoma in horses. Pigment Cell Melanoma Res 25: 28–36.

13. Sundstrom E, Imsland F, Mikko S, Wade C, Sigurdsson S, et al. (2012) Copy number expansion of the STX17 duplication in melanoma tissue from Grey horses. BMC Genomics 13: 365.

14. Salmon Hillbertz NH, Isaksson M, Karlsson EK, Hellmen E, Pielberg GR, et al. (2007) Duplication of FGF3, FGF4, FGF19 and ORAOV1 causes hair ridge and predisposition to dermoid sinus in Ridgeback dogs. Nat Genet 39: 1318–1320.

15. Flisikowski K, Venhoranta H, Nowacka-Woszuk J, McKay SD, Flyckt A, et al. (2010) A novel mutation in the maternally imprinted PEG3 domain results in a loss of MIMT1 expression and causes abortions and stillbirths in cattle (Bos taurus). PLoS One 5: e15116.

16. Venhoranta H, Pausch H, Wysocki M, Szczerbal I, Hanninen R, et al. (2013) Ectopic KIT copy number variation underlies impaired migration of primordial germ cells associated with gonadal hypoplasia in cattle (Bos taurus). PLoS One 8: e75659.

17. Liu GE, Hou Y, Zhu B, Cardone MF, Jiang L, et al. (2010) Analysis of copy number variations among diverse cattle breeds. Genome Res 20: 693–703.

18. Bickhart DM, Hou Y, Schroeder SG, Alkan C, Cardone MF, et al. (2012) Copy number variation of individual cattle genomes using next-generation sequencing. Genome Res 22: 778–790.

19. Hou Y, Liu GE, Bickhart DM, Matukumalli LK, Li C, et al. (2012) Genomic regions showing copy number variations associate with resistance or suscepti-bility to gastrointestinal nematodes in Angus cattle. Funct Integr Genomics 12: 81–92.

20. Hou Y, Bickhart DM, Chung H, Hutchison JL, Norman HD, et al. (2012) Analysis of copy number variations in Holstein cows identify potential mechanisms contributing to differences in residual feed intake. Funct Integr Genomics 12: 717–723.

21. Xu L, Hou Y, Bickhart DM, Song J, Van Tassell CP, et al. (2014) A genome-wide survey reveals a deletion polymorphism associated with resistance to gastrointestinal nematodes in Angus cattle. Funct Integr Genomics.

22. Marklund S, Kijas J, Rodriguez-Martinez H, Ronnstrand L, Funa K, et al. (1998) Molecular basis for the dominant white phenotype in the domestic pig. Genome Res 8: 826–833.

23. Giuffra E, Tornsten A, Marklund S, Bongcam-Rudloff E, Chardon P, et al. (2002) A large duplication associated with dominant white color in pigs originated by homologous recombination between LINE elements flanking KIT. Mamm Genome 13: 569–577.

24. Giuffra E, Evans G, Tornsten A, Wales R, Day A, et al. (1999) The Belt mutation in pigs is an allele at the Dominant white (I/KIT) locus. Mamm Genome 10: 1132–1136.

25. Peiffer DA, Le JM, Steemers FJ, Chang W, Jenniges T, et al. (2006) High-resolution genomic profiling of chromosomal aberrations using Infinium whole-genome genotyping. Genome Res 16: 1136–1148.

26. Winchester L, Yau C, Ragoussis J (2009) Comparing CNV detection methods for SNP arrays. Brief Funct Genomic Proteomic 8: 353–366.

27. Wang K, Chen Z, Tadesse MG, Glessner J, Grant SF, et al. (2008) Modeling genetic inheritance of copy number variations. Nucleic Acids Res 36: e138.

28. Nicholas TJ, Cheng Z, Ventura M, Mealey K, Eichler EE, et al. (2009) The genomic architecture of segmental duplications and associated copy number variants in dogs. Genome Res 19: 491–499.

29. Ramayo-Caldas Y, Castello A, Pena RN, Alves E, Mercade A, et al. (2010) Copy number variation in the porcine genome inferred from a 60 k SNP BeadChip. BMC Genomics 11: 593.

30. Liu J, Zhang L, Xu L, Ren H, Lu J, et al. (2013) Analysis of copy number variations in the sheep genome using 50K SNP BeadChip array. BMC Genomics 14: 229.

31. Wang Y, Gu X, Feng C, Song C, Hu X, et al. (2012) A genome-wide survey of copy number variation regions in various chicken breeds by array comparative genomic hybridization method. Anim Genet 43: 282–289.

32. Doan R, Cohen N, Harrington J, Veazy K, Juras R, et al. (2012) Identification of copy number variants in horses. Genome Res 22: 899–907.

33. Fadista J, Nygaard M, Holm LE, Thomsen B, Bendixen C (2008) A snapshot of CNVs in the pig genome. PLoS One 3: e3916.

34. Wang J, Jiang J, Fu W, Jiang L, Ding X, et al. (2012) A genome-wide detection of copy number variations using SNP genotyping arrays in swine. BMC Genomics 13: 273.

35. Wang J, Jiang J, Wang H, Kang H, Zhang Q, et al. (2014) Enhancing Genome-Wide Copy Number Variation Identification by High Density Array CGH Using Diverse Resources of Pig Breeds. PLoS One 9: e87571.

36. Chen C, Qiao R, Wei R, Guo Y, Ai H, et al. (2012) A comprehensive survey of copy number variation in 18 diverse pig populations and identification of candidate copy number variable genes associated with complex traits. BMC Genomics 13: 733.

37. Li Y, Mei S, Zhang X, Peng X, Liu G, et al. (2012) Identification of genome-wide copy number variations among diverse pig breeds by array CGH. BMC Genomics 13: 725.

38. Wang K, Li M, Hadley D, Liu R, Glessner J, et al. (2007) PennCNV: an integrated hidden Markov model designed for high-resolution copy number variation detection in whole-genome SNP genotyping data. Genome Res 17: 1665–1674.

39. Livak KJ, Schmittgen TD (2001) Analysis of relative gene expression data using real-time quantitative PCR and the 2(-Delta Delta C(T)) Method. Methods 25: 402–408.

40. Stankiewicz P, Lupski JR (2010) Structural variation in the human genome and its role in disease. Annu Rev Med 61: 437–455.

41. Hou Y, Liu GE, Bickhart DM, Cardone MF, Wang K, et al. (2011) Genomic characteristics of cattle copy number variations. BMC Genomics 12: 127.

42. Fadista J, Thomsen B, Holm LE, Bendixen C (2010) Copy number variation in the bovine genome. BMC Genomics 11: 284.

43. Chen W, Hayward C, Wright AF, Hicks AA, Vitart V, et al. (2011) Copy number variation across European populations. PLoS One 6: e23087.

44. de Smith AJ, Walters RG, Froguel P, Blakemore AI (2008) Human genes involved in copy number variation: mechanisms of origin, functional effects and implications for disease. Cytogenet Genome Res 123: 17–26.

45. Schrider DR, Hahn MW (2010) Gene copy-number polymorphism in nature. Proc Biol Sci 277: 3213–3221.

46. Mou L, Zhang Q, Wang Y, Zhang Q, Sun L, et al. (2013) Identification of Ube2b as a novel target of androgen receptor in mouse sertoli cells. Biol Reprod 89: 32.

47. Liu W, Shang Y, Zeng Y, Liu C, Li Y, et al. (2014) Dimeric Ube2g2 simultaneously engages donor and acceptor ubiquitins to form Lys48-linked ubiquitin chains. EMBO J 33: 46–61.

48. Murgiano L, Sacchetto R, Testoni S, Dorotea T, Mascarello F, et al. (2012) Pseudomyotonia in Romagnola cattle caused by novel ATP2A1 mutations. BMC Vet Res 8: 186.

49. De La Hera DP, Corradi GR, Adamo HP, De Tezanos Pinto F (2013) Parkinson's disease-associated human P5B-ATPase ATP13A2 increases spermi-dine uptake. Biochem J 450: 47–53.

50. De Giorgio MR, Yoshioka M, Riedl I, Moreault O, Cherizol RG, et al. (2013) Trefoil factor family member 2 (Tff2) KO mice are protected from high-fat diet-induced obesity. Obesity (Silver Spring) 21: 1389–1395.

51. Feng G, Zhang Y, Yuan H, Bai R, Zheng J, et al. (2014) DNA methylation of trefoil factor 1 (TFF1) is associated with the tumorigenesis of gastric carcinoma. Mol Med Rep 9: 109–117.

52. Xue Y, Shen L, Cui Y, Zhang H, Chen Q, et al. (2013) Tff3, as a novel peptide, regulates hepatic glucose metabolism. PLoS One 8: e75240.

53. Clark PM, Dawany N, Dampier W, Byers SW, Pestell RG, et al. (2012) Bioinformatics analysis reveals transcriptome and microRNA signatures and drug repositioning targets for IBD and other autoimmune diseases. Inflamm Bowel Dis 18: 2315–2333.

54. Wight TN, Kang I, Merrilees MJ (2014) Versican and the control of inflammation. Matrix Biol 35: 152–161.

55. Sisto M, Lisi S, D'Amore M, Lofrumento DD (2014) The metalloproteinase ADAM17 and the epidermal growth factor receptor (EGFR) signaling drive the inflammatory epithelial response in Sjogren's syndrome. Clin Exp Med.

56. Plummer NW, Squire TL, Srinivasan S, Huang E, Zawistowski JS, et al. (2006) Neuronal expression of the Ccm2 gene in a new mouse model of cerebral cavernous malformations. Mamm Genome 17: 119–128.

57. Cobrinik D, Ostrovnaya I, Hassimi M, Tickoo SK, Cheung IY, et al. (2013) Recurrent pre-existing and acquired DNA copy number alterations, including focal TERT gains, in neuroblastoma central nervous system metastases. Genes Chromosomes Cancer 52: 1150–1166.

58. Yun J, Duan Q, Wang L, Lv W, Gong Z, et al. (2014) Differential expression of leukocyte beta2 integrin signal transduction-associated genes in patients with symptomatic pulmonary embolism. Mol Med Rep 9: 285–292.

59. Coppieters F, Van Schil K, Bauwens M, Verdin H, De Jaegher A, et al. (2014) Identity-by-descent-guided mutation analysis and exome sequencing in consan-guineous families reveals unusual clinical and molecular findings in retinal dystrophy. Genet Med.

60. Zhou X, Takatoh J, Wang F (2011) The mammalian class 3 PI3K (PIK3C3) is required for early embryogenesis and cell proliferation. PLoS One 6: e16358.

61. Nuttinck F, Gall L, Ruffini S, Laffont L, Clement L, et al. (2011) PTGS2-related PGE2 affects oocyte MAPK phosphorylation and meiosis progression in cattle: late effects on early embryonic development. Biol Reprod 84: 1248–1257.

62. Endo H, Ikeda K, Urano T, Horie-Inoue K, Inoue S (2012) Terf/TRIM17 stimulates degradation of kinetochore protein ZWINT and regulates cell proliferation. J Biochem 151: 139–144.

63. Hirose K, Takizawa T, Fukawa K, Ito T, Ueda M, et al. (2011) Association of an SNP marker in exon 24 of a class 3 phosphoinositide-3-kinase (PIK3C3) gene with production traits in Duroc pigs. Anim Sci J 82: 46–51.

64. Hirose K, Ito T, Fukawa K, Arakawa A, Mikawa S, et al. (2014) Evaluation of effects of multiple candidate genes (LEP, LEPR, MC4R, PIK3C3, and VRTN) on production traits in Duroc pigs. Anim Sci J 85: 198–206.

65. Ding NS, Ren DR, Guo YM, Ren J, Yan Y, et al. (2006) Genetic variation of porcine prostaglandin-endoperoxide synthase 2 (PTGS2) gene and its association with reproductive traits in an Erhualian x Duroc F2 population. Yi Chuan Xue Bao 33: 213–219.

66. Sekulic A, Kim SY, Hostetter G, Savage S, Einspahr JG, et al. (2010) Loss of inositol polyphosphate 5-phosphatase is an early event in development of cutaneous squamous cell carcinoma. Cancer Prev Res (Phila) 3: 1277–1283.

67. Kim TO, Park J, Kang MJ, Lee SH, Jee SR, et al. (2013) DNA hypermethylation of a selective gene panel as a risk marker for colon cancer in patients with ulcerative colitis. Int J Mol Med 31: 1255–1261.

68. Yang FQ, Yang FP, Li W, Liu M, Wang GC, et al. (2014) Foxl1 inhibits tumor invasion and predicts outcome in human renal cancer. Int J Clin Exp Pathol 7: 110–122.

69. Schaefer E, Collet C, Genevieve D, Vincent M, Lohmann DR, et al. (2014) Autosomal recessive POLR1D mutation with decrease of TCOF1 mRNA is responsible for Treacher Collins syndrome. Genet Med.

70. Robinson AJ, Kunji ER, Gross A (2012) Mitochondrial carrier homolog 2 (MTCH2): the recruitment and evolution of a mitochondrial carrier protein to a critical player in apoptosis. Exp Cell Res 318: 1316–1323.

71. Moriyama M, Moriyama H, Uda J, Matsuyama A, Osawa M, et al. (2014) BNIP3 Plays Crucial Roles in the Differentiation and Maintenance of Epidermal Keratinocytes. J Invest Dermatol 134: 1627–1635.

72. Kimura J, Kudoh T, Miki Y, Yoshida K (2011) Identification of dihydropyr-imidinase-related protein 4 as a novel target of the p53 tumor suppressor in the apoptotic response to DNA damage. Int J Cancer 128: 1524–1531.

73. Kohoutek J, Blazek D (2012) Cyclin K goes with Cdk12 and Cdk13. Cell Div 7: 12.

74. Yang H, Sasaki T, Minoshima S, Shimizu N (2007) Identification of three novel proteins (SGSM1, 2, 3) which modulate small G protein (RAP and RAB)-mediated signaling pathway. Genomics 90: 249–260.

75. Matsuzaki H, Wang PH, Hu J, Rava R, Fu GK (2009) High resolution discovery and confirmation of copy number variants in 90 Yoruba Nigerians. Genome Biol 10: R125.

76. Eichler EE (2006) Widening the spectrum of human genetic variation. Nat Genet 38: 9–11.

77. Seo BY, Park EW, Ahn SJ, Lee SH, Kim JH, et al. (2007) An accurate method for quantifying and analyzing copy number variation in porcine KIT by an oligonucleotide ligation assay. BMC Genet 8: 81.

78. Johansson Moller M, Chaudhary R, Hellmen E, Hoyheim B, Chowdhary B, et al. (1996) Pigs with the dominant white coat color phenotype carry a duplication of the KIT gene encoding the mast/stem cell growth factor receptor. Mamm Genome 7: 822–830.

79. Ying J, Li H, Cui Y, Wong AH, Langford C, et al. (2006) Epigenetic disruption of two proapoptotic genes MAPK10/JNK3 and PTPN13/FAP-1 in multiple lymphomas and carcinomas through hypermethylation of a common bidirectional promoter. Leukemia 20: 1173–1175.

80. Jiang Q, Ho YY, Hao L, Nichols Berrios C, Chakravarti A (2011) Copy number variants in candidate genes are genetic modifiers of Hirschsprung disease. PLoS One 6: e21219.

15

Gene-Based Rare Allele Analysis Identified a Risk Gene of Alzheimer's Disease

Jong Hun Kim[1], Pamela Song[2], Hyunsun Lim[3], Jae-Hyung Lee[4], Jun Hong Lee[1], Sun Ah Park[5]*, for the Alzheimer's Disease Neuroimaging Initiative

1 Department of Neurology, Dementia Center, Stroke Center, Ilsan hospital, National Health Insurance Service, Goyang-shi, South Korea, 2 Department of Neurology, Inje University Ilsan Paik Hospital, Goyang-shi, South Korea, 3 Clinical Research Management Team, Ilsan hospital, National Health Insurance Service, Goyang-shi, South Korea, 4 Department of Life and Nanopharmaceutical Sciences and Department of Maxillofacial Biomedical Engineering, School of Dentistry, Kyung Hee University, Seoul, South Korea, 5 Department of Neurology, Soonchunhyang University Bucheon Hospital, Bucheon-shi, South Korea

Abstract

Alzheimer's disease (AD) has a strong propensity to run in families. However, the known risk genes excluding *APOE* are not clinically useful. In various complex diseases, gene studies have targeted rare alleles for unsolved heritability. Our study aims to elucidate previously unknown risk genes for AD by targeting rare alleles. We used data from five publicly available genetic studies from the Alzheimer's Disease Neuroimaging Initiative (ADNI) and the database of Genotypes and Phenotypes (dbGaP). A total of 4,171 cases and 9,358 controls were included. The genotype information of rare alleles was imputed using 1,000 genomes. We performed gene-based analysis of rare alleles (minor allele frequency≤3%). The genome-wide significance level was defined as meta $P<1.8\times10^{-6}$ (0.05/number of genes in human genome = 0.05/28,517). *ZNF628*, which is located at chromosome 19q13.42, showed a genome-wide significant association with AD. The association of *ZNF628* with AD was not dependent on *APOE* ε4. *APOE* and *TREM2* were also significantly associated with AD, although not at genome-wide significance levels. Other genes identified by targeting common alleles could not be replicated in our gene-based rare allele analysis. We identified that rare variants in *ZNF628* are associated with AD. The protein encoded by *ZNF628* is known as a transcription factor. Furthermore, the associations of *APOE* and *TREM2* with AD were highly significant, even in gene-based rare allele analysis, which implies that further deep sequencing of these genes is required in AD heritability studies.

Editor: Thomas Arendt, University of Leipzig, Germany

Funding: Study design, meta-analysis, and preparation of the manuscript in this study were supported by a grant from the Korea Health 21 R&D Project, Ministry of Health, Welfare, and Family Affairs, Republic of Korea (HI10C2020 and A092004). The funders had no role in study design, data collection and analysis, decision to publish, or preparation of the manuscript.

Competing Interests: The authors have declared that no competing interests exist.

* Email: sapark@schmc.ac.kr

Introduction

Alzheimer's disease (AD) is a leading cause of dementia and is known to have high heritability (as high as 60–80%) [1,2]. Genome-wide association studies (GWAS) have identified several risk genes for AD such as *ABCA7*, *BIN1*, *CD33*, *CD2AP*, *CLU*, *CR1*, *EPHA1*, *MS4A6A/MS4A4E*, and *PICALM* [3–7]. The known risk genes for AD explain only 30% of heritability [8,9]. Aside from *APOE* ε4, reported risk genes have low clinical significance because of their small effect sizes [7]. The common variant hypothesis posited common diseases are attributed to common variants and this hypothesis is base concept for GWAS [10,11]. However, similar to other common diseases, the heritability of AD cannot be fully explained by common alleles [12].

There are growing reports regarding rare variants related to complex diseases [13–17]. Contrary to the common variant hypothesis, variants with low frequency could be primary causes for common diseases, according to the rare variant hypothesis

[11,18]. The rationale of the rare variant hypothesis is that allele variants with low frequencies have a higher probability of functional significance [12]. A large scale exome sequencing study has indicated that 95.7% SNPs with functional importance are rare variants [19]. Additionally, the number of variants with loss of function showed an inverse correlation with MAF [20,21]. Considering their functional significance, rare variants may have large effect sizes. Recently, rare alleles in *TREM2*, *APP*, and *PLD3* have been reported to have association with AD [22–24]. Thus, the identification of more risk or protective rare alleles associated with AD is required.

Although rare alleles are promising targets for genetic association studies of complex diseases, the analyses of rare alleles remains challenging. For example, very large sample sizes are required to detect rare alleles that have modest effect sizes [19]. Deep sequencing of large samples is too expensive for typical researchers to perform. The mutational loads within the same genes, regions, or pathways can be alternative approach [13,25]. However, a large number of candidate rare alleles within specific

Table 1. Characteristics of studies.

Study	Genotyping platform	Case/Control with genetic data	Case/control after QC*	No of SNPs after QC	No of imputed SNPs[§]	No of imputed rare (MAF≤3%) SNPs[§]
ADNI	Illumina Human610-Quad	350/169	350/169	533479	16242208	8608819
ADNI2	Illumina GenomeStudio v2009.1	53/125	53/125	634701	14860121	7257490
GenADA	Affimetrix Mapping250K_NspMapping250K_Sty	782/806	779/803	432763	14441395	6863818
eMERGE	Illumina Human660W-Quad_v1_A	676/1843	632/1843	535401	16190257	8572925
NIA-LOAD	Illumina Human610-Quad_v1_B	2244/2320	2098/2095	542080	19568275	11943583
Framingham	Affimetrix Mapping250K_NspMapping250K_Sty	314/4711	259/4323	371114	16510848	8908801

* In addition to genotyping QC, we selected only European ancestry without missing information on age and sex.
[§] SNPs with INFO≥0.4.

regions are more difficult to obtain and interpret, than genotyping of a few loci.

Improvement of imputation methods has allowed accurate inference of rare alleles [26]. According to 1000 genomes study [20], the mean squared Pearson correlation coefficients (R^2) between rare SNPs (MAF 0.5%–5%) and imputed dosages were 0.7–0.9 in the European ancestry. Furthermore, mutational loads of rare alleles within genes obtained from imputation can confer high power [27]. In this study, we aimed to find risk genes for AD using gene-based analysis of rare alleles deduced from 1000 genomes and publicly available GWAS data.

Materials and Methods

Subjects

We used publicly available GWAS data from the Alzheimer's Disease Neuroimaging Initiative (ADNI), Genetic Alzheimer's Disease Associations (GenADA) study, Electronic Medical Records and Genomics (eMERGE), the National Institute on Aging Late Onset Alzheimer's Disease (NIA-LOAD) family study, and the Framingham study. ADNI data were obtained from https://ida.loni.ucla.edu. GenADA (dbGaP accession number: phs000219.v1) [28,29], eMERGE (dbGaP accession number: phs000234.v1), NIA-LOAD (dbGaP accession number: phs000168.v1), and the Framingham study (dbGaP accession number: phs000007.v16) data were downloaded from dbGaP (http://www.ncbi.nlm.nih.gov/gap). Subjects with European ancestry were included. After genotypic quality control (QC), missing phenotypic data exclusion, and ethnic group selection, 4171 cases and 9358 control were included in this study. Summaries about

the studies are shown in Table 1. Additional information for each study were detained in File S1. The institutional review board of Ilsan hospital approved our study. Written informed consent was given by participants. In addition patient records were anonymized prior to analysis.

Genotypic QC and imputation

We excluded alleles with low (<1%) MAF, low (<95%) call rate, and deviation of Hardy-Weinberg Equilibrium ($P<10^{-6}$). The subjects with low (<95%) call rates, too high autosomal heterozygosity (false discovery rate, FDR<1%) and too high relatedness (identical-by-state, IBS>0.95) were excluded. For genotypic QC, we used the GenABEL package, v 1.69 [30].

After estimating haplotypes using SHAPEIT, v 1.0 [31], imputation with multi-population reference panels of 1000 genomes (phase I, release Mar 2012) was executed using IMPUTE2, v 2.2 with default parameters [32,33]. We discarded imputed SNPs with INFO<0.4. The dosage data of imputation were used for further analyses. The dosage means the expected genotype score [34].

Statistical analyses

In the association study, we adjusted for age, sex, years of education, and significant principle components (PCs) of the genetic stratification (File S1). For consistency across studies, years of education were categorized as follows: **1**, ≤ 4; **2**, 4< and ≤10; **3**, 11< and ≤ 15; **4**, >15 years according to the established methods of stratifications in the GenADA study. We imputed

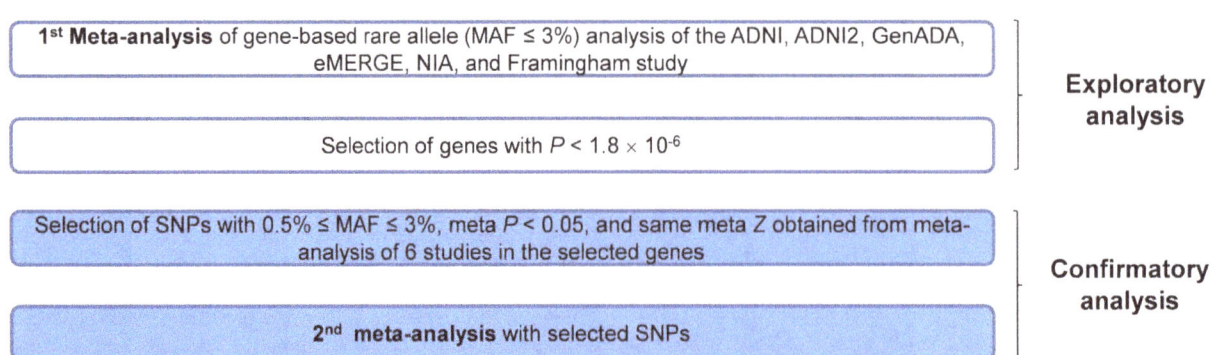

Figure 1. The overall scheme of this study.

Table 2. The highly ranked seven genes in the first meta-analyses.

Gene	CHR	Start	not adjusted for APOE ε4			adjusted for APOE ε4	
			Meta Z*	Meta P	direction§	Meta Z	Meta P
ZNF628	19	55987698	5.0	5.3×10^{-7}	++++++	5.2	1.3×10^{-7}
APOE	19	45409038	4.8	1.4×10^{-6}	+?+++-	2.3	0.023
TOMM40	19	45394477	4.1	4.0×10^{-5}	+-++++	2.3	0.018
MMP1	11	102654407	4.0	6.6×10^{-5}	-++++	4.1	4.0×10^{-5}
NAPRT1	8	144656956	3.9	7.8×10^{-5}	+++++	4.5	8.0×10^{-6}
TREM2	6	41126245	3.8	1.2×10^{-4}	+++++	4.6	3.7×10^{-6}
CBLB	3	105438891	3.8	1.5×10^{-4}	+++++	3.2	0.0015

We show the highly ranked genes (meta $P < 2.0 \times 10^{-4}$) in the first meta-analysis in this table.
Protein names: zinc finger protein 628, ZNF628; apolipoprotein E, APOE; translocase of outer mitochondrial membrane 40, TOMM40; matrix metallopeptidase 1, MMP1; nicotinate phosphoribosyltransferase domain containing 1, NAPRT1; triggering receptor expressed on myeloid cells 2, TREM2; Cbl proto-oncogene B, E3 ubiquitin protein ligase, CBLB.
* Larger absolute Z score represents smaller P and the direction of the Z score represents the direction of risk [35].
§ The signs mean those of the Z score of each study. The question mark represents missing data in the study because of low INFO or high MAF. The order of the signs is ADNI, ADNI2, GenADA, eMERGE, NIA-LOAD, and Framingham study.

missing years of education to a mean value. The years of education was regarded as a continuous variable.

We performed a weighted, Z score based, fixed-effects, meta-analysis using METAL [35]. The effect sample size (N_E) for meta-analysis is given in terms of numbers of AD (N_{AD}) and of controls (N_C), as follows [35]:

$$N_E = 4/(1/N_{AD} + 1/N_C)$$

The forest plot was drawn using 'rmeta' R package.

APOE is the strongest risk gene among the known risk genes for AD. In several genome-wide association studies for AD [3], the top ranked genes could show false associations with AD, because they are within same LD block of *APOE* ε4. In addition, the pathogenesis of AD patients might be different between carriers and noncarriers of *APOE* ε4 [36]. Therefore, we examined the dependency on *APOE* ε4 genotype status by two ways. First, the results were compared after adjustment for *APOE* ε4 genotype status – the number of *APOE* ε4 allele in each individual. eMERGE and the Framingham study did not include data on *APOE* ε4 genotype status. Therefore, we used imputed dosages of *APOE* ε4 for these two studies (Table S1 in File S1). Second, the collinearity between selected genes and *APOE* ε4 genotype status was examined.

Gene-based rare allele analysis

In this study, gene-based rare allele analysis means accumulations of rare alleles within the same coding region implemented in GRANVIL [27]. The definition of gene boundaries was based on the UCSC genome browser (build 37). The Framingham study showed inflated type I error and skewed results (Figures S1 and S2 in File S1). Therefore, we need to adjust for genetic stratification of the Framingham study using another algorithm implemented in GenABEL v1.69 and ProbABEL v0.30 [30,37] (Figure S2 in File S1). For gene-based analysis of the Framingham study, we need to make computer program for ourselves. We made a dosage of a gene (D) similar to an allele's dosage in the Framingham study, as follows [27].

$$D = \frac{\sum_{i=1}^{n} Gi}{n}$$

Where Gi is a dosage of the ith SNP and n is a number of rare alleles within a gene that were used in the analysis.

Analyses proceeded in two steps. The overall study scheme is shown in Figure 1. We performed the first meta-analysis to select genes with genome-wide significance. The genome-wide significance was defined as significance of $P < 1.8 \times 10^{-6}$ (0.05/number of genes in human genome in UCSC genome browser (build 37) = 0.05/28517). However, there are three shortcomings in the gene-based rare allele analysis using imputation. First, it is difficult to interpret if there are a lot of rare alleles in a gene. Second, by pooling risk and protective alleles, power can be decreased. However, considering such directions before selecting candidate genes, overinflation of type I error can be problematic. Third the accuracy of imputation can be decreased in rare alleles with very low MAF. We performed confirmatory analysis (the second meta-analysis) with selected SNPs We did confirmatory analysis, according to two reasons. First, if we could test genetic risk factors with a small number of SNPs, it would be more convenient for genotyping and interpretation. Therefore, we selected several risk

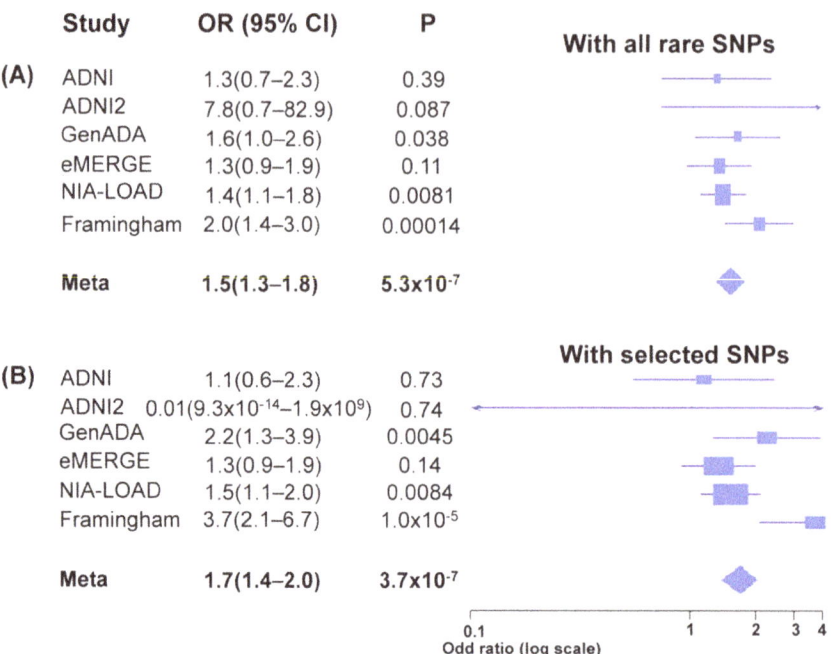

	Study	OR (95% CI)	P
(A)	ADNI	1.3(0.7–2.3)	0.39
	ADNI2	7.8(0.7–82.9)	0.087
	GenADA	1.6(1.0–2.6)	0.038
	eMERGE	1.3(0.9–1.9)	0.11
	NIA-LOAD	1.4(1.1–1.8)	0.0081
	Framingham	2.0(1.4–3.0)	0.00014
	Meta	**1.5(1.3–1.8)**	**5.3×10^{-7}**
(B)	ADNI	1.1(0.6–2.3)	0.73
	ADNI2	0.01(9.3×10^{-14}–1.9×10^{9})	0.74
	GenADA	2.2(1.3–3.9)	0.0045
	eMERGE	1.3(0.9–1.9)	0.14
	NIA-LOAD	1.5(1.1–2.0)	0.0084
	Framingham	3.7(2.1–6.7)	1.0×10^{-5}
	Meta	**1.7(1.4–2.0)**	**3.7×10^{-7}**

With all rare SNPs

With selected SNPs

Odd ratio (log scale)

Figure 2. Forest plots showing the association of *ZNF628* with AD. Results are (A) with all rare SNPs (the first meta-analysis) and (B) with only selected risk SNPs (the second meta-analysis). The weight of each study was calculated by $4/(1/N_{AD}+1/N_C)$, where N_{AD} and N_C are numbers of AD and controls, respectively [35].

SNPs in the finally selected gene according to meta *P* and meta *Z* ($P<0.05$ and Z>0) after performing classical SNP based GWAS and meta-analysis. Second, we excluded rare variants with MAF< 0.5%, because the imputation accuracy decreases in very low MAF [20].

Results

The first meta-analysis

In the meta-analysis, *ZNF628* had genome-wide significance (meta $P = 5.3\times10^{-7}$ [OR 1.5, 95% CI 1.3–1.8]) (Table 2 and Figure 2A). SNPs in *ZNF628* used in this study are summarized in Table S2 in File S1. In addition, *APOE* had also genome-wide significance (meta $P = 1.4\times10^{-6}$). Other genes with high significances, but not with genome-wide significance were *TOMM40*, *MMP1*, *NAPRT1*, *TREM2*, and *CBLB*.

Dependency on *APOE* ε4 genotype status

We examined the dependencies of the selected genes by adjusting for *APOE* ε4 (Table 2). The significance of *ZNF628* was remained, even after adjustment. However, the significance of *APOE* decreased after adjustment for *APOE* ε4 (after adjustment, *P* value of *APOE* increased to 0.023).

Additionally, the collinearity between *ZNF628* and *APOE* ε4 genotype status were examined based on the variance inflation factor (VIF, Table S3 in File S1). The VIFs of all studies were approximately 1.

Meta-analysis with selected risk SNPs (the confirmatory second analysis)

For a more applicable clinical approach, we identified significant risk SNPs by meta *P* and meta *Z* scores. Furthermore, considering the imputation accuracy [20], we selected SNPs with $0.5\% \leq MAF\leq3\%$. Two risk SNPs (dbSNP ID: rs112407198 and

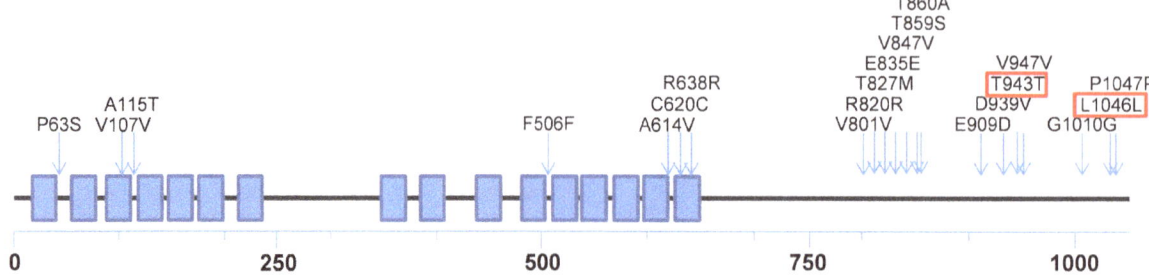

Figure 3. Schematic representation of *ZNF628* with locations of SNPs used in gene-based rare allele analysis in this study. *ZNF628* is a protein 1059 amino acids long. We briefly showed the domains (boxes) and the locations of SNPs (arrows) in a schematic linear structure of *ZNF628*. Blue boxes denote C2H2-type zinc finger domains. dbSNP ID can be found in Table S2 in File S1. SNPs within red boxes were used in the second analysis.

Table 3. Results of the second meta-analysis (confirmatory analysis).

Study	rs112407198 (position: 19:55995401)						rs73057174 (position: 19:55995710)						Gene-based	
	MAF	INFO	not adjusted for APOE ε4		adjusted for APOE ε4		MAF	INFO	not adjusted for APOE ε4		adjusted for APOE ε4		P not adjusted for APOE ε4	P adjusted for APOE ε4
			beta	P	beta	P			beta	P	beta	P		
ADNI	0.011	0.78	0.29	0.70	0.09	0.91	0.029	0.91	0.08	0.85	0.39	0.40	0.73	0.43
ADNI2	0.009	0.82	-8.7	0.79	0.50	0.87	0.031	0.95	-0.30	0.63	-0.083	0.90	0.74	0.55
GenADA	0.011	0.47	1.99	0.0052	1.93	0.0079	0.022	0.67	0.52	0.11	0.59	0.093	0.0045	0.0049
eMERGE	0.012	0.72	0.16	0.65	0.11	0.75	0.022	0.90	0.32	0.15	0.26	0.26	0.14	0.26
NIA	0.013	0.74	0.24	0.39	0.079	0.80	0.024	0.90	0.46	0.011	0.62	0.0015	0.0084	0.0042
Framingham	0.011	0.41	1.31	0.0079	1.34	0.0066	0.019	0.55	1.35	0.00026	1.34	0.00029	1.0×10^{-5}	9.7×10^{-6}
Meta				0.0043		0.019				2.8×10^{-5}		2.7×10^{-6}	3.7×10^{-7}	2.8×10^{-7}

Key: MAF, minor allele frequency; INFO, imputation quality score made by IMPUTE2.

rs73057174) selected within *ZNF628* were synonymous SNPs (Figure 3). As shown in Figure 2B and Table 3, gene-based rare allele analysis using only selected SNPs had genome-wide significance with moderately high effect size (meta $P = 3.7 \times 10^{-7}$ [OR 1.7, 95% CI 1.4–2.0]).

Gene-based rare allele analyses for the genes known to be associated with AD

Interestingly, rare alleles in *APOE* and *TREM2* showed significantly high association with AD (Table 2). Thus, we tested rare alleles of other known genes associated with AD. The most highly ranked nine genes in the AlzGene database [38] (*ABCA7, PICALM, CLU, MS4A6A/MS4AE, CD33, BIN1, CR1, and CD2AP*) were selected for the test. Based on the meta-analysis, only *BIN1* had significance (meta $P = 0.046$), but did not reach to genome-wide significance level (Table 4).

Discussion

We performed meta-analysis with publicly available genetic studies of AD with imputed rare (MAF≤3%) alleles. *ZNF628* was identified to have significant association with AD. Additionally, our rare allele analysis revealed the significant association of *APOE* and *TREM2* with AD, which suggested that our results were valid and that these genes require further study [39,40].

ZNF628 is a C2H2-zinc finger protein, a type of transcription factors [41] consisting of three exons. C2H2-type zinc finger proteins are known to be essential for normal growth and development [41]. *ZNF628* is found in mammals, but not Zebra fish or C. elegans [41]. *ZNF628* is evenly expressed in various tissues including brain [42,43]. *ZNF628* is conserved among mammals and seems to be functionally important [41]. The possible DNA binding site is the sequence motif – C/GA/TA/TGGTTGGTTGC [41]. As this time, the target proteins and related human disorders associated with *ZNF628* have not been reported. It is possible that the rare alleles in *ZNF628* change the expression levels of certain proteins related to AD pathogenesis.

In the selected allele analysis of *ZNF628* (the second confirmatory analysis), P and Z values of two SNPs (rs112407198 and rs73057174) reached the criteria of $P<0.05$ and $Z>0$. These SNPs are located outside the C2H2-type zinc finger domains and synonymous SNPs (Figure 3). The synonymous mutations are known to change the protein expression level and conformation [44] by affecting mRNA structure [45] or changing the time of cotranslational folding [46]. The altered expression levels or structure of *ZNF628* could affect the expression level of other proteins.

There were no dependencies between *ZNF628* and *APOE* ε4 genotype status. *ZNF628* is separated from *APOE* by more than 10^8 bp, although they are both located on chromosome 19. Therefore, *ZNF628* is not included in same LD block with *APOE* ε4. *ZNF628* did not lose its significance in meta-analysis even after adjustment for *APOE* ε4 genotype status. Therefore, *ZNF628* appears to be related with AD independently from *APOE* ε4. In contrast, the significance of *APOE* was affected by *APOE* ε4. The association of the rare alleles in *APOE* with AD was highly significant $(P = 1.4 \times 10^{-6})$ with AD, although this significance disappeared after adjusting for *APOE* ε4. This suggested that rare alleles in the same LD block with *APOE* ε4 conferred significant association with AD.

Other risk genes that have been found in GWAS targeting common alleles were not replicated in our gene-based rare allele

Table 4. Results of gene-based rare allele analysis top ranking genes in the AlzGene database.

Gene	CHR	start	Meta Z	Meta P	Directions*
ABCA7	19	1040101	0.2	0.8539	+++−+
PICALM	11	85668485	−0.4	0.7068	−?+−+−
CLU	8	27454450	−1.1	0.2554	−++−
MS4A6A	11	59939080	0.5	0.6377	−+−+−+
CD33	19	51728334	−0.4	0.7185	−++−
BIN1	2	127805606	2.0	0.0460	++++−
MS4A4E	11	59980567	0.5	0.6377	−+−+−+
CR1	1	207669472	0.7	0.5050	++−+−+
CD2AP	6	47445524	−1.1	0.2593	−+−+

Protein names: ATP-binding cassette, sub-family A, ABCA7; phosphatidylinositol binding clathrin assembly protein, PICALM; clusterin, CLU; membrane-spanning 4-domains, subfamily A, member 6A, MS4A6A; CD33 molecule, CD33; bridging integrator 1, BIN1; putative membrane-spanning 4-domains subfamily A member 4E, MS4A4E; complement component (3b/4b) receptor 1, CR1; CD2-associated protein, CD2AP.
* The signs represent those of the Z score of each study. The question mark represents missing data in the study because of low INFO or high MAF. The order of the signs is ADNI, ADNI2, GenADA, eMERGE, NIA, and Framingham study.

analysis. Only *TREM2*, which has been identified in previous studies targeting rare alleles, showed high significance levels [39,40]. Common alleles with small effect sizes have been explained by synthetic association of rare alleles [47,48]. Recently, however, this hypothesis was not confirmed in a large-scale study of seven common immune diseases [49]. Similarly, we could not show association of rare alleles within the known genes with AD.

There are several limitations in this study. First, a replication study with real genotyping is required. However, 1000 genomes-based imputations can enable us to find refined and novel signals [50]. Furthermore, the sample size and power can be increased by imputation [51] and meta-analysis [52]. Our gene-based rare variant analysis by imputation have comparable high power with re-sequencing analysis, especially with a large number of sample size [27]. Second, rare alleles analysis of *ZNF628* of this study was performed in White populations. Although this result should be replicated in different populations, it is difficult to identify. The two important selected SNPs of our study, rs11247198 and rs73057174, have not been reported in Asian populations, whereas higher MAF has been identified in Black populations (especially in the Bushmen). Third, current methods of rare allele analysis still have problems and need more powerful and consistent methods [53]. The simulated studies using 20 different tools did not generate consistent results [54]. Therefore, simulation studies to identify methods that generate the optimal results are required [53]. Additionally, the directions of SNPs for related diseases are not usually considered [53]. Lastly, the SNPs in introns could not be considered because of limited our computational resources.

In conclusion, we observed a noble association between *ZNF628* and AD. Considering the biological role of the *ZNF628* protein, it may contribute to AD by regulating various AD-related proteins expressions. Functional studies to elucidate its contribution to AD pathogenesis are required. Additionally, further studies addressing different populations should be replicated to assess the value of the *ZNF628* rare allele as a genetic biomarker of AD.

Acknowledgments

We used the supercomputing resource of the Korea Institute of Science and Technology Information (KISTI).

ADNI got a grant from the National Institute on Aging, the National Institute of Biomedical Imaging and Bioengineering. Additionally, ADNI was supported by the following: Alzheimer's Association; Alzheimer's Drug Discovery Foundation; BioClinica, Inc.; Biogen Idec Inc.; Bristol-Myers Squibb Company; Eisai Inc.; Elan Pharmaceuticals, Inc.; Eli Lilly and Company; F. Hoffmann-La Roche Ltd and its affiliated company Genentech, Inc.; GE Healthcare; Innogenetics, N.V.; IXICO Ltd.; Janssen Alzheimer Immunotherapy Research & Development, LLC.; Johnson & Johnson Pharmaceutical Research & Development LLC.; Medpace, Inc.; Merck & Co., Inc.; Meso Scale Diagnostics, LLC.; NeuroRx Research; Novartis Pharmaceuticals Corporation; Pfizer Inc.; Piramal Imaging; Servier; Synarc Inc.; and Takeda Pharmaceutical Company. ADNI clinical sites in Canada got funds from the Canadian Institutes of Health Research. The Foundation for the National Institutes of Health (www.fnih.org) facilitated private sector contributions. The grantee organization is the Northern California Institute for Research and Education, and the study is coordinated by the Alzheimer's Disease Cooperative Study at the University of California, Rev October 16, 2012 San Diego. The Laboratory for Neuro Imaging at the University of California, Los Angeles disseminated ADNI data. ADNI was also supported by NIH grants (P30 AG010129 and K01 AG030514). The genotypic and associated phenotypic data used in the study, "Multi-Site Collaborative Study for Genotype-Phenotype Associations in Alzheimer's Disease (GenADA)" were provided by the GlaxoSmithKline, R & D Limited. Alzheimer's Disease Patient Registry (ADPR) and Adult Changes in Thought (ACT) study was supported by a U01 from the National Institute on Aging (Eric B. Larson, PI, U01AG006781). The 3M Corporation gave a gift and it was used to expand the ACT cohort. DNA aliquots sufficient for GWAS from ADPR Probable AD cases, who had been enrolled in Genetic Differences in Alzheimer's Cases and Controls (Walter Kukull, PI, R01 AG007584) and obtained under that grant, were made available to eMERGE without charge. Genotyping, which was performed at Johns Hopkins University, was supported by the NIH (U01HG004438). GWAS were supported through a Cooperative Agreement from the National Human Genome Research Institute, U01HG004610 (Eric B. Larson, PI). The eMERGE Administrative Coordinating Center (U01HG004603) and the National Center for Biotechnology Information (NCBI) helped phenotype harmonization and genotype data cleaning. The "Genetic Consortium for Late Onset Alzheimer's Disease" was supported by the Division of Neuroscience, NIA. In NIA-LOAD study, Genetic Consortium for Late Onset Alzheimer's Disease helped phenotype harmonization and genotype cleaning, as well as with general study coordination. The Genetic Consortium for Late Onset Alzheimer's Disease includes a GWAS funded as part of the Division of Neuroscience, NIA. The Framingham Heart Study is performed and supported by the National Heart, Lung, and Blood Institute (NHLBI) in collaboration with Boston University (Contract No. N01-HC-25195). Funding for SHARe Affymetrix genotyping was provided by NHLBI Contract N02-HL-64278. The Framingham Dementia Mild Plus Incidence dataset was supported by NIH/NIA grants R01 AG08122 and R01 AG033193. The Framingham Dementia Moderate Plus Incidence dataset was supported by NIH/NIA grants R01 AG08122 and R01 AG033193.

However, we did not receive the commercial funds that are shown in this section. Although the data in our study can be publicly available, it was mandatory to show the funding sources for the studies. This does not alter our adherence to PLOS ONE policies on sharing data and materials.

This manuscript was not written in collaboration with investigators of the Framingham Heart Study and does not necessarily reflect the opinions or views of the Framingham Heart Study, Boston University, or NHLBI.

Some of data used in preparation of this article were obtained from the Alzheimer's Disease Neuroimaging Initiative (ADNI) database (adni.lo-ni.ucla.edu). As such, the investigators within the ADNI contributed to the design and implementation of ADNI and/or provided data but did not participate in analysis or writing of this report.

A complete listing of ADNI investigators can be found at:http://adni.loni.ucla.edu/wp-content/uploads/how_to_apply/ADNI_Acknowledgement_List.pdf.

Author Contributions

Conceived and designed the experiments: JHK SAP. Performed the experiments: JHK SAP. Analyzed the data: JHK HSL J-HL SAP. Contributed reagents/materials/analysis tools: JHK J-HL SAP. Contributed to the writing of the manuscript: JHK PS HSL J-HL JHL SAP. Agreement to be accountable for all aspects of the work in ensuring that questions related to the accuracy or integrity of any part of the work are appropriately investigated and resolved: JHK PS HSL J-HL JHL SAP.

References

1. Pedersen NL, Posner SF, Gatz M (2001) Multiple-threshold models for genetic influences on age of onset for Alzheimer disease: findings in Swedish twins. Am J Med Genet 105: 724–728.
2. Gatz M, Reynolds CA, Fratiglioni L, Johansson B, Mortimer JA, et al. (2006) Role of genes and environments for explaining Alzheimer disease. Arch Gen Psychiatry 63: 168–174.
3. Harold D, Abraham R, Hollingworth P, Sims R, Gerrish A, et al. (2009) Genome-wide association study identifies variants at CLU and PICALM associated with Alzheimer's disease. Nat Genet 41: 1088–1093.
4. Lambert JC, Heath S, Even G, Campion D, Sleegers K, et al. (2009) Genome-wide association study identifies variants at CLU and CR1 associated with Alzheimer's disease. Nat Genet 41: 1094–1099.
5. Hollingworth P, Harold D, Sims R, Gerrish A, Lambert JC, et al. (2011) Common variants at ABCA7, MS4A6A/MS4A4E, EPHA1, CD33 and CD2AP are associated with Alzheimer's disease. Nat Genet 43: 429–435.
6. Naj AC, Jun G, Beecham GW, Wang LS, Vardarajan BN, et al. (2011) Common variants at MS4A4/MS4A6E, CD2AP, CD33 and EPHA1 are associated with late-onset Alzheimer's disease. Nat Genet 43: 436–441.
7. Seshadri S, Fitzpatrick AL, Ikram MA, DeStefano AL, Gudnason V, et al. (2010) Genome-wide analysis of genetic loci associated with Alzheimer disease. JAMA 303: 1832–1840.
8. Bertram L (2011) Alzheimer's genetics in the GWAS era: a continuing story of 'replications and refutations'. Curr Neurol Neurosci Rep 11: 246–253.
9. Sullivan PF, Daly MJ, O'Donovan M (2012) Genetic architectures of psychiatric disorders: the emerging picture and its implications. Nat Rev Genet 13: 537–551.
10. Reich DE, Lander ES (2001) On the allelic spectrum of human disease. Trends Genet 17: 502–510.
11. Schork NJ, Murray SS, Frazer KA, Topol EJ (2009) Common vs. rare allele hypotheses for complex diseases. Curr Opin Genet Dev 19: 212–219.
12. Manolio TA, Collins FS, Cox NJ, Goldstein DB, Hindorff LA, et al. (2009) Finding the missing heritability of complex diseases. Nature 461: 747–753.
13. Asimit J, Zeggini E (2010) Rare variant association analysis methods for complex traits. Annu Rev Genet 44: 293–308.
14. van de Ven JP, Nilsson SC, Tan PL, Buitendijk GH, Ristau T, et al. (2013) A functional variant in the CFI gene confers a high risk of age-related macular degeneration. Nat Genet 45: 813–817.
15. Wheeler E, Huang N, Bochukova EG, Keogh JM, Lindsay S, et al. (2013) Genome-wide SNP and CNV analysis identifies common and low-frequency variants associated with severe early-onset obesity. Nat Genet 45: 513–517.
16. Huyghe JR, Jackson AU, Fogarty MP, Buchkovich ML, Stancakova A, et al. (2013) Exome array analysis identifies new loci and low-frequency variants influencing insulin processing and secretion. Nat Genet 45: 197–201.
17. Styrkarsdottir U, Thorleifsson G, Sulem P, Gudbjartsson DF, Sigurdsson A, et al. (2013) Nonsense mutation in the LGR4 gene is associated with several human diseases and other traits. Nature 497: 517–520.
18. Bodmer W, Bonilla C (2008) Common and rare variants in multifactorial susceptibility to common diseases. Nat Genet 40: 695–701.
19. Tennessen JA, Bigham AW, O'Connor TD, Fu W, Kenny EE, et al. (2012) Evolution and functional impact of rare coding variation from deep sequencing of human exomes. Science 337: 64–69.
20. Abecasis GR, Auton A, Brooks LD, DePristo MA, Durbin RM, et al. (2012) An integrated map of genetic variation from 1,092 human genomes. Nature 491: 56–65.
21. MacArthur DG, Balasubramanian S, Frankish A, Huang N, Morris J, et al. (2012) A systematic survey of loss-of-function variants in human protein-coding genes. Science 335: 823–828.
22. Rovelet-Lecrux A, Legallic S, Wallon D, Flaman JM, Martinaud O, et al. (2012) A genome-wide study reveals rare CNVs exclusive to extreme phenotypes of Alzheimer disease. Eur J Hum Genet 20: 613–617.
23. Jonsson T, Atwal JK, Steinberg S, Snaedal J, Jonsson PV, et al. (2012) A mutation in APP protects against Alzheimer's disease and age-related cognitive decline. Nature 488: 96–99.
24. Cruchaga C, Karch CM, Jin SC, Benitez BA, Cai Y, et al. (2014) Rare coding variants in the phospholipase D3 gene confer risk for Alzheimer's disease. Nature 505: 550–554.
25. Walsh T, McClellan JM, McCarthy SE, Addington AM, Pierce SB, et al. (2008) Rare structural variants disrupt multiple genes in neurodevelopmental pathways in schizophrenia. Science 320: 539–543.
26. Shea J, Agarwala V, Philippakis AA, Maguire J, Banks E, et al. (2011) Comparing strategies to fine-map the association of common SNPs at chromosome 9p21 with type 2 diabetes and myocardial infarction. Nat Genet 43: 801–805.
27. Magi R, Asimit JL, Day-Williams AG, Zeggini E, Morris AP (2012) Genome-Wide Association Analysis of Imputed Rare Variants: Application to Seven Common Complex Diseases. Genet Epidemiol 36: 785–796.
28. Filippini N, Rao A, Wetten S, Gibson RA, Borrie M, et al. (2009) Anatomically-distinct genetic associations of APOE epsilon4 allele load with regional cortical atrophy in Alzheimer's disease. Neuroimage 44: 724–728.
29. Li H, Wetten S, Li L, St Jean PL, Upmanyu R, et al. (2008) Candidate single-nucleotide polymorphisms from a genomewide association study of Alzheimer disease. Arch Neurol 65: 45–53.
30. Aulchenko YS, Ripke S, Isaacs A, van Duijn CM (2007) GenABEL: an R library for genome-wide association analysis. Bioinformatics 23: 1294–1296.
31. Delaneau O, Marchini J, Zagury JF (2012) A linear complexity phasing method for thousands of genomes. Nat Methods 9: 179–181.
32. Howie B, Fuchsberger C, Stephens M, Marchini J, Abecasis GR (2012) Fast and accurate genotype imputation in genome-wide association studies through pre-phasing. Nat Genet 44: 955–959.
33. Howie B, Marchini J, Stephens M (2011) Genotype imputation with thousands of genomes. G3 (Bethesda) 1: 457–470.
34. Zheng J, Li Y, Abecasis GR, Scheet P (2011) A comparison of approaches to account for uncertainty in analysis of imputed genotypes. Genet Epidemiol 35: 102–110.
35. Willer CJ, Li Y, Abecasis GR (2010) METAL: fast and efficient meta-analysis of genomewide association scans. Bioinformatics 26: 2190–2191.
36. Rhinn H, Fujita R, Qiang L, Cheng R, Lee JH, et al. (2013) Integrative genomics identifies APOE epsilon4 effectors in Alzheimer's disease. Nature 500: 45–50.
37. Aulchenko YS, Struchalin MV, van Duijn CM (2010) ProbABEL package for genome-wide association analysis of imputed data. BMC Bioinformatics 11: 134.
38. Bertram L, McQueen MB, Mullin K, Blacker D, Tanzi RE (2007) Systematic meta-analyses of Alzheimer disease genetic association studies: the AlzGene database. Nat Genet 39: 17–23.
39. Jonsson T, Stefansson H, Steinberg S, Jonsdottir I, Jonsson PV, et al. (2013) Variant of TREM2 associated with the risk of Alzheimer's disease. N Engl J Med 368: 107–116.
40. Guerreiro R, Wojtas A, Bras J, Carrasquillo M, Rogaeva E, et al. (2013) TREM2 variants in Alzheimer's disease. N Engl J Med 368: 117–127.
41. Chen GY, Muramatsu H, Ichihara-Tanaka K, Muramatsu T (2004) ZEC, a zinc finger protein with novel binding specificity and transcription regulatory activity. Gene 340: 71–81.
42. Wu C, Orozco C, Boyer J, Leglise M, Goodale J, et al. (2009) BioGPS: an extensible and customizable portal for querying and organizing gene annotation resources. Genome Biol 10: R130.
43. Derrien T, Johnson R, Bussotti G, Tanzer A, Djebali S, et al. (2012) The GENCODE v7 catalog of human long noncoding RNAs: analysis of their gene structure, evolution, and expression. Genome Res 22: 1775–1789.
44. Sauna ZE, Kimchi-Sarfaty C (2011) Understanding the contribution of synonymous mutations to human disease. Nat Rev Genet 12: 683–691.

45. Nackley AG, Shabalina SA, Tchivileva IE, Satterfield K, Korchynskyi O, et al. (2006) Human catechol-O-methyltransferase haplotypes modulate protein expression by altering mRNA secondary structure. Science 314: 1930–1933.

46. Kimchi-Sarfaty C, Oh JM, Kim IW, Sauna ZE, Calcagno AM, et al. (2007) A "silent" polymorphism in the MDR1 gene changes substrate specificity. Science 315: 525–528.

47. Dickson SP, Wang K, Krantz I, Hakonarson H, Goldstein DB (2010) Rare variants create synthetic genome-wide associations. PLoS Biol 8: e1000294.

48. Cirulli ET, Goldstein DB (2010) Uncovering the roles of rare variants in common disease through whole-genome sequencing. Nat Rev Genet 11: 415–425.

49. Hunt KA, Mistry V, Bockett NA, Ahmad T, Ban M, et al. (2013) Negligible impact of rare autoimmune-locus coding-region variants on missing heritability. Nature 498: 232–235.

50. Huang J, Ellinghaus D, Franke A, Howie B, Li Y (2012) 1000 Genomes-based imputation identifies novel and refined associations for the Wellcome Trust Case Control Consortium phase 1 Data. Eur J Hum Genet 20: 801–805.

51. de Bakker PI, Ferreira MA, Jia X, Neale BM, Raychaudhuri S, et al. (2008) Practical aspects of imputation-driven meta-analysis of genome-wide association studies. Hum Mol Genet 17: R122–128.

52. Skol AD, Scott LJ, Abecasis GR, Boehnke M (2007) Optimal designs for two-stage genome-wide association studies. Genet Epidemiol 31: 776–788.

53. Bansal V, Libiger O, Torkamani A, Schork NJ (2010) Statistical analysis strategies for association studies involving rare variants. Nat Rev Genet 11: 773–785.

54. Bansal V, Libiger O, Torkamani A, Schork NJ (2011) An application and empirical comparison of statistical analysis methods for associating rare variants to a complex phenotype. Pac Symp Biocomput: 76–87.

16

Large Genomic Region Free of GWAS-Based Common Variants Contains Fertility-Related Genes

Rong Qiu[1,2], Chao Chen[3,4]*, Hong Jiang[5], Libing Shen[6], Min Wu[1,2], Chunyu Liu[3,4,7]*

1 School of Information Science and Engineering, Central South University, Changsha, China, 2 Hunan Engineering Laboratory for Advanced Control and Intelligent Automation, Changsha, China, 3 Department of Psychiatry, University of Illinois at Chicago, Chicago, United States of America, 4 Institute of Human Genetics, University of Illinois at Chicago, Chicago, United States of America, 5 Department of Neurology, Xiangya Hospital, Central South University, Changsha, China, 6 School of Life Science, Fudan University, Shanghai, China, 7 State Key Laboratory of Medical Genetics of China, Central South University, Changsha, China

Abstract

DNA variants, such as single nucleotide polymorphisms (SNPs) and copy number variants (CNVs), are unevenly distributed across the human genome. Currently, dbSNP contains more than 6 million human SNPs, and whole-genome genotyping arrays can assay more than 4 million of them simultaneously. In our study, we first questioned whether published genome-wide association studies (GWASs) assays cover all regions well in the genome. Using dbSNP build 135 data, we identified 50 genomic regions longer than 100 Kb that do not contain any common SNPs, i.e., those with minor allele frequency (MAF)$\geq$1%. Secondly, because conserved regions are generally of functional importance, we tested genes in those large genomic regions without common SNPs. We found 97 genes and were enriched for reproduction function. In addition, we further filtered out regions with CNVs listed in the Database of Genomic Variants (DGV), segmental duplications from Human Genome Project and common variants identified by personal genome sequencing (UCSC). No region survived after those filtering. Our analysis suggests that, while there may not be many large genomic regions free of common variants, there are still some "holes" in the current human genomic map for common SNPs. Because GWAS only focused on common SNPs, interpretation of GWAS results should take this limitation into account. Particularly, two recent GWAS of fertility may be incomplete due to the map deficit. Additional SNP discovery efforts should pay close attention to these regions.

Editor: Frederick C. C. Leung, University of Hong Kong, China

Funding: This study was supported by National Basic Research Program (973 Program) (No. 2012CB944601, 2012CB517902, to Hong Jiang), New Century Excellent Talents in University (No. NCET-10-0836, to Hong Jiang), National Natural Science Foundation of China (No. 61125301 to Min Wu; No. 30971585, 30871354, 30710303061, 30400262, 81271260, to Hong Jiang). The funders had no role in study design, data collection and analysis, decision to publish, or preparation of the manuscript.

Competing Interests: The authors have declared that no competing interests exist.

* E-mail: chenchaor@gmail.com (CC); liucy@uic.edu (CL)

Introduction

The human genome contains millions of common SNPs, which are being deposited into public databases. These data have been used to design genome-wide association studies (GWASs) [1,2,3]. Common SNPs are better powered in association tests [4]. However, genomic regions not covered by common variants are neglected. Those neglected regions may contain variants with low frequencies, and should be paid more attention to because rare variants are even more likely to be functional than common ones [5].

In our study, we were interested in two questions: 1) whether the human genome is sufficiently covered by common SNPs and is sufficiently captured by common SNPs of standard GWAS platforms, and 2) whether any genes were included in those regions and their enriched biological functions.

To answer these two questions, we started with searching regions without common SNPs, called common SNP-free regions (CSFRs), regions free of both common SNPs and CNVs, called common variant-free regions (CVFRs). Next, we explored the functional enrichment of genes identified in CSFRs and CVFRs. With available personal genome sequencing data, whether these CSFRs and CVFRs contain common and rare variants were also examined.

Methods

Identification of CSFRs and CVFRs

Common SNPs (MAF$\geq$1%) in dbSNP build 135, Genome Assembly Gaps and Genome Database refGene data were downloaded from the UCSC Genome Browser (http://genome.ucsc.edu/) (Table 1). The CNV data were downloaded from the DGV (Table 1). Using the common SNP table, we calculated distances between adjacent common SNPs and subtracted regions containing the genome assembly gaps. If the remaining SNP intervals were longer than 100 kb, those intervals were defined as CSFRs. The CSFRs were further searched for CNVs. If after subtracting regions containing CNVs, the intervals were still longer than 100 kb, those intervals were defined as CVFRs. The reason we used 100 kb as bin to detect SNP free region is the SNP Linkage disequilibrium distance: several groups reported blocks of up to 100 kb in length exhibiting very strong linkage disequilibrium [6,7].

To verify our result for its impacts on GWAS, we first determined whether the CSFRs are truly missed by Affymetrix

Table 1. Data Sources Used in This Study.

Data	URL	Version	Modified date	Data description and summary statistics
Common SNP Data in HapMap	http://hgdownload.cse.ucsc.edu/goldenPath/hg19/database	Human Genome assembly hg19.	18-Dec-2011	snp135Common.txt.gz Total SNPs: 11,488,259 in chr1-chrY.
Genome Assembly Gaps data	http://hgdownload.cse.ucsc.edu/goldenPath/hg19/database	Human Genome assembly hg19.	27-Apr-2009	gap.txt.gz Total gaps, 357 in chr1-chrY.
Genomes Unzipped data	http://www.genomesunzipped.org/download/	Based on human genome hg18, upgraded to hg19	10-Oct-2010	Total of 1923 SNPs in the chrY.9 sample, 546 common SNPs with maf>1%.With data for 9 personal genome sequences.
personal genome variation data	http://hgdownload.cse.ucsc.edu/goldenPath/hg19/database/	Based on Human Genome assembly hg19.	21-Feb-2010	Total of 9 personal genomes: pgNA12878.txt.gzpgNA12891.txt.gzpgNA12892.txt.gzpgNA19240.txt.gzpgSjk.txt.gzpgVenter.txt.gz pgWatson.txt.gz pgYh1.txt.gzpgYoruban3.txt.gz
DGV data	http://hgdownload.cse.ucsc.edu/goldenPath/hg19/database/	Human Genome assembly hg19.	07-Mar-2011	dgv.txt.gz Total 101605 in chr1-chrY.
segmental duplication data	http://eichlerlab.gs.washington.edu/database.html	Human Genome assembly hg19.	27-Jun-2011	inter pairs is 22980; intra pairs is 8763
Genes	http://hgdownload.cse.ucsc.edu/goldenPath/hg19/database/refGene.txt.gz	Human Genome assembly hg19.	21-May-2012	Total number of genes is 42,742; after eliminating other chromosome, 30,332 genes in chr1-chrY remain.

Genome-Wide Human SNP Arrays. Next, we asked whether these regions included rare variations or were devoid of genetic variation. We analyzed common SNP data obtained from Genomes Unzipped (genomesunzipped.org) and Personal Genome Variation tracks from the UCSC Genome Browser. These two datasets are collections of variants that have been identified in the sequencing of personal genomes (Table 1).

Identification of genes in CSFRs and CVFRs

Gene annotation data from the Human Genome assembly hg19 UCSC refGene was used to map coding genes in the CSFRs and CVFRs (Table 2). Genes were included if their transcription regions overlapped with the CSFRs/CVFRs by at least one base pair. When a gene had multiple splicing forms, we chose the longest splicing form to define the gene region.

Pathway and functional analyses

The genes identified in the CSFRs/CVFRs were used to analyze their enrichment of biological functions through the Database for Annotation, Visualization and Integrated Discovery (DAVID, http://david.abcc.ncifcrf.gov/tools.jsp).

Isochore characterization

Isochore is a large region of DNA sequence which has a relatively uniform degree in its GC content [8]. We use 100 kb as the length of flank region and 2% GC difference as indicator to identify isochore, isochore border and unknown region among SNP free regions. All SNP free regions in this study are longer than 100 kb. CSFRs are identified as isochore if its GC content is 2% greater or lower than both right and left regions. CSFRs are identified as isochore border if the difference of GC content between two flank regions is greater than 2%, and GC-content difference between left flank and right flank region is greater than GC-content difference between CSFR and its flank regions. Unknown region means CSFR is neither isochore nor isochore border.

Results

CSFRs and CVFRs identification

We identified 50 CSFRs distributed across eight chromosomes: chr1, chr2, chr7, chr9, chr10, chr16, chrX, and chrY. The Y chromosome carried the majority of these regions–33 in total (Table 2). After excluding the CNV regions, we identified 20 CVFRs distributed across two chromosomes: chrX and chrY. The Y chromosome still carried the majority, with 18 regions (Table 3).

We checked our results in the Affymetrix SNP Array 6.0 by its annotation data. Among the CSFRs, we found 25 SNPs' information in the annotation file, and only four of them had non-zero minor allele frequency: rs11681529, rs2571764, rs2874557, and rs35516764. The other 20 are monomorphic for HapMap four populations (Caucasian, African, Chinese and Japanese). Therefore, we concluded that most of these 50 large genomic regions has not been covered properly by the Affymetrix 6.0 Array at least in those major populations investigated.

Genes in CSFRs and CVFRs and their functional enrichment

Ninety-seven genes overlapped with 28 of the 50 CSFRs (56%) (Table 2). DAVID was used to test whether the annotations of this set of genes were over presented with particular GO terms [9]. They were highly enriched with biological pathways involved with sexual reproduction, spermatogenesis, male gamete generation, gamete generation, multicellular organism reproduction, and reproductive processes in a multicellular organism (p<0.05 and FDR q<0.05, Table 4). The gene set included a number of gene previously reported to be related to reproduction, including *DAZ1* [10,11], *BPY2* [12], *TSPY2* [11], *CDY1* [13], *CDY2A* [13] and *RBMY1* [11]. A gr/gr deletion polymorphism on Y chromosome of those CSFRs has also been suggested to be a risk factor of spermatogenic impairment in some populations [14,15].

Twenty genes were overlapped with seven of the 20 CVFRs (35%) (Table 3). DAVID was also performed on these 20 genes. However, these genes were not enriched in any biological functions.

Table 2. List of 50 common SNP-free regions containing 97 genes.

Chr	CSFR_start	CSFR_end	CSFR_size	Gene_name	Isochore_type
chr1	145883118	145989503	106385	GPR89C, PDZK1P1	Isochore_border
chr2	110524226	110704031	179805	RGPD5, RGPD6, LIMS3, LIMS3-LOC440895, LIMS3 L	Isochore
chr2	111191098	111347035	155937	LIMS3-LOC440895, LIMS3, LIMS3L, RGPD6, RGPD5	Isochore_border
chr7	74765724	74866460	100736	GATSL2	Isochore
chr9	39379250	39551456	172206	LOC653501, ZNF658B	Unknown
chr9	39829606	39961804	132198	FAM75A2, FAM75A1, FAM74A1	Unknown
chr9	41497718	41635419	137701	FAM75A5, FAM75A7, LOC653501, ZNF658B	Unknown
chr9	42743905	42847394	103489	LOC286297	Isochore_border
chr10	46799214	46907775	108561	FAM35B	Isochore
chr10	48185336	48300420	115084	LOC642826, AGAP9, FAM25B, FAM25G, FAM25C, ANXA8, ANXA8 L1	Isochore_border
chr16	33142890	33293778	150888	TP53TG3, TP53TG3C, TP53TG3B	Isochore_border
chrX	52098738	52395914	297176	XAGE2, XAGE2B, XAGE1B, XAGE1A, XAGE1D, XAGE1C, XAGE1E	Unknown
chrX	52445914	52568230	122316	XAGE1A, XAGE1C, XAGE1E, XAGE1D, XAGE1B	Isochore_border
chrY	4834281	4935713	101432	PCDH11Y	Isochore_border
chrY	5012892	5205540	192648	PCDH11Y	Unknown
chrY	5274434	5421065	146631	PCDH11Y	Isochore_border
chrY	6074690	6422524	347834	TTTY23, TTTY23B, TSPY2, TTTY1B, TTTY1, TTTY2B, TTTY2, TTTY21, TTTY21B, TTTY7B, TTTY7, TTTY8B,TTTY8	Isochore
chrY	9381846	9492957	111111	RBMY3AP	Isochore
chrY	9524503	9768115	243612	TTTY8, TTTY8B, TTTY7B, TTTY7, TTTY21, TTTY21B, TTTY2B, TTTY2, TTTY1, TTTY1B, TTTY22, TTTY23,TTTY23B	Isochore
chrY	14691127	14804076	112949	TTTY15	Isochore_border
chrY	19563894	20143885	579991	FAM41AY1, FAM41AY2, LINC00230B, LINC00230A, XKRY, XKRY2, CDY2B, CDY2A	Unknown
chrY	20193885	20834702	640817	XKRY, XKRY2, LINC00230A, LINC00230B, FAM41AY1, FAM41AY2, HSFY2, HSFY1, TTTY9B, TTTY9A	Unknown
chrY	20837553	21080706	243153	TTTY9B, TTTY9A, HSFY2, HSFY1, NCRNA00185	Unknown
chrY	22564778	22665261	100483	TTTY10	Unknown
chrY	23473201	23580342	107141	RBMY2EP	Isochore_border
chrY	23634362	23838234	203872	RBMY1B, RBMY1A1, RBMY1E, RBMY1D, TTTY13	Isochore_border
chrY	23993156	24359930	366774	RBMY1A1, RBMY1D, RBMY1B, RBMY1E, PRY, PRY2, TTTY6, TTTY6B, RBMY1F, RBMY1J	Isochore_border
chrY	24500602	24620459	119857	RBMY1F, RBMY1J, TTTY6B, TTTY6	Unknown
chrY	24620459	28160890	3540431	PRY, PRY2, TTTY17B, TTTY17C,TTTY17A, TTTY4C, TTTY4B, TTTY4, BPY2B, BPY2, BPY2C, DAZ1, DAZ4, DAZ3, DAZ2, TTTY3B, TTTY3, CDY1, CDY1B, CSPG4P1Y, GOLGA2P2Y, GOLGA2P3Y	Isochore_border
chr9	42027732	42145811	118079		Isochore_border
chr9	44466205	44651655	185450		Isochore_border
chr9	45128500	45250203	121703		Isochore_border
chr9	65632583	65745692	113109		Isochore_border
chrY	3016123	3134221	118098		Isochore
chrY	3179117	3359419	180302		Isochore_border
chrY	3833777	3966707	132930		Unknown

Table 2. Cont.

Chr	CSFR_start	CSFR_end	CSFR_size	Gene_name	Isochore_type
chrY	3966708	4346934	380226		Unknown
chrY	4466077	4593373	127296		Unknown
chrY	4593411	4807708	214297		Unknown
chrY	6482140	6677618	195478		Isochore_border
chrY	7401836	7548914	147078		Unknown
chrY	8214827	8334874	120047		Isochore_border
chrY	15039955	15234829	194874		Unknown
chrY	18248698	18381734	133036		Unknown
chrY	18390543	18560004	169461		Isochore_border
chrY	19375294	19500106	124812		Unknown
chrY	22214221	22369679	155458		Isochore_border
chrY	22419679	22564743	145064		Isochore_border
chrY	23241568	23361665	120097		Isochore_border
chrY	28160891	28509481	348590		Isochore_border

SNP-free regions from personal genome sequencing and segmental duplications

We further explored those SNP-free regions in personal genome variant data. Rare variants were detected in most of the CSFRs or CVFRs. Only one region on X chromosome (chrX: 52,267,361-52,395,914) left. We also examined this region in updated dbSNP database (dbSNP137, http://www.ncbi.nlm.nih.gov/). Two more common SNPs were deteceted (rs201652812 and rs199865557). After subtract them, the left region was 105 kb (chrX: 52,290,698-52,395,914), which was the finally region not containing any

Table 3. List of 20 common variant-free regions containing 20 genes.

chr	CVFR_start	CVFR_end	CVFR_size	gene_name
chrX	52098738	52231295	132557	XAGE2, XAGE2B
chrX	52267361	52395914	128553	XAGE2, XAGE2B
chrY	4834281	4935713	101432	PCDH11Y
chrY	4935714	5205540	269826	PCDH11Y
chrY	5274434	5421065	146631	PCDH11Y
chrY	9524503	9640365	115862	TTTY8, TTTY8B, TTTY7B, TTTY7, TTTY21, TTTY21B, TTTY2B, TTTY2, TTTY1, TTTY1B TTTY22
chrY	20228333	20599266	370933	XKRY, XKRY2, LINC00230A, LINC00230B FAM41AY1, FAM41AY2
chrY	3016123	3134221	118098	
chrY	3179117	3359419	180302	
chrY	4114366	4346934	232568	
chrY	4466077	4593373	127296	
chrY	4593411	4807708	214297	
chrY	6577215	6677618	100403	
chrY	8214827	8334874	120047	
chrY	15039955	15234829	194874	
chrY	17559652	17661377	101725	
chrY	18248698	18381734	133036	
chrY	18390543	18560004	169461	
chrY	19375294	19500106	124812	
chrY	23247004	23361665	114661	

Table 4. Top 6 GO terms from the functional annotation analysis of 97 CSFR genes by DAVID.

Category	Term	Count	%	P-Value	FDR
GOTERM_BP_FAT	sexual reproduction[1]	9	14.8	0.00000003	0.000033
GOTERM_BP_FAT	Spermatogenesis[2]	8	13.1	0.000000047	0.000052
GOTERM_BP_FAT	male gamete generation[2]	8	13.1	0.000000047	0.000052
GOTERM_BP_FAT	gamete generation[2]	8	13.1	0.00000026	0.00028
GOTERM_BP_FAT	multicellular organism reproduction[2]	8	13.1	0.0000011	0.0012
GOTERM_BP_FAT	reproductive process in a multicellular organism[2]	8	13.1	0.0000011	0.0012

[1]gene included RBMY1A1, RBMY1B, RBMY1J, RBMY1F, XKRY, XKRY2, BPY2C, BPY2B, BPY2, CDY1, CDY1B, CDY2B, CDY2A, DAZ2, DAZ3, DAZ4, DAZ1, and TSPY2.
[2]gene included RBMY1A1, RBMY1B, RBMY1J, RBMY1F, BPY2C, BPY2B, BPY2, CDY1, CDY1B, CDY2B, CDY2A, DAZ2, DAZ3, DAZ4, DAZ1, and TSPY2.

known variant in all of the genome-wide sequencing data that we were able to collect. *XAGE2* and its splicing isoforms were harbored in this region.

We next tested this final region in segmental duplication database from Eichler's lab (http://eichlerlab.gs.washington.edu/database.html) [7], and found it was overlapped with one of the segmental duplication regions.

We found that 49 CSFRs did carry SNPs in the Genomes Unzipped and Personal Genome Variation tracks. And the left X chromosome region did not contain any SNPs but overlapped with segmental duplication region.

Twenty-four CSFRs are isochore borders

To dig out the sequence properties of 50 CSFRs, we characterized those regions by GC content. Different GC contents can separated DNA sequences into compositionally fairly homogeneous regions [8]. By comparing GC contents between CSFRs and their flanking regions, we found that twenty-four CSFRs belong to isochore border regions, seven belong to isochore regions, and eighteen are unknown regions (Table 2, Table S1).

Discussion

We performed a thorough search for large genomic regions that are free of common variants in dbSNP and we found 50 CSFRs and 20 CVFRs. Most of these variations free regions located on Y chromosome. Genes in the CSFRs were highly enriched for activities related to reproduction. Further investigation in the sequencing of personal genomes found most of the CSFRs (49 out of 50) did contain rare SNPs, suggesting those regions have not been covered well in the existing common variants sequencing projects, like the 1000 Genomes Project.

GWAS is one the most infusive common variants sequencing projects, but important finding might be missed because of its poor coverage of rare variants. Recently, two fertility GWAS studies were conducted but failed to find SNPs on sex chromosomes [16,17]. Both studies used Affymetrix GWAS platforms that we evaluated in this study. However, both sex chromosomes have long been implicated in infertility, specifically in spermatogenic damage in mouse models and in human candidate gene/region studies [18]. Our study found that those genomic regions free of common variants regions carrying many genes important to reproduction. With those important candidate genes missing, we must be cautious of analyzing fertility-related GWASs, which may produce false negatives.

The most reliable CVFR call contains the *XAGE2* and its isoforms, which belong to *XAGE* subfamily. *XAGE2* is strongly

expressed in normal testes, and in some tumor [19]. Because genotyping platforms cannot fully cover structural variations such as segmental duplication, we further applied structural variations filtering analysis, and observed *XAGE* region was overlapped with segmental duplication. Based on these observations, we concluded that the observation of variant free regions is more a coverage problem with the current versions of dbSNP and existing GWAS assay platforms than a lack of assayable variation. When more genomes are sequenced, we may end up with proper coverage of complete human genome by common SNPs.

We mapped our SNPs on dbSNP build 135 and regions on GRCh37.p10 (hg19) assembly reference, which is the most accurate alignment version and with all current genome knowledge available. Comparing to old versions, hg19 changed many genomic coordinates and included alternate haplotype assemblies for chr6 (7 haplotypes), chr4 (1 haplotype), and chr17 (1 haplotype). Different versions can be converted by liftOver software (http://genome.ucsc.edu/cgi-bin/hgLiftOver). More details of differences in each version are provided in NCBI (http://www.ncbi.nlm.nih.gov/genome/guide/human/release_notes.html).

Further study can focus on the sequence properties of those regions, and their conservative across species. Isochores are spatially heterogeneous in mammalian genome and varies in replication timing, gene richness, recombination rate, etc [20,21,22]. Natural selection is the most plausible explanation for formation and maintenance of isochores [20]. We observed nearly half of CSFRs are isochores and isochore border regions, which is a hint that these CSFRs may be under different selection pressure from its neighboring regions. To further test selection pressure, we mapped those regions to chimpanzee and mouse by Synteny analysis from Ensembl (http://useast.ensembl.org/Homo_sapiens/Location/Synteny?r=6:133017695-133161157), and found only 6 genes (*RGPD5, RGPD6, GATSL2, FAM25G, HSFY1, HSFY2*) can map to unique regions in the other two species. Next we applied dN/dS ratio test, the ratio of substitution rates at non-synonymous and synonymous sites, and found that human genes under more purify selection than chimpanzee genes (paired T test, p = 0.01, Table S2). Those results suggest that natural selection seems to be the major evolutionary force behind these variant-free regions.

In summary, by searching large genomic regions free of common variants for the first time, we identified tens of common variations free regions, and most of them were located on the X and Y chromosomes. The genes located in CSFRs are enriched for fertility. Incorporating personal genome data, only one region was still free of variants and harbored gene *XAGE2*, indicating most of the detections due to low coverage of rare variations. Future deep sequencing from more individuals and redesigning GWAS arrays should improve our understanding of the variability of these regions and their functional importance.

Author Contributions

Conceived and designed the experiments: CL. Performed the experiments: RQ CC. Analyzed the data: RQ LS. Contributed reagents/materials/analysis tools: CC MW HJ. Wrote the paper: RQ CC HJ.

References

1. Jiang RH, Duan JC, Windemuth A, Stephens JC, Judson R, et al. (2003) Genome-wide evaluation of the public SNP databases. Pharmacogenomics 4: 779–789.
2. Wang WYS, Barratt BJ, Clayton DG, Todd JA (2005) Genome-wide association studies: Theoretical and practical concerns. Nature Reviews Genetics 6: 109–118.
3. McCarthy MI, Abecasis GR, Cardon LR, Goldstein DB, Little J, et al. (2008) Genome-wide association studies for complex traits: consensus, uncertainty and challenges. Nature Reviews Genetics 9: 356–369.
4. Nannya Y, Taura K, Kurokawa M, Chiba S, Ogawa S (2007) Evaluation of genome-wide power of genetic association studies based on empirical data from the HapMap project. Human Molecular Genetics 16: 2494–2505.
5. Zhu QQ, Ge DL, Maia JM, Zhu MF, Petrovski S, et al. (2011) A Genome-wide Comparison of the Functional Properties of Rare and Common Genetic Variants in Humans. American Journal of Human Genetics 88: 458–468.
6. Aissani B, Perusse L, Lapointe G, Chagnon YC, Bouchard L, et al. (2006) A quantitative trait locus for body fat on chromosome 1q43 in French Canadians: linkage and association studies. Obesity (Silver Spring) 14: 1605–1615.
7. Bailey JA, Yavor AM, Massa HF, Trask BJ, Eichler EE (2001) Segmental duplications: Organization and impact within the current Human Genome Project assembly. Genome Research 11: 1005–1017.
8. Costantini M, Clay O, Auletta F, Bernardi G (2006) An isochore map of human chromosomes. Genome Res 16: 536–541.
9. Huang da W, Sherman BT, Lempicki RA (2009) Systematic and integrative analysis of large gene lists using DAVID bioinformatics resources. Nat Protoc 4: 44–57.
10. Fernandes S, Huellen K, Goncalves J, Dukal H, Zeisler J, et al. (2002) High frequency of DAZ1/DAZ2 gene deletions in patients with severe oligozoospermia. Mol Hum Reprod 8: 286–298.
11. Lardone MC, Parodi DA, Valdevenito R, Ebensperger M, Piottante A, et al. (2007) Quantification of DDX3Y, RBMY1, DAZ and TSPY mRNAs in testes of patients with severe impairment of spermatogenesis. Mol Hum Reprod 13: 705–712.
12. Choi J, Koh E, Suzuki H, Maeda Y, Yoshida A, et al. (2007) Alu sequence variants of the BPY2 gene in proven fertile and infertile men with Sertoli cell-only phenotype. Int J Urol 14: 431–435.
13. Kleiman SE, Lehavi O, Hauser R, Botchan A, Paz G, et al. (2011) CDY1 and BOULE transcripts assessed in the same biopsy as predictive markers for successful testicular sperm retrieval. Fertil Steril 95: 2297–2302, 2302 e2291.
14. Repping S, Skaletsky H, Brown L, van Daalen SK, Korver CM, et al. (2003) Polymorphism for a 1.6-Mb deletion of the human Y chromosome persists through balance between recurrent mutation and haploid selection. Nat Genet 35: 247–251.
15. Krausz C, Giachini C (2007) Genetic risk factors in male infertility. Arch Androl 53: 125–133.
16. Kosova G, Scott NM, Niederberger C, Prins GS, Ober C (2012) Genome-wide association study identifies candidate genes for male fertility traits in humans. Am J Hum Genet 90: 950–961.
17. Hu ZB, Xia YK, Guo XJ, Dai JC, Li HG, et al. (2012) A genome-wide association study in Chinese men identifies three risk loci for non-obstructive azoospermia. Nature Genetics 44: 183–186.
18. Burgoyne PS, Mahadevaiah SK, Sutcliffe MJ, Palmer SJ (1992) Fertility in Mice Requires X-Y Pairing and a Y-Chromosomal Spermiogenesis Gene-Mapping to the Long Arm. Cell 71: 391–398.
19. Chen YT, Ross DS, Chiu R, Zhou XK, Chen YY, et al. (2011) Multiple Cancer/Testis Antigens Are Preferentially Expressed in Hormone-Receptor Negative and High-Grade Breast Cancers. Plos One 6.
20. Costantini M, Cammarano R, Bernardi G (2009) The evolution of isochore patterns in vertebrate genomes. BMC Genomics 10: 146.
21. Oliver JL, Carpena P, Hackenberg M, Bernaola-Galvan P (2004) IsoFinder: computational prediction of isochores in genome sequences. Nucleic Acids Res 32: W287–292.
22. McVean GA, Myers SR, Hunt S, Deloukas P, Bentley DR, et al. (2004) The fine-scale structure of recombination rate variation in the human genome. Science 304: 581–584.

17

Longevity and Plasticity of CFTR Provide an Argument for Noncanonical SNP Organization in Hominid DNA

Aubrey E. Hill[1], Zackery E. Plyler[2], Hemant Tiwari[3], Amit Patki[3], Joel P. Tully[1,4], Christopher W. McAtee[4], Leah A. Moseley[4], Eric J. Sorscher[4,5]*

1 Department of Computer and Information Sciences, University of Alabama at Birmingham, Birmingham, Alabama, United States of America, **2** Department of Biology, University of Alabama at Birmingham, Birmingham, Alabama, United States of America, **3** Department of Biostatistics, University of Alabama at Birmingham, Birmingham, Alabama, United States of America, **4** Gregory Fleming James Cystic Fibrosis Research Center, University of Alabama at Birmingham, Birmingham, Alabama, United States of America, **5** Department of Medicine, University of Alabama at Birmingham, Birmingham, Alabama, United States of America

Abstract

Like many other ancient genes, the cystic fibrosis transmembrane conductance regulator (CFTR) has survived for hundreds of millions of years. In this report, we consider whether such prodigious longevity of an individual gene – as opposed to an entire genome or species – should be considered surprising in the face of eons of relentless DNA replication errors, mutagenesis, and other causes of sequence polymorphism. The conventions that modern human SNP patterns result either from purifying selection or random (neutral) drift were not well supported, since extant models account rather poorly for the known plasticity and function (or the established SNP distributions) found in a multitude of genes such as CFTR. Instead, our analysis can be taken as a polemic indicating that SNPs in CFTR and many other mammalian genes may have been generated—and continue to accrue—in a fundamentally more organized manner than would otherwise have been expected. The resulting viewpoint contradicts earlier claims of 'directional' or 'intelligent design-type' SNP formation, and has important implications regarding the pace of DNA adaptation, the genesis of conserved non-coding DNA, and the extent to which eukaryotic SNP formation should be viewed as adaptive.

Editor: John R. Battista, Louisiana State University and A & M College, United States of America

Funding: This work was supported by the National Institutes of Health [P30DK72482 to E. S.]; and the Cystic Fibrosis Foundation [R464 to E. S.]. The funders had no role in study design, data collection and analysis, decision to publish, or preparation of the manuscript.

Competing Interests: The authors have declared that no competing interests exist.

* Email: sorscher@uab.edu

Introduction

The classically hypothesized, random accumulation of single nucleotide polymorphisms (SNPs) through the ages presents a paradox. As a variation on the ratchet mechanism sometimes attributed to Muller [1–5] and expanded upon recently by Lynch [6–8] and Koonin [9], consider a simplistic estimate that ~1 in 1000 base pairs from our own genomes have become polymorphic after 150,000 years of human evolution. If, for argument's sake, one were to assume a similar rate of SNP accumulation among older metazoans (omitting, for the moment, the obvious contributions of negative selective pressure and identity by descent) [10], entire genomes would be rendered unrecognizable at every base pairing among vast numbers of ancient genes extant for 150,000,000 years. For still older eukaryotes, the situation would be much worse. This issue has been classically debated, but has not been addressed in the context of specific human genes or the most recent data concerning human DNA. In this report, we apply emerging knowledge from genome scale sequencing projects to view long-term DNA stability in a nontraditional way. An integrated look at a number of quotidian endpoints raises significant questions regarding purifying selection (as well as any sort of evolutionary drift or neutrality [7,8,9,11]) as explanations for the prolonged survival of genes such as CFTR.

In semblance to much of the human genome, CFTR is largely non-coding (total size approximately 190 Kb; cDNA approximately 4500 bp), and like many other human genes has been preserved across diverse species including ancient fish, amphibian, fowl, and mammalian. A great deal is known regarding the genetics and physiology attributable to homozygous or heterozygous CFTR loss in humans. Complete functional absence of one copy of CFTR occurs in 3–4% of American and European Caucasians (over ten million CFTR heterozygotes in North America alone) [12,13]. Historically, at least one CFTR mutation (F508del) likely conferred a strong selective advantage [14,15], but no deleterious effect on survival due to a single F508del allele (or any other CFTR mutation) is expected. Phenotypic findings are also absent among mice, pigs, ferrets, and rats deleted for a single CFTR [16–19]. In addition, CFTR itself is remarkably flexible and accommodates extensive polymorphism. Homozygous CF (knockout) mice lacking CFTR protein can be restored to health by insertion of a human CFTR different in coding sequence from the murine protein by approximately 30% [16]. CF manifestations can also be reversed in transgenic animals encoding CFTR with a very large (51 amino acid) deletion within the regulatory domain [20].

Mutations in CFTR or any other eukaryotic gene continue to accrue until a threshold of deleterious SNPs is reached, beyond which the profound resilience and plasticity of individual proteins, as well as their crucial epistatic effects (due to multiple loci impacting protein function) will begin to falter. In this report, we argue that over hundreds of millions of years and ongoing SNP accrual, a threshold of this sort should have been expected for CFTR long ago. Note that when individuals or organisms with severe homozygous CFTR defects are culled by purifying selection, this would not overcome a steadily accumulating mutational burden present among surviving contemporaries and their descendants, each being subject to steadily advancing numbers of SNPs over the evolutionary time scale. While overall SNP diversity within a population may fluctuate due to factors such as selection or drift, ongoing accumulation of new DNA variants is very large, and by itself suggests a number of interesting considerations. Population genomics has modeled DNA persistence and stability based on recombination (to reset the mutational ratchet) or a cumulative loss of fitness (attributable to randomly accumulating SNPs and their gene interaction networks) together with natural selection to eliminate detrimental CFTR alleles. Neither of these mechanisms, however, would overcome the continued (and potentially inexorable) accrual of SNPs among surviving members of a population. Our report furnishes recent genomic evidence that SNP accrual over vast numbers of generations could by now have left every CFTR allele so riddled with polymorphism that few, if any, would be viable (regardless of the extent of negative selection), and none would be available to recombine or restore a functional sequence. Moreover, removal of frequent unfit individuals (either an entire species or an occasional cockatrice) by natural selection would not reverse an accumulating CFTR mutational burden within surviving individuals and clades.

Statistical and population-based approaches intended to explain a species averting "mutational meltdown" have not fully addressed emerging knowledge regarding haplosufficiency of vertebrate genes, deep plasticity of the vertebrate genome, the observation that protein coding sequences such as CFTR have been remarkably conserved over hundreds of millions of years (despite an assumption of ongoing SNP accumulation) and new evidence relevant to structure/activity of eukaryotic proteins, including their redundancy and/or expendability. While acknowledging that even a modest decrease in fitness has never been established for the vast majority of random SNPs in any higher eukaryotic gene, an inference has often been that recondite evolutionary pressure somehow holds back SNP accumulation in coding sequences such as CFTR. Because this is a testable hypothesis, we developed our study to address the following questions: 1) In a survey of human populations worldwide using modern and leading-edge genomic tools, including studies conducted to minimize ascertainment bias, what can fundamental patterns of SNP accumulation in CFTR and other critical genes tell us about the production of DNA variants (and particularly the random nature of SNPs at the time of their formation)?, 2) Do these patterns appear to be either spatially or temporally neutral with respect to natural selection?, and 3) Based on a current understanding of CFTR-dependent effects on fitness, what does the analysis indicate regarding the role of purifying selection over the course of human or more ancient hominid evolution? Our findings suggest that modern human SNP compendia are not well reconciled with traditional explanations for long-term DNA persistence (including the role of purifying selection), while at the same time providing no evidence for less conventional (neutral, directional, or intelligent design type [11,21–24]) models of DNA evolution.

Results

Analysis of SNP distribution

Human exonic and intronic SNP frequencies. We began by tabulating SNP frequency within CFTR and other coding versus non-coding regions of DNA. Our expectation was that SNPs should be less prevalent (e.g., on a per 10,000 nucleotide basis) within the exome; i.e. non-coding DNA can sustain small sequence variations with minimal adaptive consequence [8,25–30]. Assumptions such as these have been challenged to some degree by recent studies indicating up to 80% of the non-coding genome subserves important regulatory function, and that point mutations within ENCODE motifs might often give rise to significant effects on fitness [31–33].

Data from dbSNP and HapMap are not optimal for addressing SNP frequency or distribution, since ascertainment bias skews these compendia towards SNPs: 1) discovered previously from selective exonic or other sequencing programs (dbSNP and HapMap), or 2) desirable from the standpoint of hapblock structure; i.e. specifically being sought as 'informative' vis-à-vis genome wide or other surveys (HapMap). On the other hand, data from 1000 Genomes provides unbiased and valuable information in this regard. We utilized complete human genomic sequences (approx. 1.8 million SNPs) from an initial 1000 Genomes release (http://pilotbrowser.1000genomes.org/index.html) that were prospective, nonbiased, and manageable in terms of computing.

Results in Figure 1 show that overall SNP frequency is diminished in exons versus intronic DNA for CFTR and 132 other genes known to cause serious human illness when disrupted (Figure 1A and Table S1). These genes were chosen because they are expected to be among the most susceptible to intense selective pressure (for many, their homozygous loss being lethal or debilitating). A significant difference in SNP frequency (exon:intron; 1:2.0, $p = 4.4 \times 10^{-46}$) was observed when critical human loci shown here were surveyed. A more expansive analysis of 4857 accessible genes indicated a similar ratio of 1:1.8 (exonic:intronic).

Is this difference in SNP frequency simply attributable to adaptive purging of deleterious exonic SNPs? If so, it becomes necessary to argue that approximately half of all single nucleotide changes across the human exome (the vast majority of which—including synonymous SNPs—would be of no known functional consequence) were instead highly significant, and that a sizeable number of these (approximately 50%) have been expunged (e.g. due to premature death or decreased fitness). If one accepts the notion that non-coding DNA is also a frequent object of selective pressure (i.e. the ENCODE analysis), even greater numbers of exonic SNPs would need to be removed to account for the findings. In addition, as discussed in detail below, the observation is anything but 'neutral' or 'random' [11,21]; the bias in favor of intronic SNPs is robust and appears to occur genome-wide.

Enhancement of synonymous versus non-synonymous SNPs in human genes. Synonymous polymorphisms are often viewed as insignificant from the standpoint of protein function, and are typically disregarded in genome scale studies of human disease (for example, GWAS or somatic SNPs responsible for cancer [27,28,34]). Among all genes—and particularly those vital to health—synonymous mutations are much better represented than their non-synonymous counterparts; examples are shown in Figure 2 and Table S2. Synonymous mutations are increased by approximately 1.6 fold among CFTR and ninety-seven other disease-associated genes with at least one exonic SNP. A more extensive test of 13,820 accessible genes (with well-defined exonic-intronic boundaries; Exon-Intron Database, human build 36.1 (http://www.utoledo.edu/med/depts/bioinfo/database.html)) in-

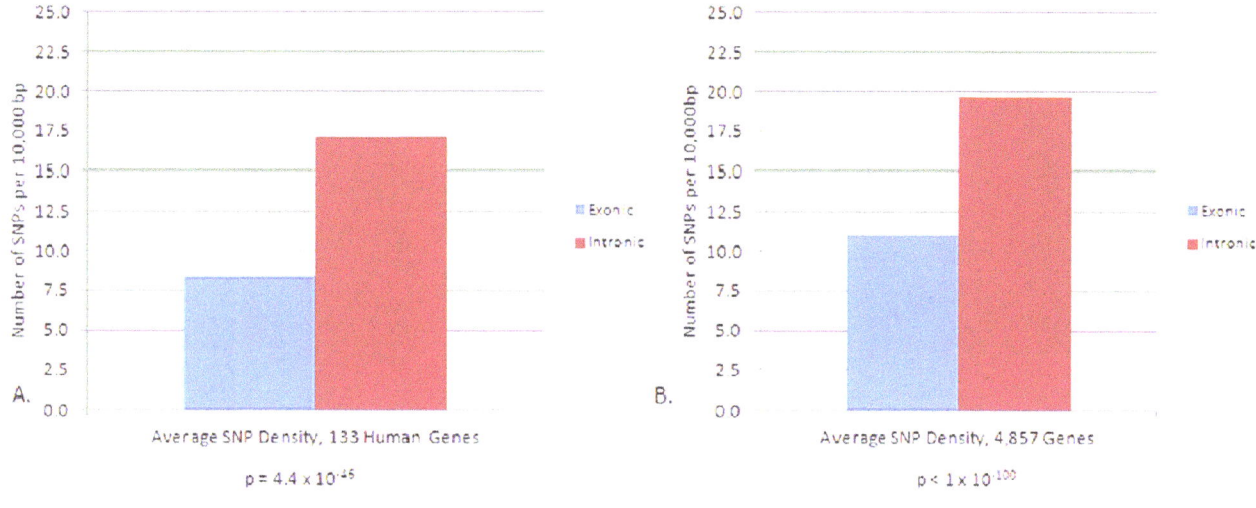

Figure 1. SNP incidence in human intronic and exonic DNA. A: SNPs in 133 human genes known to be lethal or severely debilitating if deleted [90] (Table S1); B: Survey of 4857 human genes for which intron/exon boundaries are readily definable in the Exon-Intron Database (http://www.utoledo.edu/med/depts/bioinfo/database.html) and 1000 Genomes release (http://pilotbrowser.1000genomes.org/index.html); Panel C: Composite data used to generate Panels A and B.

dicated a ratio of 1.3:1. A similar conclusion has been reached by others in a study based on dbSNP [35] and in exons exhibiting what were presumed to be accelerated rates of evolution [36,37]. Because a synonymous: non-synonymous ratio of 1:3—more than reversal of the measured frequency—is anticipated based on random nucleotide replacements within all 64 eukaryotic codons, the data suggests operation of strong selective pressures that have removed deleterious, non-synonymous mutations from human genes. This interpretation, which forms a basis for parallelism as defined by McDonald-Kreitman [22], is considered more fully in a later section.

Abundance of transition mutations in CFTR and other human genes. Intronic SNP sequences from CFTR were examined and found to have a relative paucity of transversion (A↔T, G↔C, A↔C, G↔T) in comparison with transition type polymorphisms (T↔C, A↔G, Table 1; $p = 8.5 \times 10^{-20}$) [10,38–41]. In a 1000 Genomes survey of exonic DNA from CFTR and 97 additional proteins crucial to human health with at least one SNP, a similar pattern was observed. The same was noted when spontaneous (and perhaps more recent) CFTR mutations were analyzed (Table 2) (www.genet.sickkids.on.ca). The transition-favoring aspect has been mechanistically ascribed to a failure of DNA error detection/repair, differences in misincorporation rates, or other factors [10].

Founder (ancestral) alleles account for the most common CFTR SNPs among Caucasians. In order to provide context

regarding the time frame responsible for appearance of human SNPs shown here, we reviewed haplotype block structure for CFTR and several other genes using HapMap. We utilized data for over 4 million SNPs, drawn from 270 individuals within North American Caucasian (CEU), Han Chinese (CHB), Japanese (JPT), and Yoruba (Nigeria, YRI) ethnic groups.

Figure 3 and Figure S1 describe incidence of CFTR SNPs from 45 or more subjects per ethnic background. Among JPT and CHB, virtually every SNP (by HapMap intention) is part of a major block, with disruption of the haplotype "clouds" (red circles/solid arrows) occurring due to ancestral crossover events (broken arrows). Because CFTR polymorphisms selected by HapMap were originally identified based on significant frequency in both alleles, there is reasonable agreement between SNPs shown in Figure 3 and those identified by unbiased sequencing in the 1000 Genomes database (i.e. for the specific case of a well-studied gene such as CFTR, approximately 90% of SNPs selected by the HapMap consortium were independently identified by 1000 Genomes). The majority of common CFTR SNPs in HapMap, therefore, are well represented by 1000 Genomes and suitable for the purpose described here.

SNP incidence profiles such as those shown in the CFTR minor allelic frequency (MAF) block diagram (at least for JPT, CHB) obviously cannot be explained by recent mutation in human DNA followed by purifying selection: no MAF block would otherwise be present. Such findings are attributable to ancestral haplotypes—

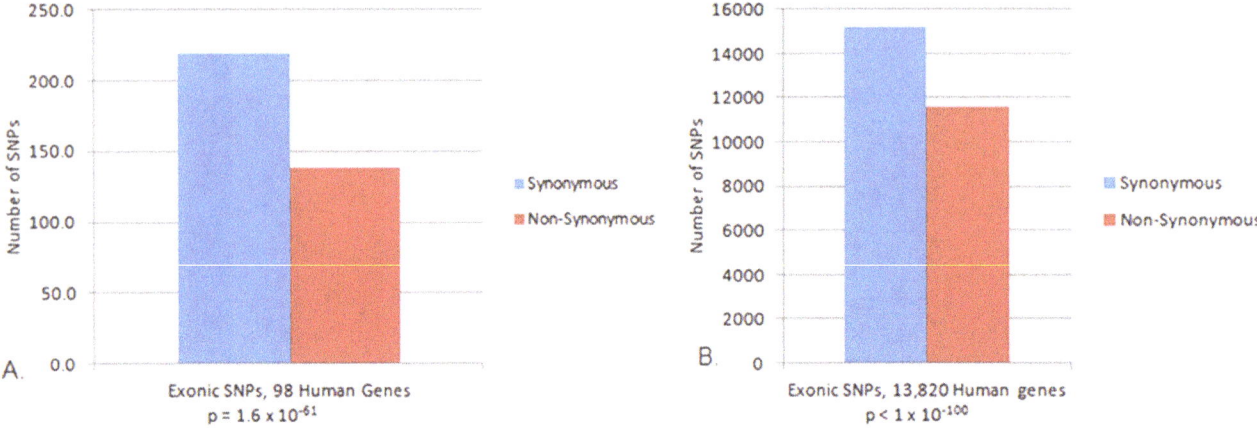

Figure 2. Synonymous and non-synonymous SNP incidence. <u>A</u>: Exonic SNPs in 98 genes known to be lethal or severely debilitating if deleted (a subset of genes in Figure 1A with at least one exonic SNP (Table S2)); <u>B</u>: Survey of 13,820 genes for which data was accessible from the Exon-Intron Database (http://www.utoledo.edu/med/depts/bioinfo/database.html) and 1000 Genomes (http://pilotbrowser.1000genomes.org/index.html); <u>C</u>: Composite data used to generate Panels A and B. All genes from Figure 1A with at least one exonic SNP were examined. Each gene was analyzed in the 1000 Genome Pilot Browser (http://pilotbrowser.1000genomes.org/index.html) including designation as synonymous vs. non-synonymous. The synonymous SNP enhancement agrees with earlier population-based studies in *Drosophila*, human, and other species [35,36,37,84]. To confirm that the ratio of synonymous to non-synonymous SNPs calculated from the set of 98 disease-associated genes was representative of the larger population, a bootstrapping analysis was conducted. Two-thousand samples of 98 genes were randomly selected from the larger gene cohort. Synonymous to non-synonymous ratios were used to determine a mean for each set of ninety-eight chosen in this manner. The overall mean of 2,000 samples was used to calculate both confidence interval and a 2-tailed t-test comparing the means of the 98 disease-associated genes and the mean derived from bootstrap sampling of the larger gene set. At the 95% confidence level, the mean synonymous to non-synonymous ratio of the 13,000 gene data set indicated a ratio between 1.37 and 1.38. A comparison to the 98 gene cohort mean yielded a p-value of 0.12.

the major source of human polymorphism—with a presumption that large haplotype blocks degenerate over evolutionary time due to recombination [42]. This same interpretation is compatible with the CEU and YRI MAF plots for CFTR, which exhibit

degenerating and abolished hapblock structure, respectively (with a caveat regarding the numbers of ancestral or founder haplotypes; see following section). Similar features are shown for NF1, a gene

Table 1. Transition Bias in Human SNPs.

SNP	CFTR Intronic	98 Human genes, Exonic
A/T	16	9
A/G	67*	150**
A/C	14	22
G/C	14	29
G/T	16	17
C/T	58*	130**

Incidence of six possible SNP configurations (transition and transversion) for CFTR intronic regions, and coding sequence from CFTR and 97 other human genes containing at least one exonic SNP (Figure 2 and Table S2). Underlined = transition mutations. The *p* values (based on an assumption of equal probability for any individual base replacement) indicate a strong bias in favor of transitions over transversions in both the human CFTR intronic DNA and the exonic sequences of 98 human genes. Transition:transversion ratio for CFTR intronic SNPs = 2.1; for exonic SNPs in 98 genes = 3.6.
*$p = 8.5 \times 10^{-20}$.
**$p = 5.7 \times 10^{-70}$.

Table 2. Transition Bias in CFTR Mutations Associated with Human Disease.

Wild Type	Disease associated mutation*	Number of Occurrences	SNP	Total observation
A	C	39		-
C	A	50	A/C	89
A	T	45		-
T	A	57	A/T	102
C	G	49		-
G	C	57	G/C	106
C	T	130		-
T	C	104	C/T	234[#]
G	A	179		-
A	G	157	A/G	336[#]
G	T	100		-
T	G	69	G/T	169
		Total = 1036		Total = 1036

Incidence of the possible SNP configurations (transition vs. transversion) among >1000 SNPs, many of which have been implicated in clinical CF (http://www.genet. sickkids.on.ca/cftr/app). p values indicate a bias towards transition based on an assumption of equal probability for any individual base replacement. Transition:transversion ratio = 1.2.
*http://www.genet.sickkids.on.ca/cftr/app.
[#]$p = 1.3 \times 10^{-56}$ for transition SNPs.

on chromosome 17 that mediates the autosomal dominant disease, neurofibromatosis (Figure 3B).

DNA variants on the Y-chromosome further indicate that human SNPs have been contributed in large measure by early ancestral alleles. The Y-chromosome furnishes an independent test of SNP derivation by minimizing contributions of ancestral alleles (anticipated to be reduced by at least 75%, since each ancestral breeding pair contributes four autosomes but only one Y-chromosome; note that if there were only one ancestral male for a specific ethnic group, there would be no SNPs on the Y attributable to founder haplotype). Results from HapMap, dbSNP, or 1000 Genomes are shown in Table 3, and indicate (as reported by others based on a variety of approaches [43–48]) a markedly diminished Y-chromosomal SNP incidence. DNA sequencing obstacles, sampling bias, background selection etc., contribute to this finding and the overall, quantitative difference is not known [44,46,48]. However, it is clear that Y-chromosomal SNPs are far fewer in number, including those within more readily sequenced regions (Table 3). The data, therefore, support early ancestral haplotypes—rather than ongoing DNA mutation—as a major contributor to SNP distributions among modern humans (i.e. SNPs of identity descent) [10,42].

Note that if point mutations in CFTR and other human genes were accounted for purely by *de novo* DNA mutation among *Homo sapiens*, rates of Y-chromosomal SNPs should be approx. 50% of autosomal SNP frequency (one Y-chromosome for every two autosomes). Anything less than 50% can be conditionally attributed to founder (ancestral) derived autosomal alleles. Because the measured SNP incidence of the Y (Table 3) is consistently less than 2% of the autosomal SNP frequency, the findings suggest that over 95% of autosomal SNPs could have been contributed by founder haplotypes. Small differences in chromosome-specific mutation rates would not significantly alter this estimate, although background selection may significantly diminish Y chromosome diversity, and complicates analyses of this kind [48].

Non-synonymous SNP rates in CFTR and other human genes are lower than expected. Standard SNP frequencies

reviewed in Figures 1 and 2 (and associated Tables) agree with findings from many laboratories investigating DNA variants among human and other species, and provide an argument against 'neutral' or 'near neutral' models of genomic evolution. Otherwise, individual alleles, large segments of the genome, as well as both coding and non-coding DNA would be required to drift in an overwhelmingly biased and uniform direction, and drift (by definition) occurs randomly [11,21,23,27]. Note that neutral models do not exclude specific genomic elements exhibiting only limited variation (e.g. hyperconserved segments), although such intervals are not believed to represent a predominant component of human DNA. By the same token, near-neutral models allow for significant numbers of deleterious variants to become fixed in small populations (i.e., tending to favor non-synonymous SNPs). In this context, therefore, it becomes informative to scrutinize the quantitative significance of a synonymous to non-synonymous SNP ratio of roughly 1.5:1 (Figure 2).

Consider, for example, a pair of extant Caucasian individuals with a common ancestor 50,000 years ago (an estimated time of hominid migration out of Africa) [49], who now differ at approximately 1 in 1,000 nucleotide positions throughout their respective genomes [50]. If exonic DNA conservatively represents ~3% of three billion human nucleotide pairs, this amounts to approximately 90 million exonic positions with ~0.1% rate of single nucleotide polymorphism, or on the order of 90,000 exonic SNPs. The data in Figure 2 indicates that upwards of 54,000 of these should be synonymous, with roughly 36,000 non-synonymous, genome-wide (i.e. synonymous: non-synonymous ratio of ~1.5).

As introduced above, in the absence of natural selection, the expected ratio of non-synonymous to synonymous polymorphisms on a full genome basis is usually taken to reflect stochastic SNP formation, since factors such as drift, shift, etc. would be minimized by random assortment. A neutral or random accrual of exonic SNPs generates a quantitative ratio of approximately 3:1 (non-synonymous to synonymous; accounting for all possible nucleotide changes in all possible human codons). The "expected"

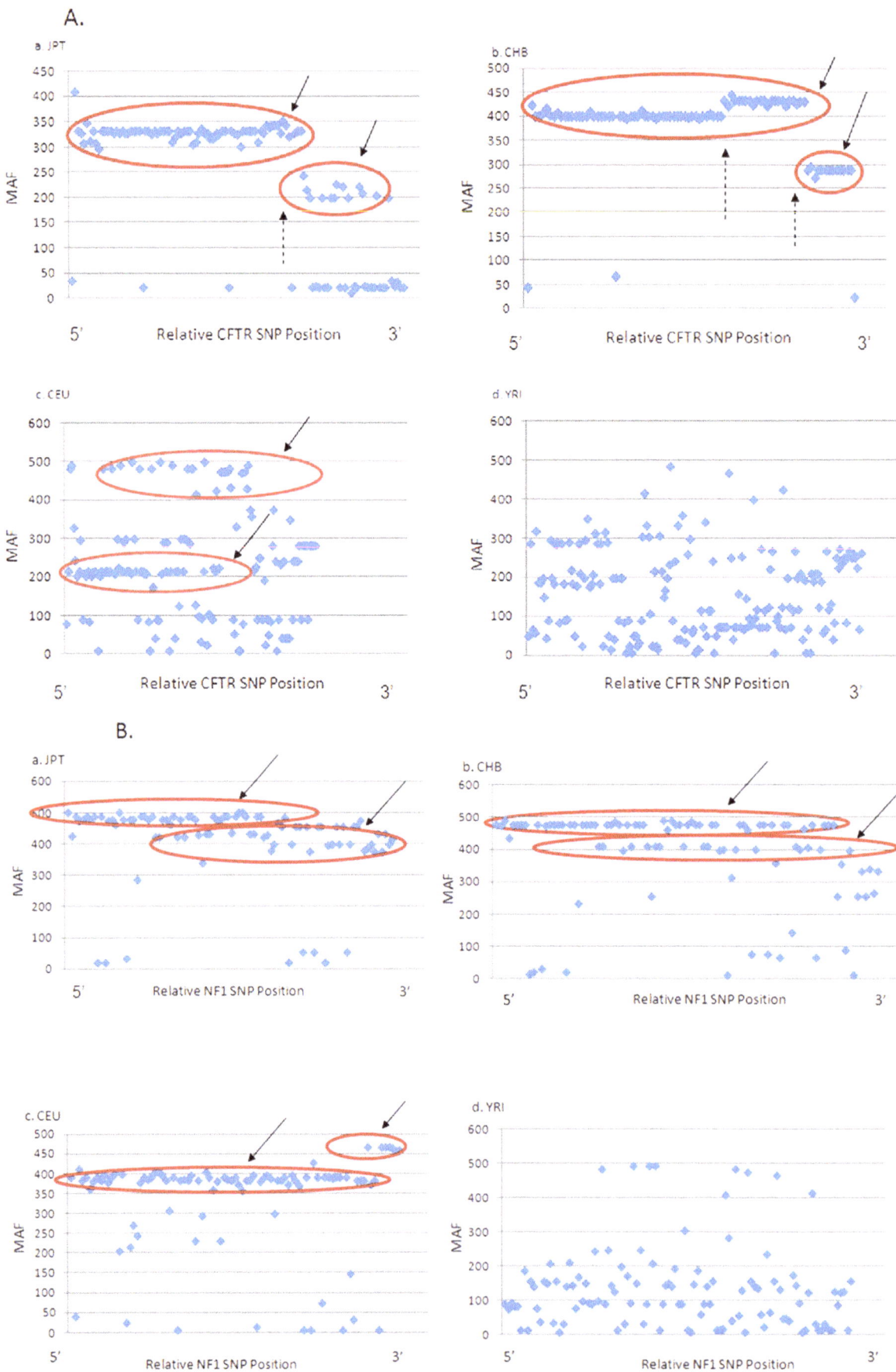

Figure 3. HapMap minor allelic frequencies (MAFs) plotted against gene sequence position. Frequency data for SNPs in CFTR (<u>Panel A</u>) or NF1 (<u>Panel B</u>) were collated for each of the ethnicities shown: JPT (Japanese in Tokyo, 45 individuals); CHB (Han Chinese in Beijing, 45 individuals); CEU (or CEPH, Utah residents with ancestry from northern and eastern Europe, 90 individuals); and YRI (Yoruba in Ibidan, Nigeria, 90 individuals). MAF refers to the relative frequency (1000 = 100% incidence) of the minor allele at each SNP position. Solid arrows/red circles depict areas indicative of a haplotype block (also referred to as MAF block) in the genes as shown; broken arrows describe sites of genomic recombination. In order to generate a MAF block diagram, allele frequency data was downloaded from UCSC genome table browser (http://genome.ucsc.edu/cgi-bin/hg:tables). After downloading, SNPs with MAF equal to zero among all four ethnicities were omitted. The remaining SNPs were then inserted into the scatter plot. Linkage disequilibrium valves for the blocks depicted here (when obtained directly from HapMap) were robust (r^2 among co:allelic SNPs shown by red circles typically = 1.0).

incidence across 90,000 exonic SNPs (omitting natural selection for the moment) is therefore at least a reversal of the "observed"— i.e. ~68,000 non-synonymous and ~22,000 synonymous SNPs. (Codon usage as a means to preserve exonic DNA is considered later in this report.)

Although fitness is an exceedingly difficult variable to quantify [51], it is problematic to imagine that negative selection could lead to such a reversal of the non-synonymous to synonymous SNP ratio over an extended period of human evolution. <u>First</u>, the mutation rate in humans ($1-3 \times 10^{-8}$ SNPs/nucleotide/generation [52–55]), and the estimated ~54,000 synonymous SNPs across ~90 million exonic positions would be predicted to require> 400,000 (not 50,000) years— i.e. most of the known synonymous SNPs must have significantly predated the ethnic founders of modern *Homo sapiens*. This agrees with conclusions from the MAF block analysis shown in Figure 3— i.e. that most human SNPs are not the result of recent DNA mutation, but attributable to ancient ancestral alleles.

<u>Second</u>, and more importantly, the synonymous versus non-synonymous SNP ratios cannot be easily reconciled with what is known about CFTR and many other human gene products. In every eukaryotic genome, SNPs continue to accumulate with each ensuing generation, and at some point would be anticipated to alter fitness. As described above, the "expected" versus "observed" SNP quota requires that under a natural selective pressure, human genomes have evolved from an expected (and stochastic) ratio of 1:3 (synonymous: non-synonymous) to the observed ratio of ~1.5:1. This would indicate that more than three of every four (approx. 78%) non-synonymous SNPs in CFTR and other genes have been deleted (or at least markedly de-enriched) from the gene pool either by classic (purifying) selection, or due to more complex, multigenic effects. The model presents a "best case" scenario, since rare mutations with a positive effect on fitness would require still higher rates of purifying selection to arrive at the same net loss of non-synonymous polymorphism. Note this analysis is quantitative and does not involve statistical or other underlying demographic assumption. (The calculation is based solely on empiric data including the directly measured (and widely accepted) genomic enrichment for synonymous vs. non-synonymous human SNPs, the expected ratio of synonymous: non-synonymous SNPs if formation was random—a value obtained from the genetic code, the best available (and readily quantified) degree of human DNA polymorphism, and size of the human genome.) The available and published findings point to a decrease from ~68,000 (predicted) to ~36,000 (observed) non-synonymous SNPs, or roughly 32,000 pre-reproductive age deaths or unfit genomes deleted or severely repressed during a finite period of human (or, in fact, pre-human) evolution.

Purifying selection and modern SNP ratios. The assertion that an estimated 32,000 single nucleotide changes (distributed over a genome with 20,000–30,000 genes) have been expunged from the gene pool of two individual humans with a common (pre-human) ancestor highlights the rarity of these mutational events. The notion that an isolated, randomly placed non-synonymous

SNP in a diploid gene such as CFTR should not only abrogate function, but also lead to death or blunt fertility of the entire organism seems incompatible with the puissant effects on fitness that would be required to explain the modern SNP distributions reviewed here. Put plainly, it seems antithetical to suppose that a single, random base substitution in any gene should disrupt activity or severely undermine fitness of an entire human, let alone that this has occurred tens of thousands of separate times over the course of an individual's evolutionary descent from an ancestral founder.

As shown in Figure 2, the synonymous to non-synonymous ratio is similar whether 98 (disease-associated) or 13,820 distinct human genes are analyzed (a value of 1.3–1.6). Because the proportion of synonymous to non-synonymous SNPs applies across numerous chromosomes, deleterious effects concentrated in a small number of genes cannot account for the finding. Moreover, if this level of non-synonymous SNP reduction represents a 'mutational burden' near to a significant (additive or fractional) effect of mutations on fitness, one could argue that many extant human genes (and individuals) should exist near some critical threshold, limping along and barely able to accommodate further polymorphism. As noted above, this is not the case for modern human CFTR, which appears strongly accommodating to polymorphism. Moreover, the notion that one random point mutation in a single allele of <u>any</u> human gene would usually (in almost 80% of cases) cause death or severely abrogate fitness is at odds with much of what has been learned about human gene and protein plasticity over the past 50 years.

Imagine that technology were available to place one exonic SNP in a single diploid gene randomly in the human genome. The likelihood that this individual SNP would be sufficient to destroy (or render less fertile) an entire individual is remote. Moreover, *Homo sapiens* is a comparatively young species. If this level of polymorphism represents a general threshold (i.e. a "tipping point") beyond which fitness is lost as genomes decompensate, it is difficult to imagine how a panoply of much more ancient genes (among far more ancient species) could have survived during an evolutionary period thousands of times more prolonged. In addition, note that in recombinant murine models, an extensive database already exists with regard to the same question. As with human, the likelihood that a single (random) non-synonymous SNP per murine gene should be a common cause of infertility or death is diminishingly small based on a vast number of transgenic animals and experimental findings. Yet the data reviewed in Figures 1 and 2 require a remarkable selective pressure of roughly this magnitude in order to account for observed SNP frequencies among the same genes in human DNA (i.e. one of every two random, exonic SNPs (including synonymous SNPs, Figure 1), or just one random, non-synonymous SNP per human gene (Figure 2) appears to be so deleterious that it typically causes death or abrogates normal reproduction).

Inurement of DNA polymorphism in the eukaryotic genome. Our earliest hominid ancestors inherited CFTR as part of a genetic legacy hundreds of millions of years old.

Table 3. Frequency of SNPs on the Y and other representative human chromosomes.

Chromosome	Size (bp)	Total SNPs			SNPs per 10,000 bp		
		db SNP	Hapmap (CEU)	1000 Genomes*	db SNP	Hapmap (CEU)	1000 Genomes*
Chr:22	49691432	399169	55941	251649	80.33	11.258	50.642
Chr:21	46944323	369905	50983	219897	78.797	10.86	46.842
Chr:20	62435964	623847	121069	396676	99.918	19.391	63.533
Chr:X	154913754	847225	122601	556264	54.69	7.914	35.908
Chr:Y	57772594	50993	722	326	8.827	0.125	0.056
Genes on the Chr Y	2173359	171	85	244	0.787	0.391	1.123

Number of SNPs is given for each of the chromosomes shown, according to data in dbSNP, HapMap, or 1000 Genomes.
*1000 Genomes Pilot Release 7.

Irrespective of whether or not modern CFTR SNP patterns are attributable to recent purifying selection, the integrity and flexibility of the human gene is well-established. How did CFTR persist without becoming riddled with polymorphism despite its ancient origins and subsequent epochs of mutation accrual? With regard to a central question posed by this report, consider hominid evolution during the past 200,000 years and a DNA mutation rate (described above) with approx. 90 new SNPs per individual per generation (for review, [46,51,56]). During 10,000 generations (assuming 20 years each), an estimated 9×10^5 mutations would be expected to distinguish a present-day individual from an early hominid ancestor. Now apply this same process for a much longer period (among significantly older metazoans) and, for the moment, omit the role of purifying selective pressure. After 250×10^6 years (the evolutionary age of many sharks), mutations would be expected in at least every codon of any sizable vertebrate genome; e.g., every codon in a 3×10^9 base pair genome would be altered in a random fashion. If a more realistic generation time (e.g. one year) is imposed, every codon in every core metabolic gene would be altered 20 times, and in a genome of 30,000,000 bp (perhaps more representative of certain diploid ancestors), each codon in every core metabolic gene would be randomly replaced nearly 2,000 times. Modestly lower rates of human mutation [e.g. 3-fold less, compare 53–55, 57] do not substantially alter this analysis. Moreover, a computer simulation conducted by our laboratory demonstrated <6% concordance of modern CFTR versus an ancient ancestor under these conditions, and that the last CFTR with >30% homology to the original CF gene product would have disappeared hundreds of millions of years ago. In other words, regardless of the type and magnitude of selective pressure that might have been applied, no CFTR would be expected today that even remotely resembles a functional protein. Moreover, even if one were to skew the analysis (and impose additional assumptions regarding population size, drift, evolutionary bottlenecks, etc.) so that CFTR somehow survived and retained its plasticity, the likelihood that an individual human with a working copy of CFTR would have also preserved 20,000–30,000 other human genes in exactly the same fashion (each gene having experienced its own stochastic mutational burden over hundreds of millions of years) does not seem compatible with genomic persistence. Regardless of population size, variation of the fitness landscape, putative valleys, drift etc. invoked so that observations better approximate the evolutionary expectation, the analysis strongly indicates that "meltdown" of human CFTR is long overdue.

In summary, based on new and emerging knowledge regarding DNA polymorphism, the classical argument that natural selection

or DNA recombination somehow reset the "ratchet" mechanism described above [58–60] does not account for significant discrepancy. The mutational burden continues to accumulate towards "meltdown" in every generation and every member of a given population. Even when natural selection removes certain disadvantageous haplotypes or enriches others, all remaining alleles continue to experience an ever-increasing SNP burden through the ages. Recombination, drift, and natural selection cannot stave off the mutational juggernaut, since every allele available for recombining continues to experience its own accumulating mutational burden, and every diploid gene that evades selection will continue to accumulate SNPs. While overall SNP diversity of a human population will fluctuate due to factors such as these, the perpetual accumulation of new DNA variants over an evolutionary time frame is very large. Moreover, even if a specific gene somehow managed to persist, its sequence could be ransacked by extensive polymorphism, and close to the threshold for dissipation. In other words, detrimental fitness effects necessary to overcome evolutionary destruction of CFTR—a eukaryotic gene that like others exhibits remarkable plasticity—account poorly for the extreme longevity of CFTR or the surrounding genome.

Concordance between SNPs in human exons and coding sequences from other species. We also compared exonic regions in human genes found permissive for SNPs (i.e. "polymorphic" by McDonald-Kreitman criteria; [22,61]) and the corresponding regions in six other chordates. An example depicting the first nine exonic SNPs reported in CFTR by 1000 Genomes is shown in Figure 4A. The complete CFTR open reading frame is approximately 50% identical among these six non-human CFTRs, including evolutionarily distant species such as chicken and frog. A corresponding but more extensive analysis is shown for CFTR and twenty-one other human genes with ≥ 50% overall concordance and at least one exonic SNP identified by 1000 Genomes (Table S3). The findings summarized in Figure 4B establish that the same DNA positions exhibiting single nucleotide polymorphism among humans also tend to be concordant with polymorphic sites from evolutionarily distant species. Similar results have been shown previously by others [22,62,63]. Notably, DNA positions polymorphic in humans (and the corresponding polymorphic positions among non-human species) are predominantly *synonymous* (Figure 4C; $p = 2.7 \times 10^{-9}$). When complete coding sequences from 629 individuals and a recent 1000 Genomes release were analyzed, approx. 69% of positions found to be polymorphic among both humans and multiple other species were synonymous

($p = 3.2 \times 10^{-13}$ versus the expected number of synonymous SNPs if accumulation were randomly distributed). Since variability is enriched in a strongly synonymous fashion, these mutations are unlikely to represent DNA positions where SNPs have been established by purifying selection. Moreover, since a robust synonymous bias occurs genome wide, neutrality or drift do not furnish a satisfactory explanation. In a related study, we observed that CFTR exonic concordance among four vertebrate species evolutionarily distant from human (horse, frog, zebrafish, and shark) is approximately 43% and that the total number of single nucleotide differences between these four species and human CFTR is 4620. The preponderance of shared SNPs among the four species was again found to be synonymous. In addition, when a SNP was applied at random in 4620 distinct instances to the 4443 base pair open reading frame of human CFTR using computer simulation, concordance between human and non-human coding sequences was much lower than observed in nature (average ~35%; $p = 6.6 \times 10^{-63}$) (Figure 4D). Observations such as these point to an important question: Why do so many synonymous DNA changes—mutations of questionable adaptive relevance—exhibit significant conservation across numerous species?

Discussion

Recent studies of CFTR sequence evolution [64], protein residue co-evolution and structure [65], selection for intronic regulatory sequences [33], inferences regarding CFTR channel gating [66], and characterization of the cystic fibrosis disease mutational spectrum [67] are predicated on a mechanism that treats CFTR exonic, intronic, synonymous, and non-synonymous SNP production as a random process. More classical aspects of genomics including DNA 'clocks' [28], polymorphic SNP formation and non-neutral evolution [22,30,36], rapidly evolving and ultra-conserved DNA [26], genesis of phenotypic complexity [68–70], and computational reconstruction of ancestral DNA [71] are likewise grounded to varying extent on SNP formation as essentially unbiased. The present study, however, suggests that inadequate attention has been paid to the non-random features of SNP formation. Based on recent knowledge regarding CFTR (and other protein) plasticity, function, and SNP distribution, our analysis indicates that approximately half of all exonic SNPs and nearly 80% of non-synonymous SNPs that should have been expected on a stochastic basis in human DNA instead were never formed in the first place.

Note that the above statement is by no means meant to imply that evolutionary selection does not purge deleterious mutation. However, insofar as human DNA is concerned, even if one adopts a very conservative estimate that a single SNP anywhere in the ~90 million nucleotides of human coding DNA has a 1 in 10 chance of causing death or undermining fitness of the entire human organism, we are left with an estimate that among SNPs expected on a random basis, 45% of all exonic SNPs and >70% of non-synonymous SNPs instead were never produced. Below we provide a summary that underscores the topics dealt with by this report.

A. Figure 1 and Table S1 establish that introns exhibit a strong increase in SNP frequency whether investigated among a selected gene set or across the entire human genome (a ratio of approx. 2:1 intronic versus exonic SNPs; $p = 4.4 \times 10^{-46}$). The assumption that this results from selective removal of exonic SNPs seems unsatisfactory, since it would require at least one of every two coding SNPs (including a preponderance of synonymous SNPs) to be markedly detrimental, leading to early death or otherwise undermining fertility, irrespective of epistasis (see also below). The likelihood that a solitary, randomly placed exonic SNP in CFTR (or other gene) should be lethal or vitiate fertility is contrasted by a substantial body of modern evidence regarding protein function and plasticity. We believe an alternative explanation has not been adequately considered; namely, that the modern SNP distributions shown here are attributable in large measure to a strong bias in their original formation. (In this context, human SNPs are not 'neutral' [13,16–21]; there is a strong and highly significant bias towards non-coding SNPs among individual genes, groups of genes, and genome wide (Figure 1), yet the observation is not directional as described by Cairns [24,72] or the result of intelligent design [73]).

B. Figure 2 describes a strong increase in synonymous SNPs compared to their non-synonymous counterparts (approx. 1.5:1) when exons are investigated from CFTR, multiple human genes and across the entire genome. Again, purifying selection does not provide a complete or satisfactory explanation, since a natural selective mechanism would require that during our evolutionary past, DNA has been so inexplicably brittle that just one new non-synonymous SNP per diploid gene routinely led to death or interrupted fertility of an entire human ancestor. The requirement that a single, randomly placed SNP would typically have such an effect on CFTR (or any protein) needs to be carefully interpreted. The number of human exonic SNP positions per gene is comparatively small, and it is not satisfactory to imbue these infrequent polymorphisms with such an overwhelming effect on fitness. Again, an alternative explanation seems to imply that synonymous SNPs were produced (at the time of their formation) in a substantially biased fashion, and at much higher frequencies than their non-synonymous counterparts. In this context, when we analyzed complete genomic sequences from 16 different murine strains (from http://www.sanger.ac.uk/cgibin/modelorgs/mousegenomes/snps.pl), heterozygous positions attributable to very recent SNP formation among congenic murine lines removed from many forms of selection (i.e. variants produced in a "minimally selective" laboratory environment with negligible predatory, pathogenic, reproductive, or environmental pressure), we measured a ratio of 1.6:1 synonymous to non-synonymous substitutions ($p = 8.3 \times 10^{-49}$) ([74] and unpublished results). This observation, as with the human data, is best explained by a strong bias favoring synonymous SNPs at the time of formation.

C. The data in Figure 4 indicate that positions of human exonic SNPs strongly resemble the corresponding sites of polymorphism among numerous evolutionarily distant species. A classical interpretation that this represents selective removal of the same detrimental point mutations across human and multiple other genomes does not account for the findings, in part because the conserved SNPs are predominantly synonymous. Instead, the results appear to suggest that SNPs across many species have been produced in a fashion that is more biased (or constrained) than classically appreciated. The observation again applies to individual eukaryotic loci such as CFTR, and a survey representing larger cohorts of genes. Examples of **specific mechanisms** that could account for pathways of this type are described later in this report.

D. Note that few (if any) studies have considered the possibility that modern human SNP patterns might depend more on the ways mutations were originally produced than the extent to

A.	Human Base Position	Human SNP	Rat	Mouse	Dog	Opossum	Chicken	Frog
	Non-Syn-593	C/T	C	C	C	C	C	C
	Non-Syn-1125	A/C	A	G	A	A	G	G
	Non-Syn-1408	G/A	C	T	A	A	T	A
	Syn-2562	T/G	G	T	T	T	C	A
	Syn-3177	A/G	A	A	A	A	A	G
	Syn-3870	A/G	A	A	A	A	T	A
	Non-Syn-3983	T/C	T	T	T	T	T	T
	Non-Syn-4121	C/G	C	C	T	T	T	A
	Syn-4389	G/A	A	A	A	A	A	A

Note: CFTR exhibits an overall 48.5% concordance across 6 non human species

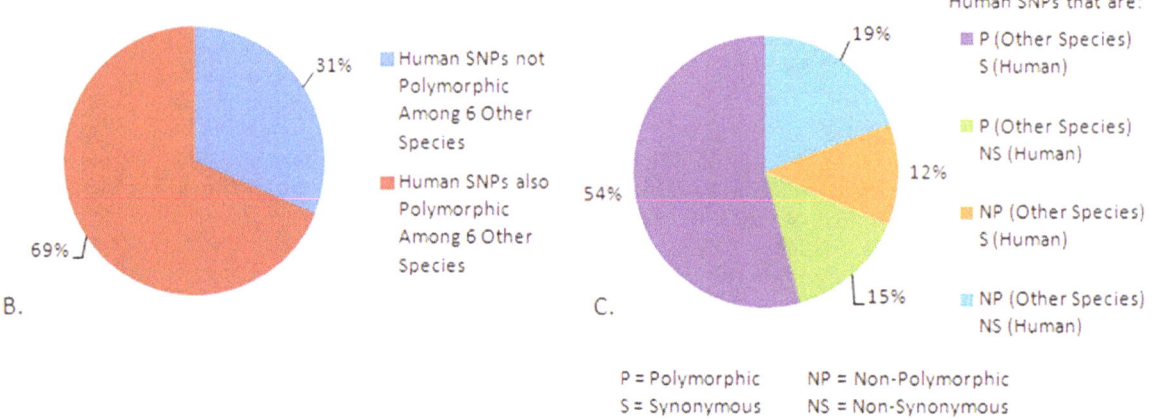

B.

Human SNPs not Polymorphic Among 6 Other Species — 31%

Human SNPs also Polymorphic Among 6 Other Species — 69%

C.

Human SNPs that are:

P (Other Species) S (Human) — 54%

P (Other Species) NS (Human) — 15%

NP (Other Species) S (Human) — 12%

NP (Other Species) NS (Human) — 19%

P = Polymorphic   NP = Non-Polymorphic
S = Synonymous   NS = Non-Synonymous

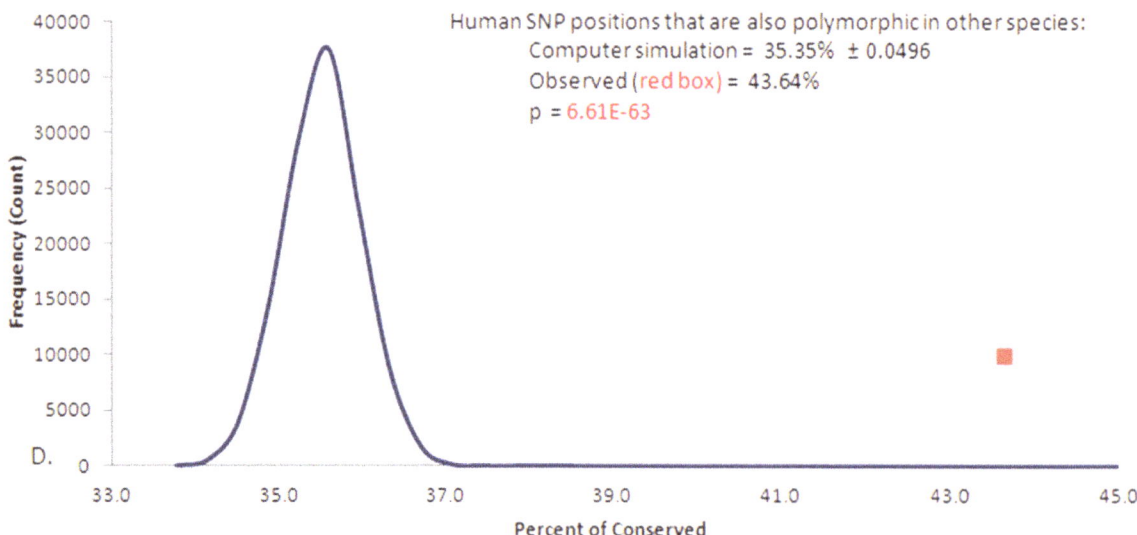

Human SNP positions that are also polymorphic in other species:
Computer simulation = 35.35% ± 0.0496
Observed (red box) = 43.64%
p = 6.61E-63

D.

Figure 4. Positions exhibiting polymorphism in human CFTR are also polymorphic among other species. A: Six of nine CFTR coding SNPs identified by unbiased analysis of individuals in 1000 Genomes were also were polymorphic among diverse species, despite approximately 50% overall nucleotide identity among the non-human CFTRs being analyzed. B, C: CFTR and 21 other genes (Table S3) were investigated in the same fashion shown in Panel A. The majority of SNPs in exonic regions found to be polymorphic were synonymous ($p = 2.7 \times 10^{-9}$, versus the stochastic ratio otherwise expected for non-synonymous to synonymous polymorphism). In order to increase stringency, only those genes in Figure 2A with ≥ 50% concordance across the six non-human species were included in the analysis. D. CFTR homologs in four evolutionarily distant species (horse, frog, zebrafish, and shark) were aligned with the human coding strand, both independently and collectively. In the collective alignment, ~43% of the coding sequence was invariant. A computer simulation was conducted and the total number of differences from human placed randomly within the human CFTR reading frame of 4443 bp. The goal was to determine in a conservative fashion whether concordance observed in a multiple species alignment could be accounted for by chance. The simulation was performed 120,000 times and the numbers of differences from human tabulated. The mean concordance (35.4%) and standard deviation (~0.05) for this set of simulation data was calculated and differed significantly from the higher level of identity observed in nature for the multiple species alignment ($p = 6.6 \times 10^{-63}$).

which they have subsequently been selected or randomly fixed. SNP formation bias is typically neglected by current models of both exonic and protein evolution. Yet our data suggest that hominid SNP *formation* has been much more frequent (on a per nucleotide basis) in non-coding (compared to coding) DNA, and that exonic SNPs are far more likely to be created as synonymous (versus non-synonymous) variants.

E. Finally, data in other Figures and Tables of the report (and throughout the text) offer a human genomic context for the well-described mutational "ratchet" scenario contemplated years ago by Muller and colleagues. One can argue that over eons of eukaryotic evolution, a mutational burden should (by now) have decimated CFTR and other diploid genes. Yet despite hundreds of millions of years of ongoing mutation accrual in core metabolic proteins, the human genome does not appear to be anywhere near "meltdown." The notion that spectacular longevity of higher eukaryotic DNA is somehow accounted for by selective elimination of detrimental SNPs does not adequately explain the findings. Although it is clear that individuals (and entire species) are constantly being expunged from the global gene pool, this would have no effect (and would not reverse) ongoing SNP accumulation in the genes or genomes of all surviving individuals (and species) whose burden of DNA mutation is not resolved simply by "weeding out" of others. SNP accrual over countless generations might be expected to leave every allele riddled with polymorphism (regardless of negative selection), and eliminate sequences that could otherwise recombine to restore a functional protein. This dilemma has not been fully considered in light of modern functional genomics; computer simulation indicates "mutational meltdown" should have occurred millions of years ago.

An alternative hypothesis

Perhaps SNP generation over the evolutionary timescale has been fundamentally less random than we typically assume. Under this arrangement, the major (high MAF) human SNPs in HapMap and 1000 Genomes would have appeared in a fashion configured to prevent overwhelming genomic attrition (for example, with SNP formation directed towards noncoding DNA and synonymous polymorphism). Such a model would help reconcile observations that human SNP populations (both synonymous and non-synonymous) are non-random (and non-neutral) in distribution (Figures 2, 3, 4, Figure S1), yet are not well explained by purifying selection (Figures 1–3; and arguments above regarding the plasticity of human genes, fitness effects necessary to account for the prevalence of non-synonymous variants, measured frequency of synonymous SNPs in human DNA, conserved silent DNA variants among multiple species, magnitude of damage to an entire organism that would be required to purge human SNPs from coding DNA, etc.) [11,23,76–78]. Moreover, we found no evidence for SNP 'directionality' as an adaptive mechanism in specific genes, or 'intelligent design' suggested by others [24,72,73,75]. The same SNP patterns were observed across numerous genes and protein functional categories, and the DNA changes were strongly synonymous. In addition, 'directional' mutations (suggested to enhance fitness) would not reverse DNA attrition, which is solely a consequence of mutation rate and time.

Implications: A mechanistic perspective

While "non-randomness" or "formation bias" could be taken prosaically to imply large numbers of irrelevant SNP "hot spots" or other physical factors that contribute to SNP accumulation, we suggest that survival of DNA and establishment of genomic polymorphism are so crucial they might not be left to chance alone [75,79–81]. For example, note that transition mutations within human ancestral alleles are very strongly favored (Tables 1 and 2), and this bias results in an exon conserving effect throughout the genome. We take the transition bias, by itself, as compelling evidence for a robust **mechanism** that opposes genetic dissipation and dictates patterns of SNP accrual that are allowable, since random replacements by transition nucleotides (as opposed to transversions) at the 3rd codon position overwhelmingly (by 94%) favor synonymous substitution. Moreover, a transition at any codon position confers a bias towards both synonymous and conservative amino acid replacement (Tables 4–5, see also [82,83]). Because the genetic code predated both eukaryotic exons and introns, these findings point to DNA transition bias as a well-organized device that evolved to favor a specific type of DNA variant. The mechanism would act to preserve crucial DNA coding sequences and delimit the types of SNPs and protein polymorphisms most likely to occur.

In unpublished studies, we recently compiled a genomic analysis of sixteen distinct strains of *Mus musculus* (http://www.sanger.ac.uk/cgi-bin/modelorgs/mousegenomes/snps.pl) [74]. We found that high prevalence DNA motifs surrounding single nucleotide polymorphisms were markedly underutilized by mammalian anticodons (Plyler, *et al.*, manuscript submitted). Because the same SNP promoting elements were otherwise conserved across murine exons, introns, and intergenic regions, this result could not manageably be attributed to ongoing purifying selection at the level of protein function. Instead, the data suggested an exon-sparing **mechanism** that regulates SNPs at the time of their formation and serves to minimize mutations within exonic DNA. The pathway was best interpreted as another device that co-evolved with both the genetic code and codon usage to help preserve the exome of higher organisms.

Intronic DNA as an exon-sparing mechanism

In contrast to exonic DNA, significant numbers of intronic and intergenic SNPs are already known to be generated in a constrained and arguably predictable manner based on transition bias, proximity to DNA recombination sites, nucleosome structure, GC rich isochores, or specific sequence contexts [10,76–82]. Non-coding SNPs provide genomic and phenotypic variation (through modification of crucial regulatory elements, microRNAs, expressed noncoding sequences, etc.) without the need for substantial transmutation of the open reading frames. Based on the analysis presented here and the vital imperative to avert "meltdown" of protein coding DNA, we speculate that the introme, itself, might not only serve as an evolutionary strategy to support the generation of diversity (through alternative splice variants, microRNAs, gene network regulation, etc.), but as a specific alternative to meddling with the exons. Non-coding DNA in this model would appear quite expendable for a given species over a few generations (as true for CF mice, innumerable other transgenic animals in which non-coding DNA has been disrupted to delete, insert, repair, or select for gene modifications, or animals in which

Table 4. Computer Simulation of SNP Accrual in the Setting of a Transition Bias Leads to Enhancement of Synonymous Variants.

Imposed Substitution Bias	Sequence	No. Runs	Muts/Run	Resulting N:S Ratio	P-Value vs. CFTR unbiased
Unbiased	CFTR	10	50	3.37	----
	CFTR GC-RICH	10	10	4.12	0.018
CFTR Mutation Database Derived Transition Bias (See Table 2)	CFTR	10	50	2.79	7.27 E-81
	CFTR GC-RICH	10	10	3.04	1.29 E-31
Exon Derived Transition Bias (See Table 1)	CFTR	10	50	2.39	1.74 E-222
	CFTR GC-RICH	10	10	2.16	2.74 E-29
Intron Derived Transition Bias (See Table 1)	CFTR	10	50	2.64	1.40 E-121
	CFTR GC-RICH	10	10	2.02	9.50 E-21

SNPs were placed randomly at computer-generated positions in the full-length CFTR sequence, or in a GC-rich region (150 base pair interval (4260–4409) of the human CFTR open reading frame) in an unbiased fashion, or with a transition bias according to the CFTR mutation database (see Table 2), or transition bias observed for either exonic or intronic SNPs from 1000 Genomes (Table 1). GC rich isochores are reported to be more likely sites of natural mutation. The ratio of resulting non-synonymous (N) to synonymous (S) SNPs is shown. The data indicates strong preference for synonymous variants in the setting of transition bias, although magnitude of the effect does not fully account for enhancement of synonymous SNPs shown in Figure 2. Transition bias may therefore represent one (perhaps among several) evolutionary mechanisms serving to augment formation of synonymous DNA polymorphism.

large expanses of the non-coding genome have been intentionally omitted without phenotypic effect [83]), but would be essential for selfish genomes attempting to cope with environmental challenge over an evolutionary timeframe. If the noncoding compartment originated in part as a strategy that helps DNA more safely 'experiment' with its own diversity (without running the risk of "meltdown"), the metabolic expense of intronic elements might be partly justified on that basis alone. In principle, by setting conservative limits regarding: 1) the range of single nucleotide mutation rate, 2) extent to which random coding mutations are

expected to disrupt protein function, and 3) likelihood that SNPs in non-coding DNA would alter gene expression, simulations to evaluate this hypothesis might be more formally undertaken in the future.

Relevance to population-type analysis

The results presented in this report should be viewed in light of earlier, population studies regarding natural selection and the consequent distribution of SNP fitness effects. For example, our finding of synonymous SNP enrichment agrees with previous data

Table 5. Computer Simulation of SNP Accrual in the Setting of a Transition Bias Leads to Enhancement of Conservative Mutations.

Imposed Substitution Bias	Sequence	No. Runs	Muts/Run	Resulting Ncon:Con	P-Value vs. Corresponding unbiased substitution
Unbiased	Artificial Sequence	10	50	2.95	----
	CFTR	10	50	1.93	----
	CFTR GC-RICH	10	10	1.75	----
CFTR Mutation Database Derived Transition Bias (See Table 2)	Artificial Sequence	10	50	2.8	2.46 E-15
	CFTR	10	50	2.04	----
	CFTR GC-RICH	10	10	1.56	2.89 E-19
Exon Derived Transition Bias (See Table 1)	Artificial Sequence	10	50	2.52	5.96 E-58
	CFTR	10	50	1.53	4.31 E-85
	CFTR GC-RICH	10	10	1.57	3.45 E-29
Intron Derived Transition Bias (See Table 1)	Artificial Sequence	10	50	2.25	8.40 E-32
	CFTR	10	50	1.99	----
	CFTR GC-RICH	10	10	1.17	6.49 E-18

SNPs were stochastically placed in 1) an artificial, assembled gene containing 1480 codons arranged randomly (i.e. random codons were used to generate a 4440 bp sequence), 2) the CFTR coding sequence (1480 codons), or 3) a GC-rich region of CFTR. The computer-generated positions to be mutated were selected randomly, and the choice of base replacement (e.g. with or without a particular transition bias) derived as above, according to the CFTR mutation database (Table 2), or rates observed for exonic or intronic SNPs (Table 1). The ratios for non-conservative (Ncon) to conservative (Con) SNPs are shown. Table 5 is the result of 10 simulation runs per sequence, indicating significant differences even after small numbers of SNP incorporation.

in *Drosophila*, human, and other species, including seminal observations from Kreitman and Colleagues [35–37,61,84–85]. On the other hand, earlier work applying data of this sort to a Poisson-type random field model in human populations has led to conclusions different from those described here [36,84,86]. For example, Boyko and colleagues [84] investigated genomic SNP ratios and reported a synonymous:non-synonymous frequency of 1.38 among Caucasians (very similar to the value of 1.3–1.6 shown in Figure 2). When a mathematical treatment was conducted (based in part on earlier quantitative work in *Drosophila* [87] and human [88]), the data fit best to a distribution in which approximately 30% of random, non-synonymous single base replacements were suggested to be highly deleterious ([84]; a finding quite different from our analysis). A study by the same group used similar quantitative methods to test natural selection among conserved non-coding elements, and concluded that synonymous SNPs in human genes are under strong (recent) positive selective pressure [86]. We note that either of these interpretations might be significantly influenced by a preference towards synonymous SNP formation. Although Poisson field experiments have utilized the best available demographic parameters, such studies require a neutral set of variants (typically synonymous SNPs) against which evolution of specific genes, genetic elements, or proteins at selected sites can be ratioed or otherwise compared [89]. Our findings suggest a strong non-neutral component during the production of new synonymous or other SNPs. We therefore believe future analyses might benefit from more prominently considering the contribution of synonymous SNP formation bias as described by the present report.

In this study, we also considered more classical, population-based analysis of natural selection utilizing McDonald-Kreitman methodology to test human CFTR variants from over 1,000 individuals and numerous ethnicities (1000 Genomes), and compared this data to an outgroup representing the most recent chimpanzee sequence (CHIMP 2.1.4) The proportion of base substitutions fixed in human versus differences between human and chimp indicated $\alpha = (-)$ 2.72, with neutrality index of 3.72 (uncorrected p value = 0.07). Based on classical interpretation, our findings could be taken to suggest a trend towards positive CFTR selection with non-neutral divergence for the human CF gene. On the other hand, evidence presented here indicates synonymous SNP formation is often non-neutral due to features such as transition bias. This aspect complicates a conventional assessment of selection intensity, which (as above) typically requires a well-defined cohort of random and fitness neutral (e.g. synonymous) SNPs for comparison to non-synonymous variants. For example, we show that CFTR (like many other genes) exhibits discrete exonic regions with high CpG content (Tables 4–5). Such coding intervals represent sites for augmented DNA methylation, increased transition bias and enhanced synonymous SNP formation. Awareness of such domains and their quantitative significance may influence the interpretation of selection intensity, including gene segments believed to undergo rapid evolution, which are otherwise predicated on largely random (and neutral) formation of non-coding or synonymous SNPs.

Concluding remarks: an adaptation to enhance adaptability

In this report, we interpret proteins and their constituent exons as 'lessons' of inestimable value picked up from iterative attempts at long term DNA survival. We maintain that despite the need for variation, lessons such as these should be viewed as far too valuable to expend—particularly when alternatives such as non-coding DNA might be utilized instead. We suggest that SNP formation in CFTR and other human genes appears configured to help preserve exons (e.g. with the majority of SNPs adhering to a set of rules that strongly bias their formation as intronic, synonymous, contextual, transitional etc.), constrained by specific molecular mechanisms such as those involving transition bias and codon usage, and should therefore be viewed as meaningful and adaptive. A precedent for the sort of adaptation proposed here has already been described for a different evolutionary mechanism, the combinatorial immune system, where antediluvian trial and error turned up a highly organized means of generating extensive diversity that is capable of responding to infectious agents not yet encountered by a species. We note that the adaptive immune response provides remarkable phenotypic variation that is regulated, has worked well through the ages, is readily explained without evoking directional evolution [24,72] or intelligent design [73], and is undoubtedly based on multiple earlier prototypes that failed or were much less effective. In the same fashion, we suggest that adaptive pathways have evolved to help regulate DNA diversification. Such mechanisms could have appeared, for example, after countless failed attempts at long term DNA survival that ended in unchecked SNP accumulation and genomic meltdown.

In summary, more attention should to be paid to ways in which production of human DNA polymorphism (i.e. at the time of SNP formation) is organized. A review of quotidian sequence data provided throughout this report indicates that genome-wide SNP patterns should be evaluated in light of long-term DNA survival and modern knowledge regarding protein plasticity. If HapMap, 1000 Genomes, and related projects in other species increasingly reveal predictable or mechanistically relevant patterns of SNP distribution as contemplated by the present study, evolution might be viewed as more regulated than has been classically interpreted. This report therefore suggests the existence of mechanisms by which DNA may regulate its own diversity, safeguard the hard-earned lessons encoded by genes, and help guide its own evolutionary path.

Materials and Methods

Exonic and synonymous SNP frequencies identified by 1000 Genomes

SNPs in 133 human genes known to be lethal or severely debilitating when deleted [90] or 4857 human genes for which intron/exon boundaries are readily definable in the Exon-Intron Database (http://www.utoledo.edu/med/depts/bioinfo/database.html), and a 1000 Genomes release (http://pilotbrowser.1000genomes.org/index.html; 6 individuals of European or African descent) were evaluated. The length of each coding sequence and the combined lengths of the intronic sequences (Exon-Intron Database), together with SNP information (1000 Genomes) were obtained. SNP totals were normalized to numbers of exonic or intronic nucleotides (exonic and intronic SNP percentages, respectively).

For evaluation of synonymous versus non-synonymous mutations, two datasets were used: SNPs in 98 of the 133 genes described above with at least one exonic variant, and 13,820 genes for which data was accessible (from http://www.utoledo.edu/med/depts/bioinfo/database.html and 1000 Genomes). Synonymous and non-synonymous SNPs were normalized to total exonic nucleotide content as above.

As a test of SNP authenticity, we manually inspected a random population of 200 coding and non-coding variants selected from 1000 Genomes using Interactive Genome Viewer (IGV) software. We tested these for features shown previously to indicate sequencing error, and found less than 2–3% of SNPs exhibited low quality score, inconsistent consensus, misalignment, artifactually high "coverage" ("pile-up") due to homologous sequences elsewhere in the genome, duplicated reads, indels, short local repeats, etc. This result provided independent confirmation for robustness of SNP data available from the 1000 Genomes resource.

HapMap based identification of minor allelic frequencies for CFTR and NF1 across four major ethnicities

Frequency data for SNPs in CFTR or NF1 were collated for each of four ethnicities: JPT (Japanese in Tokyo, 45 individuals); CHB (Han Chinese in Beijing, 45 individuals); CEU (or CEPH, Utah residents with ancestry from northern and eastern Europe, 90 individuals); and YRI (Yoruba in Ibidan, Nigeria, 90 individuals) using HapMap (http://hapmap.ncbi.nlm.nih.gov/). Minor allelic frequency (MAF) refers to the relative frequency (1000 = 100% incidence) of the minor allele at each SNP position. We focused our attention on SNPs verified by HapMap and with minor allelic frequency greater than zero in at least one population (i.e. at least one individual among 270 in the database exhibited the minor SNP). This convention allowed us to use HapMap as a less biased tool for cataloging genomic variation among very diverse individuals. In order to generate a MAF block diagram, allele frequency data was downloaded from the UCSC genome table browser (http://genome.ucsc.edu/cgi-bin/hg:tables). After downloading, SNPs with MAF equal to zero among all four ethnicities were omitted to enhance stringency. The remaining SNPs were then utilized to generate scatter plots.

Analysis of transition SNPs and Y chromosomal SNPs in human genes

SNP configurations (transition and transversion) for intronic and exonic regions using the set of 98 human genes described above with at least one exonic SNP were collated. Frequencies of CFTR SNP subtypes (transition vs. transversion) among >1000 SNPs, many of which have been implicated in clinical CF, were also tallied (http://www.genet.sickkids.on.ca/cftr/app). SNP totals for the Y and other chromosomes were obtained directly from dbSNP, HapMap, and 1000 Genomes, and presented in tabular form.

Simulation of mutation accrual in the setting of a transition bias

SNPs were placed at computer-generated positions in 1) the full-length CFTR sequence, 2) a GC-rich region of CFTR (150 base pair interval (4260–4409) of the open reading frame), or 3) an artificial, assembled gene containing 1480 codons arranged randomly (i.e. random codons to generate a 4440 bp sequence).

The site for each base substitution was determined by a random number function that served as an index into an array representing the desired base substitution ratio. For example, an unbiased base substitution array for the replacement of adenine contained cytosine, guanine and thymine in equal proportions, so that these were chosen with equal probability to replace adenine. The base substitution ratios were:

1. Unbiased – each of the four bases was replaced with equal probability by any of the other three.
2. CFTR Mutation Database – All entries in this database (http://www.genet.sickkids.on.ca/cftr/app) describe directional substitution, such as A to C, and were used to calculate base substitution frequencies.

3. Exon or Intron Base Substitution Ratios – were obtained as above from exons of 98 human genes or CFTR introns. Because the directionality of mutation is often not known (i.e. for a SNP encoding A or T, it is unknown whether A→T or T→A), and since 1) the directionality is not relevant to this particular test of transition bias, and 2) there is clearly a transition bias regardless of directionality judged by studies of more recent mutations in CFTR, the incidences of both directions were analyzed together.

Each simulation run imposed 10–50 base replacements on the human CFTR coding sequence or the artificial sequence, and 10 runs were conducted for each condition. In all instances, a site (base) was allowed to mutate more than one time, although in practice this seldom occurred. Due to the comparatively short length of the GC-rich region, 10 mutated sites were tested per simulation. Results from simulations were determined following translation and alignment with the authentic sequence.

A modified version of the above algorithm was established to model the long term consequence of mutation accrual in CFTR. Similarity to an original CFTR sequence was evaluated by translation of the resulting cDNA after several thousand generations in order to evaluate protein integrity.

Code for the above simulations was designed as follows:

1. A sequence related to CFTR as described above was represented as a character array (DNA bases: A, C, G, T).
2. Base replacement arrays were populated to reflect the particular substitution bias being simulated. For example, Table 2 lists 1036 CFTR SNPs found to be associated with human disease. From these, replacement frequency for each of the possible substitutions was calculated, and used to model SNP accrual as described below.
3. The particular position to be mutated was selected by a random number within the range of the length of the sequence to be mutated. This random number was used as an index into the array from Step 1.
4. The sequence to be mutated was read as an array using the random number from Step 3. The base at that position was determined.
5. For example, if position 5 of the sequence to be mutated contained an A, the A-substitution array (Step 2) would be consulted to determine the replacement base. A random number determined which of the three bases would replace the A. The random number was used to index the A-substitution array created and populated in Step 2.
6. The position in the sequence to be mutated randomly selected in Step 3 was overwritten with the replacement base from Step 5.
7. When the desired number of simulated mutations had been completed, the mutated sequence codons were compared with corresponding original sequence to determine the numbers of synonymous or non-synonymous mutations and (for the non-synonymous mutations) the numbers of conservative versus non-conservative amino acid replacements.

SNP concordance among evolutionarily distant species

CFTR and 21 other genes with ≥50% concordance across six non-human species (rat, mouse, dog, opossum, chicken, and frog) were obtained from the UCSC genome browser. Sequences of the non-human species were aligned using the ClustalW2 tool (http://www.ebi.ac.uk/Tools/clustalw2/index.html). The number of bases identical among the non-human species was divided by the largest total number of bases in any species to obtain the most conservative measure of percent concordance. Locations of exonic human SNPs in each gene were obtained from 1000 Genomes. Following alignment of all six non-human species, the variance among non-human genes was examined using human SNPs as a marker. For example, after observing a SNP in human CFTR at nucleotide position 593 and the corresponding position in six other species, a determination was made as to whether or not the non-human species also exhibited polymorphism at that site.

In addition, CFTR homologs in four evolutionarily distant species (horse, frog, zebrafish, and shark) were aligned with the human coding strand, both independently and collectively. In the collective alignment, ~43% of the coding sequence was invariant. A computer simulation was conducted and the total number of differences from human placed randomly within the human CFTR reading frame of 4443 bp. This simulation was performed 120,000 times, and the total number of differences from human (distributed as random point mutations) was tabulated. The goal was to determine whether concordance observed in a multiple species alignment could be accounted for by chance.

Statistical analysis

Comparison for SNP frequencies (exonic versus intronic, synonymous versus non-synonymous, transition versus transversion, observed versus predicted computer simulation frequencies, etc.) were conducted by χ^2 and contingency table analysis (2×2 tables employing Yates' correction for continuity).

Supporting Information

Figure S1   Minor allelic frequencies of prominent CFTR SNPs. The heat map emphasizes findings shown in Figure 3A; i.e. much higher minor allele frequencies (MAFs) among SNPs in CFTR for CEU, CHB, and JPT, compared with YRI. When MAFs were plotted against physical location, haplotype blocks were evident in three of four ethnicities, and underrepresented in the 'back' half of CFTR. The observation is attributable to a greater number of YRI ancestral haplotypes, additional cross-over

events among YRI, higher numbers of SNPs formed since founding of YRI, or some combination of these factors. The YRI ethnicity exhibited no MAF block structure or positional enhancement of variation, in contrast to other ethnic groups.

Table S1   SNP incidence in human intronic and exonic DNA. SNPs in 133 human genes known to be lethal or severely debilitating if deleted [90].

Table S2   Synonymous and non-synonymous SNP incidence. Exonic SNPs in 98 of the 133 genes shown in Figure 1A, specifically those containing at least one exonic SNP.

Table S3   Relationship between human SNPs and variable nucleotide positions of non-human species. From among 98 genes described in Figure 2A and Table S2, those with at least 50% concordance among six nonhuman species (rat, mouse, dog, opossum, chicken, frog) were selected for further analysis (22 genes total). The number and type of exonic SNPs in human (from 1000 Genomes) and correspondence with known polymorphic regions in other vertebrate species is shown.

Acknowledgments

The authors thank Drs. Hughes Evans, Bruce Korf, John Hartman, Elliot Lefkowitz, Michael Crowley, Robert Fischer, and Tom Broker for useful discussion and reviewing the manuscript. We are grateful to Ms. Jenny Mott, Ms. Jan Tindall, and Ms. Cheryl Owens for help preparing the paper. We also acknowledge the HapMap Consortium (The International HapMap Consortium. The International HapMap Project. *Nature* 426, 789–796 (2003)).

Author Contributions

Conceived and designed the experiments: AH ZP ES. Performed the experiments: AH ZP JT CM LM. Analyzed the data: AH ZP HT AP CM ES. Contributed reagents/materials/analysis tools: AH HT LM ES. Wrote the paper: AH ZP ES. Designed the software used in analysis: AH JT LM.

References

1. Gabriel W, Lynch M, Bürger R (1993) Muller's Ratchet and mutational meltdowns. Evolution 47: 1744–1757.
2. Haldane JBS (1937) The effect of variation on fitness. The American Naturalist LXXI(735):337–49.
3. Kimura M (1968) Evolutionary rate at the molecular level. Nature 217(5129):624–6.
4. Lynch M (1990) Mutation load and the survival of small populations. Evolution 44(7): 1725–37.
5. Muller HJ (1964) The relation of recombination to mutational advance. Mutat Res 106:2–9.
6. Sung W, Ackerman MS, Miller SF, Doak TG, Lynch M (2013) Drift does influence mutation-rate evolution. Proc Natl Acad Sci 2013;110(10):E860.
7. Sung W, Ackerman MS, Miller SF, Doak TG, Lynch M (2012) Drift-barrier hypothesis and mutation-rate evolution. Proc Natl Acad Sci 109(45):18488–92.
8. Lynch M (2010) Evolution of the mutation rate. Trends Genet 26(8):345–52.
9. Koonin EV (2012) The Logic of Chance – The natures and origin of biological evolution. New Jersey: FT Press. 516 p.
10. Jobling MA, Hurles ME, Tyler-Smith C (2004) Human evolutionary genetics: origins, peoples, and disease. Abingdon and New York: Garland Science. 458 p.
11. Kreitman M (1996) The natural theory is dead: Long live the neutral theory. BioEssays 18:678.
12. Rowe SM, Miller S, Sorscher EJ (2005) Cystic fibrosis. N Engl J Med 352(19):1992–2001.
13. Havasi V, Rowe SM, Kolettis PN, Dayangac D, Sahin A, et al. (2010) Association of cystic fibrosis genetic modifiers with congenital bilateral absence of the vas deferens. Fertil Steril 94(6):2122–7.
14. Gabriel SE, Brigman KN, Koller BH, Boucher RC, Stutts MJ (1994) Cystic fibrosis heterozygote resistance to cholera toxin in the cystic fibrosis mouse model. Science 266(5182):107–9.

15. Aeffner F, Abdulrahman B, Hickman-Davis JM, Janssen PM, Amer A, et al. (2013) Heterozygosity for the F508del Mutation in the cystic fibrosis transmembrane conductance regulator anion channel attenuates influenza severity. J Infect Dis 208(5):780–9.
16. Zhou L, Dev CR, Wert SE, DuVall MD, Frizzell RA, et al. (1994) Correction of lethal intestinal defect in a mouse model of cystic fibrosis by human CFTR. Science 266(5191):1705–8.
17. Tuggle KL, Birket SE, Cui X, Hong J, Warren J, et al. (2014) Characterization of defects in ion transport and tissue development in cystic fibrosis transmembrane conductance regulator (CFTR)-knockout rats. PLoS One 9(3):e91253.
18. Sun X, Sui H, Fisher JT, Yan Z, Liu X, et al. (2010) Disease phenotype of a ferret CFTR-knockout model of cystic fibrosis. J Clin Invest 120(9):3149–60.
19. Rogers CS, Stoltz DA, Meyerholz DK, Ostedgaard LS, Rokhlina T, et al. (2008) Disruption of the CFTR gene produces a model of cystic fibrosis in newborn pigs. Science 321(5897):1837–41.
20. Ostedgaard LS, Meyerholz DK, Vermeer DW, Karp PH, Schneider L, et al. (2011) Cystic fibrosis transmembrane conductance regulator with a shortened R domain rescues the intestinal phenotype of CFTR −/− mice. Proc Natl Acad Sci U S A 108(Abst 7).
21. Kimura M (1979) Model of effectively neutral mutations in which selective constraint is incorporated. Proc Natl Acad Sci U S A 76(7):3440–4.
22. McDonald JH, Kreitman M (1991) Adaptive protein evolution at the Adh locus in Drosophila. Nature 351(6328):652–4.
23. Hahn MW (2008) Toward a selection theory of molecular evolution. Evolution 62:255–65.
24. Cairns J, Overbaugh J, Miller S (1988) The origin of mutants. Nature 335.142–145.
25. Chen IP, Tang CY, Chiou CY, Hsu JH, Wei NV, et al. (2009) Comparative analyses of coding and noncoding DNA regions indicate that Acropora

(Anthozoa: Scleractina) possesses a similar evolutionary tempo of nuclear vs. mitochondrial genomes as in plants. Mar Biotechnol (NY) 11(1):141–52.

26. Drake JA, Bird C, Nemesh J, Thomas DJ, Newton-Cheh C, et al. (2006) Conserved noncoding sequences are selectively constrained and not mutation cold spots. Nat Genet 38(2):223–7.

27. Andolfatto P (2005) Adaptive evolution of non-coding DNA in Drosophila. Nature 437(7062):1149–52.

28. Kumar S (2005) Molecular clocks: four decades of evolution. Nat Rev Genet 6(8):654–62.

29. Thomson R, Pritchard JK, Shen P, Oefner PJ, Feldman MW (2000) Recent common ancestry of human Y chromosomes: evidence from DNA sequence data. Proc Natl Acad Sci U S A 97(13):7360–5.

30. Clark AG, Glanowski S, Nielsen R, Thomas PD, Kejariwal A, et al. (2003) Inferring nonneutral evolution from human-chimp-mouse orthologous gene trios. Science 302(5652):1960–3.

31. Maurano M, Humbert R, Rynes E, Thurman R, Haugen E, et al. (2012) Systematic localization of common disease-associated variation in regulatory DNA. Science 337(6099):1190–1195.

32. Pennisi E (2012) ENCODE project writes eulogy for junk DNA. Science 337(6099):1159–1161.

33. Ward L, Kellis M (2012) Evidence of abundant purifying selection in humans for recently acquired regulatory functions. Science 337(6102):1675–8.

34. Ley TJ, Mardis ER, Ding L, Fulton B, McLellan MD, et al. (2008) DNA sequencing of a cytogenetically normal acute myeloid leukaemia genome. Nature 456(7218):66–72.

35. Hinds DA, Stuve LL, Nilsen GB, Halperin E, Eskin E, et al. (2005) Whole-genome patterns of common DNA variation in three human populations. Science 307(5712):1072–9.

36. Bustamante CD, Fledel-Alon A, Williamson S, Nielsen R, Hubisz MT, et al. (2005) Natural selection on protein-coding genes in the human genome. Nature 437(7062):1153–7.

37. Berglund J, Pollard KS, Webster MT (2009) Hotspots of biased nucleotide substitutions in human genes. PLoS Biol 7(1):e26.

38. Doron-Faigenboim A, Pupko T (2007) A combined empirical and mechanistic codon model. Mol Biol Evol 24(2):388–97.

39. Keightley PD, Trivedi U, Thomson M, Oliver F, Kumar S, et al. (2009) Analysis of the genome sequence of three *Drosophila melanogaster* spontaneous mutation accumulation lines. Genome Res 19(7):1195–201.

40. McClellan DA, Whiting DG, Christensen R, Sailsbery J (2004) Genetic codes as evolutionary filters: subtle differences in the structure of genetic codes result in significant differences in patterns of nucleotide substitution. J Theor Biol 226(4):393–400.

41. Petrov DA, Hartl DL (1999) Patterns of nucleotide substitution in Drosophila and mammalian genomes. Proc Natl Acad Sci U S A 96(4):1475–9.

42. McClellan J, King MC (2010) Genetic heterogeneity in human disease. Cell 141(2):210–7.

43. Sachidanandam R, Weissman D, Schmidt SC, Kakol JM, Stein LD, et al. (2001) A map of human genome sequence variation containing 1.42 million single nucleotide polymorphisms. Nature 409(6822):928–33.

44. Mitchell RJ, Hammer MF (1996) Human evolution and the Y chromosome. Curr Opin Genet Dev 6(6):737–42.

45. Hammer MF (1995) A recent common ancestry for human Y chromosomes. Nature 378(6555):376–8.

46. Jobling MA, Tyler-Smith C (2003) The human Y chromosome: an evolutionary marker comes of age. Nat Rev Genet 4(8):598–612.

47. Graves JA (2006) Sex chromosome specialization and degeneration in mammals. Cell 124(5):901–14.

48. Wilson Sayres MA, Lohmueller KE, Nielsen R (2014) Natural selection reduced diversity on human Y chromosomes. PLoS Genet 10(1): e1004064.

49. Goebel T (2007) Anthropology. The missing years for modern humans. Science 315(5809):194–6.

50. Reich DE, Schaffner SF, Daly MJ, McVean G, Mullikin JC, et al. (2002) Human genome sequence variation and the influence of gene history, mutation and recombination. Nat Genet 32(1):135–42.

51. Kondrashov AS (2008) Another step toward quantifying spontaneous mutation. Proc Natl Acad Sci U S A 105(27):9133–4.

52. Xue Y, Wang Q, Long Q, Ng BL, Swerdlow H, et al. (2009) Human Y chromosome base-substitution mutation rate measured by direct sequencing in a deep-rooting pedigree. Curr Biol 19(17):1453–7.

53. Kuroki Y, Toyoda A, Noguchi H, Taylor TD, Itoh T, et al. (2006) Comparative analysis of chimpanzee and human Y chromosomes unveils complex evolutionary pathway. Nat Genet 38:158–167.

54. Conrad DF, Keebler JEM, DePristo MA, Lindsay SJ, Zhang Y, et al. (2011) Variations in genome-wide mutation rates within and between human families. Nat Genet 43(7):712–4.

55. Roach JC, Glusman G, Smith AFA, Huff CD, Hubley R, et al. (2010) Analysis of genetic inheritance in a family quartet by whole-genome sequencing. Science 328(5978):636–9.

56. Lynch M (2010) Rate, molecular spectrum, and consequences of human mutation. Proc Natl Acad Sci U S A 107(3):961–8.

57. Nachman MW, Crowell SL (2000) Estimate of the mutation rate per nucleotide in humans. Genetics 156:297–304.

58. Schön I, Martens K (2003) No slave to sex. Proc Biol Sci 270(1517):827–33.

59. Felsenstein J (1974) The evolutionary advantage of recombination. Genetics 78(2):737–56.

60. Tucker AE, Ackerman MS, Eads BD, Xu S, Lynch M (2013) Population-genomic insights into the evolutionary origin and fate of obligately asexual Daphnia pulex. Proc Natl Acad Sci 110(39):15740–5.

61. Kreitman M (1983) Nucleotide polymorphism at the alcohol dehydrogenase locus of *Drosophila melanogaster*. Nature 304:412–417.

62. Drummond DA, Wilke CO (2008) Mistranslation-induced protein misfolding as a dominant constraint on coding-sequence evolution. Cell 134(2):341–52.

63. Hodgkinson A, Eyre-Walker A (2010) The genomic distribution and local context of coincident SNPs in human and chimpanzee. Genome Biol Evol 2:547–57.

64. Mendoza JL, Schmidt A, Li Q, Nuvaga E, Barrett T, et al. (2012) Requirements for efficient correction of ΔF508 CFTR revealed by analyses of evolved sequences. Cell 148(1–2):164–74.

65. Gulyás-Kovács A (2012) Integrated analysis of residue coevolution and protein structure in ABC transporters. PLoS One 7(5):e36546.

66. Csanády L, Vergani P, Gulyás-Kovács A, Gadsby DC (2011) Electrophysiological, biochemical, and bioinformatic methods for studying CFTR channel gating and its regulation. Methods Mol Biol 741:443–69.

67. Rishishwar L, Varghese N, Tyagi E, Harvey SC, Jordan IK, et al. (2012) Relating the disease mutation spectrum to the evolution of the cystic fibrosis transmembrane conductance regulator (CFTR). PLoS One 7(8):e42336.

68. Fernandez A, Lynch M (2011) Non-adaptive origins of interactome complexity. Nature 474(7352):502–5.

69. Lynch M (2010) Scaling expectations for the time to establishment of complex adaptations. Proc Natl Acad Sci 107(38):16577–82.

70. Lynch M (2012) The evolution of multimeric protein assemblages. Mol Biol Evol 29(5):1353–66.

71. Blanchette M, Diallo AB, Green ED, Miller W, Haussler D (2008) Computational reconstruction of ancestral DNA sequences. Methods Mol Biol 422:171–84.

72. Cairns J (1995) Adaptive mutation and sex. Science 269: 289.

73. Meyer SC (2013) Darwin's Doubt: The Explosive Origin of Animal Life and the Case for Intelligent Design. New York: HarperOne. 512 p.

74. Plyler ZE, Hill AE, McAtee CW, Moseley LA, Sorscher EJ (in press) SNP distribution in the murine genome provides evidence for contextual bias and recent parallel evolution.

75. Fullerton SM, Bernardo Carvalho A, Clark AG (2001) Local rates of recombination are positively correlated with GC content in the human genome. Mol Biol Evol 18(6):1139–42.

76. Chen X, Chen Z, Chen H, Su Z, Yang J, et al. (2012) Nucleosomes suppress spontaneous mutations base-specifically in eukaryotes. Science 335(6073):1235–1238.

77. Coop G, Przeworski M (2007) An evolutionary view of human recombination. Nat Rev Genet 8(1):23–34.

78. Hodgkinson A, Ladoukakis E, Eyre-Walker A (2009) Cryptic variation in the human mutation rate. PLoS Biol 7(2):e1000027.

79. Kauppi L, Jeffreys AJ, Keeney S (2004) Where the crossovers are: recombination distributions in mammals. Nat Rev Genet 5(6):413–24.

80. Duret L, Arndt PF (2008) The impact of recombination on nucleotide substitutions in the human genome. PLoS Genet 4(5):e1000071.

81. Lercher MJ, Hurst LD (2002) Human SNP variability and mutation rate are higher in regions of high recombination. Trends Genet 18(7):337–40.

82. Rutherford SL (2003) Between genotype and phenotype: protein chaperones and evolvability. Nat Rev Genet 4(4):263–74.

83. Nobrega MA, Zhu Y, Plajzer-Frick I, Afzal V, Rubin EM (2004) Megabase deletions of gene deserts result in viable mice. Nature 431:988–993.

84. Boyko AR, Williamson SH, Indap AR, Degenhardt JD, Hernandez RD, et al. (2008) Assessing the evolutionary impact of amino acid mutations in the human genome. PLoS Genet 4(5): e1000083.

85. Yang Z, Nielsen R, Goldman N, Pedersen AM (2000) Codon-substitution models for heterogeneous selection pressure at amino acid sites. Genetics 155:431–49.

86. Torgerson DG, Boyko AR, Hernandez RD, Indap A, Hu X, et al. (2009) Evolutionary processes acting on candidate cis-regulatory regions in humans inferred from patterns of polymorphism and divergence. PLoS Genet 5 (8): e1000592.

87. Sawyer SA, Hartl DL (1992) Population genetics of polymorphism and divergence. Genetics 132:1161–1176.

88. Williamson SH, Hernandez R, Fledel-Alon A, Zhu L, Nielson R, et al. (2005) Simultaneous inference of selection and population growth from patterns of variation in the human genome. Proc Natl Acad Sci U S A 102(22):7882–7.

89. Eyre-Walker A, Keightley PD (2007) The distribution of fitness effects of new mutations. Nat Rev Genet 8(8):610–8.

90. Fortini ME, Skupski MP, Boguski MS, Hariharan IK (2000) A survey of human disease gene counterparts in the Drosophila genome. J Cell Biol 150(2):F23–30.

18

Poly-dA:dT Tracts Form an *In Vivo* Nucleosomal Turnstile

Carl G. de Boer[1], Timothy R. Hughes[1,2]*

1 Department of Molecular Genetics, University of Toronto, Toronto, Ontario, Canada, **2** Donnelly Centre for Cellular and Biomolecular Research, University of Toronto, Toronto, Ontario, Canada

Abstract

Nucleosomes regulate many DNA-dependent processes by controlling the accessibility of DNA, and DNA sequences such as the poly-dA:dT element are known to affect nucleosome binding. We demonstrate that poly-dA:dT tracts form an asymmetric barrier to nucleosome movement *in vivo*, mediated by ATP-dependent chromatin remodelers. We theorize that nucleosome transit over poly-A elements is more energetically favourable in one direction, leading to an asymmetric arrangement of nucleosomes around these sequences. We demonstrate that different arrangements of poly-A and poly-T tracts result in very different outcomes for nucleosome occupancy in yeast, mouse, and human, and show that yeast takes advantage of this phenomenon in its promoter architecture.

Editor: Mary Bryk, Texas A&M University, United States of America

Funding: This work was supported by a grant from the Canadian Institutes of Health Research (http://www.cihr-irsc.gc.ca/) to TRH (MOP-111007) and by a Natural Sciences and Engineering Research Council of Canada (http://www.nserc-crsng.gc.ca/) award to CGD (PGS D). The funders had no role in study design, data collection and analysis, decision to publish, or preparation of the manuscript.

Competing Interests: The authors have declared that no competing interests exist.

* Email: t.hughes@utoronto.ca

Introduction

In vivo, promoters are characterized by a nucleosome free region (NFR) that is followed by a periodic phasing of well-positioned nucleosomes continuing into the gene body. In yeast, this phasing is absent *in vitro*, but can be restored by the addition of a whole cell extract (WCE) and ATP, presumably a result of ATP-dependent chromatin remodelers (CRs) [1]. The promoter NFR, however, is largely preserved *in vitro* because yeast promoters contain sequences that are inherently refractory to nucleosome formation, such as low G/C content [2] and poly-dA:dT tracts [3].

Yeast promoters have a biased distribution of poly-A and poly-T elements flanking nucleosome free regions [4,5], which cannot be explained solely by the biased base content (**Figure 1**). This asymmetric poly-A/poly-T arrangement has no known function and is incongruous with the model that poly-dA:dT tracts simply exclude nucleosomes via a rigid DNA structure [6] since the DNA should resist bending equally in either orientation.

Results and Discussion

Hypothesizing that the asymmetric arrangement of these elements in promoters may have evolved to maintain promoter NFRs through some effect on nucleosome occupancy, we identified all non-overlapping poly-A sequences of exactly length five (AAAAA) in the yeast genome and analyzed the nucleosome occupancy [1] surrounding these elements (**Figure 2**). *In vitro*, both poly-A and poly-T sequences are similarly depleted of nucleosomes in an approximately symmetric fashion, both in the presence and absence of a WCE. However, upon addition of ATP, which activates CRs present in the WCE, the sequence becomes

further depleted, but in an asymmetric fashion; a nucleosome becomes well-positioned 5′ to the poly-A sequence, but not 3′, and the NFR is offset 5′ to the poly-A sequence, similar to the trend observed *in vivo* (**Figure 2**).

We next asked how nucleosomes were positioned around the three possible distinct arrangements of poly-A sequences (poly-A/poly-A, poly-A/poly-T, poly-T/poly-A). *In vivo* [7], when two poly-A elements are within ~60 bp, a strong NFR that is offset 5′ to the poly-A sequences generally results (**Figure 3A**). The poly-A/poly-T arrangement is typically much less depleted between the two motifs and yields two NFRs; one 5′ to the poly-A and the other 3′ to the poly-T (**Figure 3B**). The poly-T/poly-A combination results in the most robust NFR (**Figure 3C**), which could explain why this arrangement is preferred in yeast promoters. Further, in all cases, nucleosomes tend to be more well-positioned 5′ to poly-A sequences (3′ to poly-T). *In vitro*, in the absence of WCE and ATP [1], there is little difference between the three possible poly-A/poly-T arrangements and, in general, nucleosomes are depleted symmetrically around each poly-dA:dT element (**Figure S2** in **File S1**). We note that the occupancy bias surrounding poly-dA:dT tracts in the presence of active chromatin remodelers is unlikely to result from differences in the nucleosome isolation/quantification procedures because the same procedures were used to generate all *in vitro* data [1], but the bias occurs only when WCE and ATP are both present (**Figure 2**). Further, the nucleosome occupancy bias surrounding poly-A/poly-T combinations is consistent between *in vivo* datasets that use different approaches for crosslinking (sulfhydryl [8], formaldehyde [1,7]), cleavage (peroxide-mediate [8], MNase [1,7]), and quantification (microarray [7], sequencing [1,8]; see **Figure S3** in **File S1**).

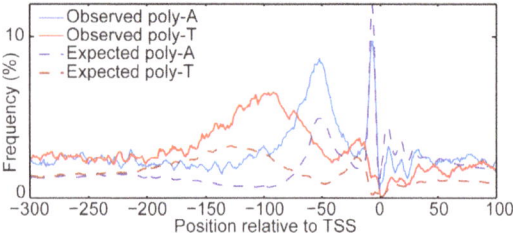

Figure 1. Yeast promoters have a biased distribution of poly-As and poly-Ts. The observed and expected frequency of poly-A and poly-T (AAAAA/TTTTT) elements across yeast promoters is shown, with expected calculated given the base content of the region. A greater number of poly-Ts and poly-As occur than expected in the $-115:-75$ and $-75:-35$ regions, respectively ($p<10^{-6}$ by simulation; see methods).

We hypothesize that the CR-dependent asymmetric arrangement of nucleosomes surrounding poly-A elements reflects differences in the nucleosome translocation efficiency from upstream vs. downstream of poly-As. It is possible that such a difference could result from the different histone-DNA contacts of the two DNA-strands. However, mouse [9] and human [10], which have nucleosomes very similar to those of yeast (84% identical in histone fold domains, between mouse and yeast), display a trend opposite to yeast (**Figure 4**); poly-A/poly-T combinations tend to be more depleted than poly-T/poly-A combinations, two consecutive poly-As generally result in 3'-biased NFRs, and, overall, there appear to be a more robust nucleosome boundaries 3' to poly-As (5' to poly-Ts). This observation suggests that specific factors (e.g. CRs) are responsible for differentiating between poly-As and poly-Ts. For example, poly-A tracts could prevent binding of CRs such that they can move a nucleosome towards poly-A sequences, but once there, the CR binds the DNA less efficiently and so cannot move it away. Indeed, previous studies have hinted that the DNA sequence could influence the repositioning of nucleosomes by CRs *in vitro*, but the mechanism, *in vivo* relevance, and sequence determinants of this phenomenon remained unknown [11,12]. More detailed studies of nucleosome positioning in the presence or absence of different CRs will be needed to determine the specificities of these CRs.

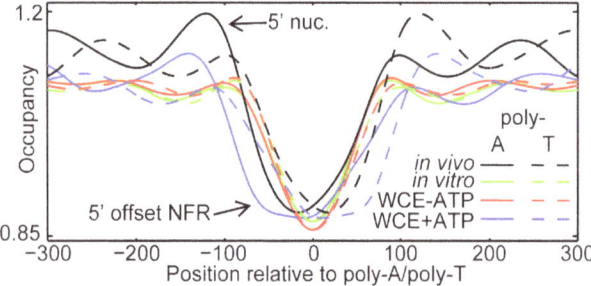

Figure 2. Nucleosomes are arranged asymmetrically around poly-dA:dT tracts. Average nucleosome occupancy surrounding poly-A and poly-T sequences (AAAAA/TTTTT) for salt gradient dialysis (*in vitro*), WCE without ATP (WCE-ATP), WCE with ATP added (WCE+ATP), as well as *in vivo* occupancy [1]. The difference in occupancy between poly-As and poly-Ts is significant only for *in vivo* and WCE+ATP (by rank sum; see **Figure S1** in **File S1**).

Our data indicate that poly-A sequences form an asymmetric barrier to CR-mediated nucleosome transit, that this asymmetry is used in yeast promoter architecture, and that the same sequences are used differently in mammals. This phenomenon helps explain part of the discrepancy between *in vitro* and *in vivo* nucleosome occupancy and indicates that the DNA sequence may play a greater role in positioning nucleosomes in the cell than previously appreciated. More complex models of nucleosome occupancy that account for CR-mediated nucleosome transit may be needed to fully explain nucleosome occupancy in the dynamic environment of the cell.

Methods

Definition of poly-A/poly-T

For **Figure 1** and **Figure 2**, we defined a poly-A element as any instance of five As in a row in the genome, with poly-T defined similarly. For **Figure 1**, we calculated the expected occurrence of poly-As and poly-Ts by using the nucleotide frequency at every base pair in the region to calculate the proportion of promoters expected to contain a poly-A or poly-T sequence at any given position. For **Figure 2**, we only considered non-overlapping instances.

The poly-A/poly-T combinations in **Figure 3**, and **Figure S2** and **S3** in **File S1** were derived by identifying all maximal poly-A and poly-T elements of at least 5 bp in the yeast genome and considering only those motif pairs whose (outer) edges lie within 500 bp and that have no additional poly-dA:dT tracts between them. **Figure 4** was created similarly, but only considering BAC-enriched regions for mouse data (i.e. regions for which high-resolution occupancy data are available) and only non-repetitive (by repeatmasker) regions of chromosome 22 for human. For **Figure S4** in **File S1**, only poly-A tracts of exactly length 5 were considered. In all cases, we used the NCBI v37 mouse genome, hg18 human genome, and R64 yeast genome.

Nucleosome occupancy normalization

For the data displayed in **Figure 2**, **Figure 4** (**D–E**), and **Figure S2** and **S3** in **File S1** (*in vitro* and *in vivo* yeast sequencing data [1,8], and MNase-digested chromatin from human granulocytes [10]), we smoothed the data within each locus (Gaussian, SD = 20 bp), while for the data in **Figure 3** (yeast microarray data [7]) and **Figure 4** (**A–C**) (mouse Th1 sequencing data, representing the centres of 147 bp fragments isolated from native, MNase-digested chromatin [9]), we performed no such smoothing. Smoothing the data in this way makes it correspond more closely to nucleosome occupancy by distributing the dyad occupancy (nucleosome centre position) over the area covered by a nucleosome. We did not smooth the mouse data because doing so obscured the poly-A/poly-T bias. We noted that the sequencing data (**Figure 2**, **Figure 4**, and **Figure S2–S4** in **File S1**) displayed significant variation in the number of reads per locus, so, for these data, we scaled each locus so that they each had a comparable numbers of reads and threw out any loci containing fewer than 40 (yeast *in vitro*; **Figure S2** and **S4A** in **File S1**), 400 (mouse *in vivo*; **Figure 4A-C** and **Figure S4C** in **File S1**), or 100 reads (human *in vivo*, **Figure 4D–F**, **Figure S4B** in **File S1**). For **Figure 3**, **Figure 4**, and **Figure S2** and **S3** in **File S1**, we also smoothed between loci to emphasize the overall occupancy trend (Gaussian, SD = 50, except for **Figure 4 A–C**, for which we used SD = 10).

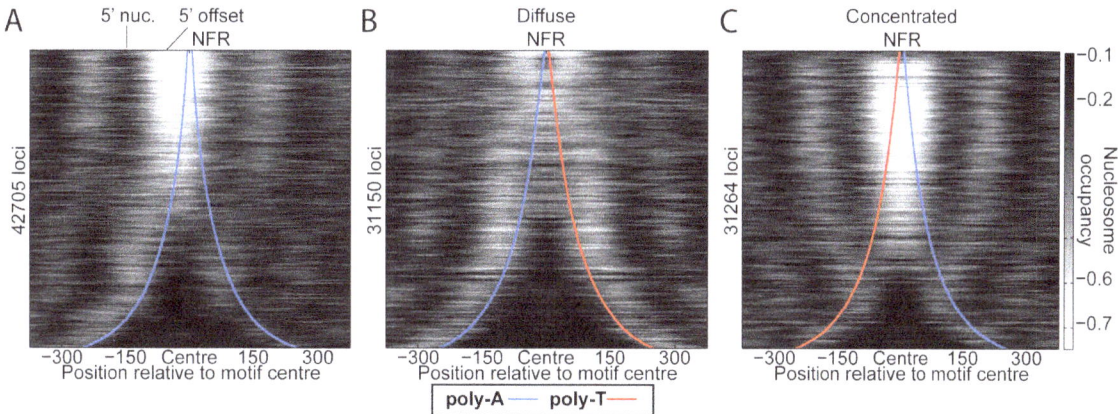

Figure 3. The different poly-A/poly-T arrangements result in vastly different nucleosome occupancy outcomes. *In vivo* nucleosome occupancy [7] (heatmap) surrounding all instances of (**A**) poly-A/poly-A, (**B**) poly-A/poly-T, and (**C**) poly-T/poly-A combinations in the yeast genome separated by no more than 500 bp. Red and blue curves represent the outer motif edges of poly-Ts and poly-As, respectively. Note that the poly-T/poly-T combination is a mirror image of the poly-A/poly-A data.

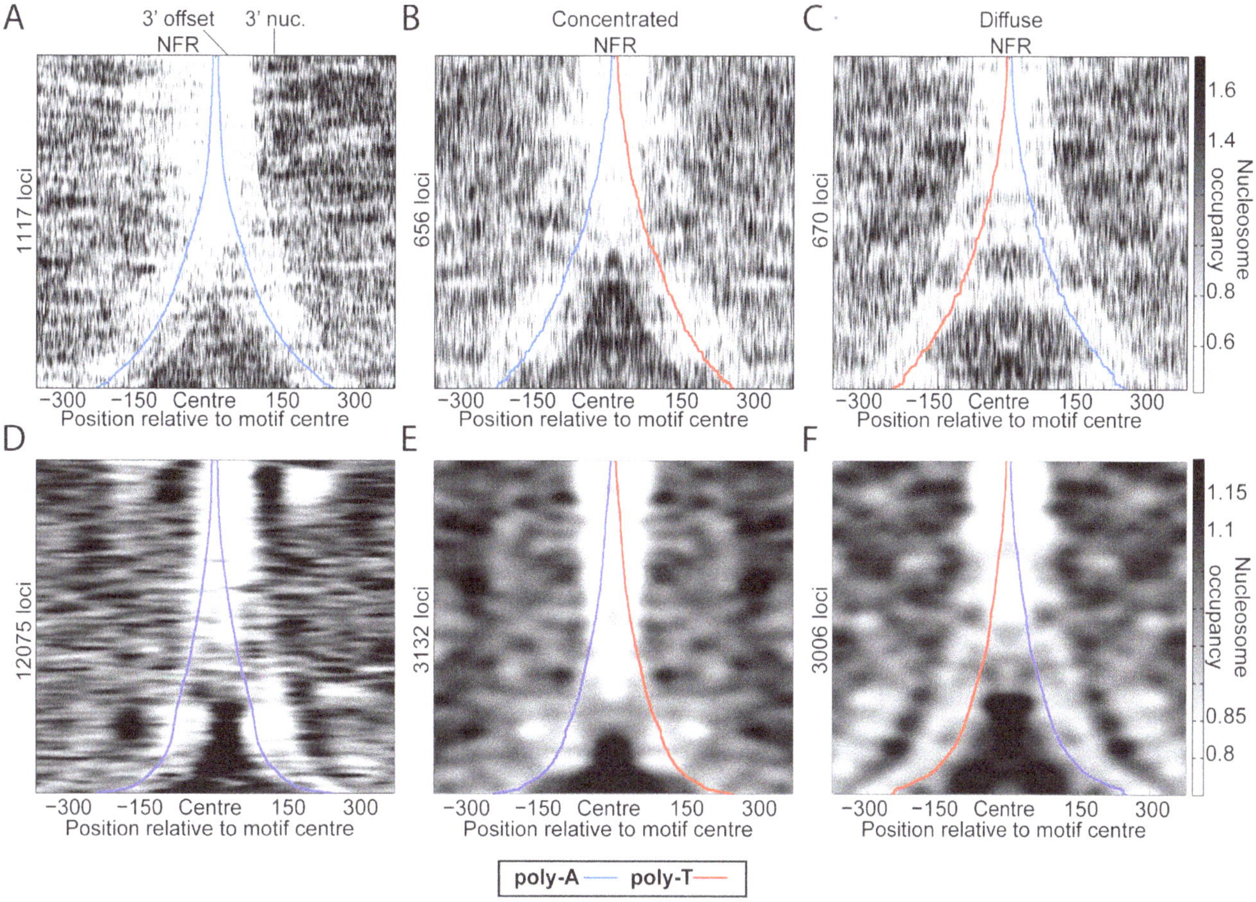

Figure 4. Mammalian nucleosome occupancy is also biased surrounding poly-As and poly-Ts, but the trend is opposite to yeast. *In vivo* nucleosome occupancy for (**A–C**) regions with available high-resolution nucleosome data from mouse Th1 cells [9] and (**D–F**) non-repetitive regions on chromosome 22, for human granulocytes [10] (heatmaps) surrounding all instances of (**A, D**) poly-A/poly-A, (**B, E**) poly-A/poly-T, and (**C, F**) poly-T/poly-A combinations. Gaussian smoothed between rows (SD = 10 and 50, for mouse and human, respectively). The distinct transitions from light to dark in the mouse data (**A-C**) result from using unsmoothed data, which corresponds roughly to nucleosome dyad occupancy (in this case the poly-A/poly-T bias was more obvious without smoothing). This distinct transition is presumably caused by the destabilization of nucleosomes as poly-dA:dT tracts are incorporated, and nucleosomes appear to be most unstable when the dyad is 69 bp from the proximal poly-dA:dT tract edge in human, mouse, and yeast (**Figure S4** in **File S1**).

Significance of poly-As and poly-Ts in promoter regions

To gauge the significance of the distribution of poly-As and poly-Ts in promoter regions, we generated "random-sequence promoters" (in equal proportion to the number of actual promoters analyzed) where, at every base position, that base had the same probability of being an A or T as the actual frequency of that base at that position. We repeated this procedure 10^6 times and each time counted the number of occurrences of 5 Ts or 5 As in a row within the $-115{:}-75$ and $-75{:}-35$ regions (relative to the TSS), respectively, but we found no randomly generated set of promoters with as extreme an occurrence of poly-As and poly-Ts in these regions as observed *in vivo* (max simulated = 1653 and 480, actual = 4919 and 2449 for A5 and T5, respectively).

Significance of nucleosome bias surrounding poly-dA:dTs

To gauge the significance of the biased distribution of nucleosomes surrounding poly-As and poly-Ts, we compared the distribution of normalized (as described above) reads surrounding these sequences within each experimental condition. We used the two-tailed (Mann-Whitney) rank sum test to gauge the significance of the difference in occupancy for poly-As compared to poly-Ts at equivalent positions relative to the poly-A/T. The result is plotted in **Figure S1** in **File S1**, along with the Bonferroni multiple hypothesis correction significance threshold.

Author Contributions

Conceived and designed the experiments: CGD. Performed the experiments: CGD. Analyzed the data: CGD. Contributed reagents/materials/analysis tools: CGD TRH. Contributed to the writing of the manuscript: CGD TRH. Supervised the research: TRH.

References

1. Zhang Z, Wippo CJ, Wal M, Ward E, Korber P, et al. (2011) A packing mechanism for nucleosome organization reconstituted across a eukaryotic genome. Science 332: 977–980.
2. Tillo D, Hughes TR (2009) G+C content dominates intrinsic nucleosome occupancy. BMC Bioinformatics 10: 442.
3. Kaplan N, Moore IK, Fondufe-Mittendorf Y, Gossett AJ, Tillo D, et al. (2009) The DNA-encoded nucleosome organization of a eukaryotic genome. Nature 458: 362–366.
4. Wu R, Li H (2010) Positioned and G/C-capped poly(dA:dT) tracts associate with the centers of nucleosome-free regions in yeast promoters. Genome Res 20: 473–484.
5. Hampson S, Kibler D, Baldi P (2002) Distribution patterns of over-represented k-mers in non-coding yeast DNA. Bioinformatics 18: 513–528.
6. Iyer V, Struhl K (1995) Poly(dA:dT), a ubiquitous promoter element that stimulates transcription via its intrinsic DNA structure. EMBO J 14: 2570–2579.
7. Lee W, Tillo D, Bray N, Morse RH, Davis RW, et al. (2007) A high-resolution atlas of nucleosome occupancy in yeast. Nat Genet 39: 1235–1244.
8. Brogaard KR, Xi L, Wang JP, Widom J (2012) A chemical approach to mapping nucleosomes at base pair resolution in yeast. Methods Enzymol 513: 315–334.
9. Yigit E, Zhang Q, Xi L, Grilley D, Widom J, et al. (2013) High-resolution nucleosome mapping of targeted regions using BAC-based enrichment. Nucleic Acids Res 41: e87.
10. Valouev A, Johnson SM, Boyd SD, Smith CL, Fire AZ, et al. (2011) Determinants of nucleosome organization in primary human cells. Nature 474: 516–520.
11. Rippe K, Schrader A, Riede P, Strohner R, Lehmann E, et al. (2007) DNA sequence- and conformation-directed positioning of nucleosomes by chromatin-remodeling complexes. Proc Natl Acad Sci U S A 104: 15635–15640.
12. van Vugt JJ, de Jager M, Murawska M, Brehm A, van Noort J, et al. (2009) Multiple aspects of ATP-dependent nucleosome translocation by RSC and Mi-2 are directed by the underlying DNA sequence. PLoS One 4: e6345.

19

Transgene Detection by Digital Droplet PCR

Dirk A. Moser[1,3], Luca Braga[2], Andrea Raso[2], Serena Zacchigna[2], Mauro Giacca[2], Perikles Simon[3*]

1 Faculty of Psychology, Genetic Psychology, Ruhr-University-Bochum, Bochum, Germany, 2 International Centre for Genetic Engineering and Biotechnology (ICGEB), Molecular Medicine, Trieste, Italy, 3 Department of Sports Medicine, Disease Prevention and Rehabilitation, Johannes Gutenberg-University Mainz, Mainz, Germany

Abstract

Somatic gene therapy is a promising tool for the treatment of severe diseases. Because of its abuse potential for performance enhancement in sports, the World Anti-Doping Agency (WADA) included the term 'gene doping' in the official list of banned substances and methods in 2004. Several nested PCR or qPCR-based strategies have been proposed that aim at detecting long-term presence of transgene in blood, but these strategies are hampered by technical limitations. We developed a digital droplet PCR (ddPCR) protocol for Insulin-Like Growth Factor 1 (*IGF1*) detection and demonstrated its applicability monitoring 6 mice injected into skeletal muscle with AAV9-*IGF1* elements and 2 controls over a 33-day period. A duplex ddPCR protocol for simultaneous detection of Insulin-Like Growth Factor 1 (*IGF1*) and Erythropoietin (*EPO*) transgenic elements was created. A new DNA extraction procedure with target-orientated usage of restriction enzymes including on-column DNA-digestion was established. *In vivo* data revealed that *IGF1* transgenic elements could be reliably detected for a 33-day period in DNA extracted from whole blood. *In vitro* data indicated feasibility of *IGF1* and *EPO* detection by duplex ddPCR with high reliability and sensitivity. On-column DNA-digestion allowed for significantly improved target detection in downstream PCR-based approaches. As ddPCR provides absolute quantification, it ensures excellent day-to-day reproducibility. Therefore, we expect this technique to be used in diagnosing and monitoring of viral and bacterial infection, in detecting mutated DNA sequences as well as profiling for the presence of foreign genetic material in elite athletes in the future.

Editor: Domenico Coppola, H. Lee Moffitt Cancer Center & Research Institute, United States of America

Funding: Work of PS and MG is carried out with the support of the World Anti-Doping Agency (WADA). The funders had no role in study design, data collection and analysis, decision to publish, or preparation of the manuscript.

Competing Interests: The authors have declared that no competing interests exist.

* Email: simonpe@uni-mainz.de

Introduction

Somatic gene therapy represents a promising tool to treat inherited or acquired diseases by transferring genetic material in order to compensate for defective genes, to produce a therapeutic substance, or to specifically trigger the immune system [1]. Despite its potential to treat life-threatening diseases, this technique might also be abused to improve physical performance [2]. Animal studies demonstrated successful viral transfer of potential performance-enhancing genes such as Insulin-Like Growth Factor 1 (*IGF1*; [3,4]) and Erythropoietin (*EPO*; [5]) but severe adverse events were also reported [6–9]. This has raised concerns about the illicit use of gene transfer technologies in elite sports (i.e. gene doping), and demonstrated the necessity to prohibit the use of gene transfer aimed at enhancing performance (WADA 2004), with the consequent need to monitor for the presence of transgenic DNA in routine gene doping tests.

The main obstacle for gene-doping detection is that the athlete's body would be enabled to produce doping substances that, in most cases, would be indistinguishable from endogenous proteins [10,11]. Therefore, currently proposed direct detection methods rely on specific sequence characteristics in transgenic DNA constructs that allow unbiased discriminability between genomic and exogenous DNA. Since skeletal muscle (the most likely target tissue for gene doping applications) would be difficult to harvest for routine doping testing, several studies explored the possibility to detect minute amounts of transgenic DNA in the blood after somatic (intramuscular) gene transfer.

In these studies different PCR-based approaches, such as nested PCR [12–14] and TaqMan qPCR [8,15–17], were applied in order to detect minute amounts of transgenic DNA in blood. These methods selectively detect and discriminate the intron-less transgene from genomic DNA by PCR. Thus, detection of transgenic DNA in whole blood samples could provide a suitable means to support conviction of unscrupulous athletes. However, current PCR-based detection approaches are either highly sensitive but require a nested PCR procedure with a laborious workflow [12–14], or they display weaknesses to sensitively detect minute amounts of the transgene in a single round qPCR, [8,15,17,18]. In addition, qPCR relies on external standard curves, which further complicates inter-lab comparability of results. It is also important to point out that DNA amplification efficiency is highly dependent on template structure in the way that circular DNA amplifies poorly compared to linear DNA [19,20]. To date, all aforementioned methods use undigested DNA as a template which can result in poor amplification of circular DNA structures and consequently in failure to detect the specific transgenic sequence in a huge background of endogenous DNA. As recombinant DNA is integrated into the host genome or persists as episomal circular supercoiled DNA, predigestion of

DNA should improve PCR sensitivity and increase the likelihood of transgene detection.

Digital droplet PCR (ddPCR) is a new method that enables the absolute measure of target DNA. Its principle is based on the portioning of PCR mixture into thousands of droplets per reaction [21]. As ddPCR can also handle huge amounts of background DNA in the reaction, it represents a convenient method to find the transgenic "needle in a haystack". Additional benefit of ddPCR over qPCR also includes absolute quantification, which does not rely on external standard curve, leading to excellent day-to-day reproducibility. This is why ddPCR is likely to become a favourite tool for accurate routine analysis, especially when performed at multi-site laboratories.

Here we describe a new digital droplet PCR assay for transgene *IGF1* detection after intramuscular AAV9 gene transfer. DNA was purified from whole blood of 6 intramuscularly AAV9-*IGF1* transduced mice and 2 uninjected control animals, which were monitored for 33 days for the detectability of transgene elements by ddPCR.

We also developed a duplex ddPCR protocol for *IGF1* and *EPO* transgene elements to simultaneously detect two candidate genes for gene doping in a single assay. We further established a method to test for ddPCR efficiency by the addition of an internal control standard (ICS) at a defined copy number to the reactions, a method already described for qPCR elsewhere [17]. This ICS differs only from the transgene in its probe binding site and can be detected in parallel to the transgene, which makes monitoring of ddPCR efficiency feasible in each reaction.

To minimize sample handling and to optimize for transgene detectability, we also developed a new DNA extraction protocol, which includes on-column DNA restriction enzyme digestion and allows the elution of fragmented high quality DNA as optimal target for ddPCR. This procedure leads to further improved sensitivity of the assay and might also find application for assays where preferential amplification of target sequences is warranted.

Thus we aim at presenting a straightforward and highly sensitive approach for DNA detection, which could lead to the next generation of transgene detection.

Methods

DNA extraction, digestion and purification

Human genomic DNA (hgDNA) for spike-in experiments was extracted at large scale using the salting-out procedure as described by Miller et al. [22]. PCR standards of the *IGF1* and *EPO* coding-sequence were generated using the primers and resulting amplicon lengths as indicated in Table 1. To generate circular standards, purified PCR products were cloned into the PCRII-TOPO vector according to the manufacturer's recommendations. All standards were Sanger-sequenced to check for correctness of the sequences. DNA was spiked with defined copy numbers of freshly prepared ddPCR quantified *IGF1-* or *EPO* standard, and digested using restriction enzymes DdeI and RsaI (2 units/μg hgDNA) in NEB buffer 2 for 1 h at 37°C followed by 20 minutes heat inactivation at 65°C.

Production, purification, and characterization of rAAV vectors

The human hepatic *IGF1-IA* (ref. seq. NM_000618.3) was PCR amplified and cloned into the pZac recombinant AAV expression vector, generating the pAAV-*IGF1* vector. Viral particles were produced by the AAV Vector Unit at ICGEB Trieste (http://www.icgeb.org/avu-core-facility.html). Methods for production and purification were previously described [23].

AAV9 titers were in the range of 1×10^{13} genome copies per milliliter.

AAV9-*IGF1* injection into the skeletal muscle

Animal care and treatments were conducted in conformity with institutional guidelines in compliance with national and international laws and policies, upon approval by the ICGEB Ethical Committee and by the Italian Minister of Health (EEC Council Directive 86/609, OJL 358, December 12, 1987). Animals were provided with housing in an enriched environment, with at least some freedom of movement, food, water and daily care and cleaning. Experiments were performed under general or local anesthesia, and with constant use of analgesics. At the end of any experiment, competent authorized persons decided the proper time and most appropriate humane method for animal sacrifice. All experiments were performed in male CD1 mice, 4–6 weeks of age. As a model of gene doping, tibialis anterior and gastrocnemius muscles were injected with 50 μl of either PBS or a viral suspension containing 10^{11} viral particles of AAV9-*IGF1*, and harvested after the indicated periods of time.

DNA extraction and digestion from mouse blood

DNA was extracted from 100 μl mouse blood using the Qiagen DNA microkit, which enables the elution of DNA in variable volumes between 20–100 μl. DNA was eluted in a volume of 25 μl and digested with 5 U DdeI and RsaI in a final volume of 30 μl at 37°C for 1 h followed by heat inactivation at 65°C for 20 minutes.

Restriction enzymes DdeI and RsaI were chosen according to the following criteria (see also Table 2):

- they do not cut the coding sequence between the primers
- they cut the *IGF1* and *EPO* coding sequence 5′ and/or 3′ of the PCR-amplicon which linearizes potential circular viral or plasmid constructs irrespective of any knowledge of the vector sequence
- they frequently cut the *IGF1* and *EPO* intronic region to prevent background amplification of their genomic locus during PCR
- in order to exclude the effects of endogenous CpG methylation on restriction enzyme activity, no CpG sites are allowed to be present in the restriction enzyme recognition sites

qPCR

IGF1-specific primers and probe were chosen to target all *IGF1* mRNA isoforms including mechano-growth factor (*MGF*) using UCSC Genome Browser (http://genome.ucsc.edu/) and Primer 3 [24]. Then, *IGF1* qPCR was optimized and tested for the limit of detection/limit of quantification (LOD/LOQ) as described by Burns et al. [25] using standard calibrators in a background of 500 ng hgDNA. The LOD was estimated by 2-fold serial dilutions between ~2000 and 1 calibrator copies per reaction (See Figure S1). We defined the LOD as the lowest copy number that gives a detectable PCR amplification product at least 95% of the time. The LOQ was defined as the lowest concentration that could be quantified with >80% accuracy, and LOD was defined as the minimum copy number for which all replicates of the same dilution could be successfully detected. qPCR mixture contained 10 μl SsoFast probes supermix (Bio-Rad) and primers and probes as indicated in Table 1. Two-step PCR protocol using CFX384 (Bio-Rad) started with 2 min at 98°C followed by 45 cycles at 95°C melting and 30 sec annealing/extension at 64°C.

Table 1. Primer and Probe sequences.

Gene	Primer and Probe sets in 5→3 orientation [concentration]	Amplicon size
Erythropoietin (EPO); NM_000799	Fw: TGAATGAGAATATCACTGTCCCAGAC [900 nM] Rev: CTTCCGACAGCAGGGCC [900 nM]; P: [Hex]AAG[+A]GG[+A]TG[+G]AG[+G]TCGG[BHQ1] [250 nM]; Sigma	114 bp
	Coding sequence primers: Fw: ATGggggtgcacgaatgt; Rv: TCAtctgtcccctgtcctg	582 bp
Insulin-like growth factor 1; (IGF1) NM_000618.3	Fw: GCTGGTGGATGCTCTTCAGTT [900 nM] Rev: TCCGACTGCTGGAGCCATAC [900 nM] P:[FAM]CTT[+T]TA[+T]TT[+C]AA[+C]AA[+G]CC[+C]AC[BHQ1] [250 nM]; Sigma	83 bp
	Coding sequence primers: Fw: ATGggaaaaatcagcagtcttc; Rv: CTAcatcctgtagttcttgtttcctg	462 bp
Internal Control Standard (ICS)	Fw: GCTGGTGGATGCTCTTCAGTT [900 nM] Rev: TCCGACTGCTGGAGCCATAC [900 nM] P: [VIC]TGCTCCAGAGAAGAAACCAC[MGB-NFQ] [250 nM]; Life Technologies	82 bp

Forward (Fw), reverse (Rev) primer-, and probe (P) sequences inclusive corresponding amplicon lengths. Start and stop codons in the coding sequence primers are capitalized.

To test for putative template effects on amplification efficiencies, two different sized plasmids containing the 462 bp *IGF1* coding sequence (pAAV9-*IGF1*-5237 bp; Topo PCR 2.1-*IGF1*-4339 bp) and a PCR-generated *IGF1* standard were subjected to qPCR with or without DdeI and RsaI double-digestion (See Figure S2).

Nested-qPCR

As a positive control, nested qPCR was performed to test for the presence of AAV9-*IGF1* elements in DNA isolated from mouse blood. The first round PCR was done using 4 µl of DNA using primers and conditions as described earlier [13]. PCR-products were 1:50 diluted in water and two microliter subjected to second round amplification (in triplicates) using the primers and conditions as described above. After qPCR, nested-qPCR products were analysed on a 2.5% agarose gel.

ddPCR

Each PCR reaction consisted of a 20 µL solution containing 10 µL ddPCR supermix for probes (Bio-Rad), 900 nM primers, 250 nM probe and 4 µl template DNA. Droplets (~20.000/reaction) were generated on the Bio-Rad QX-100 following the manufacturer's instructions. Samples were transferred on a 96 well-plate and thermal cycled to the endpoint (T100 Thermal Cycler; Bio-Rad) using a standard protocol; initial denaturation at 95°C for 10 min, followed by 40 cycles of melting at 95°C for 30 seconds and annealing/elongation at 61°C for 1 minute, before droplet stabilisation by 10 min incubation at 98°C. After cycling, the 96 well-plate was immediately transferred on a QX100 Droplet Reader (Bio-Rad) where flow cytometric analysis determined the fraction of PCR-positive droplets vs. the number of PCR-negative droplets in the original sample. Data were analysed using Poisson statistics to determine the target DNA template concentration in the original sample. Optimal annealing temper-

Table 2. Number of restriction sites for *Dde*I and *Rsa*I in the human/mouse genome and at the *IGF1* and *EPO* gene locus.

	*Dde*I (CTNAG)	*Rsa*I (GTAC)
Approximate number of cutting sites per Mbase in the human genome	4,844.1*	1,764.1*
average fragment length	206*	567*
Approximate number of cutting sites per Mbase in the mouse genome	5,291.6*	2,122.4*
average fragment length	189*	471*
IGF1		
IGF1 gene locus (human); chr12: 102789645–102874378	432	147
Igf1 gene locus (mouse); chr10: 87859056–87937047	430	129
Human IGF1 gene locus (56035 bp between primers); chr12: 102813436–102869470	293	100
IGF1 mRNA (7321 nucleotides)	28	12
IGF1 cds (462 nucleotides)	3	1
IGF1 amplicon (83 nukleotides)	0	0
EPO		
EPO gene locus (human); chr7: 100318423–100321323	19	4
Human EPO gene locus (729 bp between primers); chr7: 100319610-100320338	5	0
EPO mRNA (1340 nucleotides)	10	2
EPO cds (582 nucleotides)	1	2
EPO amplicon (114 nucleotides)	0	0

* Information taken from http://tools.neb.com.

atures were assayed carrying out gradient PCR for all primers and their specific targets (See Figure S3). Subsequently, we tested for potential inhibitory effects of restriction buffer on ddPCR and observed inhibitory effects only in the cases when more than 5 µl of DNA restriction solution (2.5 mM NaCl; 500 nM Tris-HCl; 500 nM MgCl$_2$; 50 nM DTT) were subjected to 20 µl ddPCR (Figure S4 a). Increasing amounts (up to 1500 ng) of genomic DNA in the background of the reaction were also assayed and did not show any inhibitory effects on ddPCR (Figure S4 b). Consequently, all ddPCRs were performed using a final volume of 4 µl template DNA (47 ng–1500 ng), which avoided potential salt and DNA inhibitory effects.

ddPCR detection of *IGF1* transgene from mouse blood and human spike-in experiments

For transgene detection, human/mouse genomic DNA was extracted as described above and 4 µl subjected to ddPCR using the primers and probes at concentrations as indicated in Table 1.

IGF 1 and *EPO* ddPCR-duplex assay

We also established a duplex protocol for parallel *IGF1* and *EPO* ddPCR transgene detection. *EPO*-specific primers and probe were used as described elsewhere [17]. Serially diluted standards (~5000 copies were 1:5, 1:2, 1:5, 1:2, 1:5 diluted) were assayed using ddPCR chemistry containing 900 nM *IGF1* and *EPO* forward and reverse primers, supplemented with differentially labelled *IGF1* and *EPO*-specific LNA TaqMan-probes at 250 nM final concentration (see Table 1). *EPO*-specific primers and probe were used as described elsewhere [17]. Samples were assayed in a background of 500 ng DdeI and RsaI fragmented human genomic DNA containing serial dilutions of either *IGF1* or *EPO* standards only; both standards at decreasing levels and also containing *IGF1* at decreasing and *EPO* at increasing amounts. Digital droplet PCR was performed under standard conditions as described above.

ICS (internal control standard)

To check for uniform PCR efficiency all reactions were spiked with an internal standard (similar to the internal threshold control (ITC) as described for qPCR [17]. This artificial standard (purchased and synthesized by MWG Eurofins) was designed to have the identical 5′ and 3′ sequence compared to the *IGF1* amplicon, but with the probe binding site replaced by a sequence taken from the ancestral organism Cyanobacterium stanieri (NC_019778.1). This 20 nucleotide sequence was blasted against the human genome to confirm that there was no presence of either identical or similar sequence, which could lead to false positive detection. A VIC-labelled TaqMan probe (Life Technologies) was designed to target this sequence in a duplex PCR when *IGF1* transgenic elements are also detected.

Subsequently, *IGF1* standard was assayed using ddPCR chemistry supplemented with an internal-control standard at defined concentration (100 copies/reaction).

DNA extraction by on-column digestion

Duplicates of whole blood samples of 5 human donors (100 µl) were spiked with the same amount of circular standard DNA. Using the Qiagen DNA microkit, DNA was extracted and was finally processed using 3 different procedures as follows:

a)   DNA was eluted conventionally with 30 µl H$_2$O.

b)   Water-eluted DNA was DdeI and RsaI digested (5 U each) for 1 h at 37°C.

c)   DNA was on-column digested for 1 h at 37°C with 30 µl of a solution containing 10 U DdeI and RsaI in 1 x buffer 2.

All samples were adjusted to the same salt concentration and dilution, heat inactivated for 20 min at 65°C, Nano dropped, agarose-gel visualized and subsequently ddPCR quantified under the conditions as described above.

Results

As a preliminary step towards ddPCR, qPCR experiments were performed to identify those primers and probes, which led to best PCR efficiencies, and which did not produce artefacts such as excessive by-products or false positive signals. Accordingly, primers and probes as indicated in Table 1 were used.

qPCR

We designed a new assay for *IGF1* transgene detection with primers that resulted in an 83 bp amplicon, in which the exon2/3 boundary was targeted by a 6-FAM-labelled LNA-probe. As illustrated in Figure S1, *IGF1* qPCR efficiency was 96.7% with a linearity of $r^2 = 0.98$-. The LOQ was defined as the lowest concentration that could be quantified with >80% accuracy, and set to 16 copies per reaction (See Figure S1); LOD was found to be 4 copies. We then tested for differential PCR efficiency dependent on DNA structure. We assayed serial dilutions (ranging from 10^6–10 copies) from 2 vectors carrying the *IGF1* coding sequence (pAAV9-*IGF1*-5237 bp and TOPO-*IGF1*-4397 bp) compared to linear PCR product. Improved amplification for the linear PCR standard (Ct difference of more than 3), compared to circular, supercoiled vectors was observed. Subsequently, we tested digested plasmid vs. undigested plasmid compared to linear PCR-standard and were able to amplify the digested plasmid with efficiencies close to the linear PCR-standard (See Figure S2).

ddPCR optimization

To determine best target-specific annealing temperatures, all primers used in ddPCR were initially tested by gradient ddPCR. As indicated in Figure S3, all primers worked well between 66°C–60°C with highest amplitude and best sensitivity at ~61°C. Subsequently, all ddPCR assays were conducted at 61°C annealing/extension temperature. Furthermore, we verified that genomic DNA, up to 1500 ng, in the ddPCR reaction did not affect results (See Figure S4 a), and that 1 x restriction solution, less than 5 µl, added to ddPCR did not inhibit the reaction (See Figure S4 b). Consequently, 4 µl of restriction solution were routinely used for transgene detection.

ddPCR for AAV9-*IGF1* transgene detection in mice

Our aim was to detect minute amounts of *IGF1* transgenic DNA using ddPCR, a method so far never described for gene-doping detection. From 100 µl whole blood, DNA was extracted with 25 µl H$_2$O and digested in NEB buffer 2 using the restriction enzymes DdeI and RsaI, in a final volume of 30 µl. Four µl of digested DNA were subjected to PCR in 3 independent trials, and *IGF1* was detected as illustrated in Figure 1. Results provided specific indication of *IGF1*-transgene detection in all transduced animals over a 33 day-period. As displayed in Table S1, copy numbers were in the range of 3200–164800 copies/reaction (which indicates 240–12263 viral elements/µl of whole blood) also indicating excellent ddPCR reproducibility with a tendency to lower values, due to additional thawing and freezing cycles. Control animals remained negative with one single false positive event (Fig 1-day30 and Figure S5). However, this false positive

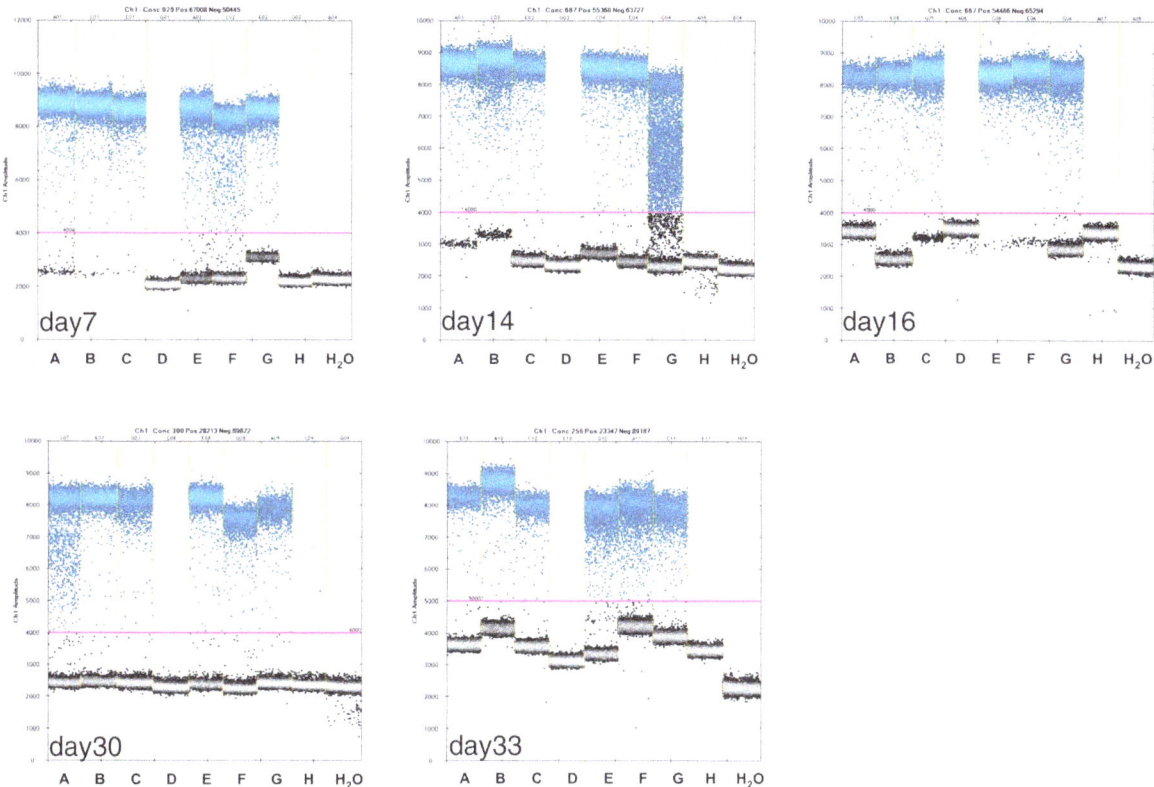

Figure 1. *IGF1* transgene detection by ddPCR. Eight mice, 6 AAV9-IGF1 muscle transduced (A, B, C, E, F, G) and two controls (D and H) were screened for 33 days by ddPCR for *IGF1* transgene detectability.

event could be clearly discriminated from true positives by manual re-adjustment of the threshold to a value that defined the lower limit of the positive control, a best practice for ddPCR, as also discussed earlier elsewhere [26].

Digital droplet PCR results were confirmed by nested-qPCR and by gel-electrophoresis of the corresponding PCR-products. PCR products showed by-products of similar size as the expected *IGF1* PCR-products, however, these by products were not detected by the *IGF1*-specific probe during qPCR (data not shown).

Spike-in duplex assay for simultaneous *IGF1* and *EPO* detection at low copy numbers

To ameliorate the use of the limited amount of DNA extracted from blood and to optimize the flow capacity of a potential routine transgene detection setting, we also developed a new duplex ddPCR assay for simultaneous *IGF1* and *EPO* detection.

Serially diluted *IGF1* and *EPO* standards were spiked into 500 ng human genomic DNA and analysed by ddPCR after DdeI and RsaI digestion. Digital droplet PCR was performed using a mixture containing *IGF1* and *EPO* primers and probes. This test should reveal sensitivity and linearity of the assay for single transgene detection and also in a duplex approach at low copy numbers (range from ~5000–10 copies/reaction). As indicated in Figure 2 (A–E), both transgenes were detected showing linearity values from 0.9997 to 1 for *IGF1* and 0.9995 to 0.9998 for *EPO* under the conditions tested. Sensitivity and linearity of the duplex assay did not differ from those obtained by the single gene assay.

IGF1 standard was also assayed in the presence of an internal control standard at ~100 copies/reaction. As indicated in

Figure 2E, ICS was clearly detected in all reactions, and could be used as a reporter for PCR efficiency in each run and for all reactions. This minimizes probability of false negatives due to potential PCR inhibitors being present in some samples.

DNA extraction by on-column digestion

We also optimized a protocol of on-column digestion prior to DNA fragmentation. This procedure led to fragmented high quality DNA, with final yields that were on average 3 times higher compared to conventionally eluted DNA as revealed by Nano-Drop 1000, agarose-gel analysis (See Figure S6).

In addition, testing DNA extracted from whole blood (a-conventionally extracted; b- conventionally extracted followed by DNA digestion; c- on-column digested) that was spiked with the same amount of plasmid standard by ddPCR revealed more sensitive detection for digested DNA, and most sensitive detection for samples subjected to on-column digestion. As indicated in Figure 3, we could achieve 1.9–2.8 fold increased ddPCR sensitivity comparing digested DNA to undigested DNA. When ddPCR was performed from the same blood after on-column digestion, ddPCR performance was further increased 2.9–19 fold compared to DNA eluted using water (Figure 3).

Discussion

Digital droplet PCR represents a new technical approach for applications where conventional PCR meets its technical limits, such as rare event detection in the presence of high amounts of genomic DNA. Compared to conventional qPCR, quantification by ddPCR does not rely on external standard curves and is more tolerant to inhibitors and variation in amplification efficiencies.

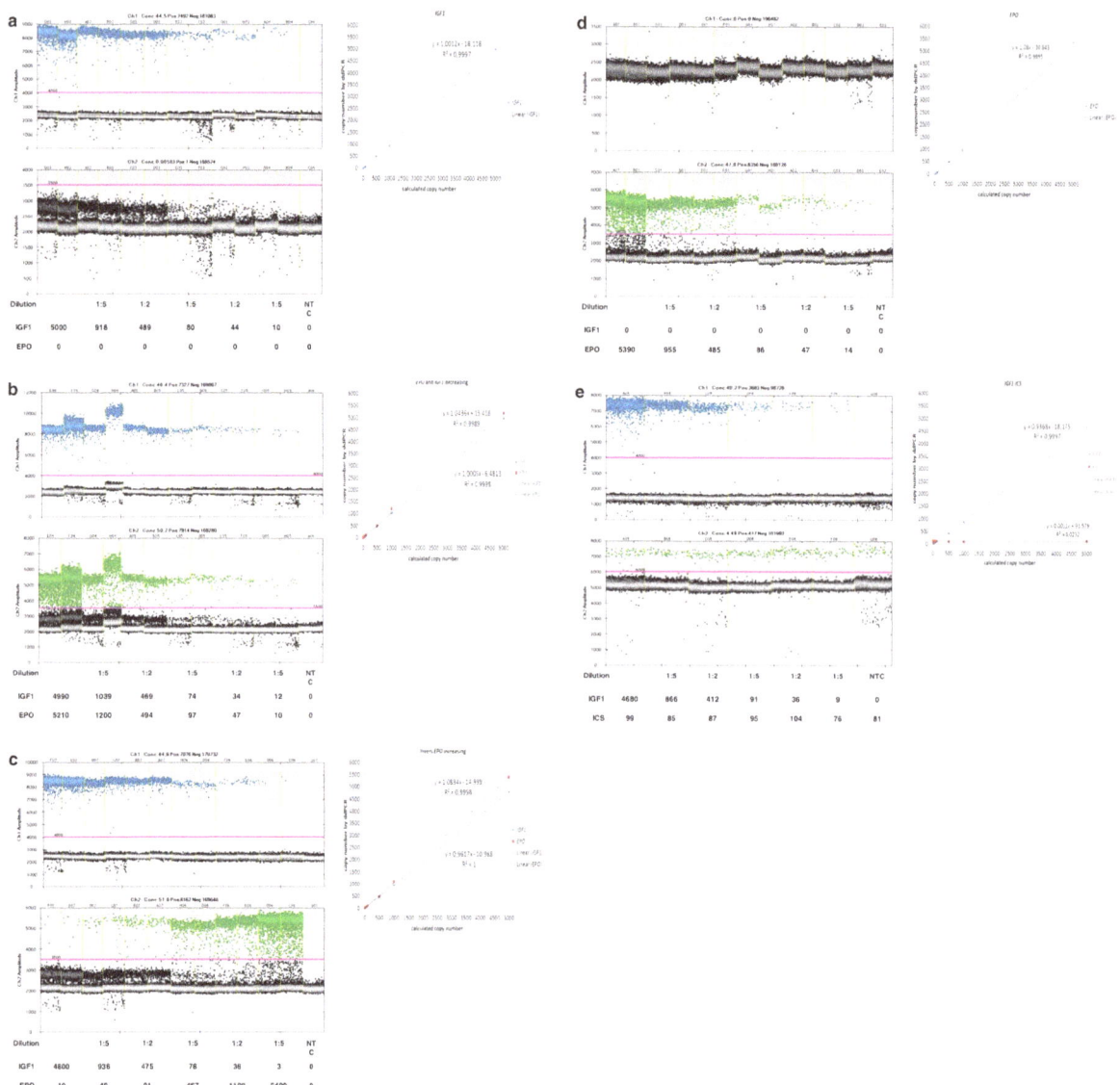

Figure 2. Duplex ddPCR for *IGF1* and *EPO* transgene elements. 500 ng human gDNA were spiked with serial dilutions (1:5; 1:2; 1:5; 1:2 and 1:5) of *IGF1*-standard (A), decreasing amounts of *IGF1* and *EPO*-standard (B), inverse amounts of *IGF1* (decreasing) and *EPO* (increasing) (C), *EPO* only (D) and *IGF1* multiplexed with ICS at ~100 copies/reaction (E). Displayed are amplitudes, copy numbers per 20 µl reaction and linearity of the signals.

Thus, ddPCR is well suited for the identification of transgenic elements with the objective of gene-doping detection.

Here we present a new ddPCR method for *IGF1* and *EPO* transgene detection with high sensitivity and reliability, in combination with an effective technique to isolate high amounts of fragmented DNA with minimal effort. The aim of this study was to test and optimize detection of transgenic elements by the use of ddPCR. Accordingly, we initially tested *IGF1*-specific primers and probes by qPCR for optimal primer annealing, amplicon generation and specific probe detection. Quantitative PCR-experiments revealed a LOQ of 16 and a LOD of 4, which indicated highly efficient PCR conditions (See Figure S1). This qPCR-assay was subsequently transferred to ddPCR and re-optimized for best annealing temperature (See Figure S3) and controlled for either salt or DNA inhibition using (See Figure S4) ddPCR specific chemistry.

Eight mice, of which 6 were transduced with AAV9-*IGF1* and 2 controls, were then monitored over a 33-day period for the detectability of transgenic elements extracted from whole blood. Strong signals could be detected for all AAV9-*IGF1* transduced mice at all points in time, with a tendency of signal reduction over time (Figure 1 and Table S1). We also created a ddPCR duplex assay aimed at the simultaneous detection of *IGF1* and *EPO* detection. As indicated in Figure 2 (A–E), both transgenes were sensitively detected showing linearity from 0.9997 to 1 for *IGF1* and 0.9995 to 0.9998 for *EPO* with copy numbers from 5400 to 3. These data are in accordance with published data describing linearity close to 1 with a limit of detection of 5 when performing duplex ddPCR assays for other genes [16].

The integration of an artificially generated internal control standard into ddPCR enabled us to control for PCR efficiency in each well. As indicated in Figure 2E, the documentation of ddPCR efficiency for each reaction represents a valuable tool for

Transgene Detection by Digital Droplet PCR

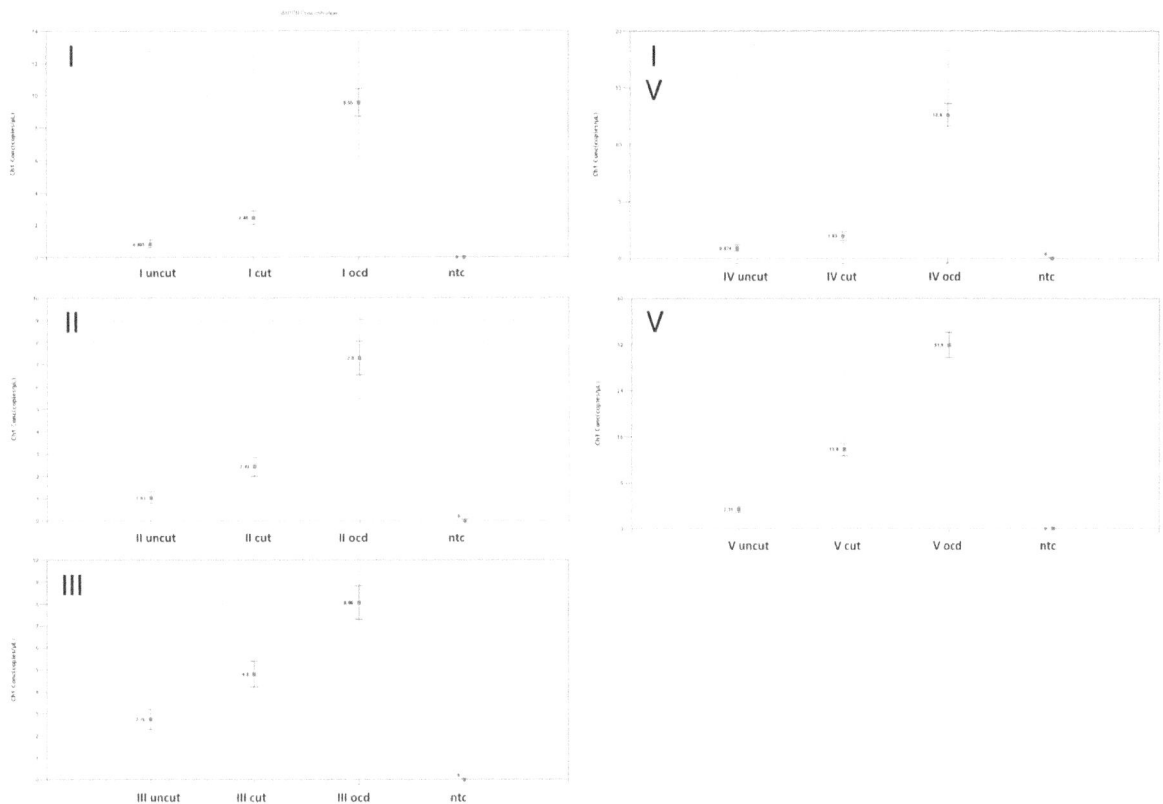

Figure 3. DdPCR results following 3 different DNA extraction procedures. Blood of 5 donors was spiked with the same amount of circular plasmid containing the *IGF1* coding sequence. DNA was extracted using 3 different procedures: uncut, Ddel and Rsal cut, and on-column Ddel and Rsal digested (ocd). Results are displayed as copies detected per µl for 5 subjects (I–V for uncut; conventionally digested (cut), on-column digested DNA (ocd), and no-template control (ntc). Results derive from two independent experiments performed in duplicates.

routine analyses as it allows for identification of false negatives, potentially originating from PCR inhibitors being present in some reactions.

On the genomic level, *IGF1* primers target exonic elements separated by a ~56 kb sequence (56035 bp), containing 293 DdeI and 138 RsaI cutting sites. Primers and probe for EPO detection were used as described by others [17] and could generate a genomic fragment of 726 bp which is cut 5 times by DdeI during digestion. DdeI and RsaI restriction enzyme digestion lead to fragmentation of the respective gene loci and inhibition of primer extension after digestion. This, and the design of *IGF1* and *EPO* specific probes which specifically target exon/exon boundaries, makes detection of false positive signals generated from *IGF1* and *EPO* genomic loci unlikely. Additionally, our data demonstrate that amplification of linearized DNA has superior PCR efficiency over circular DNA elements (Figure 3 and Figure S2), which further supports the use of restriction enzyme digestion of DNA prior to PCR. Our optimized DNA-extraction procedure includes an on column-digest at the final step of DNA extraction. This involves minimal handling of samples and results in much higher DNA yield as compared to conventional DNA extraction procedure (Figure 3 and Figure S6). This may be explained by improved elution of fragmented DNA from the column as compared to undigested high molecular DNA stretches. Comparing eluates of blood samples that were spiked with the same amount of circular *IGF1* standard after conventional DNA purification with, and without restriction to on-column digested DNA by ddPCR, we found the latter to be the most efficient

substrate for transgene detection. Accordingly, we recommend use of on-column digestion for all applications where minute amounts of DNA need to be detected, as it increases DNA yield and, thus, the probability of target elution. As displayed in Figure 3, linearization of circular target DNA improved (dd-) PCR detection efficiency, which was further enhanced by on-column digestion due to significantly improved DNA elution from the column.

In analogy to trials which aim to isolate low levels of a mutant sequence in a high background of DNA, such as ARMS-PCR (Amplification Refractory Mutation System, or allele-specific PCR; ASP [27]), we believe on-column digestion using carefully chosen enzymes may help to increase detection of specific DNA elements by PCR in the future.

We did not, however, observe a consistent Limit of Blank (LOB) for *IGF1* and *EPO* analysis, as recently described for the detection of mutated *KRAS* elements [28]. This difference might be attributed to different PCR systems, or bigger differences between wild type and transgene detection than present in mutational detection. Occasional false positives could be explained by DNA carry-over due to sample handling as previously described for other high sensitive detection methods [13].

If we compare our ddPCR assay to other transgene detection approaches, such as qPCR or agarose-gel based nested PCR, we observe similar sensitivities, as all methods can identify less than five copies per PCR. However, ddPCR offers to advantages to be independent of an artificial standard curve, with all its confounding variables, and to minimise DNA carry-over by strict reduction of DNA handling steps.

Testing for practicability and cost value of ddPCR, Morisset and co-workers [29] describe similar or improved throughput and cost-effectiveness for ddPCR when compared to qPCR. Additionally, increased usage of ddPCR can be expected to lead to further cost reduction in the future. Guidelines for standardization purposes and creation of Good Laboratory Practice (GLP) procedures for ddPCR have already been described [26] and were applied in this study.

Implementation of ddPCR for transgene detection is further encouraged by data recently presented by Strain and co-workers [30], who showed 5–20 fold improved precision comparing ddPCR to qPCR for the detection of total DNA originating from the human immunodeficiency virus (HIV) and its episomal 2-LTR (long terminal repeat) circles.

Hence, we present a new approach for transgene detection at high and minute copy amounts, including a new DNA extraction and digestion method, which delivers high quality, concentrated and fragmented DNA as an appropriate target for ddPCR.

Digital PCR as presented here, including appropriate restriction enzyme digestion of genomic DNA, represents a promising approach for the detection of minimal DNA fractions. Digital PCR will therefore most probably find implementation to monitor gene-therapy trials, and the possible abuse as gene doping, but also to screen for and monitor viral and bacterial infection, and to analyse food and feed products for xenogeneic components.

Supporting Information

Figure S1 Standard curve, amplification plot and calculation for experimental estimation of IGF1 LOD and LOQ.

Figure S2 PCR efficiency for undigested vs. digested IGF1 standards - Linearization of circular DNA leads to improved PCR efficiency and better detection at low copy numbers.

Figure S3 Temperature gradient for ddPCR detection of IGF1, EPO and ICS.

Figure S4 ddPCR efficiency under various conditions.

Figure S5 Qualitative assessment of ddPCR results.

Figure S6 Concentration, purity and integrity of DNA comparing three different DNA extraction procedures.

Table S1 IGF1 transgene copy numbers as detected by ddPCR for 6 AAV9-IGF1 transduced mice and 2 controls at 5 different days.

Acknowledgments

The authors are grateful to Dr. Pia Scheu (Bio-Rad) for her expert introduction into ddPCR technology. Furthermore, the authors thank Olga Moser and Marco Zahn for their excellent technical assistance and Thomas Beiter and Mike Bramwell for critical reading of the manuscript.

Author Contributions

Conceived and designed the experiments: DAM LB AR SZ MG PS. Performed the experiments: DAM LB AR SZ MG PS. Analyzed the data: DAM LB AR SZ MG PS. Contributed reagents/materials/analysis tools: DAM LB AR SZ MG PS. Wrote the paper: DAM LB AR SZ MG PS.

References

1. Kay MA (2011) State-of-the-art gene-based therapies: the road ahead. Nat Rev Genet 12: 316–328.
2. Schneider AJ, Friedmann T (2006) Gene doping in sports: the science and ethics of genetically modified athletes. Adv Genet 51: 1–110.
3. Barton-Davis ER, Shoturma DI, Musaro A, Rosenthal N, Sweeney HL (1998) Viral mediated expression of insulin-like growth factor I blocks the aging-related loss of skeletal muscle function. Proc Natl Acad Sci U S A 95: 15603–15607.
4. Macedo A, Moriggi M, Vasso M, De Palma S, Sturnega M, et al. (2012) Enhanced athletic performance on multisite AAV-IGF1 gene transfer coincides with massive modification of the muscle proteome. Hum Gene Ther 23: 146–157.
5. Rivera VM, Gao GP, Grant RL, Schnell MA, Zoltick PW, et al. (2005) Long-term pharmacologically regulated expression of erythropoietin in primates following AAV-mediated gene transfer. Blood 105: 1424–1430.
6. Chenuaud P, Larcher T, Rabinowitz JE, Provost N, Cherel Y, et al. (2004) Autoimmune anemia in macaques following erythropoietin gene therapy. Blood 103: 3303–3304.
7. Gao G, Lebherz C, Weiner DJ, Grant R, Calcedo R, et al. (2004) Erythropoietin gene therapy leads to autoimmune anemia in macaques. Blood 103: 3300–3302.
8. Ni W, Le Guiner C, Gernoux G, Penaud-Budloo M, Moullier P, et al. (2011) Longevity of rAAV vector and plasmid DNA in blood after intramuscular injection in nonhuman primates: implications for gene doping. Gene Ther 18: 709–718.
9. Zhou S, Murphy JE, Escobedo JA, Dwarki VJ (1998) Adeno-associated virus-mediated delivery of erythropoietin leads to sustained elevation of hematocrit in nonhuman primates. Gene Ther 5: 665–670.
10. Baoutina A, Alexander IE, Rasko JE, Emslie KR (2007) Potential use of gene transfer in athletic performance enhancement. Mol Ther 15: 1751–1766.
11. Lippi G, Guidi G (2003) New scenarios in antidoping research. Clin Chem 49: 2106–2107.
12. Beiter T, Zimmermann M, Fragasso A, Armeanu S, Lauer UM, et al. (2008) Establishing a novel single-copy primer-internal intron-spanning PCR (spiPCR) procedure for the direct detection of gene doping. Exerc Immunol Rev 14: 73–85.
13. Beiter T, Zimmermann M, Fragasso A, Hudemann J, Niess AM, et al. (2011) Direct and long-term detection of gene doping in conventional blood samples. Gene Ther 18: 225–231.
14. Moser DA, Neuberger EW, Simon P (2012) A quick one-tube nested PCR-protocol for EPO transgene detection. Drug Test Anal 4: 870–875.
15. Baoutina A, Coldham T, Bains GS, Emslie KR (2010) Gene doping detection: evaluation of approach for direct detection of gene transfer using erythropoietin as a model system. Gene Ther 17: 1022–1032.
16. Baoutina A, Coldham T, Fuller B, Emslie KR (2013) Improved Detection of Transgene and Nonviral Vectors in Blood. Hum Gene Ther Methods.
17. Ni W, Le Guiner C, Moullier P, Snyder RO (2012) Development and utility of an internal threshold control (ITC) real-time PCR assay for exogenous DNA detection. PLoS One 7: e36461.
18. Baoutina A, Coldham T, Fuller B, Emslie KR (2013) Improved detection of transgene and nonviral vectors in blood. Hum Gene Ther Methods 24: 345–354.
19. Hou Y, Zhang H, Miranda L, Lin S (2010) Serious overestimation in quantitative PCR by circular (supercoiled) plasmid standard: microalgal pcna as the model gene. PLoS One 5: e9545.
20. Lin CH, Chen YC, Pan TM (2011) Quantification bias caused by plasmid DNA conformation in quantitative real-time PCR assay. PLoS One 6: e29101.
21. Hindson BJ, Ness KD, Masquelier DA, Belgrader P, Heredia NJ, et al. (2011) High-throughput droplet digital PCR system for absolute quantitation of DNA copy number. Anal Chem 83: 8604–8610.
22. Miller SA, Dykes DD, Polesky HF (1988) A simple salting out procedure for extracting DNA from human nucleated cells. Nucleic Acids Res 16: 1215.
23. Arsic N, Zentilin L, Zacchigna S, Santoro D, Stanta G, et al. (2003) Induction of functional neovascularization by combined VEGF and angiopoietin-1 gene transfer using AAV vectors. Mol Ther 7: 450–459.
24. Untergasser A, Cutcutache I, Koressaar T, Ye J, Faircloth BC, et al. (2012) Primer3–new capabilities and interfaces. Nucleic Acids Res 40: e115.
25. Burns M, Valdivia H (2008) Modelling the limit of detection in real-time quantitative PCR. European Food Research and Technology 226: 1513–1524.
26. Huggett JF, Foy CA, Benes V, Emslie K, Garson JA, et al. (2013) The digital MIQE guidelines: Minimum Information for Publication of Quantitative Digital PCR Experiments. Clin Chem 59: 892–902.

27. Little S (2001) Amplification-refractory mutation system (ARMS) analysis of point mutations. Curr Protoc Hum Genet Chapter 9: Unit 9 8.

28. Taly V, Pekin D, Benhaim L, Kotsopoulos SK, Le Corre D, et al. (2013) Multiplex Picodroplet Digital PCR to Detect KRAS Mutations in Circulating DNA from the Plasma of Colorectal Cancer Patients. Clin Chem.

29. Morisset D, Stebih D, Milavec M, Gruden K, Zel J (2013) Quantitative analysis of food and feed samples with droplet digital PCR. PLoS One 8: e62583.

30. Strain MC, Lada SM, Luong T, Rought SE, Gianella S, et al. (2013) Highly precise measurement of HIV DNA by droplet digital PCR. PLoS One 8: e55943.

20

Genomic Assortative Mating in Marriages in the United States

Guang Guo[1,2,3]*, Lin Wang[4], Hexuan Liu[1,2], Thomas Randall[5]

1 Department of Sociology, the University of North Carolina at Chapel Hill, Chapel Hill, North Carolina, the United States of America, **2** Carolina Population Center, the University of North Carolina at Chapel Hill, Chapel Hill, North Carolina, the United States of America, **3** Carolina Center for Genome Sciences, the University of North Carolina at Chapel Hill, Chapel Hill, North Carolina, the United States of America, **4** Center for Child and Family Policy, Duke University, Durham, North Carolina, the United States of America, **5** National Institute of Environmental Health Sciences, Research Triangle Park, North Carolina, the United States of America

Abstract

Assortative mating in phenotype in human marriages has been widely observed. Using genome-wide genotype data from the Framingham Heart study (FHS; number of married couples = 989) and Health Retirement Survey (HRS; number of married couples = 3,474), this study investigates genomic assortative mating in human marriages. Two types of genomic marital correlations are calculated. The first is a correlation specific to a single married couple "averaged" over all available autosomal single-nucleotide polymorphisms (SNPs). In FHS, the average married-couple correlation is 0.0018 with $p = 3 \times 10^{-5}$; in HRS, it is 0.0017 with $p = 7.13 \times 10^{-13}$. The marital correlation among the positively assorting SNPs is 0.001 ($p = .0043$) in FHS and 0.015 ($p = 1.66 \times 10^{-24}$) in HRS. The sizes of these estimates in FHS and HRS are consistent with what are suggested by the distribution of the allelic combination. The study also estimated SNP-specific correlation "averaged" over all married couples. Suggestive evidence is reported. Future studies need to consider a more general form of genomic assortment, in which different allelic forms in homologous genes and non-homologous genes result in the same phenotype.

Editor: Margaret M. DeAngelis, University of Utah, United States of America

Funding: Funding provided by Challenge Grant RC1 DA029425-01, National Institutes of Health, http://www.nih.gov/. The funders had no role in study design, data collection and analysis, decision to publish, or preparation of the manuscript.

Competing Interests: The authors have declared that no competing interests exist.

* Email: guang_guo@unc.edu

Introduction

Assortative mating refers to a systematic departure from random mating. Positive assortative mating or homogamy occurs when mating individuals have similar traits, and negative assortative mating or heterogamy occurs when mating individuals have dissimilar traits. Human assortative mating in phenotype has been investigated for more than a century. In 1903, Pearson and colleagues report that the correlations in height, the span of arms, and the length of left forearm between husband and wife are 0.28, 0.20, and 0.20, respectively, drawing on extensive family records of 1,000 husband-wife pairs. Since Pearson's work, marriage partners have been shown to assort on a wide range of traits including race and ethnicity, age, propinquity in geography, religious belief, socio-economic status (such as educational attainment, occupation, and income), cognitive ability, anthropometric measures (such as weight, height, skin pigmentation, and other related measures), personality characteristics, mental and psychiatric conditions, and political attitudes (e.g., [1,2–13]).

If marriages are assorted to a degree by individual traits and if these traits are to a degree associated with genetic variation, it would be reasonable to hypothesize a degree of genetic assortment in human marriages. As an illustrative example, the heritability of human height is about 0.80 in developed countries [14]. Recent genome-wide association studies (GWAS) have found at least 180 independent regions of the genome that are associated with height

[15–19]. Figure 1 shows the correlation of height for different types of pairs using data from the Framingham Heart Study (FHS), with height standardized within each sex. The data show a correlation of about one half for same-sex as well as opposite-sex full-sibling pairs and parent-child pairs. The correlation for randomly paired individuals is essentially zero. The correlation for married couples in FHS after adjusting for population structure is about 0.27. This marital assortment in height likely has a major genetic component.

Genetic assortative mating may have reproductive consequences. Thiessen and Gregg [6] hypothesize that positive assortative mating outside nuclear families increases the genetic relatedness within a family, which in turn increases inclusive fitness without an extra reproductive effort. Lewontin [3] suggests that human assortative mating may play a major role in redistributing genes in contemporary times, particularly because selection through death has largely been replaced by selection through birth due to sharply-reduced mortality. If mating partners do share similar genetic variants related to, for example, obesity or psychiatric conditions, the impact of these genetic variants on the couples' offspring may be compounded. The role of genetic assortative mating may evolve with social trends. For example, college-educated Americans are increasingly more likely to marry each other rather than those with less education in comparison to a half-century ago [20]. This educational assortative mating reinforces a

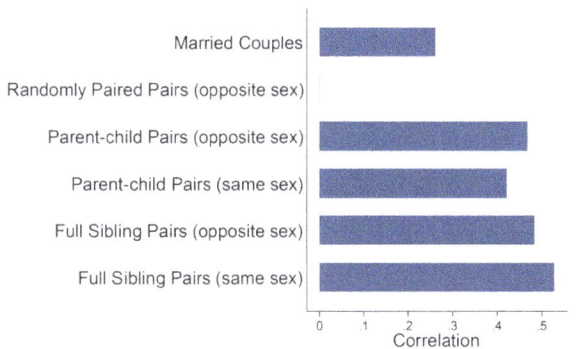

Figure 1. FHS data – the correlation of height (standardized within each sex) for married couples (N of couples = 989), opposite-sex random pairs from permuted individuals in FHS (N = 200,000), opposite-sex parent-child pairs (N = 3,447), same-sex parent-child pairs (N = 3,511), opposite-sex full sibling pairs (N = 2,815), and same-sex full sibling pairs (N = 2,898).

growing social divide between those with very low levels of education and those with more education, magnifying social class differences. This growing social divide could be partially genetic because of assortative mating.

Pearson [21] conjectures that, on average, a husband and wife are more alike than first cousins, whose coefficient of genetic relatedness is 0.125 and probably as much alike as uncle and niece, whose coefficient of genetic relatedness is 0.25, apparently basing the conjectures on the correlation findings over anthropometric measures. Pearson compares human homogamy to self-fertilization in plants; nevertheless, he realizes that human homogamy may have any degree of intensity and may be restricted to certain traits because genetic assortment can only be accomplished through phenotype.

In this project, we assess the extent to which marriage partners assort genetically using genome-wide genotype (GWAS) data from two independent studies in the United States for replication: 989 married couples in the Framingham Heart Study [22] (FHS) and 3,474 married couples in the Health and Retirement Survey (HRS). We carry out three sets of analyses: the first analysis uses 989 married couples and 287,294 SNPs in FHS; the second uses 3,474 couples and 66,526 SNPs (these 66,526 SNPs are common to both genotyping platforms used in the FHS and HRS studies); and the third analysis repeats the FHS analysis using the same 66,526 SNPs that are commonly available in FHS and HRS.

This analysis focuses on genomic assortative mating beyond race and ethnicity. It is well-known that marriages in the United States assort on race and ethnicity (e.g., [9,23]). To estimate genetic correlation within married couples net of race and ethnicity, population stratification must be controlled. In our analysis, population stratification is controlled directly in the regression models that estimate genomic assortment.

To estimate genetic assortative mating at the genomic level, we calculate two types of genome-wide marital correlations. The first is a correlation specific to a single married couple (couple correlation) "averaged" over all available autosomal SNPs. For FHS, this calculation yields 989 correlation estimates, one for each married couple averaged over 287,294 SNPs. Married-couple correlations provide a global or genomic estimate of the correlation averaged over the human genome. Such a measure is possible and attempted in this project because assortative mating

may occur over a number of human traits. Negative genomic assortment is a potential complication that may cancel negative and positive genomic assortment within a single married couple. Although assortative mating is generally considered positive, negative assortment or that opposites attract is likely to be present [1,6]. To address this issue, we estimate two additional correlations for each married couple. One is based on about half of the 287,294 SNPs that assort more positively and the other is based on the other half that assort more negatively.

The second marital correlation is a SNP correlation "averaged" over all married couples. For FHS, the SNP correlation analysis yields 287,294 correlations, one for each SNP averaged over 989 married couples. The analysis of couple correlations is quite distinct from GWAS studies. It is concerned with genetic similar within a couple averaged over the genome; it is also far more computationally demanding than a GWAS analysis. The analysis of SNP correlations appears to resemble a GWAS analysis: a GWAS study examines each SNP's association with a single phenotype in a collection of individuals and a SNP-correlation analysis estimates the average correlation over a collection of married couples with respect to a SNP. However, an important difference between the two is that married couples may assort on different phenotypes and thus assort at different genetic loci, which makes it more difficult for the analysis of SNP correlations to produce reliable estimates than a GWAS analysis.

Recent work by Domingue et al. [24] provides an estimate of genome-wide genetic similarity and an estimate of educational similarity within spousal pairs, concluding that the spousal genetic similarity over the genome is about one third or one fourth of the spousal educational similarity. Although using the same two data sources of FHS and HRS, our analysis was independently performed and reveals a number of additional insights. We use a different measure of spousal genomic similarity, calculate additional two measures of couple correlation for each married couple, and estimate SNP-correlations.

Results

Figure 2 shows the FHS distribution of couple correlation for married couples (N = 989), opposite-sex random pairs from permuted individuals in FHS (N = 200,000), opposite-sex random pairs from permuted individuals among married couples (N = 246,870), full-sibling pairs (N = 5,713), and parent-child pairs (N = 6,958). After controlling for population admixture, the married-couple correlations average 0.0018 relative to the average of randomly paired individuals (Panel 1 of Table 1). The correlation is highly significant according to both permutation tests. In contrast, the pair-specific correlations for full-sibling pairs and parent-child pairs are both centered on 0.50 with a mean of 0.503 (SD = 0.053) and 0.499 (SD = 0.007), respectively. As expected, the standard deviation of the parent-child pairs is much smaller than that of the full siblings.

Figure 3 shows the effect of controlling for population admixture via adding seven main principal components in FHS. The figure presents two estimated distributions of married-couple correlation (Panels 1 and 2) and the distribution of pair correlations estimated from random pairs (Panel 3). The results in Panels 1 and 2 are without and with control for population admixture, respectively. Once population admixture is controlled, the couple correlations that are larger than 0.02 have vanished (Panel 2).

Figure 4 shows the HRS distribution of pair correlation, for married couples (N = 3,474), opposite-sex random pairs from permuted individuals in HRS (N = 200,000), and opposite-sex

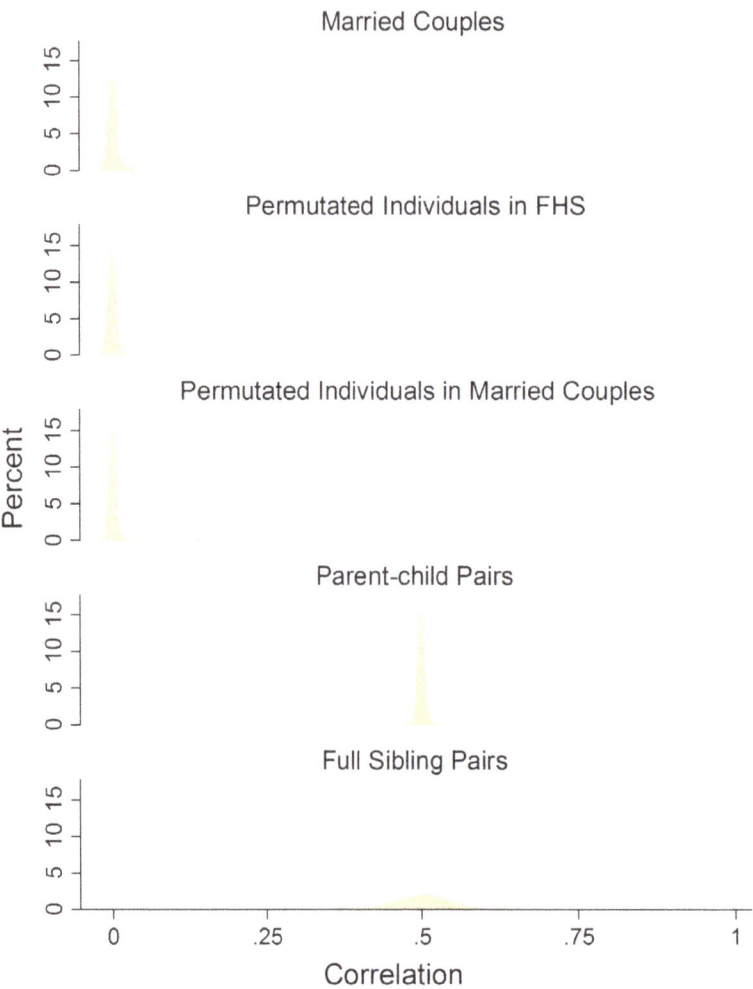

Figure 2. FHS data – the empirical density distribution of couple correlation for married-couples (N = 989), opposite-sex random pairs from permuted individuals in FHS (N = 200,000), opposite-sex random pairs from permuted individuals among married couples (N = 246,870), parent-child pairs (N = 6,958), and full sibling pairs (N = 5,713), with each mixed-model regression estimating a within-a-single-pair correlation "averaged" over 287,294 SNPs.

random pairs from permuted individuals among married couples (N = 200,000), with each mixed-model regression estimating a within-a-single-pair correlation "averaged" over the 66,526 SNPs. The results from the two permutation tests in Panel 2 of Table 1 suggest that averaged over the genome, married couples in HRS has a correlation of 0.0016–0.0017 relative to permuted random pairs. The results from both tests are highly significant. This HRS finding is similar to that from FHS.

Panel 1 of Table 2 presents FHS distribution of within-pair allelic combination for married couples, random pairs permuted among married couples, random pairs permuted among all FHS subjects, parent-child pairs, and full-sibling pairs. Large differences exist between genetically-related pairs (GRPs) and genetically nonrelated pairs (GNPs). Consistent with our hypothesis, GRPs tend to have a much higher percentage in allelic combinations of 22, 12 or 21, and 00 that contribute to positive assortment than GNPs. GNPs tend to have a much higher percentage than GRPs in allelic combinations of 02 or 20, 01 or 10, and 11 that contribute to negative assortment. Consistent with Figure 2, married couples

exhibit an allelic distribution that is almost identical to those from the two sets of random pairs. However, a careful comparison reveals that married couples have slightly higher proportions of positive-assorting SNP combinations (22, 12 or 21, and 00) than those among the two types of random pairs, suggesting that the positive genomic correlation for married couples be slightly higher than that of random pairs. For the negatively assorting combinations (02 or 20, 10 or 01, and 11), the differences between married couples and random pairs are small and the directions are mixed. Compared with random pairs, married couples have a lower proportion in 02 or 20, and 10 or 01, but a higher proportion in 11, suggesting that the negative genomic correlation for married couples be zero or extremely small.

Panel 2 of Table 2 provides the observed HRS distribution of within-pair allelic combination for different types of pairs for the 66,526 SNPs. Comparing married couples against random pairs in HRS yields a similar pattern to that in FHS: the proportions of positively assorting allelic combinations in married couples are consistently higher than those in random pairs. These allelic data

Table 1. FHS and HRS data – Two permutation tests for married-couple correlations within "negative" and "positive" SNPs: (1) permuted individuals in 989 (FHS) and 3,474 (HRS) married couples, respectively, and (2) permuted all individuals in FHS.

FHS data

		Permuted individuals in 989 married couples	Permuted all individuals in the FHS
All SNPs	Mean difference in correlation: (Married couples minus random pairs)	0.0018	0.0018
	Average p-values	0.00003	0.0001
	Proportion of p-values <0.05	99.98%	99.94%
Negative "half" of SNP combinations: 20/02, 01/10, and 11	Mean difference in correlation: (Married couples minus random pairs)	−0.000076	0.00036
	Average p-values	0.417	0.178
	Proportion of p-values <0.05	8.94%	36.14%
Positive "half" of SNP combinations: 00,12/21, and 22	Mean difference in correlation: (Married couples minus random pairs)	0.00095	0.0012
	Average p-values	0.0043	0.0088
	Proportion of p-values <0.05	98.14%	96.32%

HRS data

		Permuted individuals in 3,474 married couples	Permuted all individuals in the HRS
All SNPs	Mean difference in correlation: (Married couples minus random pairs)	0.0017	0.0016
	Average p-values	7.13×10^{-13}	8.39×10^{-12}
	Proportion of p-values <0.05	100%	100%
Negative "half" of SNP combinations: 20/02, 01/10, and 11	Mean difference in correlation: (Married couples minus random pairs)	−0.0012	−0.0012
	Average p-values	0.0023	0.0016
	Proportion of p-values <0.05	99.2%	99.3%
Positive "half" of SNP combinations: 00, 12/21, and 22	Mean difference in correlation: (Married couples minus random pairs)	0.015	0.020
	Average p-values	1.66×10^{-24}	7.75×10^{-41}
	Proportion of p-values <0.05	100%	100%

in HRS suggest that the "positive" half of the SNPs for married couples have a positive correlation while the negative correlation may be zero or extremely small. Comparing FHS and HRS, the proportion of positive assorting allelic combinations in married couples relative to random pairs appears considerably higher in HRS than in FHS, suggesting that the "positive" half of the SNPs for married couples in HRS have a larger positive correlation than those in FHS. These expectations are confirmed by regression findings.

Figure 5 provides the FHS empirical distribution of the "positive" and "negative" pair correlation, for married couples (N = 989), opposite-sex random pairs from permuted individuals in FHS (N = 200,000), and opposite-sex random pairs from permuted individuals among married couples (N = 246,870), with each mixed-model regression estimating the within a single-pair correlation "averaged" over about one half of the 287,294 SNPs.

The second half of Panel 1 of Table 1 shows the FHS results of two permutation tests for the married-couple correlations within the "negative" and "positive" SNPs. The two tests yield essentially identical findings. For the "negative" SNPs, the difference between the married-couple correlation and the random-pair correlation is small and statistically non-significant. In contrast, for

the "positive" SNPs, the average of the married-couple correlation minus the random-pair correlation is about 0.001 and statistically significant according to the average p-values (0.0043 and 0.0088).

Figure 6 presents the HRS distribution of the "positive" and "negative" pair-specific correlation, for married couples (N = 3,474), opposite-sex random pairs from permuted individuals in the HRS (N = 200,000), and opposite-sex random pairs from permuted individuals among married couples (N = 200,000), with each mixed-model regression estimating the within a single-pair correlation "averaged" over about one half of the 66,526 SNPs.

The second half of Panel 2 of Table 1 presents two permutation tests for HRS data – Two permutation tests for couple-specific correlations within "negative" and "positive" SNPs. Like in the FHS data, the two tests yield very similar findings. For the "negative" SNPs, on average, married couples have a small and statistically significant negative correlation (−0.0012, p = 0.0023; −0.0012, p = 0.0016). For the "positive" SNPs, on average married couples show a correlation of about 0.015 and 0.020, respectively, with extremely small p-values of 1.66×10^{-24} and 7.75×10^{-41}.

Panel 1 of Figure 7 plots the genome-wide SNP-specific correlation for each of the 287,294 SNPs in 989 married couples

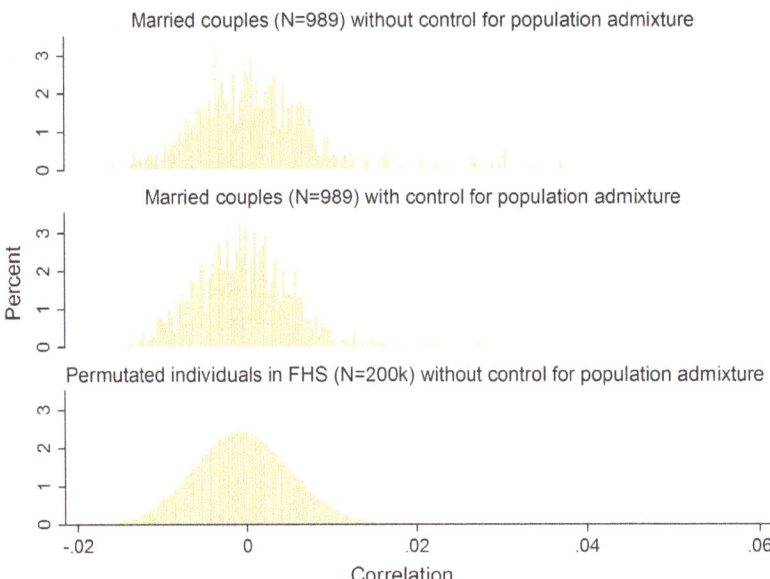

Figure 3. FHS data – the empirical density distribution of married-couple correlation over the 287,295 SNPs; (1) married couples (N = 989) without control for population admixture, (2) married couples (N = 989) with control for population admixture, and (3) opposite-sex random pairs from permuted individuals in FHS (N = 200,000). Panels (2) and (3) are the same as Panels (1) and (2) in Figure 2 and enlarged.

in FHS. The correlation was estimated using the mixed model that allows positive and negative correlations. A large majority of the SNP correlations are scattered around 0 with a range of −0.10–0.10. Panel 2 of Figure 7 parallels Panel 1 of Figure 7 except it is based on HRS with a much larger sample of 3,474 married couples. The large sample explains the much narrower ranges of estimates of SNP correlations for HRS, ranging mostly between −0.05 and 0.05.

Figure 4. HRS data – the empirical density distribution of couple correlation for married-couples (N = 3,474), opposite-sex random pairs from permuted individuals (N = 200,000), and opposite-sex random pairs from permuted individuals among married couples (N = 200,000), with each mixed-model regression estimating a within-a-single-pair correlation "averaged" over 66,526 SNPs. These 66,526 SNPs are also available in FHS.

Table 2. FHS and HRS data – the observed distribution of within-pair allelic combination for different types of pairs [%(standard deviation)] for a total of 287,294 SNPs (FHS), 66,526 SNPs (HRS), and 66,526 SNPs (FHS common to those in HRS).

Panel 1: FHS

Within-pair combination	Married Couples N = 989	Random pairs permuted among married Couples N = 246,870	Random pairs permuted among all FHS subjects N = 200,000	Parent-child Pairs N = 6,958	Full sibling Pairs N = 5,713
02/20	6.457(0.21)	6.474(0.21)	6.457(0.22)	0.024(0.02)	1.606(0.30)
01/10	32.508(0.26)	32.560(0.25)	32.578(0.26)	22.661(0.23)	19.344(1.87)
11	12.725(0.22)	12.714(0.21)	12.737(0.21)	16.260(0.23)	19.560(1.23)
00	40.061(0.32)	40.021(0.32)	40.009(0.38)	48.190(0.32)	49.050(1.10)
12/21	7.069(0.14)	7.058(0.12)	7.050(0.12)	9.922(0.16)	6.690(0.60)
22	1.180(0.06)	1.173(0.05)	1.169(0.05)	2.943(0.12)	3.757(0.41)
Total	100%	100%	100%	100%	100%

Panel 2: HRS

Within-pair combination	Married Couples N = 3,474	Random pairs permuted among married Couples N = 200,000	Random pairs permuted among all FHS subjects N = 200,000		
02/20	6.728(0.44)	7.360(1.24)	7.580(1.40)		
01/10	31.971(0.72)	32.688(1.26)	32.905(1.41)		
11	13.192(0.35)	12.956(0.60)	12.865(0.66)		
00	37.146(0.67)	36.470(1.22)	36.251(1.37)		
12/21	8.819(0.22)	8.583(0.43)	8.512(0.48)		
22	2.143(0.22)	1.943(0.33)	1.888(0.37)		
Total	100%	100%	100%		

Panel 3: HRS for a total of 66,526 SNPs in FHS that are also available in HRS

Within-pair combination	Married Couples N = 989	Random pairs permuted among married Couples N = 246,870	Random pairs permuted among all FHS subjects N = 200,000	Parent-child Pairs N = 6,958	Full sibling Pairs N = 5,713
02/20	6.708(0.22)	6.729(0.23)	6.714(0.23)	0.027(0.03)	1.669(0.30)
01/10	33.459(0.29)	33.513(0.27)	33.524(0.28)	23.390(0.25)	19.948(1.88)
11	13.233(0.23)	13.224(0.23)	13.241(0.22)	16.815(0.25)	20.246(1.26)
00	38.037(0.36)	37.993(0.35)	37.988(0.40)	46.407(0.35)	47.300(1.12)
12/21	7.338(0.15)	7.327(0.14)	7.321(0.14)	10.308(0.18)	6.947(0.62)
22	1.224(0.07)	1.215(0.06)	1.212(0.06)	3.054(0.13)	3.901(0.42)
Total	100%	100%	100%	100%	100%

To evaluate our measure of correlation, Figure 8 plots the genome-wide SNP correlation for each of the 287,294 SNPs in 5,713 full sibling pairs from FHS. Both same-sex and opposite-sex full sibling pairs are included. The large majority of the SNP correlations are scattered around 0.50 with a range of 0.40–0.60. Figure 9 presents the genome-wide SNP correlation for each of the 287,294 SNPs in 6,958 parent-child pairs. Again, both same-sex and opposite-sex parent-child pairs are included. The large majority of the SNP-specific correlations are scattered around 0.50 with a range of 0.45–0.55. As expected, the spread of the correlations for parent-child pairs is considerably narrower than that of full sibling pairs. The results in Figures 8 and 9 demonstrate that our method can produce the known patterns of genetic similarity in full sibling pairs and parent-child pairs.

Potentially problematic SNPs are those with a correlation estimate that is much less than 0.50 in the full-sibling analysis and the parent-child analysis. These SNPs do not affect our results of SNP correlations because each SNP correlation is independently calculated. In the calculation of the couple correlations where all SNPs were used in each regression, we excluded 231 out of the 287,525 SNPs. These excluded SNPs have either a full-sibling correlation less than 0.2 or greater than 0.8, or a parent-child correlation less than 0.3. The findings of couple correlations are not affected by whether these SNPs are included or excluded.

Figure 10 shows the FHS permutation tests for the SNP-specific correlations in married couples against random pairs. As will be shown in Table 3, a small number of SNPs achieve a genome-wide significance with a p-value of 5×10^{-8} or smaller. The Q–Q plot of p-values from the SNP-specific correlations is presented in Figure 11, showing that some signals remain after removing the SNPs that have genome-wide significance (Panel 2 of Figure 11).

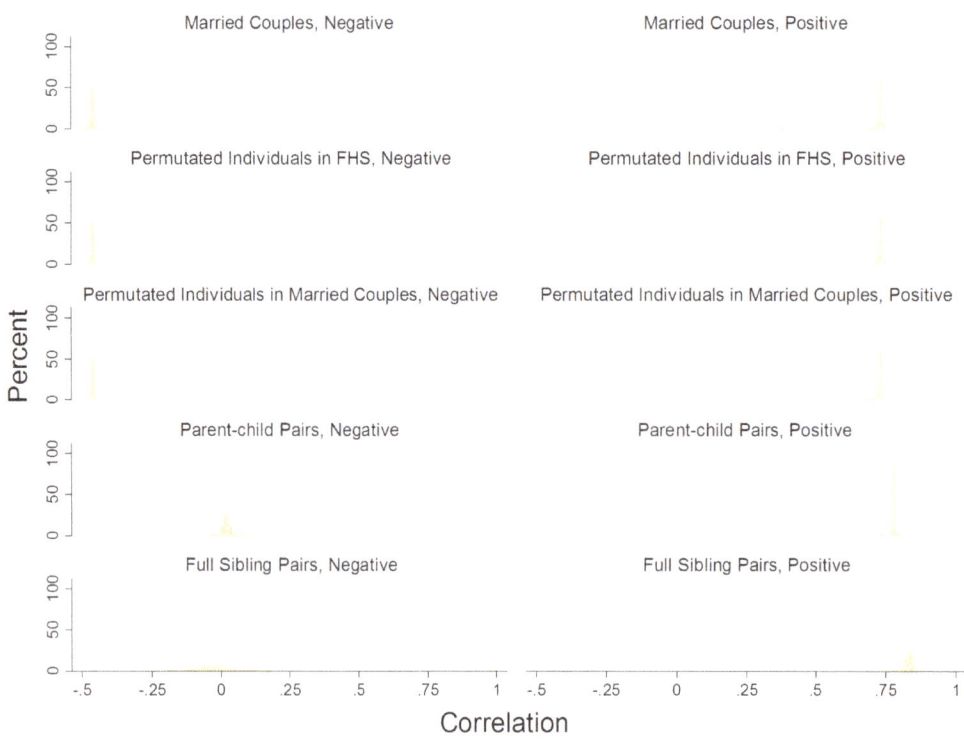

Figure 5. FHS data – the empirical density distribution of the "positive" and "negative" couple correlation, for married couples (N = 989), opposite-sex random pairs from permuted individuals in FHS (N = 200,000), opposite-sex random pairs from permuted individuals among married couples (N = 246,870), parent-child pairs (N = 6,958), and full sibling pairs (N = 5,713), with each mixed-model regression estimating the within a single-pair correlation "averaged" over about one half of the 287,294 SNPs.

Table 3 lists 10 SNPs with the smallest p-values for the SNP-specific correlations in 989 married couples out of the 287,294 SNPs from FHS. The table lists SNP name, chromosome position, gene name when available, gene location, reference allele frequency, SNP correlation for married couples and p value from the permutation test, correlation for full sibling pairs and p value, and correlation for parent-child pairs and p value. Eight SNPs have a p-value 5×10^{-8} or smaller. The largest ten correlations are all positive. The SNP correlations from full-sibling pairs and parent-child pairs are in the expected ranges.

Our replication of the top ten SNPs from FHS (Table 3) using HRS yielded two SNPs (rs16871467 and rs9483869) that are statistically significant at 0.057 and 0.050, respectively. The correlations of these two SNPs are also positive, but smaller (0.026 and 0.027, respectively) than those in FHS. Overall, three of the SNPs in the HRS analysis with 66,526 SNPs achieve a genome-wide significance with a p-value of 5×10^{-8} or smaller.

Our final analysis is an FHS-66,526-SNP analysis for couple correlation. Panel 3 of Table 2 provides the observed distribution of within-pair allelic combination for different types of pairs for these SNPs in FHS. The table indicates that the distribution is much closer to the FHS distribution based on the full set of 287,294 SNPs with the same set of individuals than that in HRS based on the exactly the same set of SNPs but a different set of individuals. The regression analysis of couple correlation of these 66,526 SNPs in FHS confirm the findings from Panel 3 of Table 2 (not shown), providing evidence that married couple correlations are predominantly determined by individuals rather than SNPs

and that the HRS 66,526-SNP analysis is likely generalizable to the full-SNP analysis.

Discussion

In FHS, the two estimates of genome-wide couple correlation are 0.0018 (p = 3×10^{-5}) and 0.0o18 (p = 10^{-4}). These couple correlation estimates in HRS are 0.0016 (p = 8.29×10^{-12}) and 0.0017 (p = 7.13×10^{-13}). The much smaller p values from HRS in these estimates as well as other estimates are likely due to the much larger samples of HRS (3,474 couples) than FHS (989 couples). These estimates of couple correlations are not threatened by multiple testing.

Consistent with the estimates of Domingue et al [24], we show positive overall similarity in genomic assortment in married couples; however, our estimates seem much smaller than theirs (0.0016–0.0018 vs. 0.02–0.045). This is the case after taking into account that the two sets of estimates are not exactly comparable. As demonstrated in this analysis (Figures 2, 8, and 9), our estimates are essentially coefficients of genetic relatedness (r) and their estimates are quartile-transformed coefficients of kinship (F) with $r = 2F$, where F is untransformed coefficient of kinship. Our estimates in spousal correlation of educational attainment or years of education with standardization within each sex are 0.59 and 0.52 for HRS and FHS, respectively. One fifth to one third of these quantities are much larger than our estimated genome-wide couple correlation of 0.0016–0.0018. The variation in couple correlation across racial/ethnic groups is examined only in HRS. Less than 1% of the couples in FHS are ethnic minorities. In HRS,

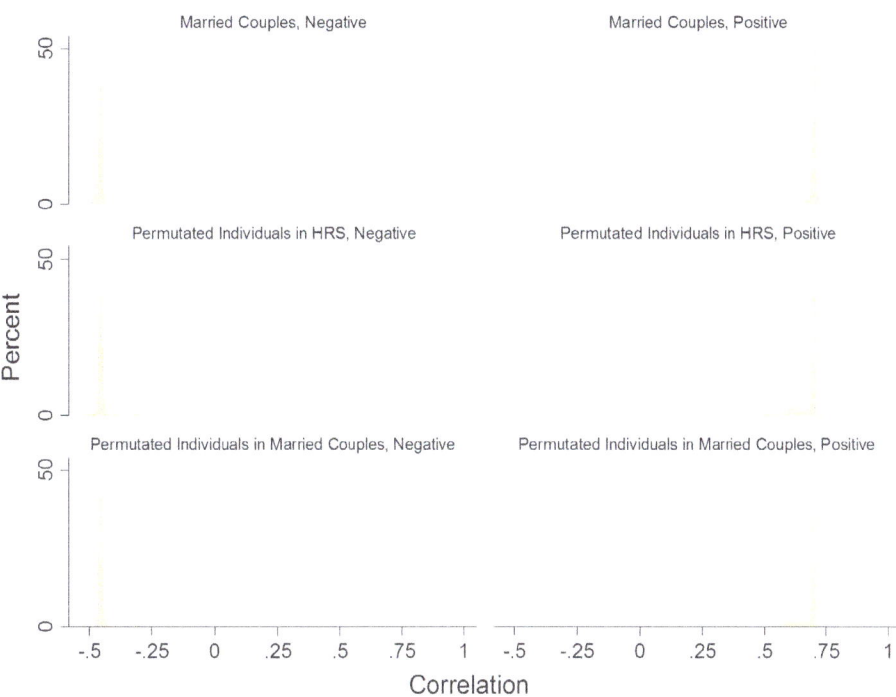

Figure 6. HRS data – the empirical density distribution of the "positive" and "negative" couple correlation for married couples (N = 3,474), opposite-sex random pairs from permuted individuals in the HRS (N = 200,000), and opposite-sex random pairs from permuted individuals among married couples (N = 200,000), with each mixed-model regression estimating the within a single-pair correlation "averaged" over about one half of the 66,526 SNPs.

Panel 1   ## Panel 2

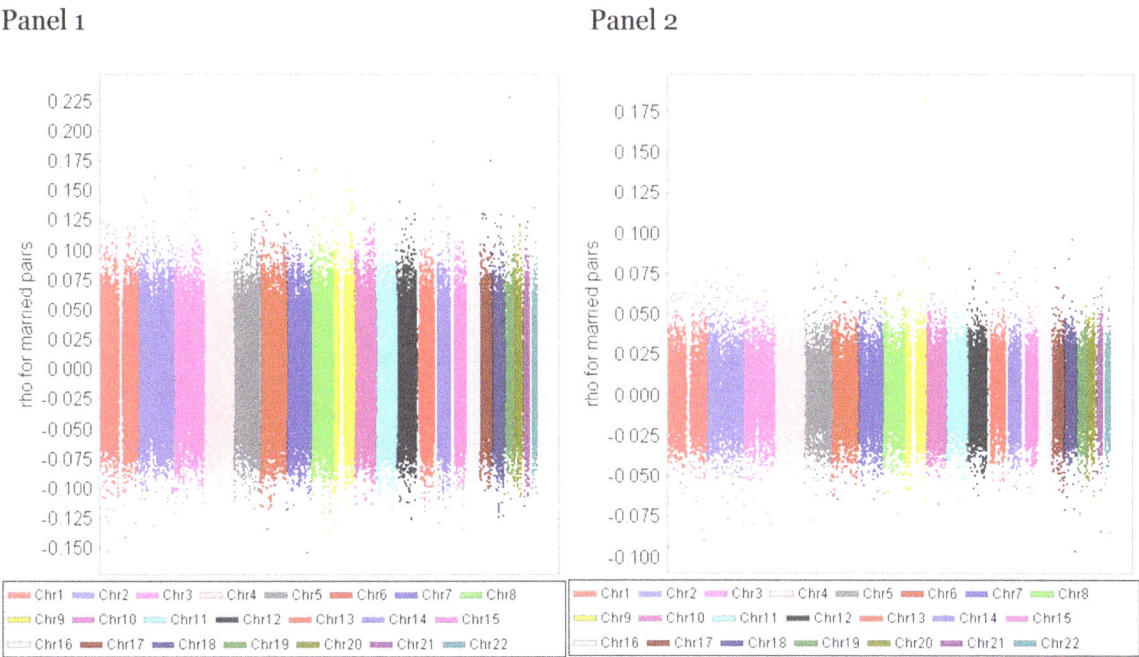

Figure 7. Panel 1: FHS data – genome-wide SNP-specific correlation for each of the 287,294 SNPs in 989 married couples. Panel 2: HRS data – genome-wide SNP-specific correlation for each of the 66,526 SNPs in 3,474 married couples (these 66,525 SNPs also available in FHS). The correlation was estimated using the mixed models with AR(1) covariance structure, controlling for population admixture.

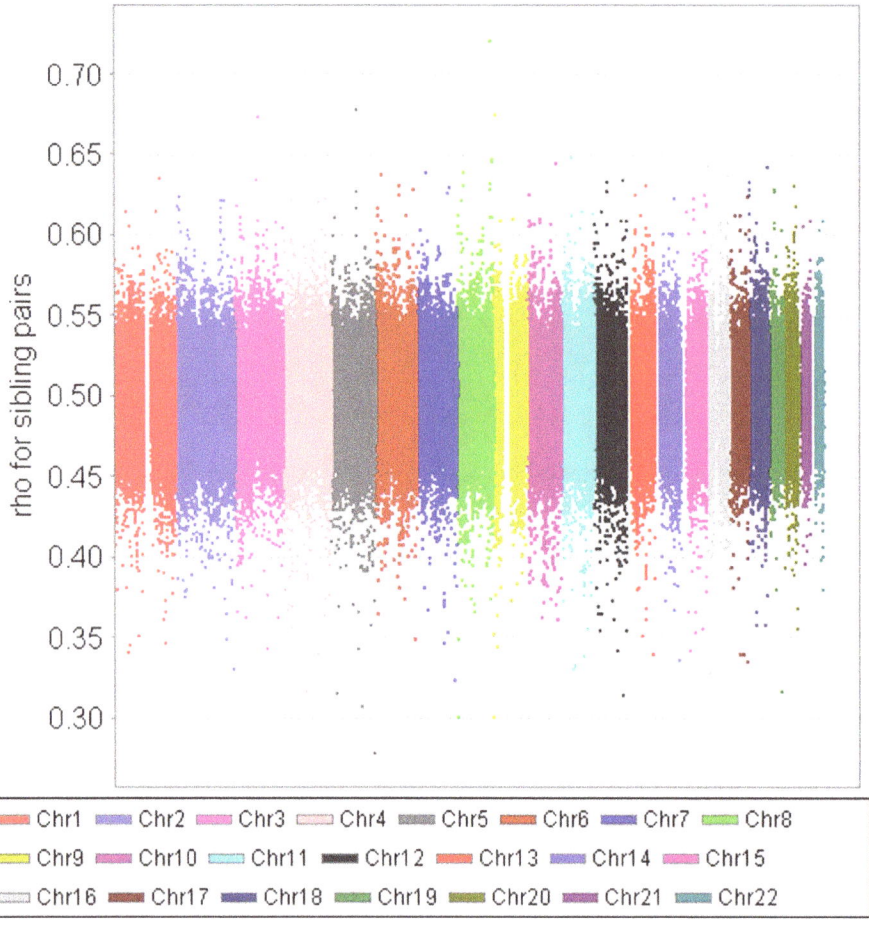

Figure 8. FHS data – genome-wide SNP-specific correlation for each of the 287,294 SNPs in 5,747 full sibling pairs. Both same-sex and opposite-sex full sibling pairs are included. The correlation was estimated using the mixed models with AR(1) covariance structure, controlling for population admixture.

constraining the sample to non-Hispanic whites yields a somewhat smaller and statistically significant couple correlation of 0.0012.

The negative couple correlations in FHS are small and statistically non-significant ($-.00008$, p = .41;.00036, p = .18). The negative marital correlations in HRS are small and statistically significant (-0.0012, p = .0023; -0.0012, p = .0016). The positive couple correlations are much larger than negative correlations in absolute values in both FHS (0.001, p = .0043; 0.0012, p = .0088) and HRS (0.015, p = 1.66×10^{-24}; 0.020, p = 7.75×10^{-41}). The sizes of these estimates in FHS and HRS are consistent with what are suggested by the distribution of the allelic combination in Panels 1 and 2 of Table 2. The data in Table 2 can be considered findings that are more closely based on raw data than those from regression analysis. In both FHS and HRS, the positive correlation is much larger and more statistically significant than the negative correlation suggesting that genetic assortative mating is primarily positive.

For the analysis of SNP-specific correlation based on FHS, of the 287,294 SNP correlations, eight have a p-value 5×10^{-8} or smaller. These SNPs are all positively correlated between married couples, with a range of 0.16–0.27. We repeated the analysis of SNP correlations for these eight SNPs using HRS data. In HRS, two of these eight SNPs (rs9483869 and rs16871467) are statistically significant at about 0.05 and also correlated positively. However, these replications are suggestive rather than definitive

because the two correlations in HRS are considerably smaller than those in FHS.

Neither rs9483869 nor rs16871467 has itself been identified as a statistically significant association in any previous GWAS analysis [25]. Rs9483869 is within an ncRNA called LINC00271, which is expressed in the brain [26]. Another SNP within LINC00271 (rs9494266) has been found to be a statistically significant hit in a GWAS on type 2 diabetes [27]. LINC00271 is in a region of high LD with the immediately adjacent gene AHI1, a gene involved in neurodevelopment and implicated in schizophrenia [27,28]. Rs16871467 is approximately 246 kb downstream of ARHGF28, a member of the Rho guanine nucleotide exchange factor family. This protein interacts with low molecular weight neurofilament mRNA and may be involved in the formation of amyotrophic lateral sclerosis neurofilament aggregates [29]. Opposite, towards the chr5 telomere, the closest defined element is the retrogene C17orf76 antisense RNA 1, approximately 36 kb away. This SNP does reside in a DNAse I hypersensitive site defined by the ENCODE project [30,31].

Genomic assortment in human marriages may vary over a number of factors. Different couples may assort on entirely different phenotypes and thus different genetic variants, which is expected to decrease the power of detecting SNP-specific correlations among couples. Genomic assortment may also be influenced by social and cultural contexts that vary across

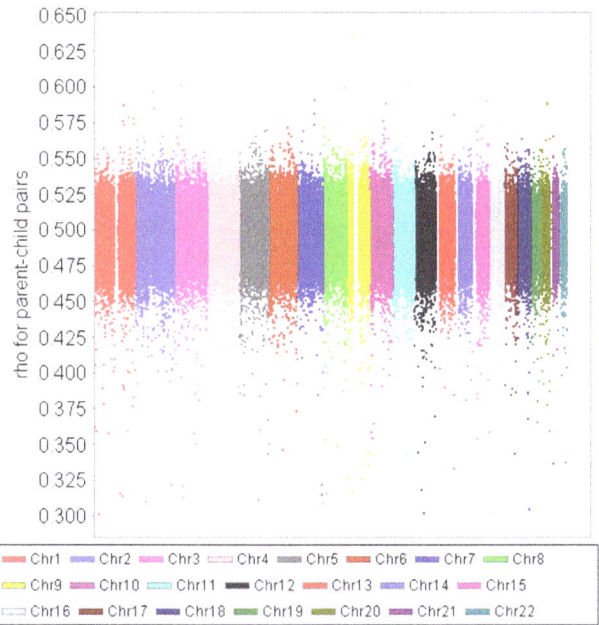

Figure 9. FHS data – genome-wide SNP-specific correlation for each of the 287,294 SNPs in 6,958 parent-child pairs. Both same-sex and opposite-sex parent-child pairs are included. The correlation was estimated using the mixed models with AR(1) covariance structure, controlling for population admixture.

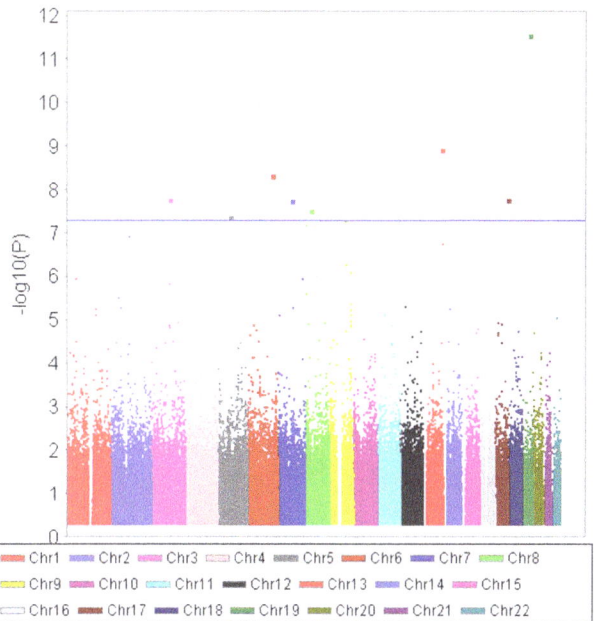

Figure 10. FHS data – the significance tests of SNP-specific correlations: the within-pair correlation of married couples against randomly-paired pairs. The tests for the 287,294 SNPs are shown in a Manhattan plot. The larger dots representing individual SNPs above the blue line indicate statistical significance at $p < 5 \times 10^{-8}$.

historical periods and geographic locations. American marriage is considerably different from marriage in other Western countries [32], not to mention marriage in non-Western countries. Pawlowski et al. [33] report an effect of World War II on mate preference in height. The advantage of taller males in the marriage market is evident among individuals born in the 1940 s, 1950 s and 1960 s, but not in the 1930 s. The authors suggest that this may be due to the relative scarcity of young men immediately after WWII. The genomic assortment may vary across geographic regions within the United States.

Overall, our data suggest a degree of genomic assortative mating at the allelic level in married couples who were born in the first half of the 20th century in the United States. Apparently, this degree of genetic assortment averaged over the human genome is much smaller than the 0.20 Pearson had conjectured based on the observed correlations in height and arm span between husband and wife. As alluded earlier, certain genetic variants such as those underlying height are likely to be heavily assorted; however, the level of overall assortment in the genome seems much less.

However, a genomic correlation of 0.015–0.02 with married couples, estimated for the "positive" assorting SNPs in HRS, can represent an important genomic assortment for at least two reasons. A married-couple correlation may be compared with genetic relatedness among biological relatives. A genomic correlation of 0.015–0.02 is close to the average genomic correlation (0.0312) among second cousins (or the genomic correlation [0.0312] of an individual with his grandfather's grandfather). While an individual passively and unselectively inherits half of his or her genes from each of the two parents, married individuals consciously or unconsciously assort on genes that play a strategic role in their reproductive marriages.

Our analysis of HRS reports a small but statistically significant negative genomic assortment, suggesting that negative genomic may, indeed, exist. This negative assortment contrasts conspicu-

ously with the only-positive assortment among genetic relatives (see Figures 2 and 4).

Our interest is in assortative mating rather than genomic similarity related to population stratification and marriages between distant relatives. The principal components included in the analysis are effective (Figure 3); nevertheless, it might be difficult to differentiate low-level genomic similarity due to assortative mating from low-level genetic similarity due to distant genetic relatives marrying each other.

There is one important methodological limitation in the current analysis. As Wright [34] pointed out decades ago, assortative mating can only be done through external phenotypes and the same phenotype may result from different DNA sequences or non-homologous genes. For example, a married couple may assort by body weight, but the body weight of the husband and the wife may depend on different sets of genes (e.g., *FTO* vs *MC4R*). Such cases of genetic assortment are missed by direct allelic comparison between homologous genes, an approach used in this analysis.

The methodological limitation underestimates a more general form of genomic assortment, in which different allelic forms cause the same phenotype within the same gene or different genes. Assortative mating may actually occur at a higher level than we estimated in this project. Only when the general form of genomic assortment is taken into account could the impact of assortative mating suggested by Lewontin [3] and Thiessen and Gregg [6] be adequately evaluated.

Methods

The Framingham Heart Study (FHS) is a community-based, prospective, longitudinal study following three generations of participants: (i) the Original Cohort enrolled in 1948 (N = 5,209); (ii) the Offspring Cohort consist of the children of the Original Cohort and their spouses, who were enrolled in 1971 (N = 5,124);

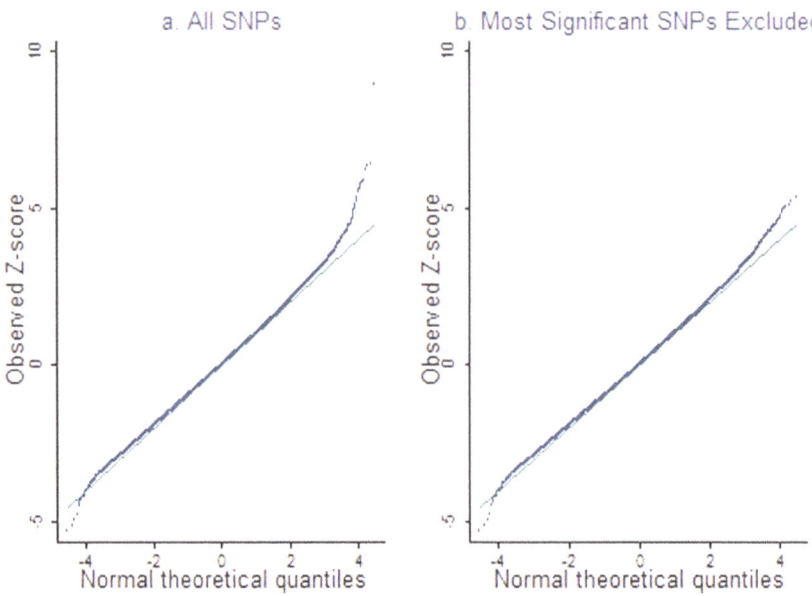

Figure 11. The QQ plot of observed Z-scores vs. expected Z-scores. The plot on the left side includes all 287,294 SNPs while the one on the right side excludes 8 SNPs with p-values smaller than 5×10^{-8}.

and (iii) the Generation Three Cohort consists of the grandchildren of the Original Cohort, who were enrolled in 2002 (N = 4,095). More information on FHS can be found online [22]. Our analysis uses the 1,978 individuals or 989 married couples whose genotype data are available. These individuals are predominantly of European origin. Less than 1% of FHS respondents were racial/ethnic minorities.

Of the 14,428 study subjects in FHS, a total of 9,237 consenting individuals have been genotyped including 4,986 women and 4,251 men. Genotyping for FHS participants was performed by Affymetrix (Santa Clara, CA, USA) using the Affymetrix 500K GeneChip array. The Y chromosome was not genotyped. The standard quality control filter is applied. Individuals with 5% or more missing genotype data are excluded from analysis. X chromosome SNPs, SNPs with a call rate ≤99% or a minor allele frequency ≤0.01 are also eliminated from analysis. The application of the quality control filter leaves 8,738 individuals with 287,525 SNPs from the 500K genotype data.

The Health and Retirement Survey (HRS), launched in 1992, is a longitudinal study, surveying more than 22,000 Americans over the age of 50 every two years and collecting information on labor force participation and health transitions. The HRS began collecting salivary DNA in 2006 and has approximately > 13,000 such DNA samples stored in repository. The genotyping for HRS was completed using the Illumina HumanOmni2.5-4v1 array, which includes more than one million SNPs. A total of 12,857 samples were genotyped and passed CIDR's quality control (QC) process. The HRS analysis used samples of 6,948 individuals or 3,474 married couples that have passed the QC. A total of 66,526 SNPs out of 287,525 SNPs used in FHS were also genotyped in HRS.

In all our analyses, the outcome variable is the dosage of minor alleles for a SNP, which is standardized with mean = 0 and SD = 1; a correlation coefficient is used to measure genetic similarity. A correlation coefficient has a range of −1 to 1 allowing measurement of positive as well as negative assortment, and was used widely in measuring phenotypic similarity in studies of assortative mating. Correlation coefficients based on dosages of

minor alleles are essentially coefficients of genetic relatedness (r). Because a coefficient of genetic relatedness is the most widely-used measurement of genetic relatedness among genetic relatives, our findings of genetic assortment among married couples can be readily understood and compared with the well-known genetic relatedness among full siblings ($r = 0.5$) and identical twins ($r = 1$).

Both married-couple-specific correlation and SNP-specific correlation are estimated by the following mixed linear model [35]:

$$Y = X\beta + \varepsilon \qquad (1)$$

where Y stands for standardized SNP dosage, X is a matrix of observed variables such as those used for controlling for population admixture, β is a coefficient vector of X including a standard intercept, and $\mathrm{Var}(Y|X) = \mathrm{Var}(\varepsilon) = \begin{pmatrix} W & \cdots & 0 \\ \vdots & \ddots & \vdots \\ 0 & \cdots & W \end{pmatrix}$

with $W = \sigma^2 \begin{pmatrix} 1 & \rho \\ \rho & 1 \end{pmatrix}$ in which ρ is either a couple correlation or a SNP correlation, depending on input data in Y. Model (1) is a special case of the auto-regressive AR(1) model. This AR(1) model allows for both positive and negative correlations, which correspond to positive and negative marital assortment.

For the couple correlation, Y_{ij} in Y is the SNP dosage for individual i and SNP j where $i = 1,2$ indexing husband and wife in a married couple and $j = 1,\ldots,287,294$ indexing the SNPs for FHS. Note that in the calculation for the couple correlation, the input data for a single mixed model FHS are a vector of SNP dosage with an extremely large dimension of $287,294 \times 2 = 574,588$. This dimension exceeds 2,000,000 if the entire set of HRS genome-wide genotype data are used for couple correlation analysis. For the SNP correlation, Y_{ij} in Y is the SNP dosage for individual i and married couple j where $i = 1,2$ indexing husband and wife in a married couple and $j = 1,\ldots,989$ indexing married couples for FHS. The mixed models for both couple correlations and SNP correlations were implemented in SAS [36].

Table 3. FHS data – ten SNPs with the smallest p-values for the SNP-specific correlation in 989 married couples out of the 287,294 SNPs, with the correlation estimated from the mixed model after controlling for population admixture.

SNP	Chromosome (position)		Gene	Location	Reference Allele (Freq.)		Married Pairs		Sibling Pairs		Parent-child Pairs	
							corr	p	corr	p	corr	p
rs16974794	19	(41499094)	CYP2B6	Intronic	A	(0.922)	0.27	0	0.47	0	0.35	0
rs1021652	17	(71403890)	SDK2	WNCG	A	(0.089)	0.20	9.30E-11	0.34	0	0.49	0
rs951954	2	(110459505)		Intergenic	T	(0.189)	0.20	1.11E-10	0.56	0	0.61	0
rs3007246	13	(105822046)		Intergenic	C	(0.928)	0.19	2.56E-10	0.56	0	0.55	0
rs16871467	5	(73484420)		Intergenic	G	(0.968)	0.18	2.43E-09	0.50	0	0.54	0
rs352416	8	(28452799)		Intergenic	C	(0.865)	0.17	9.45E-09	0.50	0	0.55	0
rs9483869	6	(136022914)	LINC00271	Noncoding RNA	T	(0.904)	0.17	1.59E-08	0.45	0	0.51	0
rs4449354	3	(105998897)		Intergenic	T	(0.090)	0.17	1.82E-08	0.49	0	0.52	0
rs16852244	3	(105993897)		Intergenic	T	(0.961)	0.16	5.15E-08	0.48	0	0.52	0
rs17155256	7	(81238711)	AY927633	WNCG	A	(0.843)	0.16	9.14E-08	0.46	0	0.53	0

Information is provided on SNP name, chromosome position, gene name when available, gene location, reference allele frequency, SNP correlation for married couples and p value from the permutation tests, correlation for full sibling pairs and p value, and correlation for parent-child pairs and p value.
WNCG: within noncoding gene&intronic.
NMDT: NMD transcript&intronic.

More intuitively, our mixed model is analogous to a multilevel model in which IQ measures of students are clustered into schools [37]. IQ measures would be equivalent to SNP dosages and schools would be equivalent to couples. In FHS, each SNP-correlation regression model estimates the correlation of a SNP averaged over 989 couples, which is equivalent to a multilevel model that estimates the intra-class or within-school correlation of an IQ measure averaged over the schools in the analysis sample. The analogy may also be applied to our couple-correlation regression where the multilevel model analyzes only one school on a large number of different cognitive measures. The multilevel model would estimate a within-school correlation averaged over the large number of cognitive measures. The model can be identified because of multiple measures of cognitive outcomes. The model makes sense because we estimate an average genomic correlation within a couple, which is similar to genomic correlation within a pair of biological siblings. In FHS, our mixed couple-correlation model estimates a correlation within a couple averaged over 287,294 SNPs. In FHS, 989 couples yielded 989 such couple estimates.

To verify that our estimated correlation coefficients are essentially coefficients of genetic relatedness, the couple correlation and SNP correlation were also performed on 5,713 pairs of full siblings and 6,958 parent-child pairs. For full-sibling pairs, each couple correlation is based on all SNPs for a single full-sibling pair and each SNP correlation is based on all sibling pairs. The parent-child estimates parallel those of full-sibling pairs. The known genetic relatedness in full siblings and parent-children can be used as a benchmark against which the genetic similarity estimates from married couples can be evaluated. The SNP correlation based on full sibling pairs and parent-child pairs can also be used to check the quality of individual SNPs. If the sibling and parent-child correlation for a specific SNP deviate severely from what is expected, the quality of that particular SNP may be questioned.

To remove the effects of race and ethnicity on genomic assortment, principal components (PCs) were estimated in FHS and in HRS by Eigensoft [38,39] and then included in regression analysis of couple and SNP correlations. Since principle components are influenced by correlation data, we excluded some of the correlated SNPs and correlated individuals when constructing PCs. To remove correlated SNPs, we used Plink to run LD-based SNP pruning and only kept the SNPs with pair-wise $r^2<0.2$. To remove the correlated individuals, we used Plink to get the pairwise identity-by-descent (IBD) estimates, and kept those with estimated genome-wide pair-wise IBD <0.1. The PCs for the subjects that were excluded for the construction of PCs were subsequently calculated using the parameter coefficients obtained from those included in the PC estimation. For both FHS and HRS, seven largest PCs were used. Previous work shows that adjusting a small number of PCs is usually sufficient to account for population admixture [38]. For FHS, 92,648 SNPs were used to construct the PCs; for HRS, the PCs were constructed on the basis of the 67,385 SNPs.

Our mixed-model approach allows controlling population stratification in the regression analysis. For the SNP correlation, the seven largest PCs were included in Equation (1) as individual predictors. For the couple correlation, the seven largest PCs were used in a regression to predict the minor allele dosage of each SNP; the resulting residuals were then used as the outcome variable in Equation (1).

The statistical significance tests for couple correlations and SNP correlations are performed following the same principles in FHS and HRS. The couple correlations are evaluated via two permutation tests. Two permutation tests based on two quite

different populations provide a robustness check for the results of significance tests. For FHS, the first permutation test is based on the individuals in the 989 married couples. We obtained 246,870 random pairs from these individuals who are genetically unrelated, unmarried, of the opposite sex, and with the male no more than 5 years older and no more than 2 years younger than the female. In the second permutation test based on all FHS individuals, we first randomly select a subset of 200,000 pairs from about 20 million possible unrelated opposite-sex pairs in FHS. A subset is selected to reduce computation. In both permutation tests, we (1) compute couple correlations for all these married couples and random pairs, (2) randomly draw 5,000 samples (N = 989) from the large pool of 200,000 (or 246,870) pairs without replacement, (3) randomly draw 5,000 samples (N = 989) from married couples with replacement, and (4) compare each of the 5,000 bootstrapped samples of married couples with the 5,000 random-pair samples using a t test.

A potential limitation of a couple correlation is that the positive and negative assortment within each married couple may cancel each other. To address this issue, we calculate two correlations for each couple, one using about half of the SNPs that contribute to the more "positive" assortment and the other using the half of SNPs that contribute to the more "negative" assortment.

The division of the entire set of the SNPs into "positive" and "negative" groups is based on the combination of minor allele dosage at each SNP for each couple. We use "02" to indicate that the minor allele dosage for a particular SNP for one spouse is "0" and for the other is "2". The combination can only take one of the six forms: 02 or 20, 01 or 10, 11, 00, 12 or 21, and 22, where 0, 1 and 2 represent a minor allele dosage. A simulation based on the observed distribution of these combinations in the married couples of FHS yields an order of 02 or 20, 01 or 10, 11, 00, 12 or 21, and 22 according to how positive a contribution each of the six combinations makes to the overall couple correlation. These simulated results were used to order the SNPs in each couple dataset.

To provide more information on the simulation, we simulated paired data with six possible combinations of 02 or 20, 01 or 10, 11, 00, 12 or 21, and 22, assuming the distribution of each combination is the same as that in the observed genome-wide genotype data. We then compared each pair of the combinations with respect to their contributions to the overall correlation. For example, when comparing the contributions of 11 and 22, we assessed the change in the overall correlation as a response to increasing the proportion of 22 and reducing the proportion of 11, while keeping the same the proportions of other combinations. Comparing all possible pairs found that increasing the proportions of 00, 12 or 21, and 22 results in an increase of the overall correlation, whereas an increase in the proportions of 20 or 02, 10 or 01, and 11 results in a decrease of the overall correlation.

For each couple, the SNPs with the combinations of 20 or 02, 10 or 01, and 11 are included in the negative group and the SNPs with the combinations of 00, 12 or 21, and 22 are included in the positive group. The statistical tests for these positive and negative correlations are performed in a similar fashion as those for the overall couple correlation.

A Z-test and its associated p-value were obtained for each SNP correlation in both FHS and HRS. For FHS, each test is a comparison of the SNP correlation based on 989 married couples against the distribution of the same-SNP correlation calculated from the 5,000 samples of randomly paired opposite-sex pairs based on the entire FHS sample. Each of the 5,000 samples has a sample size of 989 pairs.

To summarize, this study consists of three parts. The first part is an FHS analysis; it uses all available SNPs (287,294) in FHS for both couple-correlation and SNP-correlation analysis. Part-2 is an HRS analysis. Part-2 SNP-correlation analysis only uses the 10 SNPs in HRS that have the smallest P-values in FHS; and part-2 couple-correlation analysis uses 66,526 SNPs in HRS that are also available in FHS. These SNPs are the only SNPs available in both FHS and HRS. Using exactly the same set of SNPs from two independent studies offers an opportunity to replicate the findings. A non-trivial reason for not using all SNPs available in HRS in couple-correlation analysis is computational. The analysis would have to estimate an extremely large number of mixed models for permutation tests, each model using a dataset with $2 \times 2,000,000 = 4,000,000$ rows of data. Part-3 analysis is a couple-correlation analysis using the 66,526 SNPs in FHS that are available in HRS. Thus, this part-3 FHS analysis uses exactly the same set of the 66,526 SNPs that the HRS analysis of couple correlation used, but a different set of individuals in FHS to calculate couple correlations. Comparing the findings from the FHS 287,294-SNP analysis and the FHS 66,526-SNP analysis provides evidence whether the findings from the 66,526-SNP analysis in HRS can be generalized to those of the 2,000,000-SNP analysis in HRS.

Acknowledgments

Many thanks to Yunfei Wang and Qianchuan He for their invaluable support in this project.

Author Contributions

Conceived and designed the experiments: GG. Analyzed the data: LW HL. Contributed reagents/materials/analysis tools: TR. Contributed to the writing of the manuscript: GG.

References

1. Vandenberg SG (1972) Assortative mating, or who marries whom? Behavior Genetics 2: 127–157.
2. Risch N, Choudhry S, Via M, Basu A, Sebro R, et al. (2009) Ancestry-related assortative mating in Latino populations. Genome Biology 10.
3. Lewontin R, Kir D, Crow J (1968) Selective mating, assortative mating, and inbreeding: Definitions and implications. Biodemography and Social Biology 15: 141–143.
4. Ramsoy NR (1966) Assortative Mating and the Structure of Cities. American Sociological Review 31: 773–786.
5. Speakman JR, Djafarian K, Stewart J, Jackson DM (2007) Assortative mating for obesity. American Journal of Clinical Nutrition 86: 316–323.
6. Thiessen D, Gregg B (1980) Human assortative mating and genetic equilibrium: An evolutionary perspective. Ethology and Sociobiology 1: 111–140.
7. Merikangas KR (1982) Assortative Mating for Psychiatric Disorders and Psychological Traits. Arch Gen Psychiatry 39: 1173–1180.
8. Nielsen J (1964) Mental disorders in married couples (assortative mating). British Journal of Psychiatry 110: 683–697.
9. Qian Z (1998) Changes in Assortative Mating: The Impact of Age and Education, 1970–1990. Demography 35: 279–292.
10. Mare RD (1991) 5 Decades of Educational Assortative Mating. American Sociological Review 56: 15–32.
11. Torche F (2010) Educational Assortative Mating and Economic Inequality: A Comparative Analysis of Three Latin American Countries. Demography 47: 481–502.
12. Smits J, Park H (2009) Five Decades of Educational Assortative Mating in 10 East Asian Societies. Social Forces 88: 227–255.
13. Heath AC, Berg K, Eaves LJ, Solaas MH, Sundet J, et al. (1985) No Decline in Assortative Mating for Educational-Level. Behavior Genetics 15: 349–369.
14. Silventoinen K, Kaprio J, Lahelma E, Koskenvuo M (2000) Relative effect of genetic and environmental factors on body height: Differences across birth cohorts among Finnish men and women. American Journal of Public Health 90: 627–630.
15. Weedon MN, Lettre G, Freathy RM, Lindgren CM, Voight BF, et al. (2007) A common variant of HMGA2 is associated with adult and childhood height in the general population. Nature Genetics 39: 1245–1250.
16. Weedon MN, Lango H, Lindgren CM, Wallace C, Evans DM, et al. (2008) Genome-wide association analysis identifies 20 loci that influence adult height. Nature Genetics 40: 575–583.
17. Allen HL, Estrada K, Lettre G, Berndt SI, Weedon MN, et al. (2010) Hundreds of variants clustered in genomic loci and biological pathways affect human height. Nature 467: 832–838.
18. Lettre G, Jackson AU, Gieger C, Schumacher FR, Berndt SI, et al. (2008) Identification of ten loci associated with height highlights new biological pathways in human growth. Nature Genetics 40: 584–591.
19. Sanna S, Jackson AU, Nagaraja R, Willer CJ, Chen WM, et al. (2008) Common variants in the GDF5-UQCC region are associated with variation in human height. Nature Genetics 40: 198–203.
20. Schwartz CR, Mare RD (2005) Trends in Educational Assortative Marriage from 1940 to 2003. Demography 42: 621–646.
21. Pearson K (1903) Assortative mating in man. Biometrika 2: 481–489.
22. FHS (2012) Framingham Heart Study: www.framinghamheartstudy.org. Accessed 2014 Oct 17.
23. Qian ZC, Lichter DT (2007) Social boundaries and marital assimilation: Interpreting trends in racial and ethnic intermarriage. American Sociological Review 72: 68–94.
24. Domingue BW, Fletcher J, Conley D, Boardman JD (2014) Genetic and educational assortative mating among US adults. Proceedings of the National Academy of Sciences of the United States of America 111: 7996–8000.
25. Hindorff L, MacArthur J, Morales J, Junkins H, Hall P, et al. (2013) A Catalog of Published Genome-Wide Association Studies. pp. www.genome.gov/gwastudies. Accessed 2014 Oct 17.
26. Amann-Zalcenstein D, Avidan N, Kanyas K, Ebstein RP, Kohn Y, et al. (2006) AHI1, a pivotal neurodevelopmental gene, and C6orf217 are associated with susceptibility to schizophrenia. European Journal of Human Genetics 14: 1111–1119.
27. Salonen JT, Uimari P, Aalto JM, Pirskanen M, Kaikkonen J, et al. (2007) Type 2 diabetes whole-genome association study in four populations: The DiaGen consortium. American Journal of Human Genetics 81: 338–345.
28. Slonimsky A, Levy I, Kohn Y, Rigbi A, Ben-Asher E, et al. (2010) Lymphoblast and brain expression of AHI1 and the novel primate-specific gene, C6orf217, in schizophrenia and bipolar disorder. Schizophrenia Research 120: 159–166.
29. Volkening K, Leystra-Lantz C, Strong MJ (2010) Human low molecular weight neurofilament (NFL) mRNA interacts with a predicted p190RhoGEF homo-logue (RGNEF) in humans. Amyotrophic Lateral Sclerosis 11: 97–103.
30. Dunham I, Kundaje A, Aldred SF, Collins PJ, Davis C, et al. (2012) An integrated encyclopedia of DNA elements in the human genome. Nature 489: 57–74.
31. Thurman RE, Rynes E, Humbert R, Vierstra J, Maurano MT, et al. (2012) The accessible chromatin landscape of the human genome. Nature 489: 75–82.
32. Cherlin CJ (2009) The Marriage-Go-Round: The State of Marriage and the Family in America Today. New York: Alfred A. Knop.
33. Pawlowski B, Dunbar RIM, Lipowicz A (2000) Evolutionary fitness - Tall men have more reproductive success. Nature 403: 156–156.
34. Wright S (1921) Systems of mating. III. Assortative mating based on somatic resemblance. Genetics 6: 144–161.
35. Searle SR (1971) Linear Models. New York: Wiley and Sons.
36. SAS Institute Inc. (1961–2005) www.sas.com. Accessed 2014 Oct 17.
37. Goldstein H (2011) Multilevel Statistical Models. 4th ed. London: Wiley.
38. Price AL, Patterson NJ, Plenge RM, Weinblatt ME, Shadick NA, et al. (2006) Principal components analysis corrects for stratification in genome-wide association studies. Nature Genetics 38: 904–909.
39. Ma J, Amos CI (2012) Principal Components Analysis of Population Admixture. Plos One 7.

21

Genomic Analysis of *Sleeping Beauty* Transposon Integration in Human Somatic Cells

Giandomenico Turchiano[1], Maria Carmela Latella[1], Andreas Gogol-Döring[2,3], Claudia Cattoglio[4], Fulvio Mavilio[1,5], Zsuzsanna Izsvák[6], Zoltán Ivics[7], Alessandra Recchia[1]*

1 Center for Regenerative Medicine, Department of Life Sciences, University of Modena and Reggio Emilia, Modena, Italy, 2 German Centre for Integrative Biodiversity Research (iDiv) Halle-Jena-Leipzig, Leipzig, Germany, 3 Institute of Computer Science, Martin Luther University Halle-Wittenberg, Halle, Germany, 4 Howard Hughes Medical Institute, Department of Molecular and Cell Biology, University of California, Berkeley, Berkeley, California, United States of America, 5 Genethon, Evry, France, 6 Max Delbruck Center for Molecular Medicine, Berlin, Germany, 7 Division of Medical Biotechnology, Paul Ehrlich Institute, Langen, Germany

Abstract

The *Sleeping Beauty* (SB) transposon is a non-viral integrating vector system with proven efficacy for gene transfer and functional genomics. However, integration efficiency is negatively affected by the length of the transposon. To optimize the SB transposon machinery, the inverted repeats and the transposase gene underwent several modifications, resulting in the generation of the hyperactive SB100X transposase and of the high-capacity "sandwich" (SA) transposon. In this study, we report a side-by-side comparison of the SA and the widely used T2 arrangement of transposon vectors carrying increasing DNA cargoes, up to 18 kb. Clonal analysis of SA integrants in human epithelial cells and in immortalized keratinocytes demonstrates stability and integrity of the transposon independently from the cargo size and copy number-dependent expression of the cargo cassette. A genome-wide analysis of unambiguously mapped SA integrations in keratinocytes showed an almost random distribution, with an overrepresentation in repetitive elements (satellite, LINE and small RNAs) compared to a library representing insertions of the first-generation transposon vector and to gammaretroviral and lentiviral libraries. The SA transposon/SB100X integrating system therefore shows important features as a system for delivering large gene constructs for gene therapy applications.

Editor: Sebastian D. Fugmann, Chang Gung University, Taiwan

Funding: Funding was received for this study from Italian Ministry of University and Research-FIRB 2008 (AR), DEBRA international (AR) and the European Research Council (GT-SKIN) (FM). The funders had no role in study design, data collection and analysis, decision to publish, or preparation of the manuscript.

Competing Interests: The authors have declared that no competing interests exist.

* Email: alessandra.recchia@unimore.it

Introduction

The *Sleeping Beauty* (SB) transposon is a member of the Tc1/*mariner* transposon superfamily. Tc1/*mariner* elements are generally 1,300–2,400 bp in length and contain a single gene coding for the transposase that is flanked by terminal inverted repeats (IR). The IRs of SB host a pair of binding sites containing short, 15–20 bp direct repeats (DRs). Both the outer and the inner pairs of transposase-binding sites are required for transposition. The SB transposase binds the IRs in a sequence-specific manner, and mediates precise cut-and-paste transposition in a wide variety of vertebrate cells including human cells [1–3]. For this reason, the SB-based integration system is a valuable tool for functional genomics in several model organisms and represents a promising vector for human gene therapy [4,5]. However, a major bottleneck of any transposon-based application is the low transposition efficiency. Therefore, considerable effort was dedicated to improve the SB integration machinery by modifying its IRs and systematically mutating the transposase gene. In 2002, Cui et al. carefully explored the structure and functions of the IRs. They modified the outer and inner DR sites of both IRs and the spacer sequence between the DRs generating a new version of transposon IR,

called T2, with fourfold increased transposition efficiency [6]. However, the transpositional activity of this system (and that of the first-generation transposon [7]) is negatively affected by the size of transposon, resulting in an exponential drop for every kb introduced between the two IR.

In 2004, Zayed et al. constructed the "sandwich" (SA) version of the transposon vector [8]. The SA IR consists of two complete transposon elements in a head to head orientation, flanking a DNA expression cassette, thereby forming a sandwich-like arrangement. Mutation of the 5′ terminal CA nucleotides of the right IR abolishes cleavage at the innermost transposon ends; therefore, only the four terminal DRs represent the catalytic substrate for the "cut and paste" transposition. The SA transposon showed a 3.7-fold enhanced activity over first generation transposon to integrate ∼7.5 kb-DNA sequence upon SB10 transposase delivery. Five years later, a transposase 100-fold more active than SB10, named SB100X, was developed by a high-throughput, PCR-based DNA shuffling strategy [1]. The improved integration efficiency associated with SB transposition opened new avenues for its application. The hyperactive SB100X transposase was employed to obtain highly efficient germline transgenesis in pigs [9,10] rabbits [11] and rodents [12,13], stable transfer of therapeutic genes in clinical relevant cells [1,14–18], and reprogramming of mouse embryonic and human foreskin fibroblasts into iPS cells [19].

In this study, we investigated the integration efficiency of large expression cassettes mediated by the optimized SB elements: the SA transposon and the SB100X transposase. We report a side-by-side comparison between the SA and the T2 transposons carrying DNA cargo of increasing length. We performed a deep molecular characterization of SA-mediated integrants in epithelial cell lines and in primary immortalized keratinocytes stressing the SB system with cargos up to 18 kb. These data provide evidence for stability of SB-mediated integration and the reproducibility of the cut-and-paste mechanism even with large transposons embedded between two double IRs. Moreover, clonal analysis reveals a linear correlation between transposon copies harboured into the genomic DNA and their expression, an important characteristic for gene therapy application. Finally, high-resolution, genome-wide mapping of SA integrations in human keratinocytes revealed a close-to-random integration pattern with respect to genes and chromosomes, highlighting a relative low risk of genotoxicity as previously reported for SB transposition in cell lines [20–23]. Interestingly, the high-throughput analysis of SA integration sites showed an overrepresentation of integration events into repetitive elements (RE) of the human genome, in particular satellite, small RNA and LINE elements.

Materials and Methods

Cell culture

HeLa cells were cultured using DMEM medium (Lonza) added with 10% Fetal Bovine Serum (FBS), 1% L-Glutamine (L-Gln) and 1% Penicillin-Streptomycin (Pen/Strep). For each experiment, an aliquot of cryo-preserved HeLa cells was thawed and plated on 8 cm dishes. Upon reaching 80–90% of confluency, cells were re-plated on 6-wells culture plates at a concentration of 2×10^5 cells/well. After 24 h, cultures in each well were at 70–80% confluency, ready to be transfected.

Mouse NIH3T3 fibroblast cell line was maintained in Dulbecco's Modified Eagle's medium (Euroclone), supplemented with 10% bovine serum.

We have used SV40 immortalized keratinocytes derived from a patient affected by generalized atrophic benign epidermolysis bullosa (GABEB) produced by Borradori et al. [24] and kindly provided by J.W. Bauer. GABEB cells were cultivated in EpiLife medium supplemented with human keratinocyte growth supplement (HKGS) (Invitrogen, US). EpiLife is a serum-free keratinocyte culture medium with a low calcium (0.06 mM) concentration supplemented with HKGS which results in a final concentration of 0.2% (v/v) BPE, 5 lg/mL bovine insulin, 0.18 *lg/mL* hydrocortisone, 5 lg/mL bovine transferrin and 0.2 ng/mL human EGF. Upon reaching 80–90% of confluency, cells were re-plated on 6-wells culture plates at a concentration of 2.3×10^5 cells/well. After 24 h, cultures in each well were at 70–80% confluency, ready to be transfected.

Plasmid constructs

The plasmid carrying the T2 IRs including a Venus reporter gene driven by the chicken β actin promoter fused to CMV early enhancer element (CAGGS) and the construct coding for the SB100X were described in Mates et al. [1]; the SA transposon IRs were described in Zayed et al. [8]. The CAGGS Venus expression cassette was *Dra III* excised from pT2 3.2 and introduced into *EcoRV* digested pSA to obtain pSA 5.7. pT2 3.2 and pD28 [25]

were digested with *XbaI* to clone a non coding DNA of 2.7 kb from pD28 into the transposon.

Two fragments of the first intron of the HPRT gene were PCR amplified and cloned into the pCR 2.1 (TOPO cloning kit, Invitrogen) plasmid. The pT2 10 plasmid was cloned ligating the pT2 CAGGS Venus *SpeI* with *NheI* fragment of the amplified HPRT intron 1. The pT2 14 plasmid derives from pT2 10 digested with *ClaI* ligated to the *NotI* fragment of the amplified HPRT intron 1. Finally, pT2 18 was obtained by ligating a third sequence amplified from the HPRT intron 1 with pT2 14 through *EcoRI* restricted ends. The pSA 5.7 plasmid was digested with *NheI* and ligated to the *NheI* non coding fragment of the HPRT gene to obtain the pSA 9.7. Then the pSA 9.7 was digested with *PmeI* enzyme and ligated with a *PvuII* fragment of the HPRT intron 1 to obtain the pSA 14. To enlarge the pSA14, a sequence amplified from the intron 3 of the Lamb3 gene was introduced by *EcorV* compatible ends to obtain pSA 18.

Transfection-based transposition and calculation of transposition efficiency

HeLa and GABEB cells were both transfected with FugeneHD transfection reagent (Roche). For each sample 2 μg of DNA were added to 100 μl of either DMEM (for HeLa) or EpiLife (for GABEB). The media used for this transfection reaction mix were not added with FBS, L-Gln or Pen/Strep.

The transposon/transposase amounts of plasmid DNA were calculated to respect the stoichiometric ratio of 1:1 or, for transposon >10 kb, 2:1, in a total quantity of 2 μg. 2 μg of transposon-only plasmid were used for non-transposed control.

Each transfection reaction mix was complexed with 6 μl of FugeneHD (10 μl with SA and T2 18 in GABEB cells) and subsequently mixed by pulse-vortexing for a few seconds. The mixes were thereafter left at room temperature for 10′ in order to allow the formation of lipoplexes. After the 10′ had expired, each mix was added drop-by-drop to a cell culture sample, which was subsequently incubated at 37°C.

HeLa cells were transfected with Calcium Phosphate method using 15 μg of 14- or 18 kb transposons mixed with the plasmid carrying the transposase expression cassette.

The percentage of Venus+ cells was determined 2 and 20–30 days post-transfection via flow cytometry and the transposition efficiency was calculated as: Venus+ cells at 20–30 days post transfection/Venus+ cells at Day 2×100. Cells that were only transfected with the transposon plasmid represented the control for background integration events.

Transposed clones were analysed via flow cytometry to determine the presence of doublets and the Venus mean fluorescence intensity (MFI).

Isolation of single cell clones

GABEB cells were limiting diluted to obtain a concentration of 0.5 cell/well, plated onto lethally irradiated NIH3T3 cells and cultured in keratinocyte growth medium, a DMEM and Ham's F12 media mixture (2:1) containing FCS (10%), penicillin-streptomycin (1%), glutamine (2%), insulin (5 μg/ml), adenine (0.18 mM), hydrocortisone (0.4 μg/ml), cholera toxin (0.1 nM), and triiodothyronine (2 nM). After 1 week, the medium was replaced by EpiLife medium supplemented with HKGS. After 2 weeks GABEB cells were trypsinised at subconfluence and re-plated without the NIH 3T3 feeder-layer in EpiLife HKGS medium.

HeLa cells were seeded to obtain a concentration of 0.3 cells/well in a 96 well plate in DMEM medium complemented with 10% FBS.

Southern blot analysis

Ten μg of genomic DNA, extracted from $1-5\times10^6$ cells by a QIAmp DNA Mini kit (Qiagen), were digested overnight with *Nhe*I (SA 9.7-derived clones) and *Afl*II (T2 10-derived clones) to verify the copy number of the transposed cassette, or with *Nco*I (SA 9.7-derived clones) and *Mfe*I plus *Nde*I (T2 10-derived clones) to verify the integrity of the transposed cassette. Digested gDNA was run on a 0,8% agarose gel, transferred to a nylon membrane (Duralon, Stratagene) by Southern capillary transfer and probed with 2×10^7 cpm ^{32}P-labeled Venus probe according to standard techniques [26].

PCR screening for episomal SB vectors

About 100 ng of template gDNA were used in a PCR reaction. Primers capable to amplify the Amp resistance gene or the SB100X transposase (**Table S1**) were used to detect genomic integrations of SA 9.7 backbone and SB100X, respectively. PCR conditions were as follows: 30′′ at 94°C, 30′′ at 58°C and 30′′ at 72°C for 30 cycles.

LM-PCR and bioinformatic analysis

Integration sites were amplified by Linker Mediated PCR (LM-PCR), as described [27]. Briefly, genomic DNA was extracted from $0.5-5\times10^6$ transposed cells and digested with *Mse*I and *Xho*I enzyme to prevent amplification from internal mutated IR fragments. An *Mse*I double-stranded linker was then ligated and LM-PCR performed with nested primers specific for the linker and SA IR/DR (**Table S1**).

LM-PCR derived amplicons were run on a Roche/454 GS FLX using titanium chemistries by GATC Biotech AG Next Gen Lab. A valid integration contained: the TAGpSAIR nested primer and the entire SA IR/DR sequence up to a TA dinucleotide.

Alignment pipeline. 31,603 sequencing reads were tested for the presence of the SA IR sequence and TA dinucleotide. The SA IR and any primer sequences were trimmed, and the remaining reads starting with TA dinucleotides were mapped to the human genome (hg19) using NCBI BLAST (blastn with default parameters). We kept only reads which were mapped to a single genomic site with at least 90% sequence identity and an E-value of at most 0.05. Only reads which could be mapped from their 5′ end onwards were considered for further analysis. Redundant reads mapping to identical genomic positions were collapsed. This way we got 2019 unique SA integration sites.

For the statistical analysis we generated 10,000 control sites in-silico taking into account the bias introduced by LM-PCR techniques. We first generated artificial reads starting with TA dinucleotide of the human genome in a way that the control sequences had both the length and the frequency of *Mse*I restriction sites (TTAA) as observed in real sequencing reads. The artificial reads were then processed by the same mapping criteria used for the SA sites.

RM blast analysis. Analyses of repetitive element were performed with RepeatMasker Blast (http://repeatmasker.org) [28]. To achieve reliable and comparable results we processed the raw sequences trimming out the primer sequences used in LM-PCR, the IR/LTR/linker specific sequences following the primers. Resulting reads were further trimmed till the 40th nucleotide discarding every sequence with less than 40 nucleotides. Finally, we collapsed the reads that were either identical or with one mismatch. A two-sample test for proportions was used for pairwise comparison of the RE within the different datasets.

For statistical analysis we created control sets as follows. We first randomly sampled 1 Million sequences 49 bp in length from the human reference genome (hg19). Then we discarded all sequences

not starting with TA. The resulting set of 65,826 TA-weighted sequences was used as a background for T*neo* SB integrations. For a second random control set we first randomly sampled 10 Million sequences of length 120 bp from the genome. Then we discarded all sequences not starting with TA, or either not containing the *Mse*I restriction motif TTAA or having a TTAA within the first 39 bp of the sequence. After removing the part of the sequences following the first occurrence of TTAA, we received 292,917 sequences of lengths between 40 bp and 120 bp, which were weighted for TA and *Mse*I and could be used as a background for SA integrations. We passed the generated sequences through the same filtering/trimming pipeline as the actual integration reads.

A third random control set of 45,235 genomic sequences weighted for *Mse*I was adapted from Cattoglio et al. [29] and used as a background for MLV and HIV integrations.

Bidirectional PCR mapping on GABEB clones

Transposon integrations in GABEB clones were amplified by LM-PCR as described. PCR products were shotgun-cloned (TOPO TA cloning kit, Invitrogen) and then sequenced. Sequences between the TA and the linker primers were mapped onto the human genome by the BLAT genome browser (UCSC Human Genome hg19). Sequences featuring a unique best hit with ≥90% identity to the human genome were considered genuine integration sites. To confirm the genuine integration in both directions we design primers on the genomic region hit and performed a direct PCR in conjunction with the pSAIR specific primer for the SA IR sequence (**Table S1**). The derived amplicons were loaded on agorose gel and checked for the expected length.

Results

Efficiency of T2 and SA transposons

The sandwich (SA) transposon vector has superior ability to transpose >10 kb transgenes with respect to the first-generation transposon when SB10 transposase was provided [8]. Nevertheless, the T2 transposon, resulting from site-specific mutations in the IR sequences and insertion of double TA flanking each IR, has been demonstrated to have a four-fold enhanced activity over the first-generation transposon construct [6]. A side-by-side comparison of SA and T2 transposon was needed to address the transposition efficiency of increasing DNA cargoes and to verify their molecular behaviours once integrated into the human genome.

We generated SA- and T2-based plasmids (SA 5.7 and T2 3.2 **Figure 1**) keeping the Venus reporter gene as standard expression cassette. Increasing sizes of a non-coding human stuffer DNA (4-, 8.3- and 12.3 kb in the SA plasmid; 6.8-, 10.8- and 14.8 kb in the T2 plasmid) were introduced between the two IR/DR to produce transposons of comparable length. For the sake of simplicity, we named these plasmids with the transposon construct type and the size of the transposable cassette expressed in kilobases (**Figure 1**).

Transposition experiments were performed in HeLa cells and in immortalized primary keratinocytes derived from patients affected by Generalized Atrophic Benign Epidermolysis Bullosa (GABEB), an inherited skin adhesion defect. All the experiments aimed at the identification of the integration efficiency of the IR-flanked transgene were measured by long-term Venus fluorescence in the absence of selective pressure. We co-transfected the SB100X transposase-expressing plasmid together with transposon plasmids in two different molar ratios (1:1 or 1:2) depending on the transposon length. Larger cargos required more transposon DNA to reach good transfection efficiency.

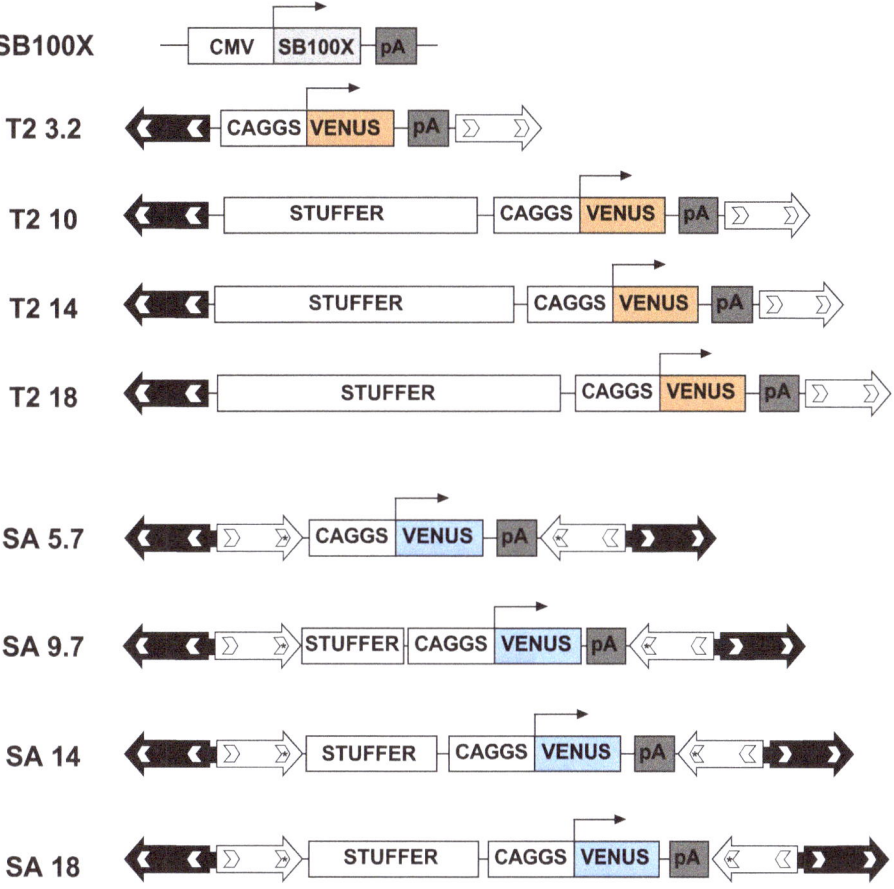

Figure 1. Transposon fleet. Schematic representation of the generated plasmids. SB100X carries the Hyperactive *Sleeping Beauty* transposase coding sequence placed under the control of the CMV promoter and followed by an SV40 poly-Adenylation (pA) signal. The transposons T2 and SA possess the expression cassette consisting of the CAGGS promoter, VENUS reporter gene and SV40 pA signal. The stuffer DNA represented has variable increasing size. The arrows represent the IR/DR ends recognised by the transposase. SA constructs are characterized by the presence of two complete IR/DR at each ends (white and black arrows) and the asterisks underline the IR mutated site not recognized as a catalytic substrate by the transposase. Numbers following T2 or SA abbreviation indicate the size in kilobases of the transposed cassette.

At least three independent experiments for each cell type and transposon were performed in order to reduce variability due to the transfection procedure. Mock-transfected HeLa and GABEB cells, and cells transfected with the T2 or SA Venus constructs alone were used as controls (in the absence of transposase, no transposition event should occur and residual reporter gene expression after long periods would only be attributable to noise or to rare random plasmid integration events). Transgene expression all along the culture period (up to 31 days) was measured via flow cytometry to follow the trend of the signal that persists in presence of SB100X and drops without the transposase (**Figure S1**).

The transposition efficiency was normalized by transfection efficiency (numbers of cells that received the plasmids after transfection) and calculated as the ratios between the percentage of Venus$^+$ cells at the endpoint (20–31 days) and the percentage of transfected cells 2–3 days after DNA delivery to the cells. The endpoint of each experiment is achieved when the percentage of Venus$^+$ cells in the sample transfected with the transposon alone stabilized to less than ~0.5%.

Figure 2A and 2B show the transposition rate obtained in HeLa and GABEB cells. As previously reported [1,8], the transposition efficiency was inversely proportional to the transposon size. In HeLa cells, the transposition efficiency dropped 7.8 fold (from 58.5% to 7.5%) when increasing the cargo payload from 3.2 kb to 18 kb, independently of the transposon structure (T2 or SA). Interestingly, this size-dependent effect was less pronounced in GABEB cells. In this cell type the decrease was of 1.8 fold (from 44% to 24%) for T2 and SA and the transposition rate for 18 kb transposons remained approximately 24% compared to the 7.5% in HeLa cells.

Clonal molecular analysis

Although we performed a molecular characterization of almost all T2 and SA vectors in HeLa or GABEB cells (**Table S2**.), we focused our genomic analysis on a relatively large T2 and SA transposons cassette (10 kb) and on GABEB keratinocytes. Bulk populations of transposed cells were sorted for Venus expression 20–35 days post transfection and cloned by limiting dilution. Genomic DNA extracted from each clone was first investigated by PCR for the presence of the transposon backbone and SB100X expressing plasmid. Notably, we scored 14.8% of clones (8 out of 54) positive for the Ampicillin sequence present within the transposon backbone about 60 days post transfection, while few (2 out 54) of the analysed clones were positive for the SB100X sequence (**Table S2**).

A

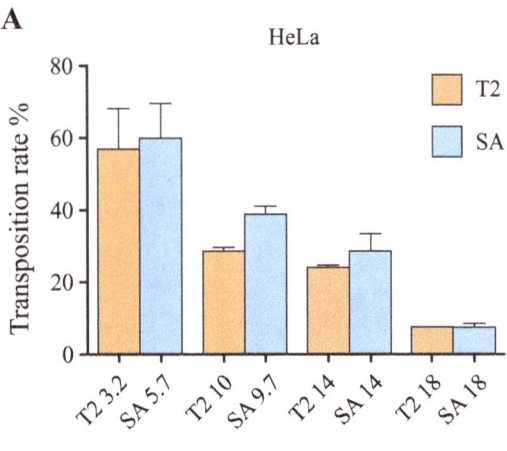

B

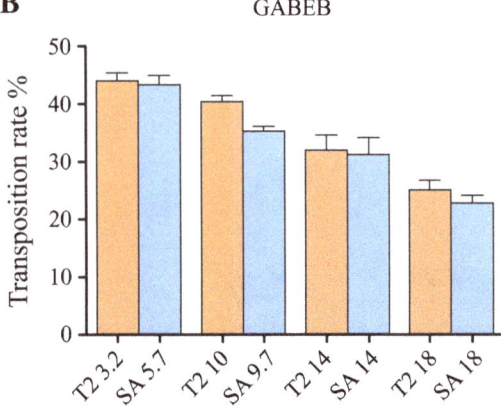

Figure 2. Transposition efficiency. HeLa (A) and GABEB (B) cells were co-transfected with the T2 and SA transposons- and transposase-carrying plasmids. The transposition rate, on the Y axis, is derived by the ratio between the percentage of Venus+ cells at about 20 and 2 days post transfection. Data are representative of three independent experiments (mean ± SEM; $n = 3$).

We next performed Southern blotting on the genomic DNA of 16 clones for each transposon type to determine the transgene copies harboured in the genome and their integrity. To this end, we digested the genomic DNA with *AflII* (T2 clones) or *NheI* (SA clones) that release fragments longer than 3.4 and 4.2 kb. Hybridization with a Venus-specific probe showed that most of the SA treated samples (13 out of 16) carry a single integrated transposon, only 1 clone (#26) had 3 copies, and 2 out of 16 clones contained 2 copies (#8, #13) resulting in an average copy number of 1.3. Surprisingly, 16 GABEB clones obtained with T2, harbour 1 to 7 copies with an average of 3 integrated transposons per clone **(Figure 3A). In general we observed that the mean copy number is more affected by the transfection efficiency (Table S2) respect to the size and type of transposons.**

Further restriction analysis performed with *MfeI* and *NdeI* on 9 T2 clones and with *NcoI* on 8 SA clones showed that all clones harbour the full-length transposon cassette **(Figure 3B)**. Among the 21 integrated transposons in the 9 T2 clones, only one, belonging to clone #3, is shorter than expected. None of the 13 integrated transposons in the 8 SA clones was rearranged.

To unequivocally prove that all the integration events mediated by SA transposition resulted from a genuine "cut and paste" mechanism, we mapped the insertion site at both transposon ends using an adapted version of Linker-Mediated PCR (LM-PCR)

[27]. Ten Venus-expressing GABEB clones, derived from transposition of the SA 5.7 plasmid, were examined. Six integrants (#1, 4, 7, 13, 14, 16) belonging to 5 clones were bi-directionally mapped by LM-PCR. Additional 21 integrants were revealed by LM-PCR and confirmed by specific PCR on the genomic region flanking the opposite IR **(Figure 3C)**. Importantly, almost all the integration events occurred without genomic rearrangements, deletions or insertions, in the target sites. Only 2 out of 27 integrations (#26 and #27 belonging to clone 34) could not bi-directionally confirmed.

Finally, we correlated the expression level of the reporter gene with the copy number of the transposon. The positional effect variegation primarily observed with retroviral and lentiviral vectors [30] could lead to the silencing of the therapeutic gene delivered by the vector. We asked weather the SB integrations would be affected by this phenomenon. We correlated the expression of Venus protein, measured by Mean Fluorescence Intensity (M.F.I.), with copy number of either the SA and T2 transposon, as determined by Southern blot or q-PCR analyses of 62 GABEB clones. For comparison, we analysed the M.F.I of a GFP reporter gene, driven by the human Keratin 14 promoter, in 70 HaCaT clones isolated upon LV transduction. A linear correlation curve was traced to retrieve the R^2 coefficient of determination. Transposon samples show an $R^2 = 0.759$ with a statistically defined correlation between two variables ($P_N = 0.6$). LV samples display an $R^2 = 0.001$ with a null defined correlation **(Figure 4)**. Independent analysis of transposed clones obtained in different cells (HaCaT and GABEB) and carrying a reporter gene driven by PGK or CAGGS promoter showed comparable results indicating common directly correlation between MFI and copy number (data not shown). We conclude that SB integrants tend to express their cargo faithfully, and multi-copy integrants express in a copy-number dependent manner, consistent with earlier observations [31].

Integration pattern analysis

In the last few years, several papers described the integration profile of the SB, *piggyBac* (PB), and Tol2 transposons [20–23,32–36]. Here we report the integration profile and preference of the sandwich compared with the first-generation SB transposon [20] in human epithelial cells. To generate a library of SA integration events, we transfected 20 million GABEB cells with SA transposon- and SB100X-carrying plasmids. The 20% of Venus-positive cells were sorted three days after transfection to enrich the population expressing the reporter gene. A 90%-pure sorted population was kept in culture for 3 weeks to dilute the un-integrated SA vector reaching a stable 78% Venus+ bulk population. We used LM-PCR and pyrosequencing to generate 6,084 non-redundant SA-linked genomic sequences in human immortalized GABEB keratinocytes. The Blast alignment retrieved 2,019 unambiguously mapped integration sites. As a control, 10,000 random unique sequences were generated in silico balancing the biases introduced by the LM-PCR (amplicon lenght and *MseI* proximity) and the availability of the TA dinucleotides in the genome. In the analysis we also annotated a large dataset (59,169 hits) generated in HeLa cells transposed with the first-generation T*neo* transposon and selected for 2 weeks with neomycin [20]. The integration sites and control sites were annotated as transcriptional start site (TSS)-proximal when mapping in the ±2.5 kb window around a TSS, intragenic when mapping within a transcription unit, and intergenic in all other cases. Among SA integrations, 58.6% were in an intergenic position, 38.9% were within the transcribed portion of at least 1 gene, and 2.5% was within a 5 kb window encompassing the TSS

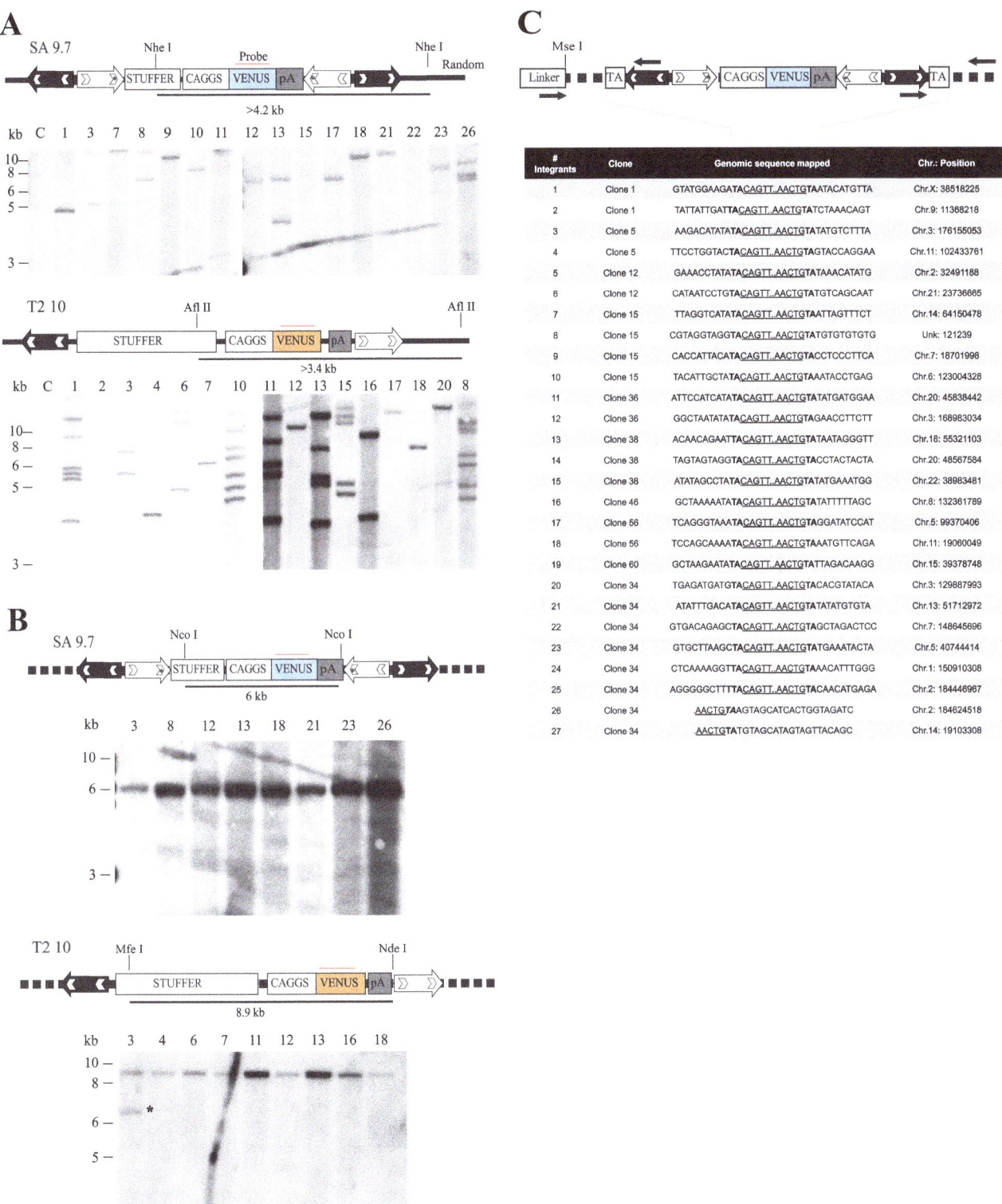

Figure 3. Molecular characterization of the *Sleeping Beauty*-mediated integration events in GABEB cell clones. (A) Southern blot analysis of genomic DNA from GABEB cell clones digested with *Nhe*I (SA clones) or *Afl*II (T2 clones), single cutter in the transposon cassette, and hybridized to a Venus probe. A single band higher than 4.2 kb (SA clones) and 3.4 kb (T2 clones) indicates integration of one copy of the transposon into the genome. Multiple Venus-specific bands correspond to repeated integration events. (B) Southern Blot analysis of genomic DNA from 8 (SA) and 9 (T2) clones digested with *Nco*I (SA clones) or *Mfe*I and *Nde*I (T2 clones). The expected Venus-specific band corresponding to 6 kb for SA and 8.9 kb for T2 transposon indicates the correct integration of the transposons into the genome. C, mock-transfected cells; red bars, Venus-specific probe. Clone showing rearrangement of the transposon cassette is highlighted by black asterisk. (C) Bi-directional mapping of the junctions between transposon and genomic DNA. The table summarizes 27 integrations belonging to 10 single clones. For each integrant, the underlined sequence

represents a portion of the transposon IRs, left (CAGTT) and right (AACTG) separated by dots; TA dinucleotide (in bold) is the target site correctly duplicated after transposition. Hit chromosomes and positions are reported. UnK, unknown region of the human genome based on UCSC hg19 assembly.

(**Figure 5A**; the complete list of sequences is available in GenBank database with the accession number SRP047118). In general, the distribution of the SB integrants in both datasets is fairly random and resembles the composition of the human genome showing no statistical differences compared to their relative controls, i.e. all p-values (both two-sample tests for proportions and Fisher's Exact Tests) were $>10^{-2}$.

We then analysed the frequency of human repetitive elements in the transposon libraries, SA and T*neo* [20], and their relative weighted controls availing of the RM Blast browser [28]. For comparison we also analysed two viral-derived integration datasets (MLV and HIV) generated in human CD34$^+$ multipotent hematopoietic progenitor cells (HPCs) [29] and their control library weighted for *MseI* restriction site distribution. The raw data generated by deep sequencing of the LM-PCR (applied to SA, MLV and HIV treated cells) and LAM-PCR (applied in [20]) products were filtered and trimmed in order to rescue the genuine integration events (see materials and methods). After filtering and trimming we retrieved 6,084 and 165,887 unique sequences in SA and first-generation vector libraries, respectively, and 37,873 and 31,204 unique sequences from MLV and HIV datasets, respectively. We generated large control datasets taking into account the bias introduced by the respective technique. In particular from the hg19 genome database we retrieved 45,235 control reads weighted for *MseI*, 65,570 sequences weighted for the presence of TA dinucleotide hit by the SB transposons, and 209,913 sequences *MseI*- and TA-double weighted.

The RM Blast analysis revealed an overrepresentation of REs in the SA integrations (34%) with respect to the TA and *MseI*-weighted control (14%) and to all the other datasets analysed (**Figure 5B**). In particular, Satellite, small RNA and LINE elements were enriched in the SA library (24-, 7.6- and 3.5-fold increase over the background, respectively) whereas in the first-

generation vector library only a slight increase in the satellite and simple repeats elements was measurable (3.5- and 2.6-fold over the background, respectively); comparable LINE frequency was detected.

Besides the higher frequency in the satellite elements, the two SB transposon datasets share a slight under-representation of SINE, LTR and DNA transposable elements in comparison with their random control libraries. We introduced MLV and HIV libraries to compare the frequency of integration into RE generated by a retroviral integrase-mediated integration mechanism. The RM Blast analysis pointed out that viral vectors disfavour integration in RE (14–16% vs 24%), and, in particular, satellite, LTR and LINE elements are underrepresented. These data clearly confirm a difference in the integration site selection between viral vectors and SB transposons and identify new signatures in the SA integrome that should be taken into consideration when using them as tools for genetic manipulation.

Discussion

The SB transposon IRs were mutated to improve their capacity to be mobilized and, to date, there is not a direct comparison that define genetic characteristics of the T2 and SA IRs [8]. In this study, HeLa cells and GABEB keratinocytes [24] were transfected with a panel of T2 or SA transposons carrying size-increasing Venus expression cassette in combination with SB100X plasmid (Figure 1). Transfection rate was higher in HeLa than in GABEB cells (**Figure S1**) and the transposition efficiency was inversely proportional to the transposon size (**Figure 2**). Interestingly, HeLa cells were severely affected by the transposon size compared to primary immortalized cells. These results suggest that the transposase activity could be favoured by some cellular factor differentially expressed in GABEB and HeLa cells. Nonetheless, T2 and SA constructs carrying cargos of comparable size showed similar transposition efficiency in both cell lines. From these data we can conclude that the T2 IR construct is interchangeable with the SA construct with some advantages: T2 has shorter IRs thereby it could accept a larger cargo cassette.

Transposed GABEB and HeLa populations were subjected to limiting dilution to obtain a single cell derived expansion. The derived clones were employed to characterize several molecular parameters: transposon-independent insertion, copy number, genomic stability, faithful transposition activity, correlation between copy number and expression of the integrated cassette. The SB100X sequence was retrieved in 6 out of 211 analysed clones while almost 14% of the clones (30 clones) were found positive for the transposon backbone sequences (**Table S2**). We hypothesize that the plasmid backbone carrying the transposon could have some advantages to remain episomal or to integrate in the genome. The transposon excision step from the plasmid leaves the backbone with a double strand break that induce recruitment of the endogenous repair machinery and integration into the cell genome. We also analysed the copy number of the clones. **Figure 3A** shows an average of 1.3 SA copies/clone while T2 copy number spans from 1 to 7 transposons with an average of 3 copies. This difference mostly depends on the transfection efficiency as confirmed by the analysis of the other transposed cell populations generated in this study. Therefore, it is possible to fine tune this parameter by adjusting the ratios of the two SB

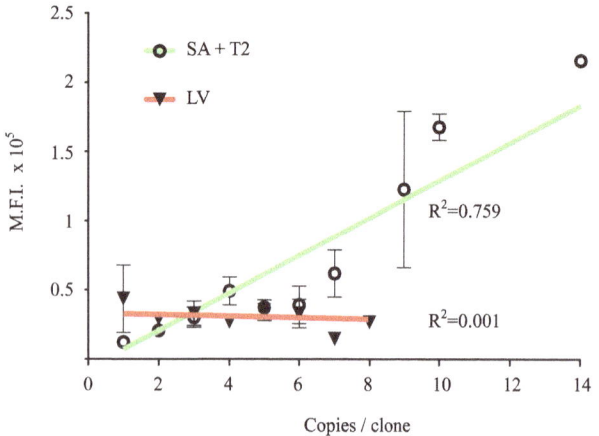

Figure 4. Correlation between copy number and expression of the integrated cassette. Mean fluorescence intensity (M.F.I.) of 62 GABEB clones positive for Venus expressing SB transposons are represented as circles; triangles indicate the M.F.I. of 70 GFP$^+$ HaCaT clones transduced with a lentiviral vector (LV). Standard deviation bars are present for those clones carrying the same copy number. R^2 coefficients of determination were extracted from the linear regression plot, green line for SA and T2 transposons and red line for LV.

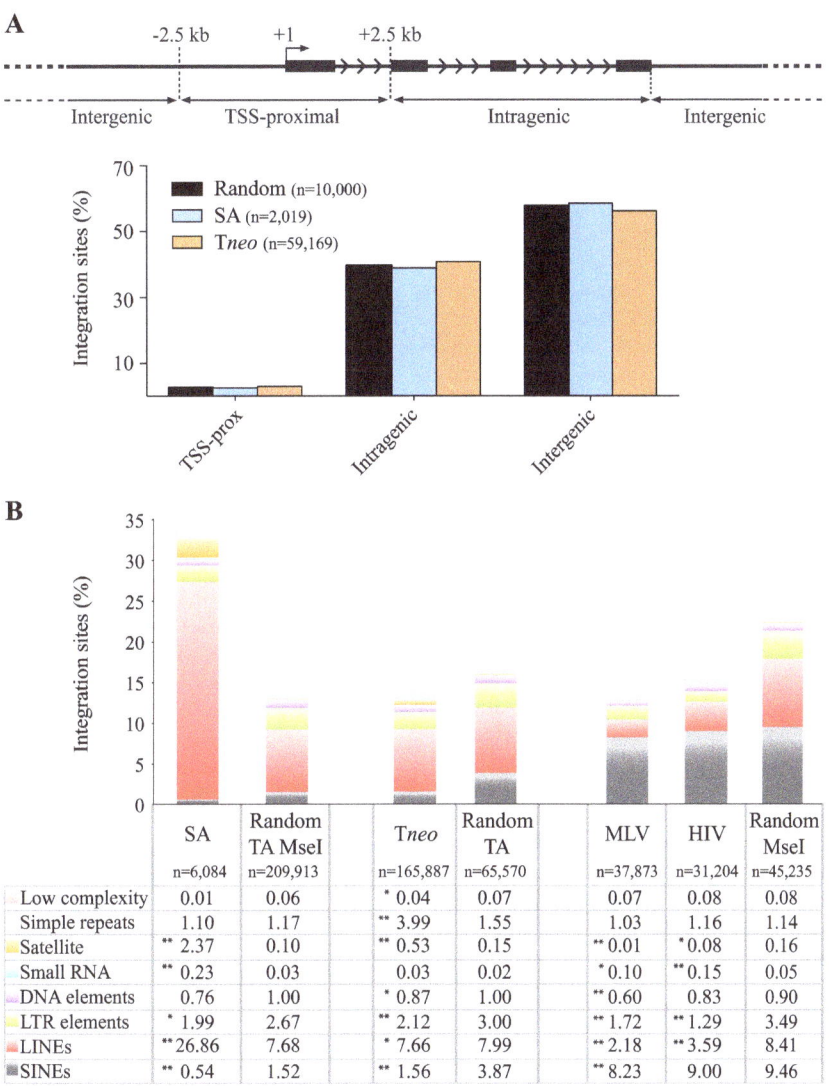

Figure 5. Integration pattern analysis. (A) Integration sites were annotated as "TSS-proximal" when occurring within a distance of ±2.5 kb from the gene's TSS, as "Intragenic" when occurring in a gene body and as "Intergenic" in all other cases. Black bars represent exons of a schematic gene, arrowhead indicates the direction of transcription. Distribution of SA, T*neo* and random integration sites in the genome is plotted accordingly to defined annotations. (B) Distribution of Repetitive Elements in SB SA and T*neo* libraries, in MLV and HIV libraries. Relative weighted random libraries were reported: TA and *Mse*I-weighted for SA, TA-weighted for T*neo* and *Mse*I-weighted for MLV and HIV libraries. **$p \leq 10^{-3}$, *$p \leq 10^{-2}$.

components used for transfection or, as previously reported, bypass the transfection procedure through the viral delivery of transposase and transposon by adenoviral vector [37], integration defective lentiviral vector [38,39], retroviral particle [40] and adeno-associated vectors [41].

We were able to associate copy number of the transposon with the expression level of the Venus fluorescence gene. Mean Fluorescence Intensity does follow a direct proportion with the copies harboured (**Figure 4**). In contrast, expression of the reporter gene in lentiviral-mediated integrants does not correlate with copy number and is more subjected to the activity of surrounding genomic sequences [42,43].

Next, the integrated transposons in these clones were also analysed for their integrity via Southern Blot. Retroviral and lentiviral vectors can rearrange during the reverse transcription step resulting in partially-deleted integrated proviruses, a frequent occurrence in transgene hosting repetitive sequences [44,45]. The

SB mediated integration, by contrast, does not require reverse transcription and thus is expected to preserve the integrity of the transgene. Ninety-eight percent of the integrants, resulting from T2 and SA transposition, have a correct size (**Figure 3 B**).

The sandwich transposon has a doubled IR/DR structure at both ends with 8 transposase binding sites in total. In principle, every transposase unit, bound to one DR site, could interact with the others to create different chiasm geometries (also described in [6]); some of these conformations could modify the integration activity resulting in chromosomal aberrations. To investigate the fidelity of the transposition process 10 GABEB clones were mapped bi-directionally by LM-PCR and transposon-genome junction was amplified by site-specific PCR. Twenty-five integrations, out of 27 (92.6%), were validated for a canonical transposition event with the TA target site duplication signature at both ends (**Figure 3C**). Two integrations mapped by LM-PCR were not confirmed in the opposite transposon end suggesting rearrangements probably caused by the repair mechanism occurred in the transposition break.

LM-PCR was also employed to derive a high-definition map of SA/SB100X integration sites in the genome of a transposed GABEB bulk population. This analysis is commonly applied to integrating vectors (i.e. retroviral and lentiviral vectors) because it allows to evaluate genotoxicity [46,47] and to understand molecular mechanism driving the integration towards specific regions of the genome [29,48–50]. The technique returned 2,019 SA unambiguously mappable integration sites randomly distributed throughout the human genome, in accordance with previously published data on first-generation transposon [20,23] (**Figure 5A**). For gene therapy purposes, the SB system results in a safer integration profile compared to other integrating vector such as Tol2, PB transposon and retroviral vectors [20–23,32,33], which favor TSS-proximal regions or gene body sequences.

Although the integration site distribution in relation to genes was found close to random, the RM Blast analysis shows a significant bias distribution of SA integrations in repetitive elements (RE), particularly in satellite, LINE and small RNA genes. It could be that these genomic regions are favourable for integration due to their base composition (TA-richness) or there might be molecular mechanisms that actively recruit the transposon/transposase complex at specific RE sites [51–55].

Curiously, the frequency of RE elements in the first-generation transposon library and its weighted control were comparable. Differently from the SA (obtained in 80% Venus expressing immortalized keratinocytes without selective pressure), the first-generation transposon library derives from transposed HeLa cells [56] selected for two weeks by antibiotic resistance. This culture condition could negatively select those integrations landing into poorly expressed genomic loci or into heterochromatin regions [35]. Nevertheless, the first-generation transposon integrations were slightly increased into satellite regions and SINE, whereas LTR and DNA elements were underrepresented compared to the background.

These data identify some common features in SB datasets. Conversely, MLV and HIV-derived viral vectors disfavour integration in RE (satellite, LTR and LINE accordingly also to [57]) suggesting an active role of viral integrase in the selection of integration sites that could better support the expression, replication and survival of the viral progeny. The genomic features newly identified in the SA integrome raise an interesting matter that needs to be deeply investigated for future application.

Supporting Information

Figure S1 Transposition trend. Expression of Venus fluorescence protein was detected by cytofluorimetric analyses at different time points in HeLa cells (A) and in GABEB cells (B). The days post transfection (p.t.) are plotted on the X axis, while the percentage of Venus+ cells are represented on the Y axis. The maximum expression from a transfected reporter gene was achieved 2 days p.t (black vertical dotted line). The plot shows the samples co-transfected with SB100X and transposon plasmids: T2 3.2 or the SA 5.7 (blue), T2 10 or SA 9.7 (green), T2 14 or SA 14 (purple), and the 18 kb transposons (orange). SA constructs represented with dashed lines, T2 with continuous lines. In gold are represented the negative controls transfected with T2 3.2 or SA 5.7 alone, without the SB100X plasmid.

Table S1 List of primers used for plasmid episomial amplification, LM-PCR, and site-specific amplification of the SA-genome junctions.

Table S2 Transposed clones were analysed to show the following parameters: number of retrieved Venus$^+$ clones for each bulk; percentage of Venus$^+$ cells in bulk populations 48 hours p.t.; percentage of stable Venus expressing cells in bulk populations; percentage of clones positive for the Ampicillin or SB100X sequence carried by transfected plasmids; mean copy number retrieved by Southern blot analysis; recombinant events detected in transposed clones by Southern blot analysis.

Acknowledgments

We thank Davide Pietrobon for compiling the scripts used for filtering and trimming the raw sequences from SB and viral libraries.

Author Contributions

Conceived and designed the experiments: GT MCL. Performed the experiments: GT MCL. Analyzed the data: GT MCL CC AGD FM Z. Izsvák Z. Ivics AR. Wrote the paper: GT AR.

References

1. Mates L, Chuah M, Belay E, Jerchow B, Manoj N, et al. (2009) Molecular evolution of a novel hyperactive Sleeping Beauty transposase enables robust stable gene transfer in vertebrates. Nature genetics 41: 753–761.

2. Ivics Z, Izsvak Z, Minter A, Hackett PB (1996) Identification of functional domains and evolution of Tc1-like transposable elements. Proc Natl Acad Sci U S A 93: 5008–5013.

3. Ivics Z, Hackett P, Plasterk R, Izsvak Z (1997) Molecular reconstruction of Sleeping Beauty, a Tc1-like transposon from fish, and its transposition in human cells. Cell 91: 501–510.

4. Hackett PB, Largaespada DA, Cooper LJ (2010) A transposon and transposase system for human application. Mol Ther 18: 674–683.

5. Ivics Z, Kaufman C, Zayed H, Miskey C, Walisko O, et al. (2004) The Sleeping Beauty transposable element: evolution, regulation and genetic applications. Current issues in molecular biology 6: 43–55.

6. Cui Z, Geurts A, Liu G, Kaufman C, Hackett P (2002) Structure-function analysis of the inverted terminal repeats of the sleeping beauty transposon. Journal of molecular biology 318: 1221–1235.

7. Izsvak Z, Ivics Z, Plasterk R (2000) Sleeping Beauty, a wide host-range transposon vector for genetic transformation in vertebrates. Journal of molecular biology 302: 93–102.

8. Zayed H, Izsvak Z, Walisko O, Ivics Z (2004) Development of hyperactive sleeping beauty transposon vectors by mutational analysis. Mol Ther 9: 292–304.

9. Ivics Z, Garrels W, Mates L, Yau TY, Bashir S, et al. (2014) Germline transgenesis in pigs by cytoplasmic microinjection of Sleeping Beauty transposons. Nat Protoc 9: 810–827.

10. Garrels W, Holler S, Taylor U, Herrmann D, Niemann H, et al. (2014) Assessment of fetal cell chimerism in transgenic pig lines generated by sleeping beauty transposition. PLoS One 9: e96673.

11. Ivics Z, Hiripi L, Hoffmann OI, Mates L, Yau TY, et al. (2014) Germline transgenesis in rabbits by pronuclear microinjection of Sleeping Beauty transposons. Nat Protoc 9: 794–809.

12. Katter K, Geurts AM, Hoffmann O, Mates L, Landa V, et al. (2013) Transposon-mediated transgenesis, transgenic rescue, and tissue-specific gene expression in rodents and rabbits. FASEB J 27: 930–941.

13. Ivics Z, Mates L, Yau TY, Landa V, Zidek V, et al. (2014) Germline transgenesis in rodents by pronuclear microinjection of Sleeping Beauty transposons. Nat Protoc 9: 773–793.

14. Jin Z, Maiti S, Huls H, Singh H, Olivares S, et al. (2011) The hyperactive Sleeping Beauty transposase SB100X improves the genetic modification of T cells to express a chimeric antigen receptor. Gene Ther 18: 849–856.

15. Liu L, Sanz S, Heggestad AD, Antharam V, Notterpek L, et al. (2004) Endothelial targeting of the Sleeping Beauty transposon within lung. Mol Ther 10: 97–105.

16. Belur LR, Frandsen JL, Dupuy AJ, Ingbar DH, Largaespada DA, et al. (2003) Gene insertion and long-term expression in lung mediated by the Sleeping Beauty transposon system. Mol Ther 8: 501–507.

17. Zhu J, Kren B, Park C, Bilgim R, Wong P, et al. (2007) Erythroid-specific expression of beta-globin by the sleeping beauty transposon for Sickle cell disease. Biochemistry 46: 6844–6858.

18. Wilber A, Linehan JL, Tian X, Woll PS, Morris JK, et al. (2007) Efficient and stable transgene expression in human embryonic stem cells using transposon-mediated gene transfer. Stem Cells 25: 2919–2927.

19. Grabundzija I, Wang J, Sebe A, Erdei Z, Kajdi R, et al. (2013) Sleeping Beauty transposon-based system for cellular reprogramming and targeted gene insertion in induced pluripotent stem cells. Nucleic Acids Res 41: 1829–1847.

20. Ammar I, Gogol-Doring A, Miskey C, Chen W, Cathomen T, et al. (2012) Retargeting transposon insertions by the adeno-associated virus Rep protein. Nucleic acids research 40: 6693–6712.

21. Huang X, Guo H, Tammana S, Jung Y-C, Mellgren E, et al. (2010) Gene transfer efficiency and genome-wide integration profiling of Sleeping Beauty, Tol2, and piggyBac transposons in human primary T cells. Mol Ther 18: 1803–1813.

22. Huang X, Wilber A, Bao L, Tuong D, Tolar J, et al. (2006) Stable gene transfer and expression in human primary T cells by the Sleeping Beauty transposon system. Blood 107: 483–491.

23. Voigt K, Gogol-Doring A, Miskey C, Chen W, Cathomen T, et al. (2012) Retargeting sleeping beauty transposon insertions by engineered zinc finger DNA-binding domains. Mol Ther 20: 1852–1862.

24. Borradori L, Chavanas S, Schaapveld R, Gagnoux-Palacios L, Calafat J, et al. (1998) Role of the bullous pemphigoid antigen 180 (BP180) in the assembly of hemidesmosomes and cell adhesion–reexpression of BP180 in generalized atrophic benign epidermolysis bullosa keratinocytes. Experimental cell research 239: 463–476.

25. McCormack W, Seiler M, Bertin T, Ubhayakar K, Palmer D, et al. (2006) Helper-dependent adenoviral gene therapy mediates long-term correction of the clotting defect in the canine hemophilia A model. Journal of thrombosis and haemostasis 4: 1218–1225.

26. Sambrook J, Russell DW (2001) Molecular cloning: a laboratory manual. CSHL press 1.

27. Schmidt M, Hoffmann G, Wissler M, Lemke N, Mussig A, et al. (2001) Detection and direct genomic sequencing of multiple rare unknown flanking DNA in highly complex samples. Hum Gene Ther 12: 743–749.

28. Smit AFA, Hubley R, Green P (1996–2010) RepeatMasker Open-3.0. Available: http://repeatmasker.org. Accessed 2014 Oct 23.

29. Cattoglio C, Facchini G, Sartori D, Antonelli A, Miccio A, et al. (2007) Hot spots of retroviral integration in human CD34+ hematopoietic cells. Blood 110: 1770–1778.

30. Cavazza A, Cocchiarella F, Bartholomae C, Schmidt M, Pincelli C, et al. (2013) Self-inactivating MLV vectors have a reduced genotoxic profile in human epidermal keratinocytes. Gene Ther 20: 949–957.

31. Garrels W, Mates L, Holler S, Dalda A, Taylor U, et al. (2011) Germline transgenic pigs by Sleeping Beauty transposition in porcine zygotes and targeted integration in the pig genome. PLoS One 6: e23573.

32. Hackett P, Largaespada D, Switzer K, Cooper L (2013) Evaluating risks of insertional mutagenesis by DNA transposons in gene therapy. Translational research.

33. Yant S, Wu X, Huang Y, Garrison B, Burgess S, et al. (2005) High-resolution genome-wide mapping of transposon integration in mammals. Molecular and cellular biology 25: 2085–2094.

34. Zhang W, Muck-Hausl M, Wang J, Sun C, Gebbing M, et al. (2013) Integration profile and safety of an adenovirus hybrid-vector utilizing hyperactive sleeping beauty transposase for somatic integration. PLoS One 8: e75344.

35. de Jong J, Akhtar W, Badhai J, Rust AG, Rad R, et al. (2014) Chromatin landscapes of retroviral and transposon integration profiles. PLoS Genet 10: e1004250.

36. Wang Y, Wang J, Devaraj A, Singh M, Jimenez Orgaz A, et al. (2014) Suicidal autointegration of sleeping beauty and piggyBac transposons in eukaryotic cells. PLoS Genet 10: e1004103.

37. Yant S, Ehrhardt A, Mikkelsen J, Meuse L, Pham T, et al. (2002) Transposition from a gutless adeno-transposon vector stabilizes transgene expression in vivo. Nature biotechnology 20: 999–1005.

38. Moldt B, Miskey C, Staunstrup N, Gogol-Doring A, Bak R, et al. (2011) Comparative genomic integration profiling of Sleeping Beauty transposons mobilized with high efficacy from integrase-defective lentiviral vectors in primary human cells. Mol Ther 19: 1499–1510.

39. Field AC, Vink C, Gabriel R, Al-Subki R, Schmidt M, et al. (2013) Comparison of lentiviral and sleeping beauty mediated alphabeta T cell receptor gene transfer. PLoS One 8: e68201.

40. Galla M, Schambach A, Falk C, Maetzig T, Kuehle J, et al. (2011) Avoiding cytotoxicity of transposases by dose-controlled mRNA delivery. Nucleic acids research 39: 7147–7160.

41. Zhang W, Solanki M, Muther N, Ebel M, Wang J, et al. (2013) Hybrid adeno-associated viral vectors utilizing transposase-mediated somatic integration for stable transgene expression in human cells. PLoS one 8.

42. Moiani A, Paleari Y, Sartori D, Mezzadra R, Miccio A, et al. (2012) Lentiviral vector integration in the human genome induces alternative splicing and generates aberrant transcripts. The Journal of clinical investigation 122: 1653–1666.

43. Cesana D, Sgualdino J, Rudilosso L, Merella S, Naldini L, et al. (2012) Whole transcriptome characterization of aberrant splicing events induced by lentiviral vector integrations. The Journal of clinical investigation 122: 1667–1676.

44. Titeux M, Pendaries V, Zanta-Boussif MA, Decha A, Pironon N, et al. (2010) SIN retroviral vectors expressing COL7A1 under human promoters for ex vivo gene therapy of recessive dystrophic epidermolysis bullosa. Mol Ther 18: 1509–1518.

45. Holkers M, Maggio I, Liu J, Janssen JM, Miselli F, et al. (2013) Differential integrity of TALE nuclease genes following adenoviral and lentiviral vector gene transfer into human cells. Nucleic Acids Res 41: e63.

46. Aiuti A, Cassani B, Andolfi G, Mirolo M, Biasco L, et al. (2007) Multilineage hematopoietic reconstitution without clonal selection in ADA-SCID patients treated with stem cell gene therapy. The Journal of clinical investigation 117: 2233–2240.

47. Biffi A, Bartolomae C, Cesana D, Cartier N, Aubourg P, et al. (2011) Lentiviral vector common integration sites in preclinical models and a clinical trial reflect a benign integration bias and not oncogenic selection. Blood 117: 5332–5339.

48. Bushman F, Lewinski M, Ciuffi A, Barr S, Leipzig J, et al. (2005) Genome-wide analysis of retroviral DNA integration. Nature reviews Microbiology 3: 848–858.

49. Bushman F (2007) Retroviral integration and human gene therapy. The Journal of clinical investigation 117: 2083–2086.

50. Montini E, Cesana D, Schmidt M, Sanvito F, Bartholomae C, et al. (2009) The genotoxic potential of retroviral vectors is strongly modulated by vector design and integration site selection in a mouse model of HSC gene therapy. The Journal of clinical investigation 119: 964–975.

51. Liu G, Geurts AM, Yae K, Srinivasan AR, Fahrenkrug SC, et al. (2005) Target-site preferences of Sleeping Beauty transposons. J Mol Biol 346: 161–173.

52. Vigdal T, Kaufman C, Izsvak Z, Voytas D, Ivics Z (2002) Common physical properties of DNA affecting target site selection of sleeping beauty and other Tc1/mariner transposable elements. Journal of molecular biology 323: 441–452.

53. Olson W, Zhurkin V (2011) Working the kinks out of nucleosomal DNA. Current opinion in structural biology 21: 348–357.

54. Foltz D, Jansen L, Black B, Bailey A, Yates J, et al. (2006) The human CENP-A centromeric nucleosome-associated complex. Nature cell biology 8: 458–469.

55. Masumoto H, Nakano M, Ohzeki J-I (2004) The role of CENP-B and alpha-satellite DNA: de novo assembly and epigenetic maintenance of human centromeres. Chromosome research 12: 543–556.

56. De Luca M, Pellegrini G, Mavilio F (2009) Gene therapy of inherited skin adhesion disorders: a critical overview. The British journal of dermatology 161: 19–24.

57. Carteau S, Hoffmann C, Bushman F (1998) Chromosome structure and human immunodeficiency virus type 1 cDNA integration: centromeric alphoid repeats are a disfavored target. J Virol 72: 4005–4014.

22

Indexes of Large Genome Collections on a PC

Agnieszka Danek[1], **Sebastian Deorowicz**[1]*, **Szymon Grabowski**[2]

1 Institute of Informatics, Silesian University of Technology, Gliwice, Poland, **2** Institute of Applied Computer Science, Lodz University of Technology, Łódź, Poland

Abstract

The availability of thousands of individual genomes of one species should boost rapid progress in personalized medicine or understanding of the interaction between genotype and phenotype, to name a few applications. A key operation useful in such analyses is aligning sequencing reads against a collection of genomes, which is costly with the use of existing algorithms due to their large memory requirements. We present MuGI, Multiple Genome Index, which reports all occurrences of a given pattern, in exact and approximate matching model, against a collection of thousand(s) genomes. Its unique feature is the small index size, which is customisable. It fits in a standard computer with 16–32 GB, or even 8 GB, of RAM, for the 1000GP collection of 1092 diploid human genomes. The solution is also fast. For example, the exact matching queries (of average length 150 bp) are handled in average time of 39 μs and with up to 3 mismatches in 373 μs on the test PC with the index size of 13.4 GB. For a smaller index, occupying 7.4 GB in memory, the respective times grow to 76 μs and 917 μs. Software is available at http://sun.aei.polsl.pl/mugi under a free license. Data S1 is available at PLOS One online.

Editor: Stephen Moore, University of Queensland, Australia

Funding: The work was supported by the Polish Ministry of Science and Higher Education under the project DEC-2013/09/B/ST6/03117 and European Social Fund project UDA-POKL.04.01.01-00-106/09. The work was performed using the infrastructure supported by POIG.02.03.01-24-099/13 grant: "GeCONiI---Upper Silesian Center for Computational Science and Engineering". The funders had no role in study design, data collection and analysis, decision to publish, or preparation of the manuscript.

Competing Interests: The authors have declared that no competing interests exist.

* Email: sebastian.deorowicz@polsl.pl

Introduction

About a decade ago, thanks to breakthrough ideas in succinct indexing data structures, it was made clear that a full mammalian-sized genome can be stored and used in indexed form in main memory of a commodity workstation (equipped with, e.g., 4 GB of RAM). Probably the earliest such attempt, by Sadakane and Shibuya [1], resulted in approximately 2 GB sized compressed suffix array built for the April 2001 draft assembly by Human Genome Project at UCSC. (Obtaining low construction space, however, was more challenging, although later more memory frugal, or disk-based, algorithms for building compressed indexes appeared, see, e.g., [2] and references therein.). Yet around 2008, only a few sequenced human genomes were available, so the possibility to look for exact or approximate occurrences of a given DNA string in a (single) genome was clearly useful. Nowadays, when repositories with a thousand or more genomes are easily available, the life scientists' goals are also more ambitious, and it is desirable to search for patterns in large genomic collections. One application of such a solution could be simultaneous alignment of sequencing reads against multiple genomes [3]. Other applications are discussed in the last section.

Interestingly, this is a largely unexplored area yet. On one hand, toward the end of the previous decade it was noticed that the "standard" compressed indexes (surveyed in [4]), e.g. from the FM or CSA family, are rather inappropriate to handle large collections of genomes of the same species, because they cannot exploit well the specific repetitiveness. On a related note, standard compres-sion methods were inefficient for a simpler problem of merely compressing multiple genomes. Since around 2009 we can observe a surge of interest in practical, multi-sequence oriented DNA compressors [5–15], often coupled with random access capabilities and sometimes also offering indexed search. The first algorithms from 2009 were soon followed by more mature proposals, which will be presented below, focusing on their indexing capabilities. More information on genome data compressors and indexes can be found in the recent surveys [16–18].

Mäkinen *et al.* [19] added index functionalities to compressed DNA sequences: *display* (which can also be called the random access functionality) returning the substring specified by its start and end position, *count* telling the number of times the given pattern occurs in the text, and *locate* listing the positions of the pattern in the text. Although those operations are not new in full-text indexes (possibly also compressed), the authors noticed that the existing general solutions, paying no attention to long repeats in the input, are not very effective here and they proposed novel *self-indexes* for the considered problem.

Claude *et al.* [7] pointed out that the full-text indexes from [19], albeit fast in counting, are rather slow in extracting the match locations, a feature shared by all compressed indexes based on the Burrows–Wheeler transform (BWT) [4]. They proposed two schemes, one basically an inverted index on q-grams, the other being a grammar-based self-index. The inverted index offers interesting space-time tradeoffs (on real data, not in the worst case), but can basically work with substrings of fixed length q. The grammar-based index is more elegant and can work with any substring length, but uses significantly more space, is slower and needs a large amount of RAM in the index build phase. None of these solutions can scale to large collections of mammalian-sized genomes, since even for 37 sequences of S. cerevisiae totaling 428 Mbases the index construction space is at least a few gigabytes.

While a few more indexes for repetitive data were proposed in recent years (e.g., [20–23]), theoretically superior to the ones presented above and often handling approximate matches, none of them can be considered a breakthrough, at least for bioinformatics, since none of them was demonstrated to run on multi-gigabyte genomic data.

A more ambitious goal, of indexing 1092 human genomes, was set by Wandelt et al. [24]. They obtained a data structure of size 115.7 GB, spending 54 hours on a powerful laptop. The index (loaded to RAM for a single chromosome at a time), called RCSI, allows to answer exact matching queries in about 250 μs, and in up to 2 orders of magnitude longer time for k-approximate matching queries, depending on the choice of k (up to 5).

Sirén et al. [25] extended the BWT transform of strings to acyclic directed labeled graphs, to support path queries as an extension to substring searching. This allows, e.g., for read alignment on an extended BWT index of a graph representing a pan-genome, i.e., reference genome and known variants of it. The authors built an index over a reference genome and a subset of variants from the dbSNP database, of size less than 4 GB and allowing to match reads in less than 1 ms in the exact matching mode. The structure, called GCSA, was built in chromosome-by-chromosome manner, but unfortunately, they were unable to finish the construction for a few "hard" chromosomes even in 1 TB of RAM! We also note that a pan-genome contains less information than a collection of genomes, since the knowledge about variant occurrences in individual genomes is lost.

A somewhat related work, by Huang et al. [26], presents an alignment tool, BWBBLE, working with a multi-genome (which is basically synonymous with pan-genome in the terminology of [25]). BWBBLE follows a more heuristic approach than GCSA and can be constructed using much more humble resources. Its memory use, however, is over $16n \log_2 n$ bits, where n is the multi-genome length. This translates to more than 200 GB of memory needed to build a multi-genome for a collection of 1092 human genomes. Both BWBBLE and GCSA need at least 10 ms to find matches with up to 3 errors.

The recently proposed journaled string tree (JST) [27] takes a different approach, providing an online scan over the reference sequence, but also keeping track of coverages of variants falling into the current window over the reference. Each individual is represented as a journal string, that is, a referentially compressed version of the original sequence; segments of journal strings, together with helper data, are stored in a journal string tree. The JST approach allows to generically speed up many sequential pattern matching algorithms (for exact or approximate search) when working on a collection of similar sequences. A drawback of this approach is that search times are never better than of an online scan over a single (reference) sequence.

Also recently, Durbin [28] presented an interesting data structure dubbed Positional Burrows–Wheeler Transform (PBWT), to find long matches between sequences within a given collection, or between a new test sequence and sequences from the collection. PBWT provides very compact representation of the dataset being searched, yet its application is different to ours: only binary information about variant occurrences are kept (not even their position in a reference sequence), which means that handling standard locate queries (given a string, report all its match

positions in the relevant sequences in the indexed collection) is impossible in this way.

Aligning sequencing reads to a genome with possible variants was also recently considered in theoretical works, under the problem name of indexing text with wildcard positions [29,30], where the wildcards represent SNPs. No experimental validation of the results was presented in the cited papers.

Most of the listed approaches are traditional string data structures, in the sense that they can work with arbitrary input sequences. The nowadays practice, however, is to represent multi-genome collections in repositories as basically a single reference genome, plus a database of possible variants (e.g., SNPs), plus information on which of the variants from the database actually occur in each of the individual genomes. The popular VCF (Variant Call Format) format allows to keep more information about a sequenced genome than listed here, but this minimal collection representation is enough to export each genome to its FASTA form. Dealing with input stored in such compact form should allow to build efficient indexes much more easily than following the standard "universal" way, not to say about tremendous resource savings in the index construction.

This modern approach was initiated in compression-only oriented works [5,13,14], and now we propose to adapt it in construction of a succinct and efficient index. According to our knowledge, this is the first full-text index capable to work on a scale of thousand(s) of human genomes on a PC, that is, a small workstation equipped with 16–32 GB of RAM. What is more, for a price of some slow-down the index can be used even on an 8 GB machine. No matter the end of the space-time tradeoff we are, the index is capable of handling also approximate matching queries, that is, reporting patterns locations in particular genomes from the collection with tolerance for up to 5 mismatches. As said, the index is not only compact, but also fast. For example, if up to 3 errors are allowed, the queries are handled in average time of 373 μs on the test PC and the index takes 13.4 GB of memory, or in 917 μs when the index is of size 7.4 GB. The current version of our index requires more resources (from 38 GB to 47 GB of RAM, depending on the index settings) in the construction phase; a drawback which may be eliminated in a future work, as discussed in the last section of this paper.

Materials and Methods

Datasets

We are indexing large collection of genomes of the same species, which are represented as the reference genome in FASTA format together with the VCF [31] file, describing all possible reference sequence variations and the genotype information for each of the genome in the dataset. We are only interested in details allowing for the recovery of the DNA sequences, all non-essential fields are ignored. Therefore, the data included in the VCFmin format, used in [14], are sufficient. Each line describes a possible variant that may be a single nucleotide polymorphism (SNP), a deletion (DEL), an insertion (INS) or a structural variation (SV), which is typically a combination of a very long deletion and an insertion. The genotype of each genome is specified in one designated column with information if each of the variant is found in this genome. In case of diploid and phased genotypes this information concerns two basic, haploid chromosome sets for each genome and treats them independently. Thus for any phased diploid genome, its DNA sequence is twice the size the reference sequence.

In our experiments we used the data available from Phase 1 of the 1000 Genomes Project [32] describing the collection of 1092 phased human genomes. We concatenated the available 24 VCF files (one for each chromosome), to get one combined VCF file, which—together with the reference sequence—is the input of our algorithm building the index.

The general idea

Our tool, Multiple Genome Index (MuGI), performs fast approximate search for input patterns in an indexed collection of genomes of the same species. The searched patterns can be provided in a text file (one pattern per line), or in FASTA or FASTQ format. The index is built based on the reference genome and the VCF file describing the set. The search answers the locate query—the result consists of all positions of the pattern with respect to the reference genome along with the list of all individuals in which it can be found.

The basic search regime is exact matching. Its enhanced version allows for searching with mismatches. Both modes use the seed-and-extend scheme. The general mechanism is to quickly find a substring of the pattern and then extend this seed to verify if it answers the query.

The index has one construction-time parameter, k, which is the maximum possible length of the seed. The match can be found directly in the reference genome and/or in its modified form, with some of the variations introduced. To find the seed we build an array of all possible k-length sequences (k-mers) occurring in all genome sequences. In the space-efficient version only a part of the array is kept. The extension step is done using the reference and the available database of variants, checking which combination of possible variations introduced, if any, allows to find the full pattern.

To know individuals in which the match can be found, we have to identify all variants whose occurrence, or absence of, have impact on the match, and list only the genomes with such combination of variants.

Building the index

To build the index, we process the input data to create the following main substructures, described in detail in the successive paragraphs:

- the reference sequence (REF),
- the Variant Database (VD),
- the Bit Vectors (BVs) with information about variants in all genomes,
- the k-Mer Array (kMA) for all unique k-length sequences in the set.

REF is stored in compact form, where 4 bits are used to (conveniently) encode a single character.

VD contains details about all possible variations. For each variant, the following items are stored: type (1 byte), preceding position (4 bytes) and alternative information (4 bytes). (Note that we keep the preceding positions to be able to manage the variants INSs, DELs and SVs, as this convention conforms to their description in VCF files.) The last item indicates alternative character in case of SNP, length of the deletion in case of DEL and position in the additional arrays of bytes (VD-aux) in case of INS and SV. VD-aux holds insertion length (4 bytes) and all inserted characters (1 byte each), if any, for every INS and SV. For SV it also stores length of the deletion (4 bytes). The variants are ordered by the preceding position and a lookup table is created to accelerate search for a variant by its location. VD together with REF can be used to decode the modified sequence from some given position to the right, by introducing certain variants. To be able to decode the sequence to the left, an additional list of all

deletions (SVs and DELs), ordered by the resulting position, is created. The list, VD-invDel, stores for each variant its number in the main VD (4 bytes) and the resulting position, that is, the position in the reference after the deletion (4 bytes).

There is one BV for each variant, each of size of the number of genomes in the collection (2 times the number of genomes for diploid organisms). Value 1 at some j th position in this vector means that the current variant is found in the j th haploid genome. To reduce the required size, while preserving random access, we keep the collection of these vectors in compressed form, making use of the fact that spatially close variant configurations are often shared across different individuals. The compression algorithm makes use of a dictionary of all possible unique 192-bit chunks (the size chosen experimentally). Each BV is thus represented as a concatenation of $\lceil no\_haploid\_genomes/192 \rceil$ 4-byte tokens (vocabulary IDs).

kMA keeps information about each k-length sequence (k-mer) occurring in the whole collection of genomes. The k-mer sequence itself is not kept. Instead, only the minimum information needed to retrieve it with help of REF and VD is stored. Based on the amount of details necessary to keep, we partition k-mers into four groups, each stored in one of the four subarrays of kMA: kMA^0, kMA^1, kMA^2 or kMA^3. The entries in each subarray are sorted according to the lexicographical order of k-mers they represent. All k-mers beginning with the unknown character (i.e., N or n) are filtered out.

All k-mers found in REF are kept in kMA^0. Only the preceding position $\langle pos\_ref \rangle$ (4 bytes) is stored for each such k-mer, as it is enough (using REF) to retrieve its sequence. These k-mers are present in all genomes with no variants introduced in the corresponding segment.

The k-mers that are obtained by applying some variant to the reference sequence are stored in kMA^1/kMA^2/kMA^3. They are produced with going through the reference genome and checking for each position p if there is any possible variant with the preceding position in the range from p to $p+k-1$. If the check is positive, we decode the k-mer. The decoding process takes into account all possible *paths*. By *path* we understand any combination of occurrence of subsequent variants, influencing the decoded sequence. For example, if SNP is possible at current position (i.e., it is listed in VD), two paths are considered: when it is found and when it is absent, resulting in two decoded sequences, differing in the last inspected character. Thus, starting from a single preceding position, many resulting sequences may be obtained. To decode most k-mers, it is enough to store the preceding position plus flags about the presence/absence of following variants. This evidence list (evList) is stored as a bit vector, where 1 means that the corresponding variant is present. For any k-mer starting inside an insertion (INS or SV) it is also necessary to store the *gap* from the beginning of the inserted string to the first character of the k-mer.

The k-mer with no *gap* and at most 32 evidences about consecutive variants from VD in the evList is stored in kMA^1, where each entry is defined as $\langle pos\_ref, evList \rangle$ $(4+4$ bytes). If there is also a *gap* involved, such k-mer goes to kMA^2, defining each entry as $\langle pos\_ref, gap, evList \rangle$ $(4+4+4$ bytes). All k-mers with more than 32 evidences in the evList or with evidences about nonconsecutive (with respect to VD) variants are kept in kMA^3, where each k-mer is represented by four fields: $\langle pos\_ref, gap, evSize, evList \rangle$ $(4+4+4+evSize \times 4$ bytes). The representative example of the latter case is a k-mer with SV introduced and many variants in VD placed within the deleted region. Keeping track of these variants, not altering the resulting sequence, is pointless.

Table 1. Pseudocode of the basic search algorithm.

Algorithm 1 exactSearch(P)
{kMA, $vtList$ and $evList$ are global variables}
1   $p \leftarrow \min(\lvert P \rvert, k)$
2   $S \leftarrow \mathrm{substring}(P, 0, p-1)$ {Retrieving the seed S}
3   **for** $i \leftarrow 0$ **to** 3 **do**
4   $(\ell, r) \leftarrow \mathrm{binSearch}(kMA^i, S)$ {Locating the seed S}
5   **for** $j \leftarrow \ell$ **to** r **do**
6   $(vtList, evList, pos\_curr, vt\_curr) \leftarrow \mathrm{partDecode}(kMA^i[j], p)$
7   $\mathrm{extend}(P, p, kMA^i[j].pos\_ref, pos\_curr, vt\_curr)$ {Extending the seed S to find P locations}

Any k-mer is kept in kMA only if there is at least one haploid genome that includes it, that is, has the same combination of occurring variants. It is checked with help of BV. Recall that the k-mers in each subarray kMA^i, $i \in \{0,1,2,3\}$, are sorted lexicographically. To speed up the binary search (by narrowing down the initial search interval), a lookup table, taking into account the first 12 characters, is created for each subarray.

The basic search algorithm

The pseudocode of the basic search algorithm is presented as Algorithm 1 in Table 1. It looks for all exact occurrences of the pattern P in the compressed collection, using the seed-and-extend scheme. The undetermined nucleotides (i.e., N or n) occurring in P are encoded differently than in REF, so they never match any character in the collection. The seed S is chosen to be a substring of P, precisely its first k characters, or the full pattern, if $\lvert P \rvert < k$ (lines 1–2).

The first step is to scan kMA for all k-mers whose prefixes (or simply full sequences, if $\lvert S \rvert = k$) match S. It is done with binary search in each subarray kMA^i, $i \in \{0,1,2,3\}$, separately (lines 3–4). Next, each found seed is partly decoded and then extended (lines 5–7). The partial decoding, done by the partDecode function, starts from pos_ref of the current k-mer and move $p = \lvert S \rvert$ characters forward, according to the k-mer's details (i.e., there may be a need to introduce some found variant). Character-by-character matching is not performed, as it is already known that $\lvert S \rvert$-length prefix of the k-mer matches S. Function partDecode returns the seed's succeeding position (pos_curr) and variant (vt_curr) in the reference, along with the list of encountered variants ($vtList$) and the list of evidences about their presence or absence ($evList$). The latter is a vector of 0 s in case of kMA^0 and a copy of k-mer's $evList$ (or its part) for other subarrays. The first variant (the one with preceding position greater than or equal to the preceding position of the k-mer) is found with binary search in VD. It is not shown in the pseudocode, but for each seed also the preceding SVs and DELs are taken into account when creating the initial $vtList$ and $evList$.

The seed S is recursively extended according to all possible combinations of variants, that is, as long as succeeding characters match the characters in P and found occurrences of P are reported (line 7). The pseudocode of the algorithm extending the seed and reporting the results is presented as Algorithm 2 in Table 2. Maintained variables are: full pattern P, ch (number of decoded characters), pre and pos (the preceding position of the seed and the current position, both in relation to the reference), and vt (next variant from VD). Also REF, BV, the current $vtList$ and $evList$ are available. If position of vt ($vt.pos$) is greater than pos (lines 2–5), no variant is introduced and the next character is taken from REF. If it does not match the related character in P, the extension is stopped, as the current path is not valid. If vt is encountered at pos (lines 6–11), it is added to the $vtList$ and two paths are checked—when it is introduced (new bit in $evList$ is set to 1) and when it is not (new bit in $evList$ is set to 0). The first path is not taken if vt does not match P. It can happen for SNPs and inserted characters (from INS or SV). If $vt.pos$ is less than pos

Table 2. Pseudocode of the algorithm extending the found seed in the basic search.

Algorithm 2 extend(P, ch, pre, pos, vt)
{REF, BV, $vtList$ and $evList$ are global variables}
1   **while** $ch < \lvert P \rvert$ **do**
2   **if** $vt.pos > pos$ **then**   {No variant at pos}
3   **if** $REF[pos] = P[ch]$ **then**
4         $pos \leftarrow pos + 1$; $ch \leftarrow ch + 1$
5   **else** report **false**   {Invalid path}
6   **else if** $vt.pos = pos$ **then**
7       $vtList.add(vt)$; $evList.add(1)$;
8   **if** vt matches P **then**
9         $new \leftarrow pos + vt.delLen$
10        $\mathrm{extend}(pre, new, ch + vt.len, vt + 1)$
11       $evList.setLast(0)$; $vt \leftarrow vt + 1$
12   **else**   {$vt.pos < pos$}
13       $new \leftarrow vt.pos + vt.delLen$
14   **if** $new > pos$ **then**
15        $vtList.add(vt)$; $evList.add(1)$;
16   **if** vt matches P **then**
17          $\mathrm{extend}(pre, new, ch + vt.len, vt + 1)$
18        $evList.setLast(0)$
19      $vt \leftarrow vt + 1$
20   $R \leftarrow 1^{noHaploidGenomes}$ {a bit-vector of noHaploidGenomes bits 1}
21   **for** $i \leftarrow 1$ **to** $vtList.size$ **do**
22   **if** $evList[i]$ **then** $R \leftarrow R \mathrel{\&} BV[i]$
23   **else** $R \leftarrow R \mathrel{\&} {\sim}BV[i]$
24   **if** $R = 0$ **then** report **false**   {Invalid path}
25   **else** report (pre, R)   {P found}

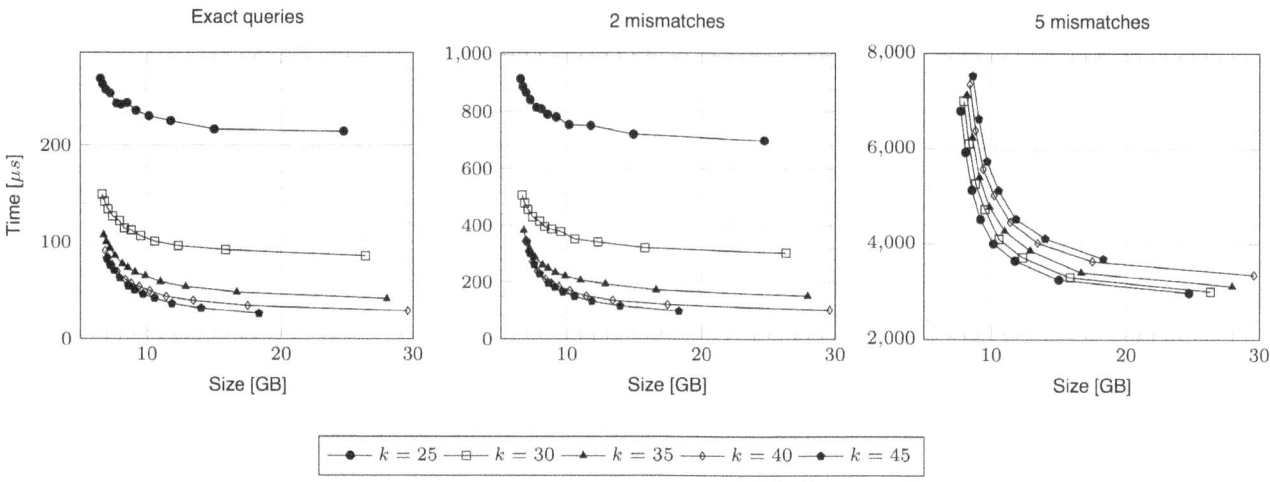

Figure 1. Average query times vs. index sizes. Simulated reads were used.

(lines 12–19), it means *vt* is placed in region previously deleted by other variant. The only possibility that *vt* is taken into account is if it deletes characters beyond previous deletion. Otherwise it is skipped.

When the extension reaches the end of the pattern *P*, it is checked in which individuals, if in any, the relevant combination of variants (track kept in *vtList*) is found (lines 20–25). The bit vector *R* is initialized to be the size of the number of haploid genomes. The value 1 at *j* th position means that *j* th haploid genome contains the found sequence. The vector *R* is set to all 1 s at the beginning, because if *vtList* is empty, the sequence is present in all genomes. To check which genomes have the appropriate combination of variants, the bitwise AND operations are performed between all BVs related to variants from the *vtList*, negating all BVs with 0s at the corresponding position in the *evList*. If *R* contains any 1 s, pattern *P* is reported to be found with the preceding position *pre* (in relation to the reference genome) and vector *R* specifies genomes containing such sequence.

The space-efficient version

To reduce the required space, while still being able to find all occurrences of the pattern, we make use of the idea of sparse suffix array [33]. This data structure stores only the suffixes with preceding position being a multiple of *s* (*s* > 1 is a construction-time parameter). In our scheme, the two largest subarrays, kMA^0 and kMA^1, are kept in sparse form, based on preceding positions of *k*-mers. For kMA^1, it is also necessary to keep all *k*-mers that begin with deletion or insertion (the first variant has the same preceding position as the *k*-mer).

The search algorithm has to be slightly modified. Apart from looking for the *k*-length prefix of the pattern (i.e., $P[0 \ldots k-1]$) in *kMA*, also *k*-length substrings starting at positions $1 \ldots s-1$ must be looked for in kMA^0, kMA^1, and kMA^3 (as some specific seeds may be present only in kMA^3). The substrings, if found in one of mentioned subarrays, must be then decoded to the left, to check if their prefix (from 1 to $s-1$ characters, depending on the starting position) matches the pattern *P*. The VD-invDel substructure is

Table 3. Index sizes.

Sparsity	Size [GB]				
	k = 25	*k* = 30	*k* = 35	*k* = 40	*k* = 45
1	24.7	26.3	27.9	29.6	31.2
2	15.0	15.8	16.6	17.5	18.3
3	11.8	12.3	12.9	13.4	14.0
4	10.2	10.6	11.0	11.4	11.8
5	9.2	9.5	9.9	10.2	10.5
6	8.5	8.8	9.1	9.4	9.7
7	8.1	8.3	8.6	8.8	9.1
8	7.7	7.9	8.2	8.4	8.6
10	7.2	7.4	7.6	7.8	8.0
12	6.9	7.1	7.2	7.4	7.5
14	6.7	6.8	7.0	7.1	7.2
16	6.5	6.6	6.8	6.9	7.0

Table 4. Query times for various variants of indexes for simulated data.

k	sparsity	size [GB]	Max. allowed mismatches					
			0	1	2	3	4	5
25	1	24.7	214.2	450.8	699.5	971.5	1,438.3	2,976.8
25	3	11.8	225.0	481.2	751.6	1,024.3	1,599.9	3,647.1
25	4	10.2	229.8	493.1	754.0	1,050.9	1,676.4	4,004.2
25	8	7.7	243.1	528.3	814.6	1,158.5	2,341.4	6,790.4
25	12	6.9	257.2	558.3	868.0	1,337.8		
25	16	6.5	268.8	588.6	916.6	1,787.9		
30	1	26.3	85.4	193.4	303.0	456.0	1,036.4	3,004.6
30	3	12.3	95.7	220.8	340.6	520.4	1,258.0	3,716.2
30	4	10.6	100.4	227.8	351.5	544.0	1,376.5	4,104.5
30	8	7.9	121.8	267.0	414.5	713.6	2,215.2	6,994.5
30	12	7.1	134.0	291.4	456.4	959.4		
30	16	6.6	149.2	319.3	506.8	1,490.4		
35	1	27.9	41.4	98.0	152.0	301.8	1,033.4	3,114.6
35	3	12.9	53.6	121.2	193.0	380.4	1,280.8	3,861.2
35	4	11.0	58.6	130.2	206.3	419.2	1,411.7	4,277.6
35	8	8.2	77.2	166.1	260.3	608.3	2,224.4	7,120.4
35	12	7.2	93.3	196.2	314.6	905.3		
35	16	6.8	107.0	222.2	382.4	1,506.4		
40	1	29.6	28.8	65.2	102.3	291.0	1,109.9	3,348.8
40	3	13.4	39.4	85.5	136.1	372.5	1,334.0	4,021.0
40	4	11.4	43.4	94.4	151.4	412.2	1,471.1	4,461.1
40	8	8.4	61.0	128.9	210.3	615.4	2,297.9	7,350.3
40	12	7.4	76.3	160.0	271.8	917.0		
40	16	6.9	90.4	184.4	344.3	1,514.1		
45	2	18.3	25.9	56.2	97.9	329.6	1,207.0	3,687.2
45	3	14.0	31.3	67.7	116.5	375.5	1,353.3	4,115.0
45	4	11.8	36.3	77.2	132.6	421.2	1,490.9	4,525.3
45	8	8.6	54.2	112.4	196.2	625.8	2,394.4	7,523.9
45	12	7.5	70.4	142.9	262.5	942.0		
45	16	7.0	82.1	168.4	342.3	1,531.9		
GEM mapper		5.0	24.0	50.6	64.9	86.4	131.0	217.3

All times in μs. We do not provide times for large sparsities and more errors than 3, since in such cases the internal queries would be for very short sequences and in turn result in numerous matches and significant times; thus, we do not recommend to use MuGI in such parameter configurations.

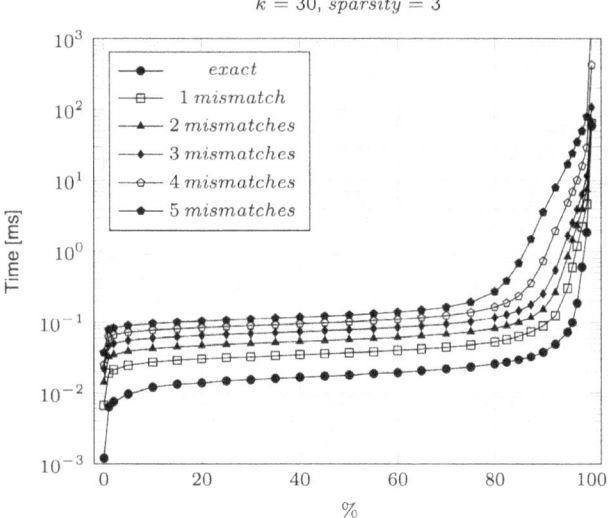

$k = 30$, $sparsity = 3$

Figure 2. Query time percentiles for exact and approximate matching, for max error up to 5. For example, the 80th percentile for 1 error equal to 0.52 ms means that 80% of the test patterns were handled in time up to 0.52 ms each, allowing for 1 mismatch. Simulated reads were used.

used for the process. The rest of search is the same as in the basic search algorithm.

The approximate search algorithm

The approximate search algorithm looks for all occurrences of the given pattern with some maximum allowed number of mismatches. According to the well-known property, for any sequence of length ℓ with m mismatches at least one of the consecutive substrings of length $q = \left\lfloor \dfrac{\ell}{m+1} \right\rfloor$ is the same as in the original sequence. Therefore, the approximate search begins with dividing the string to $m+1$ substrings of length q. Next, the exact search algorithm is used to look for each of the substrings. If a substring is found in the collection, it is further decoded to the right and to the left, similarly as in the exact search, but allowing for at most m differences between the decoded sequence and the searched sequence. Expanding to the left is done with aid of the same auxiliary substructure as in the space-efficient version (VD-

invDel). The list of genomes in which the found sequences are present is obtained in the same way as in the exact searching.

Test data

To evaluate the algorithm, we first used a similar methodology as the one in [24]. To this end, we generated a file with 100K queries, where each pattern is a modified excerpt of length $\ell = 120 \ldots 170$ (uniformly random value) from a randomly selected genome from the collection, starting at a randomly selected position. Excerpts containing undetermined nucleotide (i.e., N) were rejected. The modifications consisted in introducing random nucleotides in place of x existing nucleotides, where x is a randomly selected integer number from the $[0, 0.05 \times \ell]$ range.

Additionally, we use real reads from the 1000GP repository. There are 140K reads chosen randomly in such a way that each of 14 human populations is represented with 10K reads. Their length varies between 100 and 120 bp. Both data sets are available at project home page.

As the index construction costs are not that small (as mentioned earlier), we provide an exemplary index over the 1000GP data at our software page.

Results

All experiments were performed on a PC with Intel Core i7 4770 3.4 GHz CPU (4 cores with hyperthreading), equipped with 32 GB of RAM, running Windows 7 OS. The C++ sources were compiled using GCC 4.7.1 compiler.

The index was built on another machine (2.4 GHz Quad-Core AMD Opteron CPU with 128 GB RAM running Red Hat 4.1.2-46) and required more RAM: from 38 GB (for $k = 25$) to 47 GB (for $k = 45$). The corresponding build times were 15 hours and 72 hours, respectively. The index build phase was based on parallel sort (using Intel TBB and OpenMP libraries), while all the queries in our experiments were single-threaded. The correctness of obtained query results, in exact and approximate matching mode, was experimentally verified with a set of patterns, for which a naïve (sequential) scan over all the sequences was run.

From Table 3 we can see that the fastest index version (i.e., with sparsity 1, which translates to standard k-mer arrays) may work on the test machine even for the seed maximum length of 40 symbols. Significant savings in the index size are however possible if sparsity of 3 or more is set, making the index possible to operate on a commodity PC with 16 GB of RAM. If one (e.g., a laptop user) requires even less memory, then the sparsity set to 16 makes it possible to run the index even in 8 GB of RAM. Naturally, using

Table 5. Query times for simulated data for $k = 40$ and $sparsity = 3$ (size 13.4 GB).

Percentile	Max. allowed mismatches					
	0	1	2	3	4	5
10%	12.4	27.7	44.0	62.1	81.1	100.0
25%	15.4	32.2	50.3	69.9	90.7	111.8
50%	18.7	38.3	58.8	81.4	105.8	131.4
75%	23.8	48.5	73.5	103.7	141.9	199.5
90%	35.9	76.5	116.0	188.0	707.0	3,747.5
95%	57.0	112.7	182.9	718.1	4,972.8	17,619.4
average	39.4	85.5	136.1	372.5	1,334.0	4,021.0

All times in μs.

Table 6. Query times for various variants of indexes for real data.

k	sparsity	size [GB]	Max. allowed mismatches				
			0	1	2	3	4
25	1	24.7	214.1	483.2	741.6	1,123.3	3,566.8
25	3	11.8	227.6	509.4	791.1	1,271.6	4,575.3
25	4	10.2	238.8	520.9	795.0	1,375.8	5,134.4
25	8	7.7	259.1	569.2	865.3	2,180.1	9,634.0
25	12	6.9	282.0	605.6	960.4	4,052.4	
25	16	6.5	292.6	644.0	1,264.0	9,888.8	
30	1	26.3	93.9	206.2	342.4	938.8	3,652.4
30	3	12.3	105.8	234.5	384.8	1,191.8	4,691.9
30	4	10.6	111.2	241.7	386.8	1,310.5	5,310.1
30	8	7.9	131.3	283.9	485.4	2,183.5	9,862.3
30	12	7.1	149.2	316.0	674.2	4,075.0	
30	16	6.6	161.1	343.7	1,051.7	10,128.0	
35	1	27.9	51.5	109.5	200.2	977.1	3,782.0
35	3	12.9	62.6	132.4	255.7	1,267.1	5,224.2
35	4	11.0	75.2	156.7	307.5	1,491.0	5,977.3
35	8	8.2	94.3	193.3	463.7	2,434.8	9,893.8
35	12	7.2	98.9	206.7	630.0	4,099.5	
35	16	6.8	113.2	230.7	1,018.4	10,267.9	
40	1	29.6	34.1	68.1	191.4	1,004.6	3,793.7
40	3	13.4	43.3	90.6	250.3	1,248.2	4,878.4
40	4	11.4	49.3	100.2	280.1	1,375.3	5,491.2
40	8	8.4	67.2	134.4	426.2	2,282.5	10,236.1
40	12	7.4	80.7	165.3	645.2	4,240.8	
40	16	6.9	95.1	193.3	1,055.9	10,497.4	
45	2	18.3	30.0	61.3	219.2	1,116.7	4,327.8
45	3	14.0	35.5	72.7	259.0	1,281.7	4,988.0
45	4	11.8	40.7	82.2	283.5	1,401.1	5,598.8
45	8	8.6	58.3	118.4	432.1	2,320.6	10,420.5
45	12	7.5	73.4	149.9	657.3	4,289.5	
45	16	7.0	86.4	182.4	1,072.4	10,647.0	
GEM mapper		5.0	22.1	56.5	78.5	126.2	221.6

All times in μs. We do not provide times for large sparsities and more errors than 3, since in such cases the internal queries would be for very short sequences and in turn result in numerous matches and significant times; thus, we do not recommend to use MuGI in such parameter configurations.

Table 7. Comparison of MuGI and JST on simulated and real data, both over 1092 individual sequences of chr1.

Algorithm	Max. allowed mismatches						RAM usage [GB]
	0	1	2	3	4	5	
Simulated data							
JST-Horspool	8.0 s	—	—	—	—	—	2.58
JST-Myers	22.5 s	24.0 s	23.9 s	24.3 s	24.5 s	24.9 s	2.58
MuGI, $k=30$, sparsity = 1	8.2 μs	14.8 μs	22.7 μs	33.6 μs	69.6 μs	176.0 μs	1.84
MuGI, $k=30$, sparsity = 3	10.7 μs	21.7 μs	32.4 μs	47.9 μs	90.7 μs	239.0 μs	0.98
Real data							
JST-Horspool	6.9 s	—	—	—	—	—	2.58
JST-Myers	18.4 s	19.1 s	19.2 s	20.0 s	20.3 s	20.3 s	2.58
MuGI, $k=30$, sparsity = 1	7.6 μs	14.3 μs	22.1 μs	53.4 μs	172.3 μs	476.5 μs	1.84
MuGI, $k=30$, sparsity = 3	12.3 μs	23.1 μs	35.3 μs	72.0 μs	238.2 μs	617.9 μs	0.98

MuGI was executed for parameters $k=30$ and sparsities: 1 (index size 2.0 GB), 3 (index size 1.0 GB). Results for JST are averages from only 100 queries due to very long running times. JST times include block generation (blocks of 100K SNPs were used), but in our experiments they are at least an order of magnitude lower than pattern searching. JST-Horspool uses the Boyer–Moore–Horspool exact matching algorithm, while JST-Myers uses Myers' bit-parallel approximate matching algorithm, handling the Levenshtein distance (k-differences). The JST index size was 468 MB, in addition to the 253 MB of the reference sequence. Note its memory use during the search is significantly higher than the index size and depends on the block size (e.g., its memory use grows to about 13 GB with blocks of 1 M SNPs).

larger sparsities comes at a price of slower searches; in Fig. 1, each series of results for a given value of k corresponds to sparsities from $\{1,2,\ldots,8,10,12,14,16\}$ (sparsities of 1 correspond to the rightmost points, with the exception of the case of $k=45$, for which the sparsities start from 2). Still, this tradeoff is not very painful: even the largest allowed sparsity value (16) slows down the fastest (for sparsity of 1) queries by factor about 2 on average, in most cases.

Costlier, in terms of query times, is handling mismatches. In particular, allowing 4 or 5 mismatches in the pattern requires at least an order of magnitude longer query times than in the exact matching mode. Yet, even for 5 allowed errors the average query time was below 10 ms in all tests. This translates, for example, to 224 mapped reads per second allowing up to 5 mismatches and 10,593 mapped reads per second with up to 1 mismatch, at index size of 11.4 GB ($k=40$, sparsity of 4, simulated reads; cf. Table 4).

Apart from the average case, one is often interested also in the pessimistic scenario. Our search algorithms do not have interesting worst-case time complexities, but fortunately pathological cases are rather rare. To measure this, for each test scenario a histogram of query times over 100K patterns was gathered, and the time percentiles are shown in Fig. 2. Note that the easy cases dominate: for all maximum errors allowed, for 90% test patterns the query time is below the average. Yet, there are a few percent of test patterns for which the times are several times longer, and even a fraction of a percent of patterns with query times exceeding 100 ms (at least for approximate matching). More details exposing the same phenomenon are presented in Table 5.

While we cannot directly compare our solution to RCSI by Wandelt et al. [24], as their software is not public, we can show some comparison. Their index was built over twice less data (haploid human genomes vs. diploid genomes in our data). We handle exact matches much faster (over 6 times shorter reported average times, but considering the difference in test computers this probably translates to factor about 4). Roughly similar differences can be observed for the approximate matching scenario, but RCSI handles the Levenshtein distance, while our scheme handles (so far) only mismatches. Finally, and perhaps most importantly, our index over 1092 diploid human genomes can be run on a standard PC, equipped with 32 or 16 GB of RAM (or even 8 GB, for the price of more slow-down), while RCSI requires a machine with 128 GB (unless searches are limited to one chromosome, when a portion of the index may be loaded into memory).

We were are not able to run GCSA [25] or BWBBLE [26], due to their large memory requirements in the construction phase.

We did, however, ran a preliminary comparison of MuGI against GEM [34], one of the fastest single genome read mappers. We ran it on 1 CPU core, for mismatches only, in the all-strata mode, in which all matches with $0,1,\ldots,max\_mismatches$ errors are reported, in arbitrary order. Table 4 contains a detailed rundown of the results on simulated reads. For example, we can see that GEM performed exact matching in 24.0 μs, found matches with up to 1 mismatch in 50.6 μs, matches with up to 3 mismatches in 86.4 μs, and matches with up to 5 mismatches in 217.3 μs. The memory use was 5.0 GB. This means that, depending on chosen options of our solution, GEM was only about twice faster in the exact matching mode and 15–20 times faster when 5 mismatches were allowed. On real reads (Table 6) GEM is about 1.5–5 times faster with exact matching and about 10–20 times faster with 3 allowed mismatches. The major scenario difference is however that GEM performs mapping to a single (i.e., our reference) genome, so to obtain the same mapping results GEM would have to be run 2×1092 times, once per haploid genome. We thus consider these preliminary comparative results very promising.

Table 8. Sizes of the index components.

k	sparsity	REF	VD	BV	kMA^0	kMA^1	kMA^2	kMA^3	Total
25	1	1,548	698	2,704	11,502	8,021	84	123	24,680
25	3	1,548	698	2,704	3,879	2,731	84	123	11,767
25	4	1,548	698	2,704	2,926	2,070	84	123	10,153
25	8	1,548	698	2,704	1,496	1,078	84	123	7,732
25	12	1,548	698	2,704	1,020	747	84	123	6,925
25	16	1,548	698	2,704	782	582	84	123	6,521
30	1	1,548	698	2,704	11,502	9,634	85	137	26,307
30	3	1,548	698	2,704	3,879	3,270	85	137	12,320
30	4	1,548	698	2,704	2,926	2,474	85	137	10,571
30	8	1,548	698	2,704	1,496	1,281	85	137	7,948
30	12	1,548	698	2,704	1,020	883	85	137	7,074
30	16	1,548	698	2,704	782	684	85	137	6,637
35	1	1,548	698	2,704	11,502	11,254	85	151	27,942
35	3	1,548	698	2,704	3,879	3,810	85	151	12,875
35	4	1,548	698	2,704	2,926	2,880	85	151	10,992
35	8	1,548	698	2,704	1,496	1,484	85	151	8,167
35	12	1,548	698	2,704	1,020	1,019	85	151	7,225
35	16	1,548	698	2,704	782	786	85	151	6,754
40	1	1,548	698	2,704	11,502	12,881	86	166	29,584
40	3	1,548	698	2,704	3,879	4,354	86	166	13,434
40	4	1,548	698	2,704	2,926	3,288	86	166	11,415
40	8	1,548	698	2,704	1,496	1,689	86	166	8,387
40	12	1,548	698	2,704	1,020	1,156	86	166	7,377
40	16	1,548	698	2,704	782	889	86	166	6,872
45	1	1,548	698	2,704	11,502	14,515	86	181	31,234
45	2	1,548	698	2,704	5,784	7,303	86	181	18,305
45	4	1,548	698	2,704	2,926	3,697	86	181	11,840
45	8	1,548	698	2,704	1,496	1,894	86	181	8,608
45	12	1,548	698	2,704	1,020	1,293	86	181	7,531
45	16	1,548	698	2,704	782	993	86	181	6,992

All sizes in MBs.

Finally, in Table 7 we compare MuGI against a recent tool JST by Rahn *et al.* [27]. As we can see, MuGI is usually 5–6 orders of magnitude faster at somewhat less memory consumption. This huge gap in performance can be explained with two different search "philosophies": sequential scan over the reference sequence in JST vs. fully indexed search in MuGI. As in this test we used only chr1 data (1092 sequences), the performance gap would probably be larger with the full human collection. On the other hand, we admit that JST performance with growing k (the maximum allowed number of errors) remains unchanged (which is a property of Myers' algorithm), therefore this scheme might be a satisfactory choice for a collection of short and highly-varied genomes.

Discussion

We presented an efficient index for exact and approximate searching over large repetitive genomic collections, in particular: multiple genomes of the same species. This has a natural application in aligning sequencing reads against a collection of genomes, with expected benefits for, e.g., personalized medicine and deeper understanding of the interaction between genotype and phenotype. Experiments show that the index built over a collection of 2×1092 human genomes fits a PC machine with 16 GB of RAM, or even half less, for the price of some slow-down. According to our knowledge, this is the first feat of this kind. The obtained solution is capable of finding all pattern occurrences in the collection in much below 1 ms in most use scenarios.

We point out that representing a "true" genome as a linear sequence over the ACGT(N) alphabet is inherently imperfect, since our knowledge about these sequences is (and will likely remain in the near future) limited. Every sequencing technology introduces its errors, therefore storing qualities (i.e., *estimated* correctness probabilities) together with the DNA symbols would convey more information useful for read mapping, yet we are unable to imagine an index over large collections based on such information not requiring huge amount of resources (especially main memory) in its runtime and construction stages. Moreover, large discrepancies between the reference and a given genome, e.g., long indels, result in reads that cannot be usually mapped, which implies incomplete variant information in the built VCF. Basically for those reasons the application of MuGI (and related software, like RCSI or BWBBLE) for mapping sequencing reads trades some accuracy for performance and reasonable memory use, yet with improving sequencing technologies the obtained mapping results should also be more valuable.

On the other hand, we should stress that MuGI is an index rather than a full-fledged read mapper. Aligning reads to multiple genomes is one of its possible applications. Another example could

be searching for nullomers, that is, k-mers with no occurrences in a given genome (or, in our scenario, genome collection). To apply MuGI here, we may generate random strings of specified length (e.g., 20) in a loop and check if they have any occurrence; we may also force the mimimum distance to any 20-mer in the genome to be 2 or 3, with running the MuGI engine in the approximate matching mode, to minimize the impact of noisy data in a genome, at still acceptable search speed. Also a closely related problem of finding the minimal absent word was investigated in the literature, and it can be solved with MuGI with a systematic scan over its component structures. Nullomers/minimal absent words can be used for studies of population genetics, drug discovery and development, evolution studies, design of molecular barcodes or specific adaptors for PCR primers [35,36]. Other (or more general) areas for application of our algorithm may include comparative genomics and personalized medicine.

Several aspects of the presented index require further development. The current approximate matching model comprises mismatches only; it is desirable to extend it to edit distance. The pathological query times could be improved with extra heuristics (even if it is almost irrelevant for large bulk queries). A more practical speedup idea is to enhance the implementation with multi-threading. Some tradeoffs in component data structures (cf. Table 8) may be explored, e.g., the reference genome may be encoded more compactly but at a cost of somewhat slower access. A soft spot of the current implementation is the index construction phase, which is rather naïve and can be optimized especially towards reduced memory requirements. We believe that existing disk-based suffix array creation algorithms (e.g., [37]) can be adapted for this purpose. Alternatively, we could build our indexing data structure separately for each chromosome (with memory use for the construction reduced by an order of magnitude) and then merge those substructures, onto disk, using little memory. The sparse suffix array may be replaced with a sampled suffix array variant [38], for a hopefully faster search at a similar space consumption. Finally, experiments on other collections should be interesting, particularly on highly-polymorphic ones.

Author Contributions

Conceived and designed the experiments: AD SD SG. Performed the experiments: AD SD. Analyzed the data: AD SD SG. Contributed reagents/materials/analysis tools: AD SD. Wrote the paper: AD SD SG.

References

1. Sadakane K, Shibuya T (2001) Indexing huge genome sequences for solving various problems. Genome Informatics Series 12: 175–183.
2. Hon WK, Sadakane K, Sung WK (2009) Breaking a time-and-space barrier in constructing full-text indices. SIAM Journal of Computing 38: 2162–2178.
3. Schneeberger K, Hagmann J, Ossowski S, Warthmann N, Gesing S, et al. (2009) Simultaneous alignment of short reads against multiple genomes. Genome Biology 10: Article no.R98.
4. Navarro G, Mäkinen V (2007) Compressed full-text indexes. ACM Computing Surveys 39: Article no.2.
5. Christley S, Lu Y, Li C, Xie X (2009) Human genomes as email attachments. Binformatics 25: 274–275.
6. Brandon M, Wallace D, Baldi P (2009) Data structures and compression algorithms for genomic sequence data. Bioinformatics 25: 1731–1738.
7. Claude F, Fariña A, Martínez-Pietro M, Navarro G (2010) Compressed q-gram indexing for highly repetitive biological sequences. In: Proceedings of the 10th IEEE Conference on Bioinformatics and Bioengineering. pp. 86–91.
8. Kuruppu S, Puglisi S, Zobel J (2010) Relative Lempel–Ziv compression of genomes for large-scale storage and retrieval. LNCS 6393: 201–206.
9. Kuruppu S, Puglisi S, Zobel J (2011) Optimized relative Lempel–Ziv compression of genomes. In: Proceedings of the ACSC Australasian Computer Science Conference. pp. 91–98.
10. Deorowicz S, Grabowski S (2011) Robust relative compression of genomes with random access. Bioinformatics 27: 2979–2986.
11. Kreft S, Navarro G (2013) On compressing and indexing repetitive sequences. Theoretical Computer Science 483: 115–133.
12. Yang X, Wang B, Li C, Wang J, Xie X (2013) Efficient direct search on compressed genomic data. In: Proceedings of the IEEE 29th International Conference on Data Engineering. pp. 961–972.
13. Pavlichin D, Weissman T, Yona G (2013) The human genome contracts again. Bioinformatics 29: 2199–2202.
14. Deorowicz S, Danek A, Grabowski S (2013) Genome compression: a novel approach for large collections. Bioinformatics 29: 2572–2578.

15. Wandelt S, Leser U (2014) FRESCO: Referential compression of highly-similar sequences. IEEE/ACM Transactions on Computational Biology and Bioinformatics 10: 1275–1288.

16. Vyverman M, De Baets B, Fack V, Dawyndt P (2012) Prospects and limitations of full-text index structures in genome analysis. Nucleic Acids Research 40: 6993–7015.

17. Deorowicz S, Grabowski S (2013) Data compression for sequencing data. Algorithms for Molecular Biology 8: Article no.25.

18. Giancarlo R, Rombo S, Utro F (2014) Compressive biological sequence analysis and archival in the era of high-throughput sequencing technologies. Briefings in Bioinformatics 15: 390–406.

19. Mäkinen V, Navarro G, Sirén J, Välimäki N (2010) Storage and retrieval of highly repetitive sequence collections. Journal of Computational Biology 17: 281–308.

20. Huang S, Lam T, Sung W, Tam S, Yiu S (2010) Indexing similar DNA sequences. LNCS 6124: 180–190.

21. Gagie T, Gawrychowski P, Puglisi S (2011) Faster approximate pattern matching in compressed repetitive texts. LNCS 7074: 653–662.

22. Do H, Jansson J, Sadakane K, Sung WK (2014) Fast relative Lempel-Ziv self-index for similar sequences. Theoretical Computer Science 532: 14–30.

23. Ferrada H, Gagie T, Hirvola T, Puglisi S (2014) Hybrid indexes for repetitive datasets. Philosophical Transactions of The Royal Society A 372: Article no.2016.

24. Wandelt S, Starlinger J, Bux M, Leser U (2013) RCSI: Scalable similarity search in thousand(s) of genomes. Proceedings of the VLDB Endowment 6: 1534–1545.

25. Sirén J, Välimäki N, Mäkinen V (2014) Indexing graphs for path queries with applications in genome research. IEEE/ACM Transactions on Computational Biology and Bioinformatics 11: 375–388.

26. Huang L, Popic V, Batzoglou S (2013) Short read alignment with populations of genomes. Bioinformatics 29: i361–i370.

27. Rahn R, Weese D, Reinert K (2014) Journaled string tree—a scalable data structure for analyzing thousands of similar genomes on your laptop. Bioinformatics : doi: 10.1093/bioinformatics/btu438.

28. Durbin R (2014) Efficient haplotype matching and storage using the Positional Burrows-Wheeler transform (PBWT). Bioinformatics 30: 1266–1272.

29. Thachuk C (2013) Compressed indexes for text with wildcards. Theoretical Computer Science 483: 22–35.

30. Hon WK, Ku TH, Shah R, Thankachan S, Vitter J (2013) Compressed text indexing with wildcards. Journal of Discrete Algorithms 19: 23–29.

31. Danecek P, Auton A, Abecasis G, Albers C, Banks E, et al. (2011) The variant call format and VCFtools. Bioinformatics 27: 2156–2158.

32. Consortium TGP (2012) An integrated map of genetic variation from 1,092 human genomes. Nature 491: 56–65.

33. Kärkkäinen J, Ukkonen E (1996) Sparse suffix trees. LNCS 1090: 219–230.

34. Marco-Sola S, Sammeth M, Guigó R, Ribeca P (2012) The GEM mapper: fast, accurate and versatile alignment by filtration. Nature Methods 9: 1185–1188.

35. Hampikian G, Andersen T (2007) Absent sequences: nullomers and primes. In: Pacific Symposium on Biocomputing. volume 12, pp. 355–366.

36. Garcia S, Pinho A, Rodrigues J, Bastos C, Ferreira P (2011) Minimal absent words in prokaryotic and eukaryotic genomes. PloS ONE 6: e16065.

37. Kärkkäinen J (2007) Fast BWT in small space by blockwise suffix sorting. Theoretical Computer Science 387: 249–257.

38. Grabowski S, Raniszewski M (2014) Sampling the suffix array with minimizers. arXiv preprint http://arxiv.org/abs/14062348.

Permissions

The contributors of this book come from diverse backgrounds, making this book a truly international effort. This book will bring forth new frontiers with its revolutionizing research information and detailed analysis of the nascent developments around the world.

We would like to thank all the contributing authors for lending their expertise to make the book truly unique. They have played a crucial role in the development of this book. Without their invaluable contributions this book wouldn't have been possible. They have made vital efforts to compile up to date information on the varied aspects of this subject to make this book a valuable addition to the collection of many professionals and students.

This book was conceptualized with the vision of imparting up-to-date information and advanced data in this field. To ensure the same, a matchless editorial board was set up. Every individual on the board went through rigorous rounds of assessment to prove their worth. After which they invested a large part of their time researching and compiling the most relevant data for our readers.

The editorial board has been involved in producing this book since its inception. They have spent rigorous hours researching and exploring the diverse topics which have resulted in the successful publishing of this book. They have passed on their knowledge of decades through this book. To expedite this challenging task, the publisher supported the team at every step. A small team of assistant editors was also appointed to further simplify the editing procedure and attain best results for the readers.

Apart from the editorial board, the designing team has also invested a significant amount of their time in understanding the subject and creating the most relevant covers. They scrutinized every image to scout for the most suitable representation of the subject and create an appropriate cover for the book.

The publishing team has been an ardent support to the editorial, designing and production team. Their endless efforts to recruit the best for this project, has resulted in the accomplishment of this book. They are a veteran in the field of academics and their pool of knowledge is as vast as their experience in printing. Their expertise and guidance has proved useful at every step. Their uncompromising quality standards have made this book an exceptional effort. Their encouragement from time to time has been an inspiration for everyone.

The publisher and the editorial board hope that this book will prove to be a valuable piece of knowledge for researchers, students, practitioners and scholars across the globe.

List of Contributors

Sterling Sawaya and Neil Gemmell
Centre for Reproduction and Genomics, Department of Anatomy, and Allan Wilson Centre for Molecular Ecology and Evolution, University of Otago, Dunedin, New Zealand

Andrew Bagshaw
Department of Pathology, University of Otago, Christchurch, New Zealand

Emmanuel Buschiazzo
School of Natural Sciences, University of California Merced, Merced, California, United States of America

Pankaj Kumar
G. N. Ramachandran Knowledge Centre for Genome Informatics, Delhi, India

Shantanu Chowdhury
G. N. Ramachandran Knowledge Centre for Genome Informatics, Delhi, India
Proteomics and Structural Biology Unit, Institute of Genomics and Integrative Biology, Council of Scientific and Industrial Research, Delhi, India

Michael A. Black
Department of Biochemistry, University of Otago, Dunedin, New Zealand

Serdar Bozdag
Neuro-Oncology Branch, National Cancer Institute, National Institute of Neurological Disorders and Stroke, National Institutes of Health, Bethesda, Maryland, United States of America
Department of Mathematics, Statistics, and Computer Science, Marquette University, Milwaukee, Wisconsin, United States of America

Aiguo Li, Gregory Riddick, Margaret C. Cam, Svetlana Kotliarova and Mehmet Baysan
Neuro-Oncology Branch, National Cancer Institute, National Institute of Neurological Disorders and Stroke, National Institutes of Health, Bethesda, Maryland, United States of America

Yuri Kotliarov
Neuro-Oncology Branch, National Cancer Institute, National Institute of Neurological Disorders and Stroke, National Institutes of Health, Bethesda, Maryland, United States of America
Center for Human Immunology, Autoimmunity and Inflammation, National Heart Lung and Blood Institute, National Institutes of Health, Bethesda, Maryland, United States of America

Fabio M. Iwamoto
Neuro-Oncology Branch, National Cancer Institute, National Institute of Neurological Disorders and Stroke, National Institutes of Health, Bethesda, Maryland, United States of America
The Neurological Institute of New York, College of Physicians and Surgeons, Columbia University, New York, New York, United States of America

Howard A. Fine
Neuro-Oncology Branch, National Cancer Institute, National Institute of Neurological Disorders and Stroke, National Institutes of Health, Bethesda, Maryland, United States of America
New York University Cancer Institute, New York University Langone Medical Center, New York, New York, United States of America

Yan P. Yu, Amantha Michalopoulos and Jian-Hua Luo
Department of Pathology, University of Pittsburgh School of Medicine, Pittsburgh, Pennsylvania, United States of America

Ying Ding and George Tseng
Department of Statistics, University of Pittsburgh School of Medicine, Pittsburgh, Pennsylvania, United States of America

José Ignacio Lucas Lledó
Institut de Biotecnologia i de Biomedicina, Universitat Autónoma de Barcelona, Bellaterra, Barcelona, Spain

Mario Cáceres
Institut de Biotecnologia i de Biomedicina, Universitat Autónoma de Barcelona, Bellaterra, Barcelona, Spain
Institució Catalana de Recerca i Estudis Avançats, Barcelona, Spain

Christophe Verbeurgt
Department of Otorhinolaryngology, Erasme University Hospital, Brussels, Belgium

Françoise Wilkin and Pierre Chatelain
ChemCom S.A., Brussels, Belgium

Maxime Tarabichi and Jacques E. Dumont
Institute of Interdisciplinary Research in human and molecular Biology, Free University of Brussels, Brussels, Belgium

Françoise Gregoire
Laboratory of Pathophysiological and Nutritional Biochemistry, Department of Biochemistry, Free University of Brussels, Brussels, Belgium

Shamsul Mohd Zain and Zahurin Mohamed
The Pharmacogenomics Laboratory, Faculty of Medicine, University of Malaya, Kuala Lumpur, Malaysia
Department of Pharmacology, Faculty of Medicine, University of Malaya, Kuala Lumpur, Malaysia

Rosmawati Mohamed, Sanjiv Mahadeva and Wah-Kheong Chan
Department of Medicine, Faculty of Medicine, University of Malaya, Kuala Lumpur, Malaysia

David N. Cooper
Institute of Medical Genetics, School of Medicine, Cardiff University, Cardiff, United Kingdom

Rozaimi Razali, Arif Anwar and Nurul Shielawati Mohamed Rosli
Sengenics Sdn Bhd, High Impact Reseach Building, University of Malaya, Kuala Lumpur, Malaysia

Sanjay Rampal
Julius Centre University of Malaya, Department of Social and Preventive Medicine, Faculty of Medicine, University of Malaya, Kuala Lumpur, Malaysia

Anis Shafina Mahfudz
Medical Imaging Unit, Faculty of Medicine, University of Technology MARA, Sungai Buloh Campus, Selangor, Malaysia

Phaik-Leng Cheah
Department of Pathology, Faculty of Medicine, University of Malaya, Kuala Lumpur, Malaysia

Roma Choudhury Basu
Clinical Investigation Centre, University Malaya Medical Centre, Kuala Lumpur, Malaysia

Dimitris Polychronopoulos
Institute of Biosciences and Applications, National Center for Scientific Research "Demokritos", Athens, Greece
Department of Biochemistry and Molecular Biology, Faculty of Biology, National and Kapodistrian University of Athens, Athens, Greece

Diamantis Sellis
Department of Biology, Stanford University, Stanford, California, United States of America

Greg Elgar
Systems Biology, MRC National Institute for Medical Research, Mill Hill, London, United Kingdom

Yannis Almirantis
Institute of Biosciences and Applications, National Center for Scientific Research "Demokritos", Athens, Greece

Chuan-Le Xiao, Zhi-Biao Mai, Xin-Lei Lian, Jia-Yong Zhong, Jing-jie Jin, Qing-Yu He and Gong Zhang
Key Laboratory of Functional Protein Research of Guangdong Higher Education Institutes, Institute of Life and Health Engineering, College of Life Science and Technology, Jinan University, Guangzhou, China

Dilrini R. De Silva
Systems Biology, MRC National Institute for Medical Research, Mill Hill, London, United Kingdom
School of Biological and Chemical Sciences, Queen Mary University of London, London, United Kingdom

Richard Nichols
School of Biological and Chemical Sciences, Queen Mary University of London, London, United Kingdom

Hashem Koohy
The Babraham Institute, Babraham Research Campus, Cambridge, United Kingdom
Wellcome Trust Sanger Institute, Wellcome Trust Genome Campus, Cambridge, United Kingdom

Thomas A. Down and Tim Hubbard
The Babraham Institute, Babraham Research Campus, Cambridge, United Kingdom

Mikhail Spivakov
Wellcome Trust Sanger Institute, Wellcome Trust Genome Campus, Cambridge, United Kingdom

Jesús M. Siqueiros-García and Andrea Robina-Galatas
Ethical, Legal and Social Studies Department, National Institute of Genomic Medicine, Mexico City, D.F., Mexico

Enrique Hernández-Lemus
Computational Genomics Department, National Institute of Genomic Medicine, Mexico City, D.F., Mexico
Computational Genomics Department, National Institute of Genomic Medicine, Mexico City, D.F., Mexico

Rodrigo García-Herrera
Complexity in Systems Biology, Center for Complexity Sciences, National Autonomous University of Mexico Mexico City, D.F., Mexico

Seyhan Yazar and George E. C. Gooden
Centre for Ophthalmology and Visual Science, University of Western Australia, Lions Eye Institute, Perth, Western Australia, Australia

David A. Mackey
Centre for Ophthalmology and Visual Science, University of Western Australia, Lions Eye Institute, Perth, Western Australia, Australia
School of Medicine, Menzies Research Institute Tasmania, University of Tasmania, Hobart, Tasmania, Australia

Alex W. Hewitt
Centre for Ophthalmology and Visual Science, University of Western Australia, Lions Eye Institute, Perth, Western Australia, Australia
School of Medicine, Menzies Research Institute Tasmania, University of Tasmania, Hobart, Tasmania, Australia
Centre for Eye Research Australia, University of Melbourne, Department of Ophthalmology, Royal Victorian Eye and Ear Hospital, Melbourne, Victoria, Australia

Thomas Kvist, Line Sondt-Marcussen and Marie Just Mikkelsen
Samplix ApS, Ballerup, Denmark

Yanan Wang, Yaqi Sun, Hongyang Wang, Chao Wang, Shaobo Yu, Jing Liu, Yu Zhang, Bin Fan and Bang Liu
Lab of Molecular Biology and Animal Breeding, Key Laboratory of Agricultural Animal Genetics, Breeding and Reproduction of Ministry of Education, Huazhong Agricultural University, Wuhan, PR China

Zhonglin Tang and Kui Li
Key Laboratory of Farm Animal Genetic Resources and Germplasm Innovation of Ministry of Agriculture, Institute of Animal Science, Chinese Academy of Agricultural Sciences, Beijing, PR China

Jong Hun Kim and Jun Hong Lee
Department of Neurology, Dementia Center, Stroke Center, Ilsan hospital, National Health Insurance Service, Goyang-shi, South Korea

Pamela Song
Department of Neurology, Inje University Ilsan Paik Hospital, Goyang-shi, South Korea

Hyunsun Lim
Clinical Research Management Team, Ilsan hospital, National Health Insurance Service, Goyang-shi, South Korea

Sun Ah Park
Department of Neurology, Soonchunhyang University Bucheon Hospital, Bucheon-shi, South Korea

Libing Shen
School of Life Science, Fudan University, Shanghai, China

Jae-Hyung Lee
Department of Life and Nanopharmaceutical Sciences and Department of Maxillofacial Biomedical Engineering, School of Dentistry, Kyung Hee University, Seoul, South Korea

Rong Qiu and Min Wu
School of Information Science and Engineering, Central South University, Changsha, China

Hunan Engineering Laboratory for Advanced Control and Intelligent Automation, Changsha, China

Chao Chen
Department of Psychiatry, University of Illinois at Chicago, Chicago, United States of America
Institute of Human Genetics, University of Illinois at Chicago, Chicago, United States of America

Hong Jiang
Department of Neurology, Xiangya Hospital, Central South University, Changsha, China

Chunyu Liu
Department of Psychiatry, University of Illinois at Chicago, Chicago, United States of America
Institute of Human Genetics, University of Illinois at Chicago, Chicago, United States of America
State Key Laboratory of Medical Genetics of China, Central South University, Changsha, China

Carl G. de Boer
Department of Molecular Genetics, University of Toronto, Toronto, Ontario, Canada

Timothy R. Hughes
Department of Molecular Genetics, University of Toronto, Toronto, Ontario, Canada
Donnelly Centre for Cellular and Biomolecular Research, University of Toronto, Toronto, Ontario, Canada

Dirk A. Moser
Faculty of Psychology, Genetic Psychology, Ruhr-University-Bochum, Bochum, Germany,
Department of Sports Medicine, Disease Prevention and Rehabilitation, Johannes Gutenberg-University Mainz, Mainz, Germany

Luca Braga, Andrea Raso, Serena Zacchigna and Mauro Giacca
International Centre for Genetic Engineering and Biotechnology (ICGEB), Molecular Medicine, Trieste, Italy

Perikles Simon
Department of Sports Medicine, Disease Prevention and Rehabilitation, Johannes Gutenberg-University Mainz, Mainz, Germany

Zsuzsanna Izsvák
Max Delbruck Center for Molecular Medicine, Berlin, Germany

Guang Guo
Department of Sociology, the University of North Carolina at Chapel Hill, Chapel Hill, North Carolina, the United States of America
Carolina Population Center, the University of North Carolina at Chapel Hill, Chapel Hill, North Carolina, the United States of America
Carolina Center for Genome Sciences, the University of North Carolina at Chapel Hill, Chapel Hill, North Carolina, the United States of America

Lin Wang
Center for Child and Family Policy, Duke University, Durham, North Carolina, the United States of America

Hexuan Liu
Department of Sociology, the University of North Carolina at Chapel Hill, Chapel Hill, North Carolina, the United States of America
Carolina Population Center, the University of North Carolina at Chapel Hill, Chapel Hill, North Carolina, the United States of America

Thomas Randall
National Institute of Environmental Health Sciences, Research Triangle Park, North Carolina, the United States of America

Giandomenico Turchiano, Maria Carmela Latella and Alessandra Recchia
Center for Regenerative Medicine, Department of Life Sciences, University of Modena and Reggio Emilia, Modena, Italy

Andreas Gogol-Döring
German Centre for Integrative Biodiversity Research (iDiv) Halle-Jena-Leipzig, Leipzig, Germany
Institute of Computer Science, Martin Luther University Halle-Wittenberg, Halle, Germany

Claudia Cattoglio
Howard Hughes Medical Institute, Department of Molecular and Cell Biology, University of California, Berkeley, Berkeley, California, United States of America

Fulvio Mavilio
Center for Regenerative Medicine, Department of Life Sciences, University of Modena and Reggio Emilia, Modena, Italy
Genethon, Evry, France

Zoltán Ivics
Division of Medical Biotechnology, Paul Ehrlich Institute, Langen, Germany

Agnieszka Danek and Sebastian Deorowicz
Institute of Informatics, Silesian University of Technology, Gliwice, Poland

Szymon Grabowski
Institute of Applied Computer Science, Lodz University of Technology, Łódź, Poland

Index

www.ingramcontent.com/pod-product-compliance
Lightning Source LLC
Chambersburg PA
CBHW080408190526
45161CB00003B/173